Cellular Lipid Metabolism

Christian Ehnholm
Editor

Cellular Lipid Metabolism

Editor
Prof. Christian Ehnholm
National Public Health Institute
Mannerheimintie 166
00300 Helsinki
Finland
christian.ehnholm@thl.fi

ISBN: 978-3-642-00299-1 e-ISBN: 978-3-642-00300-4
DOI: 10.1007/978-3-642-00300-4

Library of Congress Control Number: 2009922260

Cover design: WMXDesign GmbH. Heidelberg, Germany

Printed on acid-free paper

Springer is part of Springer Science+Business Media (www.springer.com)

Preface

The key to every biological problem must in the end be sought in the cell and yet, although we know a lot about the mechanism by which cells operate, there is still a shortage in our understanding of how lipids affect cell biology. For years lipids have fascinated cell biologists and biochemists because they have profound effects on cell function. Encoded within lipid molecules is the ability to spontaneously form macroscopic, two-dimensional membrane systems. In addition to their function as physical and chemical barriers separating aqueous compartments, membranes are involved in many regulatory processes, such as secretion, endocytosis, and signal transduction. The functional interaction between lipids and proteins is essential for such membrane activities.

Lipids serve as one of the major sources of energy, both directly and when stored in adipose tissues. They also act as thermal insulators in the subcutaneous tissues and serve as electrical insulators in myelinated nerves, allowing the rapid propagation of waves of depolarization. Some lipids act as biological modulators and signal transducers (e.g., pheromones, prostaglandins, thromboxanes, leukotrienes, steroids, platelet-activating factor, phosphatidylinositol derivates) and as vehicles for carrying fat-soluble vitamins.

Research on cell biology is at present in a very active phase and molecular genetics is helping us to recognize and exploit the unity of all living systems and to reveal the fundamental mechanisms by which the cell operates.

The challenge in composing a book on *Cellular lipid metabolism* has been to select concepts that are important for our understanding in areas that have changed or in which new concepts have emerged. Recognizing that it is impossible to be comprehensive, I have tried to ensure that this book provides a survey of cell biology in areas that I consider important.

This book was planned to be a resource for scientists at post-doctoral level and above, in other words, a rather specific publication to highlight recent findings in cell biology and biochemistry but also to include important findings made in the past and give a good overview. I contacted the best experts in 13 fields and the chapters represent their specialized contributions. They represent analyses at the molecular level and reveal the principles by which cellular lipid metabolism functions.

There are still large areas of ignorance in cell biology and numerous intriguing observations that cannot be explained. In this volume we try to expose them and to stimulate readers to contemplate and discover ways of solving the open questions.

I hope this book will be of interest to all reasearchers in the area of cell biology, lipid metabolism and atherosclerosis, providing a useful review of accomplishments and a stimulating guide for future studies.

Helsinki, February 2009 Christian Ehnholm

Contents

Contributors

Martin Adiels
The Wallenberg Laboratory, Sahlgrenska Center for Metabolic and Cardiovascular Research, Department of Molecular and Clinical Medicine, Sahlgrenska University Hospital, Sahlgrenska Academy, University of Göteborg, 41345 Göteborg, Sweden

Linda Andersson
The Wallenberg Laboratory, Sahlgrenska Center for Metabolic and Cardiovascular Research, Department of Molecular and Clinical Medicine, Sahlgrenska University Hospital, Sahlgrenska Academy, University of Göteborg, 41345 Göteborg, Sweden

Ulrike Beisiegel
Institute for Biochemistry and Molecular Biology II: Molecular Cell Biology, University Medical Center Hamburg – Eppendorf, Martinistrasse 52, 20246 Hamburg, Germany
beisiegel@uke.uni-hamburg.de

Nora Bijl
Department of Medical Biochemistry (Room K1-106), Academic Medical Centre, University of Amsterdam, Meibergdreef 9,1105 AZ Amsterdam
The Netherlands

Jan Borén
The Wallenberg Laboratory, Sahlgrenska Center for Metabolic and Cardiovascular Research, Department of Molecular and Clinical Medicine, Sahlgrenska University Hospital, Sahlgrenska Academy, University of Göteborg, 41345 Göteborg, Sweden

Pontus Boström
The Wallenberg Laboratory, Sahlgrenska Center for Metabolic and Cardiovascular Research, Department of Molecular and Clinical Medicine, Sahlgrenska University Hospital, Sahlgrenska Academy, University of Göteborg, 41345 Göteborg, Sweden

Christopher J. Fielding
Cardiovascular Research Institute, Box 0130, University of California San Francisco, 4th & Parnassus, San Francisco, CA 94143, USA
christopher.fielding@ucsf.edu

Bernhard Föger
Department of Internal Medicine, Landeskrankenhaus Bregenz, Carl-Pedenz-Strasse 2, 6900 Bregenz, Austria

Albert K. Groen
Department of Medical Biochemistry (Room K1-106), Academic Medical Centre, University of Amsterdam, Meibergdreef 9,1105 AZ Amsterdam, The Netherlands; a.k.groen@amc.uva.nl

Oreste Gualillo
NEIRID (Neuroendocrine Interactions in Rheumatic and Inflammatory Diseases) Laboratory, University Clinical Hospital, Travesía Choupana s/n, 15706 Santiago de Compostela, Spain
oreste.gualillo@usc.es

Joerg Heeren
Institute for Biochemistry and Molecular Biology II: Molecular Cell Biology, University Medical Center Hamburg – Eppendorf, Martinistrasse 52, 20246 Hamburg, Germany
heeren@uke.uni-hamburg.de

Sander Kersten
Nutrition, Metabolism and Genomics Group, Division of Human Nutrition, Wageningen University, Bomenweg 2, 6703 HD Wageningen, the Netherlands; and Nutrigenomics Consortium, TI Food and Nutrition, Nieuwe Kanaal 9A, 6709 PA Wageningen, the Netherlands
sander.kersten@wur.nl

Jens Lagerstedt
The Wallenberg Laboratory, Sahlgrenska Center for Metabolic and Cardiovascular Research, Department of Molecular and Clinical Medicine, Sahlgrenska University Hospital, Sahlgrenska Academy, University of Göteborg, 41345 Göteborg, Sweden

Francisca Lago
Cellular and Molecular Cardiology Research Laboratory, University Clinical Hospital, Travesía Choupana s/n, 15706 Santiago de Compostela, Spain
mfrancisca.lago@usc.es

Lu Li
The Wallenberg Laboratory, Sahlgrenska Center for Metabolic and Cardiovascular Research, Department of Molecular and Clinical Medicine, Sahlgrenska University Hospital, Sahlgrenska Academy, University of Göteborg, 41345 Göteborg, Sweden

Dieter Lütjohann
Department of Clinical Pharmacology, Zentrallabor, Bonn University, Sigmund-Freud-Strasse 25, 53105 Bonn, Germany
dieter.luetjohann@ukb.uni-bonn.de

Yves L. Marcel
Lipoprotein and Atherosclerosis Research Group, University of Ottawa Heart Institute, 40 Ruskin Street, Ottawa, K1Y 4W7, Ontario, Canada
ylmarcel@ottawaheart.ca

Monique Mulder
Department of Internal Medicine, Division of Pharmacology, Vascular and Metabolic Diseases, Erasmus Medical Center, 3015 CE Rotterdam, The Netherlands
m.t.mulder@erasmusmc.nl

Gunilla Olivecrona
Department of Medical Biosciences/Physiological Chemistry, Bldg 6M, 3rd Floor, Umeå University, 90187 Umeå, Sweden
gunilla.olivecrona@medbio.umu.se

Thomas Olivecrona
Department of Medical Biosciences/Physiological Chemistry, Bldg 6M, 3rd Floor, Umeå University, 90187 Umeå, Sweden
thomas.olivecrona@medbio.umu.se

Vesa M. Olkkonen
Department of Molecular Medicine, National Public Health Institute, Biomedicum, P.O. Box 104, 00251 Helsinki, Finland; and Institute of Biomedicine/Anatomy, P.O. Box 63, 00014 University of Helsinki, Finland
vesa.olkkonen@ktl.fi

Sven-Olof Olofsson
The Wallenberg Laboratory, Sahlgrenska Center for Metabolic and Cardiovascular Research, Department of Molecular and Clinical Medicine, Sahlgrenska University Hospital, Sahlgrenska Academy, University of Göteborg, 41345 Göteborg, Sweden
Sven-Olof.Olofsson@wlab.gu.se

Mireille Ouimet
Lipoprotein and Atherosclerosis Research Group, University of Ottawa Heart Institute, 40 Ruskin Street, Ottawa, K1Y 4W7, Ontario, Canada

Josef R. Patsch
Department of Internal Medicine, Innsbruck Medical University, Anichstrasse 35, 6020 Innsbruck, Austria

Jeanna Perman
The Wallenberg Laboratory, Sahlgrenska Center for Metabolic and Cardiovascular Research, Department of Molecular and Clinical Medicine, Sahlgrenska University Hospital, Sahlgrenska Academy, University of Göteborg, 41345 Göteborg, Sweden

Andreas Ritsch
Department of Internal Medicine, Innsbruck Medical University, Anichstrasse 35, 6020 Innsbruck, Austria

Lucia Rohrer
Institut für Klinische Chemie, Universitätsspital Zürich, Rämistrasse 100, 8091 Zürich, Switzerland

Mikael Rutberg
The Wallenberg Laboratory, Sahlgrenska Center for Metabolic and Cardiovascular Research, Department of Molecular and Clinical Medicine, Sahlgrenska University Hospital, Sahlgrenska Academy, University of Göteborg, 41345 Göteborg, Sweden

Ivan Tancevski
Department of Internal Medicine, Innsbruck Medical University, Anichstrasse 35, 6020 Innsbruck, Austria
ivan.tancevski@i-med.ac.at

Astrid van der Velde
Department of Medical Biochemistry (Room K1-106), Academic Medical Centre, University of Amsterdam, Meibergdreef 9,1105 AZ Amsterdam, The Netherlands

Tim Vanmierlo
Department of Internal Medicine, Division of Pharmacology, Vascular and Metabolic Diseases, Erasmus Medical Center, 3015 CE Rotterdam, The Netherlands

Arnold von Eckardstein
Institut für Klinische Chemie, Universitätsspital Zürich, Rämistrasse 100, 8091 Zürich, Switzerland
arnold.voneckardstein@usz.ch

Ming-Dong Wang
Lipoprotein and Atherosclerosis Research Group, University of Ottawa Heart Institute, 40 Ruskin Street, Ottawa, K1Y 4W7, Ontario, Canada

Andreas Wehinger
Department of Internal Medicine, Landeskrankenhaus Bregenz, Carl-Pedenz-Strasse 2, 6900 Bregenz, Austria

Daniel Wüstner
Department of Biochemistry and Molecular Biology, University of Southern Denmark, Campusvej 55, 5230 Odense M, Denmark
wuestner@bmb.sdu.dk

Chapter 1
The Lipid Droplet: a Dynamic Organelle, not only Involved in the Storage and Turnover of Lipids

Sven-Olof Olofsson, Pontus Boström, Jens Lagerstedt, Linda Andersson, Martin Adiels, Jeanna Perman, Mikael Rutberg, Lu Li, and Jan Borén

Abstract Neutral lipids such as triglycerides are stored in cytosolic lipid droplets. These are dynamic organelles and consist of a core of neutral lipids surrounded by amphipathic lipids and proteins. The surface is complex and contains proteins involved in lipid biosynthesis and turnover and proteins involved in sorting and trafficking events in the cell. Lipid droplets are formed at microsomes as primordial droplets, which increase in size by fusion. In this chapter, we review the assembly and fusion of lipid droplets. We also discuss a possible mechanism to explain the link between lipid accumulation in muscle cells and the development of insulin resistance. Triglycerides are secreted as milk globules from the epithelial cells of the mammary glands, as chylomicrons from enterocytes, and as very low-density lipoproteins (VLDL) from hepatocytes. We review the processes involved in the formation of milk globules and VLDL, and we discuss the clinical consequences of overproduction of VLDL.

Abbreviations ADRP, adipocyte differentiation related protein; ATGL, adipose triglyceride lipase; DGAT, diacylglycerol acyltransferase; ER, endoplasmic reticulum; ERGIC, ER Golgi intermediate compartment; ERK2, extracellular signal regulated kinase 2; GPAT, glycerol-3-phosphate acyltransferase; HDL, high-density lipoprotein; LDL, low-density lipoprotein; NSF, N-ethylmaleimide-sensitive factor; PA, phosphatidic acid; PAP, phosphatidic acid phosphohydrolase; PC, phosphatidylcholine; PLD1, phospholipase D1; SNAP23, synaptosomal-associated protein of 23 kDa; α-SNAP, α-soluble NSF adaptor protein; SNARE, SNAP receptor; VAMP, vesicle-associated protein; VLDL, very low-density lipoprotein

---

S.-O. Olofsson, P. Boström, J. Lagerstedt, L. Andersson,
M. Adiels, J. Perman, M. Rutberg, L. Li, and J. Borén
The Wallenberg Laboratory, Sahlgrenska Center for Metabolic and Cardiovascular Research, Department of Molecular and Clinical Medicine, Sahlgrenska University Hospital, Sahlgrenska Academy, University of Göteborg, 41345 Göteborg, Sweden
e-mail: Sven-Olof.Olofsson@wlab.gu.se

C. Ehnholm (ed.), *Cellular Lipid Metabolism*,
DOI 10.1007/978-3-642-00300-4_1, © Springer-Verlag Berlin Heidelberg 2009

1.1 Introduction

Neutral lipids, such as triglycerides and cholesterol esters, are stored in the cells within so-called cytosolic lipid droplets. The neutral lipids form the core of the lipid droplet and are surrounded by an outer layer of amphipathic lipids, such as phospholipids and cholesterol (Brown 2001; Martin and Parton 2006; Fig. 1.1). The surface of the lipid droplet is generally considered to be a monolayer of lipids (Robenek et al. 2005).

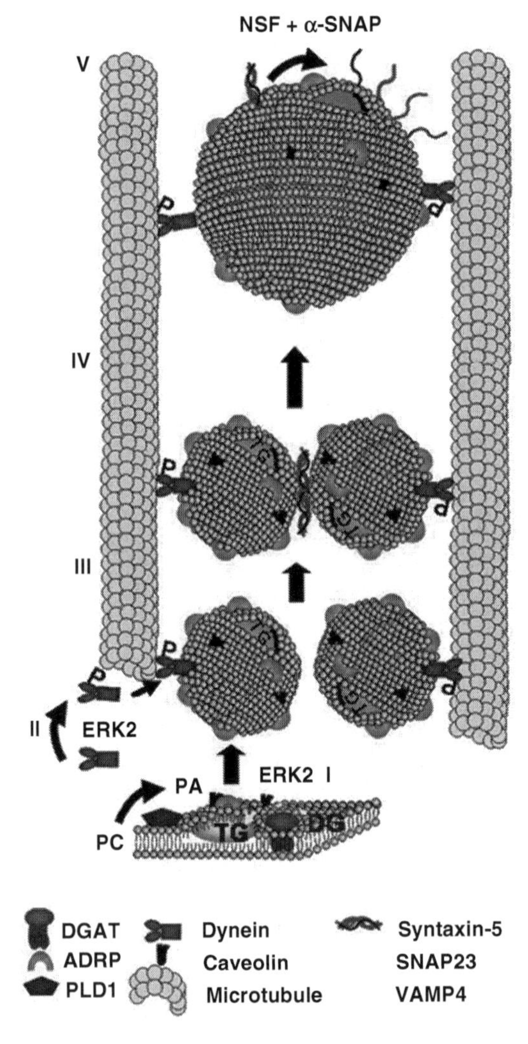

Fig. 1.1 A model for the formation of lipid droplets. Diglycerides (*DG*) are catalyzed by DGAT to form triglycerides (*TG*) in the microsomal membrane. TG have limited solubility in the amphipathic monolayer, and therefore oil out between the leaflets of the membrane to form a TG lens, which will become the core of the lipid droplet (*I*). Lipid droplet assembly also requires the production of phosphatidic acid (*PA*) from phosphatidylcholine (*PC*) catalyzed by PLD1 and an active ERK2 (*II*).

Specific proteins that are essential for the formation, structure and function of the lipid droplet are bound to its surface (Brasaemle 2007; Martin and Parton 2006). As a result of recent advances in our knowledge about the structure and function of lipid droplets, they are now considered to be dynamic organelles that can interact with other organelles and have a key role in the cellular turnover of lipids (Martin and Parton 2006).

In today's increasingly overweight society, the problems associated with excess levels of triglycerides are now well recognized. Accumulation of triglycerides, particularly in the liver and muscles, is highly correlated with the development of insulin resistance and type 2 diabetes, which are important risk factors for arteriosclerosis and cardiovascular diseases (Taskinen 2003). Triglycerides are stored to a variable extent in most cells, but they are only efficiently secreted by certain organs, i.e. liver and intestine (Olofsson and Boren 2005) and mammary glands (McManaman et al. 2007). In this article, we review the storage and secretion of triglycerides.

1.2 Lipid Droplets Form as Primordial Structures at Microsomal Membranes

The nature of the site of assembly of lipid droplets has not been conclusively determined. Results from a cell-free system indicate that they can be formed from a microsomal fraction enriched in markers for the endoplasmic reticulum (ER) and Golgi apparatus but lacking makers for the plasma membrane (Marchesan et al. 2003). An ER localization of the assembly of lipid droplets is also suggested by results showing that lipid droplets are associated with adipocyte differentiation related protein (ADRP)-enriched regions of the ER (Robenek et al. 2006). However, other results (Ost et al. 2005) indicate that the plasma membrane may be a source of droplets: triglycerides accumulate in the plasma membrane of adipocytes and this accumulation seems to be a precursor for the formation of cytosolic lipid droplets.

1.2.1 Microsomal Membrane Proteins Involved in Lipid Droplet Formation

The formation of lipid droplets is highly linked to the biosynthesis of triglycerides (Marchesan et al. 2003). Glycerol-3-phosphate acyltransferase (GPAT) catalyzes the first step, i.e. the formation of lyso-phosphatidic acid. GPAT exists in several

Fig. 1.1 (continued) The assembly process first forms a primordial droplet with a diameter <0.5 μm. ERK2 phosphorylates the motor protein dynein, which is then sorted to droplets allowing them to transfer on microtubules (*II*). This allows long-distance transport of the droplet in the cell and is also required for lipid droplet fusion (*III–V*). The fusion process is catalyzed by the SNAREs SNAP23, syntaxin-5 and VAMP4 (*IV*). After the fusion, the four-helix bundle formed by the SNARE domains of these three SNAREs is recognized by α-SNAP which, together with the ATPase NSF, unwinds the bundle, allowing new fusions to occur (*V*)

isoforms. GPAT1 and GPAT2 are present on mitochondria and GPAT3 and GPAT4 are present on ER (Gonzalez-Baro et al. 2007; for a review, see Coleman et al. 2000). The mitochondrial isoforms of GPAT were first identified and cloned and most information is from studies of these isoforms. Overexpression of GPAT1 has been shown to increase the accumulation of triglycerides in the cell and promote the formation of steatoses (Gonzalez-Baro et al. 2007). GPAT3 has also been cloned (Cao et al. 2006). This isoform is highly upregulated during adipocyte differentiation and overexpression leads to lipid accumulation in the cell (Cao et al. 2006).

The formation of phosphatidic acid from lyso-phosphatidic acid is catalyzed by 1-acylglycerol-3-phosphate O-acyltransferase (AGPAT). This enzyme exists in several isoforms of which 1 and 2 are confined to microsomes. AGPAT is a membrane-spanning protein that catalyzes the reaction on the cytosolic side of the ER (for reviews with references, see for example Agarwal and Garg 2003; Leung 2001).

When associated with microsomal membranes, the amphipathic enzyme phosphatidic acid phosphohydrolase hydrolyzes phosphatidic acid, forming diacylglycerol (for reviews, see Carman and Han 2006; Coleman et al. 2000). Diacylglycerol acyltransferase (DGAT), an integral membrane protein of microsomes (Stone et al. 2006), then catalyzes the conversion of diacylglycerol to triglycerides. There are two mammalian forms of DGAT: DGAT1 and DGAT2. DGAT1 is a multifunctional enzyme (Yen et al. 2005) whereas DGAT2 has been shown to be more potent and specific for triglyceride synthesis (Stone et al. 2004). In addition to its localization on the ER membrane, DGAT2 has also been identified on lipid droplets (Kuerschner et al. 2008). However, DGAT2 is a membrane protein that spans the bilayer twice; and it remains to be clarified how it is integrated into the amphipathic monolayer that surrounds lipid droplets. Alternatively, there may be a very tight interaction between lipid droplets and the ER allowing the DGAT2 product formed in the microsomal membrane to enter into the droplets. Such a tight interaction has been demonstrated and shown to be dependent on the GTPase Rab18 (for a review, see Martin and Parton 2006).

Lipid droplet assembly is also dependent on phospholipase D (PLD) activity and the formation of phosphatidic acid (Marchesan et al. 2003). Using intact cells, we showed that the active isoform is PLD1 and not PLD2 (Andersson et al. 2006), which is consistent with the localization of the two isoforms: PLD1 is present in ER and Golgi membranes (Andersson et al. 2006; Freyberg et al. 2001) while PLD2 is confined to the plasma membrane (Andersson et al. 2006; Du et al. 2004).

The observation that most of the identified enzymes associated with lipid droplet assembly are localized on microsomes supports the idea that lipid droplets are formed at the ER and/or Golgi apparatus.

1.2.2 Model for the Assembly of Lipid Droplets

The lipid droplets formed at the isolated microsomal membranes in a cell-free system have a diameter of 0.1–0.4 μm (Marchesan et al. 2003). This corresponds well to the size of the smallest droplets observed in cells by electron microscopy

(Marchesan et al. 2003). The newly formed droplets recovered from the cell-free system contain ADRP and are rich in caveolin and vimentin; and we propose that they represent the first primordial structures formed during the assembly process (Marchesan et al. 2003).

Although no experimental results have been obtained to date demonstrating how lipid droplets are formed, a tentative model for their assembly has been proposed (see for example Brown 2001; Fig. 1.1). Triglycerides (formed from diglycerides and acyl-CoA by the DGAT reaction in the microsomal membranes) are highly hydrophobic and have limited solubility in the monolayer of the membrane. The formed triglycerides will therefore "oil out" as a separate phase between the two leaflets, forming a lens structure that is the core of the lipid droplets. One problem is that the formed triglycerides may rapidly diffuse laterally in the ER and Golgi membranes and saturate these organelles before the oiling out occurs. However, this could be prevented if the regions of triglyceride synthesis are sealed off from the rest of the organelle.

1.3 Lipid Droplet Size Increases by Fusion

We have shown that droplets can increase in size by a fusion process, which is independent of triglyceride biosynthesis (Bostrom et al. 2005; Fig. 1.1). Approximately 15% of all droplets in the cells are engaged in fusion events at any given time (Bostrom et al. 2005) and thus fusion is a frequently occurring event that represents an important mechanism by which lipid droplets increase in size.

Lipid droplets are transported relatively long distances on microtubules (Welte et al. 1998) and motor proteins such as dynein have been shown to be present on droplets (Bostrom et al. 2005). We demonstrated that dynein is sorted to the droplets following phosphorylation by the cytosolic protein extracellularly regulated kinase 2 (ERK2; Andersson et al. 2006; Fig. 1.1). Both dynein and microtubules are essential for the fusion between droplets (Andersson et al. 2006; Bostrom et al. 2005).

1.3.1 SNAREs are Involved in Lipid Droplet Fusion

We have shown that the fusion between lipid droplets is catalyzed by N-ethylmaleimide-sensitive factor adaptor protein receptors (SNAREs), the synaptosomal-associated protein of 23 kDa (SNAP 23), syntaxin-5 and vesicular-associated protein 4 (VAMP4; Fig. 1.1). In addition, the fusion requires the ATPase N-ethylmaleimide-sensitive factor (NSF) and α-soluble NSF adaptor protein (α-SNAP; Fig. 1.1).

The role of these proteins has been extensively described for the fusion process between transport vesicles and target membranes (see for example Jahn and Scheller 2006). The SNAREs present on the target membrane (t- or Q-SNAREs) interact with a SNARE on the transport vesicle (v- or R-SNARE) to form a SNARE complex that causes fusion. A central feature in this process is the formation of a superhelix bundle, formed by four α-helical SNARE domains from the different SNAREs. The formation of the four-helix bundle forces the two membranes together, promoting their fusion.

After the fusion, the SNARE complex is unwound by NSF and α-SNAP (for reviews of the SNARE system, see Hong 2005; Jahn and Scheller 2006).

The four-helix bundle is mostly stabilized by hydrophobic interactions, except in the zero-plane, where an arginine side-chain from one SNARE domain (R-SNARE) interacts with the glutamine side-chains from the other three SNARE domains (Q-SNAREs). On the basis of the homology of the SNARE domains, Q-SNARES are subdivided into Qa, Qb and Qc; and a complete and functioning SNARE complex has the structure QabcR. The most well known t-SNARE complex is formed by a syntaxin (Qa-SNARE) and SNAP25 or SNAP23 (Qbc SNAREs). The R-SNARE is present on the transport vesicle and belongs to the VAMP family of SNAREs (for reviews, see Hong 2005; Jahn and Scheller 2006).

1.3.2 Model for the Fusion Between Lipid Droplets

As oils in water fuse spontaneously, unprotected triglycerides would fuse to form large hydrophobic regions that may influence the function of the cell. We hypothesize that this spontaneous fusion is reduced by protecting the triglycerides with amphipathic structures such as phospholipids and proteins. We also propose that the SNARE system could restore the fusion capacity of intact droplets and, moreover, provide a way to control the fusion process.

SNAP23 is a covalently palmitoylated SNARE and thus can anchor in the monolayer surrounding the lipid droplet by the palmitic acid residues (Bostrom et al. 2007). Syntaxin-5 and VAMP4 are tail-anchored proteins (High and Abell 2004). Such proteins are synthesized on ribosomes present in the cytosol and inserted into bilayer membranes through interactions with chaperone proteins (High and Abell 2004). It is reasonable to assume that syntaxin-5, which has a very hydrophobic C-terminus, could similarly be inserted into the surface of lipid droplets. However, the end of the C-terminus of VAMP4 is less hydrophobic (Bostrom et al. 2007) and thus its incorporation into the surface of lipid droplets may require other mechanisms (e.g. interaction with specific proteins, or the formation of hair-pin structures in which the hydrophobic regions dip down into the hydrophobic part of the droplet). It has been suggested that bilayer regions in the lipid droplet surface could allow the focal insertion of SNARE proteins (Sollner 2007).

The surface of the droplet is an amphipathic monolayer, whereas a vesicle is surrounded by a bilayer. Thus, it is likely that there are differences between the fusion of lipid droplets and the fusion of a transport vesicle with a target membrane (Fig. 1.2). The stalk hypothesis has been proposed to describe the fusion process between bilayers as an ordered sequence of transition states (Jahn and Scheller 2006; Fig. 1.2). We postulate that fusion between lipid droplets requires fewer steps and is complete at a stage equivalent to the creation of a "fusion stalk", i.e. when the two outer monolayers of the bilayers have fused and there is a continuum between the hydrophobic portions of the two membranes. For lipid droplets, this would

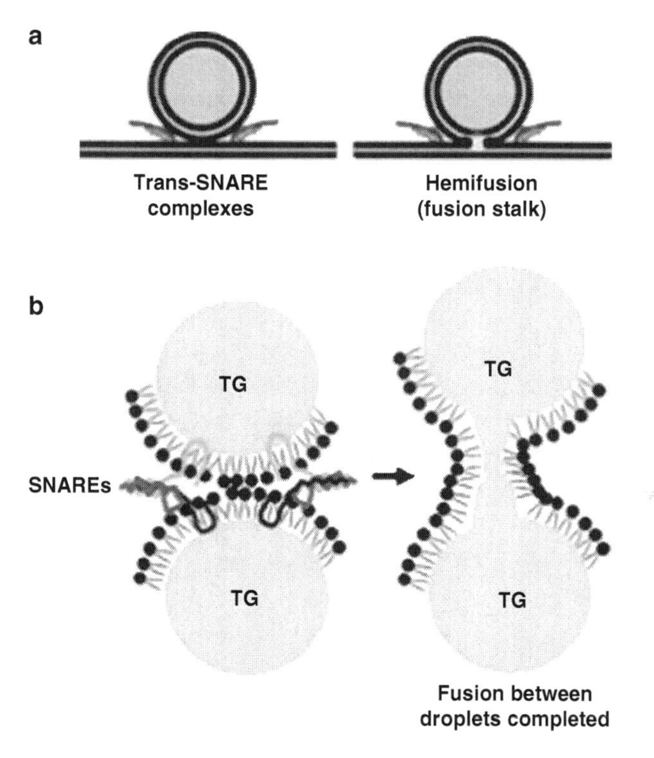

Fig. 1.2 A model for the fusion between lipid droplets. We propose that the fusion between lipid droplets (which involves two monolayers of amphipathic structures) corresponds to the first step in the fusion between transport vesicles and target membranes (which involves two bilayers). **a** According to the stalk hypothesis, a 'fusion stalk' is formed by a fusion of the outer leaflets, which allows the hydrophobic parts of the two membranes to contact each other (adapted from Jahn and Scheller 2006). **b** Formation of a four-helix bundle between syntaxin-5 (*Qa*), SNAP23 (*Qbc*) and VAMP4 (*R*) forces the two monolayers of the lipid droplets to fuse with each other, which results in a connection between the hydrophobic phases. This is equivalent to the creation of a fusion stalk in **a** (adapted from Bostrom et al. 2007)

correspond to a fusion of the monolayers surrounding the two droplets connecting the two hydrophobic cores (Bostrom et al. 2007; Fig. 1.2).

1.4 Lipid Droplets and the Development of Insulin Resistance

The accumulation of lipids in muscle (Falholt et al. 1988; Krssak et al. 1999; Machann et al. 2004) is highly correlated with the development of insulin resistance and type 2 diabetes (for reviews, see Goossens 2007; Kovacs and Stumvoll 2005; Sell et al. 2006; Yki-Jarvinen 2002; Yu and Ginsberg 2005). The accumulation of triglycerides in muscle occurs when the inflow of fatty acids exceeds the capacity of the cell to use them (by oxidation or biosynthesis).

The glucose transporter GLUT4 is of central importance for the insulin-regulated uptake of glucose in skeletal muscle (for reviews, see Dugani and Klip 2005; Huang and Czech 2007; Pilch 2008; Watson and Pessin 2006, 2007). GLUT4 exists mainly in intracellular compartments and is translocated to the plasma membrane in response to insulin. Thus, insulin results in an increase in the plasma membrane pool of GLUT4 and a subsequent increase in glucose uptake.

The sorting of GLUT4 in the cell is complex and not fully understood. It has been the subject of recent reviews (see for example Dugani and Klip 2005; Huang and Czech 2007; Pilch 2008; Watson and Pessin 2006, 2007) and the reader is referred to these reviews for details and references. One important step is the accumulation of GLUT4 in GLUT4-specific vesicles (GSV). Insulin promotes the transport of these vesicles to the plasma membrane and their subsequent fusion with the plasma membrane. The fusion process involves three SNARE proteins: syntaxin-4, VAMP2 and SNAP23 – the protein involved in lipid droplet fusion.

We showed that incubation of muscle cells with fatty acids results in an increased formation of lipid droplets and increased insulin resistance, measured as a reduced response to insulin for both glucose uptake and GLUT4 translocation (Bostrom et al. 2007). Furthermore, we found that fatty acid treatment decreases the plasma membrane pool of SNAP23 and increases the intracellular pool of SNAP23, in part due to a sequestering of SNAP23 on the increasing lipid droplet pool (Bostrom et al. 2007). Moreover, the fatty acid-induced insulin resistance can be reversed by increasing the pool of SNAP23 in the cell (Bostrom et al. 2007). These results indicate that SNAP23 has a central role in the development of lipid-induced insulin resistance. They also indicate that it is not the lipid droplets per se that promote the development of insulin resistance, but the influence of the fatty acids on SNAP23. The effect of fatty acids on the insulin signal is well established and is proposed to be mediated by fatty acid metabolites, such as diglycerides (for reviews, see Morino et al. 2006; Savage et al. 2007), ceramides (reviewed by Summers 2006) and partially oxidized fatty acids (Koves et al. 2008). However, the exact mechanisms involved are not known.

1.5 Lipid Droplet-Associated Proteins

1.5.1 PAT Proteins

In addition to those discussed above, numerous proteins have been identified in association with lipid droplets. The quantitatively most important are the so-called PAT proteins (Dalen et al. 2007; Londos et al. 1999; reviewed by Brasaemle 2007), named after the three first identified species of the family: *p*erilipin, *A*DRP and tail interacting protein 47 (TIP47). The name PAT could also reflect the existence of a *p*erilipin *a*mino-*t*erminal domain with high degree of homology between the family members. There are two recent additions to this family: lipid storage droplet protein 5 (LSDP5), also known as OXPAT, myocardial lipid droplet protein (MLDP) or PAT-1 (Dalen et al. 2007; Wolins et al. 2006), and S3-12 (Dalen et al. 2004; Wolins et al. 2003).

In addition to the highly homologous N-terminal PAT domain, there are reports of a C-terminal PAT domain, which shows very low homology between the family members (Lu et al. 2001; Miura et al. 2002). The PAT domains do not seem to be important for the binding of the protein to lipid droplets, but ADRP mutation studies indicate that the middle domain – including α-helical regions between amino acids 189 and 205 – directs ADRP to droplets (Nakamura and Fujimoto 2003).

1.5.1.1 Perlipin

Perilipin was initially identified as a protein kinase A (PKA)-phosphorylated protein in lipolytically stimulated adipocytes (Greenberg et al. 1991, 1993). The dominating sites of expression are adipocytes and steriogenic cells (Londos et al. 1995). There are three isoforms (perilipin A, B, C), which are formed from a single gene through alternative splicing (Lu et al. 2001). Perilipin A is by far the most abundant and best investigated isoform (Londos et al. 1999). The gene for human perilipin is located on chromosome 15q26 at a locus that has been linked to diabetes, hypertriglyceridemia and obesity (for a review, see Tai and Ordovas 2007).

Perilipin has an important and dualistic role in the turnover of triglycerides in lipid droplets. It protects the degradation of triglycerides when expressed in cells that lack natural expression of the protein (Brasaemle et al. 2000). Moreover, perilipin A knockout mice show a substantial increase in basal lipolysis and reduction in adipose mass, and are resistant to diet-induced obesity (Martinez-Botas et al. 2000; Tansey et al. 2001). Perilipin also promotes triglyceride degradation: β-adrenergic stimulation does not promote lipolysis in perilipin A knockout mice; and cells derived from these mice fail to show translocation of hormone-sensitive lipase (HSL) to the lipid droplet (Martinez-Botas et al. 2000; Tansey et al. 2001; reviewed by Londos et al. 2005). A proposed model, based on these and more direct results (Granneman et al. 2007), states that concomitant phosphorylation of perilipin and HSL results in translocation of HSL to the lipid droplet where it catalyzes the hydrolyzation of triglycerides (Granneman et al. 2007).

In addition to its effect on HSL, perilipin seems to have an important role in the activation of the first step in the degradation of triglycerides, i.e. the step catalyzed by the newly discovered adipose triglyceride lipase (ATGL) and its co-activator, a perilipin-interacting protein CGI-58 (or Abhd5; Lass et al. 2006). It is suggested that phosphorylation of perilipin A results in its dissociation from CGI-58, which then associates with ATGL on lipid droplets to allow lipolysis (Granneman et al. 2007; Miyoshi et al. 2007).

1.5.1.2 ADRP

Although perilipin has a central role in the turnover of triglycerides in adipocytes, there is no experimental evidence to indicate that it participates in the assembly of lipid droplets. Indeed, ADRP seems to play a central role in the assembly process even in perilipin-expressing cells such as the adipocyte. Thus, ADRP is expressed in increasing amounts early in adipocyte differentiation, but is later replaced by perilipin (Brasaemle et al. 1997).

ADRP was identified as a protein related to adipocyte differentiation (Jiang and Serrero 1992) and was originally thought to be a fatty acid-binding protein (Serrero et al. 2000). However, it later proved to be one of the major PAT proteins with a striking homology to perilipin in the N-terminus (Brasaemle 2007; Londos et al. 1999). In contrast to perilipin, ADRP is expressed ubiquitously (Brasaemle et al. 1997). This expression is highly related to the amount of neutral lipid in the cell (Heid et al. 1998) and overexpression of ADRP results in an increased formation of droplets (Imamura et al. 2002; Magnusson et al. 2006; Wang et al. 2003). The regulation of ADRP levels in the cell is complex. ADRP is regulated at the transcriptional level by peroxisome proliferator-activated receptor α (PPARα; Dalen et al. 2006; Edvardsson et al. 2006; Targett-Adams et al. 2003), but also through post-translational degradation by the proteasomal system (Masuda et al. 2006; Xu et al. 2005), which occurs when there are low levels of lipids in the cell. Thus, accumulation of intracellular triglycerides appears to stabilize ADRP and prevents it from being sorted to degradation.

Knockout of ADRP results in a rather modest phenotype (a reduced amount of triglycerides in the liver and resistance to diet-induced hepatosteatosis; no effect on adipocyte differentiation and lipolysis; Chang et al. 2006). One reason for this modest phenotype is that TIP47 is directed to the droplets and replaces ADRP (Sztalryd et al. 2006). ADRP–/– cells treated with siRNA against TIP47 retain the ability to form lipid droplets, although to a lesser extent, and added fatty acids are to a greater extent directed to phospholipid biosynthesis (Sztalryd et al. 2006). Interestingly, the combined knockdown of ADRP and TIP47 in cultured liver cells results in large lipid droplets with high turnover of triglycerides and insulin resistance (Bell 2006).

1.5.1.3 TIP47

TIP47 was initially described as a ubiquitously expressed cytosolic and endosomal 47-kDa protein involved in the intracellular transport of mannose 6-phosphate receptors between the *trans*-Golgi and endosomes (Diaz and Pfeffer 1998; Krise et al. 2000). TIP47 is believed to act as an effector for the Rab9 protein in this process, causing budding of vesicles directed to lysosomes (Carroll et al. 2001). TIP47 is also present on lipid droplets (Wolins et al. 2001). In contrast to ADRP, which is always associated with droplets and is degraded in the absence of neutral lipid, cytosolic TIP47 is shifted to lipid droplets in the presence of increased levels of fatty acids (Wolins et al. 2001).

The C-terminus of TIP47 has been crystallized and its structure determined (Hickenbottom et al. 2004). It has an α/β domain of novel topology and four helix bundles resembling the low-density lipoprotein (LDL) receptor-binding domain of apolipoprotein E (Hickenbottom et al. 2004) and the N-terminal domain of apolipoprotein A-I (Ajees et al. 2006; Lagerstedt et al. 2007). These results suggest an analogy between PAT proteins and plasma apolipoproteins.

1.5.1.4 LSDP5 and S3-12

LSDP5 is mainly expressed in tissues with high rates of β-oxidation, such as muscle, heart, liver and brown adipose tissue (Wolins et al. 2006), and shows a high degree of homology with ADRP and TIP47 (Dalen et al. 2007; Scherer et al. 1998). Forced overexpression of LSDP5 in cultured cells results in a substantial increase in the accumulation of triglycerides in response to fatty acid treatment (Wolins et al. 2006), which might be explained by a decrease in both basal and stimulated lipolysis (Dalen et al. 2007). Thus, LSDP5 seems to protect the triglyceride core of lipid droplets from degradation in a similar manner to perilipin. LSDP5 is transcriptionally regulated by PPARα in striated muscle and liver and by PPARγ in white adipose tissue (Dalen et al. 2007; Wolins et al. 2006; Yamaguchi et al. 2006).

S3-12 is mainly expressed in white adipose tissue; and it shares only a weak sequence homology with the PAT proteins (Wolins et al. 2003). S3-12 expression is transcriptionally regulated by PPARγ (Dalen et al. 2004).

1.5.2 Other Lipid Droplet-Associated Proteins

Vimentin has been shown to be present as cages around the lipid droplets (Franke et al. 1987). Knockdown by anti-sense RNA has been shown to result in a decrease in the formation of droplets (Lieber and Evans 1996). However, this finding was not verified in vimentin –/– mice (Colucci-Guyon et al. 1994).

Caveolin is also present on lipid droplets protein (Brasaemle et al. 2004; Liu et al. 2004) and, again, its function is not fully understood. One suggestion is that it may reflect an involvement of caveolin-rich plasma membranes in the assembly of lipid droplets (Ost et al. 2005). Caveolin has also been linked to triglyceride lipolysis in the droplets (for a review, see Martin and Parton 2006).

A number of other proteins have been identified on the lipid droplets by proteomics (Brasaemle et al. 2004; Cermelli et al. 2006; Liu et al. 2004). These include proteins involved in lipid biosynthesis (e.g. acyl-CoA synthetase, lanosterol synthetase) and turnover (e.g. ATGL, CGI-58, HSL) as well as in sorting and trafficking events (e.g. the Rab proteins; Grosshans et al. 2006). In addition, there are several other proteins with functions that are yet to be elucidated.

1.6 Lipid Droplets and the Secretion of Triglycerides from the Cell

Triglycerides are secreted as milk globules from the epithelial cells in mammary glands, as chylomicrons from enterocytes in the intestine and as very low-density lipoproteins (VLDL) from hepatocytes in the liver. The mechanism involved in the secretion of milk globules is very different from that involved in the secretion of chylomicrons and VLDL; and here we illustrate the two different mechanisms by focusing on the secretion of milk globules and VLDL.

1.6.1 The Assembly and Secretion of Milk Globules

Milk globules have been extensively reviewed (e.g. Heid and Keenan 2005; Mather and Keenan 1998; McManaman and Neville 2003; McManaman et al. 2007). The secreted milk globule consists of a core of triglycerides covered by a membrane that has a tripartite structure. The outer portion, or primary membrane, has a typical bilayer structure and seems to originate from the apical plasma membrane of the mammary epithelial cell. This is supported by the finding that secreted milk globules contain plasma membrane-specific proteins such as the plasma membrane calcium-transporting ATPase 2 (Reinhardt and Lippolis 2006). The material that covers the triglyceride core originates from the ER and has the appearance of a monolayer of lipid and proteins (Heid and Keenan 2005; Mather and Keenan 1998; McManaman and Neville 2003; McManaman et al. 2007).

The formation of milk globules starts in the ER by the formation of ADRP- and TIP47-containing lipid droplets, most likely by the same mechanism as other droplets (Fig. 1.3). The primordial milk globule has a diameter of less than 0.5 μm and it is thought that these newly formed droplets increase in size by fusion (Fig. 1.3; see reviews cited above). A recent proteomics study showed that milk globules appear to contain several SNARE proteins, such as SNAP23, VAMP2, syntaxin-3 and Ykt6 (Reinhardt and Lippolis 2006). The presence of significant

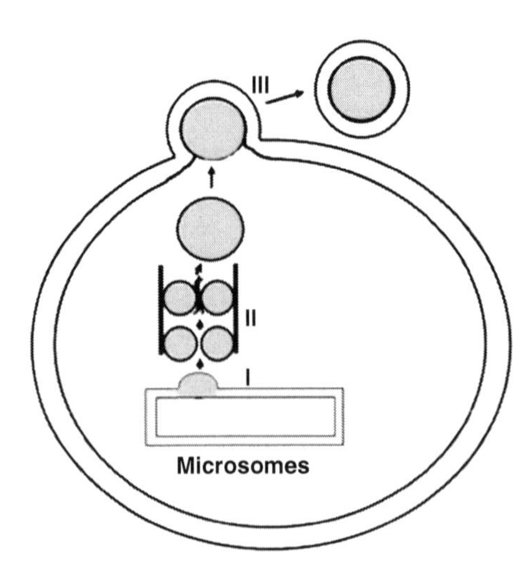

Fig. 1.3 The formation and secretion of milk globules from mammary glands. Primordial lipid droplets <0.5 μm in diameter are formed at microsomal membranes (*I*), transported on microtubules and increase in size by fusion (*II*) The mature lipid droplet is enveloped by the bi-layer of the plasma membrane and is then secreted (*III*)

levels of SNAP23 was confirmed by Western blot experiments; and thus the fusion process may be the same as that described for droplets in other cells. However, several of these proteins are abundant in the plasma membrane of the cell and the view that the outer bilayer of the milk globule structure is derived from the apical plasma membrane (see discussion below) could explain the presence of the SNARE proteins on milk globules. The mechanism by which lipid droplets are transported to the plasma membrane requires clarification.

Milk globules are secreted via a mechanism that differs completely from that used for the other types of triglyceride secretion; and two potential mechanisms have been proposed. First, it has been suggested that secretory vesicles fuse on the surface of the milk globule, leading to an intracytoplasmic vacuole containing both casein micelles and lipid droplets enveloped with secretory vesicle membrane. The content of such vacuoles is then released from the cell by exocytosis. Mather and Keenan (1998) discuss concerns about the data supporting this mechanism. The alternative and currently favored mechanism is that the milk globule approaches the plasma membrane where it is enveloped and, ultimately, pinched off (see Fig. 1.3; reviewed by Heid and Keenan 2005). The mechanism by which the milk globule interacts with the plasma membrane has not been elucidated in detail. However, butyrophilin and xanthine oxidoreductase are present both in milk globules and on the apical region of the plasma membrane of mammary gland epithelial cells; and it has been suggested that these proteins together with ADRP form a tripartite structure that is of importance for this interaction and the secretion of the milk globule (for reviews, see Heid and Keenan 2005; McManaman et al. 2007). There is still debate about the molecular details for the secretion of milk globules and indeed whether the three proteins interact (for a discussion, see McManaman et al. 2007).

1.6.2 ApoB100: the Structural Protein of VLDL

Liver and intestinal cells express apolipoprotein B (apoB), which is essential for the formation of triglyceride-containing lipoproteins and triglyceride secretion (Davidson and Shelness 2000b; Gibbons et al. 2004; Olofsson and Asp 2005; Olofsson et al. 1999, 2000). ApoB is a large amphipathic protein (Segrest et al. 2001), which exists in two forms: apoB100 and apoB48. In humans, apoB100 is expressed in the liver and is required for the formation of VLDL whereas apoB48 is expressed in the intestine and is lipidated to form chylomicrons. Here, we focus on apoB100 and the assembly of VLDL.

ApoB100 has a pentapartite structure consisting of one globular N-terminal structure, two domains of amphipathic β-sheets and two domains of amphipathic α-helices (Segrest et al. 2001). The N-terminal domain is of vital importance for the formation of VLDL as it interacts with microsomal triglyceride transfer protein (MTP), which catalyzes the transfer of lipids to apoB during the formation of lipoproteins (see below; Dashti et al. 2002).

ApoB differs from other apolipoproteins in that it is nonexchangeable, i.e. it cannot equilibrate between different lipoproteins but remains bound to the particle on which it was secreted into plasma. This is generally thought to be explained by the presence of antiparallel β-sheets with a width of approximately 30 Å, which form very strong lipid-binding structures (Segrest et al. 2001).

The three-dimensional structure of apoB100 is not known in detail, but its overall organization on LDL has been elucidated (Chatterton et al. 1995). ApoB has an elongated structure encompassing the entire particle. The C-terminus folds back over the preceding structure and crosses it at amino acid residue 3500 (arginine). The arginine binds to a tryptophan (residue 4396) preventing the C-terminus from sliding over the binding site for the LDL receptor (residues 3359–3369; Boren et al. 1998, 2001). A number of known mutations involving these amino acids break the arginine–tryptophan interaction and result in reduced binding of LDL to its receptor (Boren et al. 1998, 2001).

1.6.3 ApoB100 and the Secretory Pathway

Secretory proteins such as apoB are synthesized on ribosomes attached to the surface of the ER (Fig. 1.4). During its biosynthesis, the "nascent" polypeptide is translocated through a channel (for reviews, see Johnson and Haigh 2000; Johnson and van Waes 1999) to the lumen of the ER, where it is folded into its correct structure with the help of chaperone proteins. Correctly folded proteins are sorted into exit sites to leave the ER by transport vesicles. If the correct tertiary structure is not achieved, the protein is retained in the ER, retracted through the membrane channel and sorted to proteasomal degradation (Ellgaard and Helenius 2001, 2003; Ellgaard et al. 1999; Johnson and Haigh 2000; Johnson and van Waes 1999; Kostova and Wolf 2003; Lippincott-Schwartz et al. 2000).

ApoB100 exits the ER in vesicles that bud off from specific sites on the ER membrane and form the ER Golgi intermediate compartment (ERGIC), which is involved in protein sorting. ER-specific proteins are returned to the ER from the ERGIC and thereby prevented from entering the later part of the secretory pathway to be secreted. The ERGIC matures into *cis*-Golgi, which undergoes "cisternal maturation" to form the medial and *trans*-Golgi apparatus. During this maturation, the proteins that will be secreted are transferred through the Golgi stack. Finally, the proteins are transported from the *trans*-Golgi to the plasma membrane for secretion (for reviews, see Elsner et al. 2003; Kartberg et al. 2005).

1.6.4 The Assembly of VLDL

The assembly of VLDL involves three types of particles: a primordial lipoprotein (pre-VLDL), a triglyceride-poor form of VLDL (VLDL2) and a triglyceride-rich atherogenic form of VLDL (VLDL1; Stillemark-Billton et al. 2005).

1.6.4.1 The Primordial Lipoprotein

The assembly process starts when the growing apoB100 is co-translationally lipidated by MTP in the lumen of the ER (Fig. 1.4). By analyzing the content of the secretory pathway, we revealed a dense form of an apoB100-containing lipoprotein (Boström et al. 1988; Stillemark-Billton et al. 2005). This lipoprotein is not secreted from the cell but is a precursor to VLDL2 and VLDL1. We propose that it represents a partially lipidated form of apoB100; and we refer to it as a primordial lipoprotein (or pre-VLDL). It differs from the VLDL2 analog formed by apoB48 (see below), which is a mature particle that is avidly secreted from the cell (Stillemark-Billton et al. 2005).

The appearance of the apoB100 primordial lipoprotein is highly dependent on the C-terminal region of apoB100 (Stillemark-Billton et al. 2005). Moreover, the

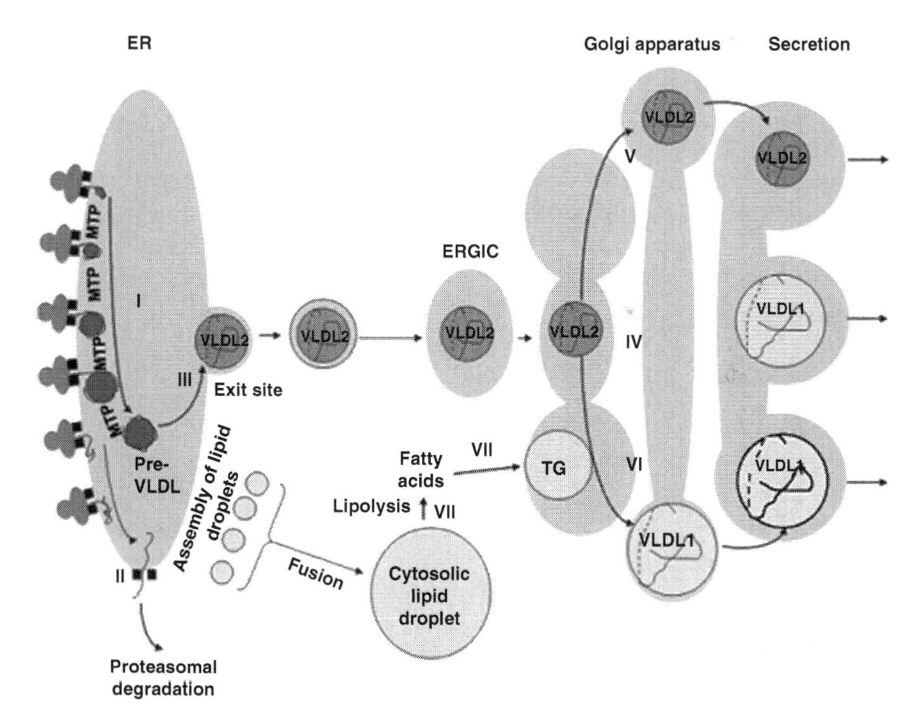

Fig. 1.4 The assembly and secretion of VLDL. ApoB100 is co-translationally lipidated in the ER by the transfer protein MTP to form a partially assembled primordial particle (pre-VLDL; *I*). If apoB100 is not co-translationally lipidated, it is retracted to the cytosol and sorted to proteasomal degradation (*II*). Pre-VLDL is either retained and degraded, or further lipidated to form a small triglyceride-poor form of VLDL (VLDL2; *III*). VLDL2 reaches the Golgi apparatus (*IV*) where it can either be secreted (*V*) or converted to the large triglyceride-rich form, VLDL1, by a bulk addition of triglycerides (*VI*). The cytosolic lipid droplets supply the apoB with triglycerides. The triglycerides in the droplets are thought to be hydrolyzed (*VII*) and the released fatty acids are re-esterified into new triglycerides that are added onto apoB (*VIII*)

particle is highly associated with chaperone proteins such as BiP (binding proteins) and PDI (protein disulfide isomerise) A potential explanation is that regions in the C-terminal of apoB100 sense the degree of lipidation and anchor the partially lipidated particle to chaperones, which retain the particle in the ER. Once the level of lipidation is sufficient to allow apoB100 to fit on the particle and the C-terminal to fold correctly, the chaperones dissociate and the particle can transfer out of the ER. Thus, the primordial lipoprotein is either retained in the cell and degraded (if not sufficiently lipidated) or further lipidated to form VLDL2 (Fig. 1.4; Olofsson and Asp 2005; Olofsson et al. 1999, 2000; Stillemark-Billton et al. 2005).

1.6.4.2 VLDL2

VLDL2 formation is highly dependent on the size of apoB. Bona fide VLDL2 is only formed with apoB100; and truncated forms of apoB result in the formation of denser particles (Stillemark et al. 2000). Indeed, there is an inverse relation between the density of the lipoproteins formed and the size of apoB. The lipoprotein formed by apoB48 (i.e. the apoB expressed in the intestine) has the density of a high-density lipoprotein (HDL; Stillemark-Billton et al. 2005) and we propose that it is a VLDL2 analog. Both bona fide VLDL2 and VLDL2 analogs can either be secreted directly or further lipidated to form VLDL1 and secreted (Fig. 1.4).

1.6.4.3 VLDL1

VLDL1 is formed from VLDL2. Thus a precursor product relationship between VLDL2 and VLDL1 is seen in pulse-chase experiments in cell cultures (Stillemark-Billton et al. 2005) and also in turnover experiments in humans (Adiels et al. 2007).

The formation of VLDL1 involves a second type of lipidation in which VLDL2 or the VLDL2 analog receives a bulk load of lipids in the Golgi apparatus (Stillemark-Billton et al. 2005). In contrast to the formation of VLDL2, apoB only needs to have a minimum size of apoB48 to allow the conversion to VLDL1 (Stillemark-Billton et al. 2005). Our recent results indicate that a sequence very close to the C-terminal end of apoB48 is required for this lipidation (Beck et al., personal communication).

The formation of VLDL1 is dependent on the GTPase ADP ribosylation factor 1 (ARF-1), a protein that is required in the anterograde transport from ERGIC to *cis*-Golgi (Asp et al. 2000). This is consistent with results showing that VLDL1 formation occurs in the Golgi apparatus (Stillemark et al. 2000) and indicates that the formation of VLDL1 requires a transfer of apoB100 from the ER to the Golgi apparatus (Fig. 1.4). Thus, one would expect a time delay of approximately 15 min between the biosynthesis of apoB100 and the major addition of lipids to form the VLDL1 particle. Indeed, in turnover studies in humans, we confirmed the presence of two steps in the assembly of VLDL1 with a 15 min difference between the secretion of newly formed apoB100 and newly formed triglycerides (Adiels et al. 2005b).

It is not known how lipids are added to VLDL2 during the formation of VLDL1, but it has been suggested that the formation of lipid droplets in the lumen of the secretory pathway plays a central role in VLDL assembly, i.e. lipid droplets fuse with apoB to form VLDL (Alexander et al. 1976). This is an interesting hypothesis as it links the formation of the core of VLDL to the process by which the core of a cytosolic lipid droplet is assembled. As discussed above, it has been proposed that the assembly of a droplet starts in the hydrophobic portion of the microsomal membrane with the formation of a triglyceride lens, which is then released to the cytosol. A triglyceride lens may also bud into the lumen of the secretory pathway, thereby giving rise to a luminal droplet that becomes the core of VLDL. This hypothesis remains to be tested experimentally (Brown 2001; Murphy and Vance 1999; Olofsson et al. 1987).

Several authors have demonstrated that fatty acids used for the biosynthesis of VLDL triglycerides are derived from triglycerides stored in cytosolic lipid droplets (Gibbons et al. 2000; Salter et al. 1998; Wiggins and Gibbons 1992) and that the enzymes involved in the release of such fatty acids have an influence on the formation of VLDL (Dolinsky et al. 2004a, b; Gilham et al. 2003; Lehner and Vance 1999; Trickett et al. 2001).

1.6.5 Regulation of VLDL Assembly

1.6.5.1 Insulin

Several possibilities to explain why insulin inhibits VLDL1 formation have been summarized by Taskinen (2003). One potential reason is that insulin inhibits lipolysis in adipose tissue and thereby decreases the inflow of fatty acids to the liver. The assembly of VLDL and in particular VLDL1 is highly dependent on the triglyceride level in the hepatocytes; and thus an insulin-induced reduction in the triglyceride level results in a reduced formation of VLDL1. This is supported by the observation that the insulin-dependent inhibition of VLDL1 assembly is not working in patients with high amount of lipids in the liver (Adiels et al. 2007). Further support for an influence on the lipolysis in the adipose tissue was obtained from studies using nicotinic acid analogs (reviewed by Taskinen 2003).

1.6.5.2 Lipid Droplets

Insulin is known to promote the formation of lipid droplets (Andersson et al. 2006). Thus, it is possible that insulin diverts triglycerides from the bulk lipidation of VLDL2 to the formation of lipid droplets. Indeed, we have observed other situations where promoting the formation of lipid droplets inhibits VLDL1 assembly. For example, increased levels of ADRP in the cell result in increased storage of neutral lipids in lipid droplets and reduce their entry into the assembly pathway (Magnusson et al. 2006). Such a manipulation results in decreased VLDL1 production, despite increased levels of triglycerides in the cell. In addition, increasing the rate of fusion between droplets (e.g. by treatment with epigallocatechin gallate) results in decreased VLDL secretion and an increased number of droplets in the liver cell (Li et al. 2005).

1.6.5.3 ApoB100 Degradation

It is generally believed that the secretion of apoB is regulated post-transcriptionally by co- and post-translational degradation. It has long been known that apoB100 undergoes intracellular degradation (for reviews, see Davidson and Shelness 2000a; Olofsson et al. 1999; Shelness and Sellers 2001). The degradation is dramatically reduced when the supply of fatty acids (and the biosynthesis of triglycerides) is increased (Borén et al. 1993; Boström et al. 1986). The intracellular degradation of apoB100 occurs at three different levels (Fisher and Ginsberg 2002; Fisher et al. 2001): (i) close to the biosynthesis of apoB (co- or post-translationally) by a mechanism that involves retraction of the apoB molecule from the lumen of the ER to the cytosol (through the same channel as it entered during its biosynthesis), ubiquitination and subsequent proteasomal degradation (Fisher et al. 2001; Liang et al. 2000; Mitchell et al. 1998; Pariyarath et al. 2001), (ii) post-translationally by an unknown mechanism that seems to occur in a compartment separate from the rough ER and is referred to as post-ER presecretory proteolysis (PERPP; Fisher et al. 2001) and (iii) by reuptake from the unstirred water layer around the outside of the plasma membrane (Williams et al. 1990) via the LDL receptor. The LDL receptor has been shown to have an important role in regulation of the secretion of apoB100-containing lipoproteins (Horton et al. 1999; Twisk et al. 2000).

The intracellular degradation of apoB seems to be a consequence of a failure to form the correct particle. To avoid degradation, apoB100 needs to form pre-VLDL during translation and pre-VLDL must be converted to VLDL2 (see Sect. 1.6.4.1). Both are depending on the amount of lipids that are loaded on to apoB. The formation of VLDL1 is not necessary for apoB100 secretion but allows increased secretion of triglycerides from the liver. The importance of triglycerides for the assembly and secretion of VLDL1 is supported by our observations from turnover studies in vivo, which have demonstrated that the secretion of VLDL1 apoB100 increases with increasing concentrations of liver lipids (Adiels et al. 2005a)

1.6.6 Clinical Implications of VLDL1 Production

Insulin resistance is a major risk factor for the development of premature atherosclerosis. In part, this could be explained by the increased production of VLDL1. Overproduction of VLDL1 has a central role in the development of the atherogenic dyslipidemia of diabetes (Taskinen 2003), which is characterized by low levels of high-density lipoproteins, the appearance of small dense LDL (sdLDL) and high levels of plasma triglycerides and apoB.

A model has been proposed in which LDL, in particular sdLDL, is retained by the intercellular substance of the arterial wall (Tabas et al. 2007; Williams and Tabas 1995, 1998). Through modification, this LDL is targeted to endocytosis into macrophages via scavenger receptors (Tabas et al. 2007). This results in the accumulation of neutral lipids [primarily cholesterol esters (Mattsson et al. 1993) but also triglycerides (Mattsson et al. 1993)] within lipid droplets in these cells. Such lipid-loaded macrophages, or foam cells, are a characteristic feature of atherosclerotic lesions and are highly involved in the progression of such lesions.

1.7 Conclusions

Triglycerides are stored in lipid droplets, dynamic organelles equipped with a protein machinery that allows their participation in sorting and trafficking processes. Here, we have presented recent breakthroughs in our understanding of how lipid droplets are assembled. Additional studies are required to determine how and where in the cell the primordial lipid droplets are formed. Of key importance is the identification of a link between the accumulation of lipid droplets and the development of insulin resistance, but further research is required to clarify the precise mechanism and proteins involved. We propose that these investigations may identify targets that could be modulated to reduce the accumulation of lipid droplets and hence reverse the associated complications.

Triglycerides can be secreted both as milk globules and lipoproteins (chylomicrons, VLDL). We have reviewed the assembly and secretion of milk globules. Further research is required to clarify how a newly assembled milk globule interacts with the plasma membrane during the budding of the globule. We have also reviewed the assembly of VLDL and described a model for its assembly. As overproduction of VLDL1 is a key feature of insulin resistance and type 2 diabetes, it is important to further elucidate how VLDL formation is regulated in humans and to clarify the role of cytosolic lipid droplets in the formation of VLDL1.

Acknowledgements We thank Dr Rosie Perkins for expert editing of the manuscript. This work was supported by grants from the Swedish Research Council, the Swedish Foundation for Strategic Research, the Swedish Heart and Lung Foundation, the NovoNordic Foundation, The Swedish Diabetes Society and the EU program NACARDIO.

References

Adiels M, Boren J, Caslake MJ, Stewart P, Soro A, Westerbacka J, Wennberg B, Olofsson SO, Packard C, Taskinen MR (2005a) Overproduction of VLDL1 driven by hyperglycemia is a dominant feature of diabetic dyslipidemia. Arterioscler Thromb Vasc Biol 25:1697–1703

Adiels M, Packard C, Caslake MJ, Stewart P, Soro A, Westerbacka J, Wennberg B, Olofsson SO, Taskinen MR, Boren J (2005b) A new combined multicompartmental model for apolipoprotein B-100 and triglyceride metabolism in VLDL subfractions. J Lipid Res 46:58–67

Adiels M, Westerbacka J, Soro-Paavonen A, Hakkinen AM, Vehkavaara S, Caslake MJ, Packard C, Olofsson SO, Yki-Jarvinen H, Taskinen MR, et al. (2007) Acute suppression of VLDL1 secretion rate by insulin is associated with hepatic fat content and insulin resistance. Diabetologia 50:2356–2365

Agarwal AK, Garg A (2003) Congenital generalized lipodystrophy: significance of triglyceride biosynthetic pathways. Trends Endocrinol Metab 14:214–221

Ajees AA, Anantharamaiah GM, Mishra VK, Hussain MM, Murthy HM (2006) Crystal structure of human apolipoprotein A-I: insights into its protective effect against cardiovascular diseases. Proc Natl Acad Sci USA 103:2126–2131

Alexander CA, Hamilton RL, Havel RJ (1976) Subcellular localization of B apoprotein of plasma lipoproteins in rat liver. J Cell Biol 69:241–263

Andersson L, Bostrom P, Ericson J, Rutberg M, Magnusson B, Marchesan D, Ruiz M, Asp L, Huang P, Frohman MA, et al. (2006) PLD1 and ERK2 regulate cytosolic lipid droplet formation. J Cell Sci 119:2246–2257

Asp L, Claesson C, Borén J, Olofsson S-O (2000) ADP-ribosylation factor 1 and its activation of phospholipase D are important for the assembly of very low density lipoproteins. J Biol Chem 275:26285–26292

Bell M, Fo S, Londos C, Sztalryd C (2006) ADRP and Tip-47 down-regulation results in abnormal lipid droplet morphology, lipid metabolism and insulin-stimulated Akt phosphorylation in cultured liver cells. Diabetes 55:A326

Borén J, Rustaeus S, Wettesten M, Andersson M, Wiklund A, Olofsson S-O (1993) Influence of triacylglycerol biosynthesis rate on the assembly of apoB-100-containing lipoproteins in Hep G2 cells. Arterioscler Thromb 13:1743–1754

Boren J, Lee I, Zhu W, Arnold K, Taylor S, Innerarity TL (1998) Identification of the low density lipoprotein receptor-binding site in apolipoprotein B100 and the modulation of its binding activity by the carboxyl terminus in familial defective apo-B100. J Clin Invest 101:1084–1093

Boren J, Ekstrom U, Agren B, Nilsson-Ehle P, Innerarity TL (2001) The molecular mechanism for the genetic disorder familial defective apolipoprotein B100. J Biol Chem 276:9214–9218

Boström K, Wettesten M, Borén J, Bondjers G, Wiklund O, Olofsson S-O (1986) Pulse-chase studies of the synthesis and intracellular transport of apolipoprotein B-100 in Hep G2 cells. J Biol Chem 261:13800–13806

Boström K, Borén J, Wettesten M, Sjöberg A, Bondjers G, Wiklund O, Carlsson P, Olofsson S-O (1988) Studies on the assembly of apo B-100-containing lipoproteins in HepG2 cells. J Biol Chem 263:4434–4442

Bostrom P, Rutberg M, Ericsson J, Holmdahl P, Andersson L, Frohman MA, Boren J, Olofsson SO (2005) Cytosolic lipid droplets increase in size by microtubule-dependent complex formation. Arterioscler Thromb Vasc Biol 25:1945–1951

Bostrom P, Andersson L, Rutberg M, Perman J, Lidberg U, Johansson BR, Fernandez-Rodriguez J, Ericson J, Nilsson T, Boren J, et al. (2007) SNARE proteins mediate fusion between cytosolic lipid droplets and are implicated in insulin sensitivity. Nat Cell Biol 9:1286–1293

Brasaemle DL (2007) Thematic review series: adipocyte biology The perilipin family of structural lipid droplet proteins: stabilization of lipid droplets and control of lipolysis. J Lipid Res 48:2547–2559

Brasaemle DL, Barber T, Wolins NE, Serrero G, Blanchette-Mackie EJ, Londos C (1997) Adipose differentiation-related protein is an ubiquitously expressed lipid storage droplet-associated protein. J Lipid Res 38:2249–2263

Brasaemle DL, Rubin B, Harten IA, Gruia-Gray J, Kimmel AR, Londos C (2000) Perilipin A increases triacylglycerol storage by decreasing the rate of triacylglycerol hydrolysis. J Biol Chem 275:38486–38493

Brasaemle DL, Dolios G, Shapiro L, Wang R (2004) Proteomic analysis of proteins associated with lipid droplets of basal and lipolytically stimulated 3T3-L1 adipocytes. J Biol Chem 279:46835–46842

Brown DA (2001) Lipid droplets: proteins floating on a pool of fat. Curr Biol 11:R446–R449

Cao J, Li JL, Li D, Tobin JF, Gimeno RE (2006) Molecular identification of microsomal acyl-CoA:glycerol-3-phosphate acyltransferase, a key enzyme in de novo triacylglycerol synthesis. Proc Natl Acad Sci USA 103:19695–19700
Carman GM, Han GS (2006) Roles of phosphatidate phosphatase enzymes in lipid metabolism. Trends Biochem Sci 31:694–699
Carroll KS, Hanna J, Simon I, Krise J, Barbero P, Pfeffer SR (2001) Role of Rab9 GTPase in facilitating receptor recruitment by TIP47. Science 292:1373–1376
Cermelli S, Guo Y, Gross SP, Welte MA (2006) The lipid-droplet proteome reveals that droplets are a protein-storage depot. Curr Biol 16:1783–1795
Chang BH, Li L, Paul A, Taniguchi S, Nannegari V, Heird WC, Chan L (2006) Protection against fatty liver but normal adipogenesis in mice lacking adipose differentiation-related protein. Mol Cell Biol 26:1063–1076
Chatterton JE, Phillips ML, Curtiss LK, Milne R, Fruchart J-C, Schumaker VN (1995) Immunoelectron microscopy of low density lipoproteins yields a ribbon and bow model for the conformation of apolipoprotein B on the lipoprotein surface. J Lipid Res 36:2027–2037
Coleman RA, Lewin TM, Muoio DM (2000) Physiological and nutritional regulation of enzymes of triacylglycerol synthesis. Annu Rev Nutr 20:77–103
Colucci-Guyon E, Portier M-M, Dunia I, Paulin D, Pournin S, Babinet C (1994) Mice lacking vimentin develop and reproduce without obvious phenotype. Cell 79:679–694
Dalen KT, Schoonjans K, Ulven SM, Weedon-Fekjaer MS, Bentzen TG, Koutnikova H, Auwerx J, Nebb HI (2004) Adipose tissue expression of the lipid droplet-associating proteins S3–12 and perilipin is controlled by peroxisome proliferator-activated receptor-gamma. Diabetes 53:1243–1252
Dalen KT, Ulven SM, Arntsen BM, Solaas K, Nebb HI (2006) PPARalpha activators and fasting induce the expression of adipose differentiation-related protein in liver. J Lipid Res 47:931–943
Dalen KT, Dahl T, Holter E, Arntsen B, Londos C, Sztalryd C, Nebb HI (2007) LSDP5 is a PAT protein specifically expressed in fatty acid oxidizing tissues. Biochim Biophys Acta 1771:210–227
Dashti N, Gandhi M, Liu X, Lin X, Segrest JP (2002) The N-terminal 1000 residues of apolipoprotein B associate with microsomal triglyceride transfer protein to create a lipid transfer pocket required for lipoprotein assembly. Biochemistry 41:6978–6987
Davidson NO, Shelness GS (2000a) Apolipoprotein B: mRNA editing, lipoprotein assembly, and presecretory degradation. Annu Rev Nutr 20:169–193
Davidson NO, Shelness GS (2000b) APOLIPOPROTEIN B: mRNA editing, lipoprotein assembly, and presecretory degradation. Annu Rev Nutr 20:169–193
Diaz E, Pfeffer SR (1998) TIP47: a cargo selection device for mannose 6-phosphate receptor trafficking. Cell 93:433–443
Dolinsky VW, Douglas DN, Lehner R, Vance DE (2004a) Regulation of the enzymes of hepatic microsomal triacylglycerol lipolysis and re-esterification by the glucocorticoid dexamethasone. Biochem J 378:967–974
Dolinsky VW, Gilham D, Alam M, Vance DE, Lehner R (2004b) Triacylglycerol hydrolase: role in intracellular lipid metabolism. Cell Mol Life Sci 61:1633–1651
Du G, Huang P, Liang BT, Frohman MA (2004) Phospholipase D2 localizes to the plasma membrane and regulates angiotensin II receptor endocytosis. Mol Biol Cell 15:1024–1030
Dugani CB, Klip A (2005) Glucose transporter 4: cycling, compartments and controversies. EMBO Rep 6:1137–1142
Edvardsson U, Ljungberg A, Linden D, William-Olsson L, Peilot-Sjogren H, Ahnmark A, Oscarsson J (2006) PPARalpha activation increases triglyceride mass and adipose differentiation-related protein in hepatocytes. J Lipid Res 47:329–340
Ellgaard L, Helenius A (2001) ER quality control: towards an understanding at the molecular level. Curr Opin Cell Biol 13:431–437
Ellgaard L, Helenius A (2003) Quality control in the endoplasmic reticulum. Nat Rev Mol Cell Biol 4:181–191
Ellgaard L, Molinari M, Helenius A (1999) Setting the standards: quality control in the secretory pathway. Science 286:1882–1888

Elsner M, Hashimoto H, Nilsson T (2003) Cisternal maturation and vesicle transport: join the band wagon. Mol Membr Biol 20:221–229

Falholt K, Jensen I, Lindkaer Jensen S, Mortensen H, Volund A, Heding LG, Noerskov Petersen P, Falholt W (1988) Carbohydrate and lipid metabolism of skeletal muscle in type 2 diabetic patients. Diabet Med 5:27–31

Fisher EA, Ginsberg HN (2002) Complexity in the secretory pathway: the assembly and secretion of apolipoprotein B-containing lipoproteins. J Biol Chem 277:17377–17380

Fisher EA, Pan M, Chen X, Wu X, Wang H, Jamil H, Sparks JD, Williams KJ (2001) The triple threat to nascent apolipoprotein B Evidence for multiple, distinct degradative pathways. J Biol Chem 276:27855–27863

Franke WW, Hergt M, Grund C (1987) Rearragement of vimentin cytoskeleton durin adipose conversion: formation of an intermediate filament cage around lipid globules. Cell 49:131–141

Freyberg Z, Sweeney D, Siddhanta A, Bourgoin S, Frohman M, Shields D (2001) Intracellular localization of phospholipase D1 in mammalian cells. Mol Biol Cell 12:943–955

Gibbons GF, Islam K, Pease RJ (2000) Mobilisation of triacylglycerol stores. Biochim Biophys Acta 1483:37–57

Gibbons GF, Wiggins D, Brown AM, Hebbachi AM (2004) Synthesis and function of hepatic very-low-density lipoprotein. Biochem Soc Trans 32:59–64

Gilham D, Ho S, Rasouli M, Martres P, Vance DE, Lehner R (2003) Inhibitors of hepatic microsomal triacylglycerol hydrolase decrease very low density lipoprotein secretion. FASEB J 17:1685–1687

Gonzalez-Baro MR, Lewin TM, Coleman RA (2007) Regulation of triglyceride metabolism. II. Function of mitochondrial GPAT1 in the regulation of triacylglycerol biosynthesis and insulin action. Am J Physiol Gastrointest Liver Physiol 292:G1195–G1199

Goossens GH (2007) The role of adipose tissue dysfunction in the pathogenesis of obesity-related insulin resistance. Physiol Behav 94:206–218

Granneman JG, Moore HP, Granneman RL, Greenberg AS, Obin MS, Zhu Z (2007) Analysis of lipolytic protein trafficking and interactions in adipocytes. J Biol Chem 282:5726–5735

Greenberg AS, Egan JJ, Wek SA, Garty NB, Blanchette-Mackie EJ, Londos C (1991) Perilipin, a major hormonally regulated adipocyte-specific phosphoprotein associated with the periphery of lipid storage droplets. J Biol Chem 266:11341–11346

Greenberg AS, Egan JJ, Wek SA, Moos MC Jr, Londos C, Kimmel AR (1993) Isolation of cDNAs for perilipins A and B: sequence and expression of lipid droplet-associated proteins of adipocytes. Proc Natl Acad Sci USA 90:12035–12039

Grosshans BL, Ortiz D, Novick P (2006) Rabs and their effectors: achieving specificity in membrane traffic. Proc Natl Acad Sci USA 103:11821–11827

Heid HW, Keenan TW (2005) Intracellular origin and secretion of milk fat globules. Eur J Cell Biol 84:245–258

Heid HW, Moll R, Schwetlick I, Rackwitz HR, Keenan TW (1998) Adipophilin is a specific marker of lipid accumulation in diverse cell types and diseases. Cell Tissue Res 294:309–321

Hickenbottom SJ, Kimmel AR, Londos C, Hurley JH (2004) Structure of a lipid droplet protein; the PAT family member TIP47. Structure (Camb) 12:1199–1207

High S, Abell BM (2004) Tail-anchored protein biosynthesis at the endoplasmic reticulum: the same but different. Biochem Soc Trans 32:659–662

Hong W (2005) SNAREs and traffic. Biochim Biophys Acta 1744:493–517

Horton JD, Shimano H, Hamilton RL, Brown MS, Goldstein JL (1999) Disruption of LDL receptor gene in transgenic SREBP-1a mice unmask hyperlipidemia resulting from production of lipid-rich VLDL. J Clin Invest 103:1067–1076

Huang S, Czech MP (2007) The GLUT4 glucose transporter. Cell Metab 5:237–252

Imamura M, Inoguchi T, Ikuyama S, Taniguchi S, Kobayashi K, Nakashima N, Nawata H (2002) ADRP stimulates lipid accumulation and lipid droplet formation in murine fibroblasts. Am J Physiol Endocrinol Metab 283:E775–E783

Jahn R, Scheller RH (2006) SNAREs – engines for membrane fusion. Nat Rev Mol Cell Biol 7:631–643

Jiang HP, Serrero G (1992) Isolation and characterization of a full-length cDNA coding for an adipose differentiation-related protein. Proc Natl Acad Sci USA 89:7856–7860

Johnson AE, Haigh NG (2000) The ER translocon and retrotranslocation: is the shift into reverse manual or automatic. Cell 102:709–712
Johnson AE, van Waes MA (1999) The translocon: a dynamic gateway at the ER membrane. Annu Rev Cell Dev Biol 15:799–842
Kartberg F, Elsner M, Froderberg L, Asp L, Nilsson T (2005) Commuting between Golgi cisternae – mind the GAP. Biochim Biophys Acta 1744:351–363
Kostova Z, Wolf DH (2003) For whom the bell tolls: protein quality control of the endoplasmic reticulum and the ubiquitin–proteasome connection. EMBO J 22:2309–2317
Kovacs P, Stumvoll M (2005) Fatty acids and insulin resistance in muscle and liver. Best Pract Res Clin Endocrinol Metab 19:625–635
Koves TR, Ussher JR, Noland RC, Slentz D, Mosedale M, Ilkayeva O, Bain J, Stevens R, Dyck JR, Newgard CB, et al. (2008) Mitochondrial overload and incomplete fatty acid oxidation contribute to skeletal muscle insulin resistance. Cell Metab 7:45–56
Krise JP, Sincock PM, Orsel JG, Pfeffer SR (2000) Quantitative analysis of TIP47-receptor cytoplasmic domain interactions: implications for endosome-to-trans Golgi network trafficking. J Biol Chem 275:25188–25193
Krssak M, Falk Petersen K, Dresner A, DiPietro L, Vogel SM, Rothman DL, Roden M, Shulman GI (1999) Intramyocellular lipid concentrations are correlated with insulin sensitivity in humans: a 1H NMR spectroscopy study. Diabetologia 42:113–116
Kuerschner L, Moessinger C, Thiele C (2008) Imaging of lipid biosynthesis: how a neutral lipid enters lipid droplets. Traffic 9:338–352
Lagerstedt JO, Budamagunta MS, Oda MN, Voss JC (2007) Electron paramagnetic resonance spectroscopy of site-directed spin labels reveals the structural heterogeneity in the N-terminal domain of apoA-I in solution. J Biol Chem 282:9143–9149
Lass A, Zimmermann R, Haemmerle G, Riederer M, Schoiswohl G, Schweiger M, Kienesberger P, Strauss JG, Gorkiewicz G, Zechner R (2006) Adipose triglyceride lipase-mediated lipolysis of cellular fat stores is activated by CGI-58 and defective in Chanarin–Dorfman Syndrome. Cell Metab 3:309–319
Lehner R, Vance DE (1999) Cloning and expression of a cDNA encoding a hepatic microsomal lipase that mobilizes stored triacylglycerol. Biochem J 343:1–10
Leung DW (2001) The structure and functions of human lysophosphatidic acid acyltransferases. Front Biosci 6:D944–D953
Li L, Stillemark-Billton P, Beck C, Bostrom P, Andersson L, Rutberg M, Ericsson J, Magnusson B, Marchesan D, Ljungberg A, et al. (2005) Epigallocatechin gallate promotes the formation of cytosolic lipid droplets and causes a subsequent reduction in the assembly and secretion of VLDL-containing apoB-100. J Lipid Res 47:67–77
Liang J-S, Wu X, Fisher EA, Ginsberg HN (2000) The amino-terminal domain of apolipoprotein B does not undergo retrograde translocation from the endoplasmic reticulum to the cytosol. Proteasomal degradation of nascent apolipoprotein B begins at the carboxyl terminus of the protein, while apolipoprotein B is still in its original translocon. J Biol Chem 275:32003–32010
Lieber JG, Evans RM (1996) Disruption of the intermediate filament system during adipose conversion of 3T3-L1 cells inhibits lipid droplet accumulation. J Cell Sci 109:3047–3058
Lippincott-Schwartz J, Roberts TH, Hirschberg K (2000) Secretory protein trafficking and organelle dynamics in living cells. Annu Rev Cell Dev Biol 16:557–589
Liu P, Ying Y, Zhao Y, Mundy DI, Zhu M, Anderson RG (2004) Chinese hamster ovary K2 cell lipid droplets appear to be metabolic organelles involved in membrane traffic. J Biol Chem 279:3787–3792
Londos C, Brasaemle DL, Gruia-Gray J, Servetnick DA, Schultz CJ, Levin DM, Kimmel AR (1995) Perilipin: unique proteins associated with intracellular neutral lipid droplets in adipocytes and steroidogenic cells. Biochem Soc Trans 23:611–615
Londos C, Brasaemle DL, Schultz CJ, Segrest JP, Kimmel AR (1999) Perilipins, ADRP, and other proteins that associate with intracellular neutral lipid droplets in animal cells. Semin Cell Dev Biol 10:51–58
Londos C, Sztalryd C, Tansey JT, Kimmel AR (2005) Role of PAT proteins in lipid metabolism. Biochimie 87:45–49

Lu X, Gruia-Gray J, Copeland NG, Gilbert DJ, Jenkins NA, Londos C, Kimmel AR (2001) The murine perilipin gene: the lipid droplet-associated perilipins derive from tissue-specific, mRNA splice variants and define a gene family of ancient origin. Mamm Genome 12:741–749

Machann J, Haring H, Schick F, Stumvoll M (2004) Intramyocellular lipids and insulin resistance. Diabetes Obes Metab 6:239–248

Magnusson B, Asp L, Bostrom P, Ruiz M, Stillemark-Billton P, Linden D, Boren J, Olofsson SO (2006) Adipocyte differentiation-related protein promotes fatty acid storage in cytosolic triglycerides and inhibits secretion of very low-density lipoproteins. Arterioscler Thromb Vasc Biol 26:1566–1571

Marchesan D, Rutberg M, Andersson L, Asp L, Larsson T, Boren J, Johansson BR, Olofsson SO (2003) A phospholipase D-dependent process forms lipid droplets containing caveolin, adipocyte differentiation-related protein, and vimentin in a cell-free system. J Biol Chem 278:27293–27300

Martin S, Parton RG (2006) Lipid droplets: a unified view of a dynamic organelle. Nat Rev Mol Cell Biol 7:373–378

Martinez-Botas J, Anderson JB, Tessier D, Lapillonne A, Chang BH, Quast MJ, Gorenstein D, Chen KH, Chan L (2000) Absence of perilipin results in leanness and reverses obesity in Lepr(db/db) mice. Nat Genet 26:474–479

Masuda Y, Itabe H, Odaki M, Hama K, Fujimoto Y, Mori M, Sasabe N, Aoki J, Arai H, Takano T (2006) ADRP/adipophilin is degraded through the proteasome-dependent pathway during regression of lipid-storing cells. J Lipid Res 47:87–98

Mather IH, Keenan TW (1998) Origin and secretion of milk lipids. J Mamm Gland Biol Neoplasia 3:259–273

Mattsson L, Johansson H, Ottoson M, Bondjers G, Wiklund O (1993) Expression of lipoprotein lipase mRNA and secretion in macrophages isolated from human atherosclerotic aorta. J Clin Invest 92:1759–1765

McManaman JL, Neville MC (2003) Mammary physiology and milk secretion. Adv Drug Deliv Rev 55:629–641

McManaman JL, Russell TD, Schaack J, Orlicky DJ, Robenek H (2007) Molecular determinants of milk lipid secretion. J Mamm Gland Biol Neoplasia 12:259–268

Mitchell DM, Zhou M, Pariyarath R, Wang H, Aitchison JD, Ginsberg HN, Fisher EA (1998) Apoprotein B 100 has a prolonged interaction with the translocon during which its lipidation and translocation change from dependence on the microsomal triglyceride transfer protein to independence. Proc Natl Acad Sci USA 95:14733–14738

Miura S, Gan JW, Brzostowski J, Parisi MJ, Schultz CJ, Londos C, Oliver B, Kimmel AR (2002) Functional conservation for lipid storage droplet association among Perilipin, ADRP, and TIP47. (PAT)-related proteins in mammals, *Drosophila*, and *Dictyostelium*. J Biol Chem 277:32253–32257

Miyoshi H, Perfield JW 2nd, Souza SC, Shen WJ, Zhang HH, Stancheva ZS, Kraemer FB, Obin MS, Greenberg AS (2007) Control of adipose triglyceride lipase action by serine 517 of perilipin A globally regulates protein kinase A-stimulated lipolysis in adipocytes. J Biol Chem 282:996–1002

Morino K, Petersen KF, Shulman GI (2006) Molecular mechanisms of insulin resistance in humans and their potential links with mitochondrial dysfunction. Diabetes 55[Suppl 2]:S9–S15

Murphy DJ, Vance J (1999) Mechanisms of lipid-body formation. Trends Biochem Sci 24:109–115

Nakamura N, Fujimoto T (2003) Adipose differentiation-related protein has two independent domains for targeting to lipid droplets. Biochem Biophys Res Commun 306:333–338

Olofsson S-O, Asp L (2005) The assembly of very low density lipoproteins. (VLDL) in the liver. In: Packard CJ, Rader D. (eds) Lipid and atherosclerosis: advances in translational medicine. Taylor and Francis, London, pp 1–15

Olofsson SO, Boren J (2005) Apolipoprotein B: a clinically important apolipoprotein which assembles atherogenic lipoproteins and promotes the development of atherosclerosis. J Intern Med 258:395–410

Olofsson S-O, Bjursell G, Boström K, Carlsson P, Elovson J, Protter AA, Reuben MA, Bondjers G (1987) Apolipoprotein B: structure, biosynthesis and role in the lipoprotein assembly process. Atherosclerosis 68:1–17

Olofsson S-O, Asp L, Borén J (1999) The assembly and secretion of apolipoprotein B-containing lipoproteins. Curr Opin Lipidol 10:341–346
Olofsson S-O, Stillemark-Billton P, Asp L (2000) The intracellular assembly of VLDL – a process that consists of two major steps that occur in separate cell compartments. Trends Cardiovasc Med 10:338–345
Ost A, Ortegren U, Gustavsson J, Nystrom FH, Stralfors P (2005) Triacylglycerol is synthesized in a specific subclass of caveolae in primary adipocytes. J Biol Chem 280:5–8
Pariyarath R, Wang H, Aitchison JD, Ginsberg HN, Welch WJ, Johnson AE, Fisher EA (2001) Co-translational interactions of apoprotein B with the ribosome and translocon during lipoprotein assembly or targeting to the proteasome. J Biol Chem 276:541–550
Pilch PF (2008) The mass action hypothesis: formation of Glut4 storage vesicles, a tissue-specific, regulated exocytic compartment. Acta Physiol (Oxford) 192:89–101
Reinhardt TA, Lippolis JD (2006) Bovine milk fat globule membrane proteome. J Dairy Res 73:406–416
Robenek H, Robenek MJ, Buers I, Lorkowski S, Hofnagel O, Troyer D, Severs NJ (2005) Lipid droplets gain PAT family proteins by interaction with specialized plasma membrane domains. J Biol Chem 280:26330–26338
Robenek H, Hofnagel O, Buers I, Robenek MJ, Troyer D, Severs NJ (2006) Adipophilin-enriched domains in the ER membrane are sites of lipid droplet biogenesis. J Cell Sci 119:4215–4224
Salter AM, Wiggins D, Sessions VA, Gibbons GF (1998) The intracellular triacylglycerol/fatty acid cycle: a comparison of its activity in hepatocytes which secrete exclusively apolipoprotein (apo) B100 very-low-density lipoprotein (VLDL) and in those which secrete predominantly apoB48 VLDL. Biochem J 332:667–672
Savage DB, Petersen KF, Shulman GI (2007) Disordered lipid metabolism and the pathogenesis of insulin resistance. Physiol Rev 87:507–520
Scherer PE, Bickel PE, Kotler M, Lodish HF (1998) Cloning of cell-specific secreted and surface proteins by subtractive antibody screening. Nat Biotechnol 16:581–586
Segrest JP, Jones MK, De Loof H, Dashti N (2001) Structure of apolipoprotein B-100 in low density lipoproteins. J Lipid Res 42:1346–1367
Sell H, Dietze-Schroeder D, Eckel J (2006) The adipocyte–myocyte axis in insulin resistance. Trends Endocrinol Metab 17:416–422
Serrero G, Frolov A, Schroeder F, Tanaka K, Gelhaar L (2000) Adipose differentiation related protein: expression, purification of recombinant protein in Escherichia coli and characterization of its fatty acid binding properties. Biochim Biophys Acta 1488:245–254
Shelness GS, Sellers JA (2001) Very-low-density lipoprotein assembly and secretion. Curr Opin Lipidol 12:151–157
Sollner TH (2007) Lipid droplets highjack SNAREs. Nat Cell Biol 9:1219–1220
Stillemark P, Borén J, Andersson M, Larsson T, Rustaeus S, Karlsson K-A, Olofsson S-O (2000) The assembly and secretion of apolipoprotein-B48-containing very low density lipoproteins in McA-RH7777 cells. J Biol Chem 275:10506–10513
Stillemark-Billton P, Beck C, Boren J, Olofsson SO (2005) Relation of the size and intracellular sorting of apoB to the formation of VLDL 1 and VLDL 2. J Lipid Res 46:104–114
Stone SJ, Myers HM, Watkins SM, Brown BE, Feingold KR, Elias PM, Farese RV Jr. (2004) Lipopenia and skin barrier abnormalities in DGAT2-deficient mice. J Biol Chem 279:11767–11776
Stone SJ, Levin MC, Farese RV Jr. (2006) Membrane topology and identification of key functional amino acid residues of murine acyl-CoA:diacylglycerol acyltransferase-2. J Biol Chem 281:40273–40282
Summers SA (2006) Ceramides in insulin resistance and lipotoxicity. Prog Lipid Res 45:42–72
Sztalryd C, Bell M, Lu X, Mertz P, Hickenbottom S, Chang BH, Chan L, Kimmel AR, Londos C (2006) Functional compensation for adipose differentiation-related protein (ADFP) by Tip47 in an ADFP null embryonic cell line. J Biol Chem 281:34341–34348
Tabas I, Williams KJ, Boren J (2007) Subendothelial lipoprotein retention as the initiating process in atherosclerosis: update and therapeutic implications. Circulation 116:1832–1844
Tai ES, Ordovas JM (2007) The role of perilipin in human obesity and insulin resistance. Curr Opin Lipidol 18:152–156

Tansey JT, Sztalryd C, Gruia-Gray J, Roush DL, Zee JV, Gavrilova O, Reitman ML, Deng CX, Li C, Kimmel AR, et al. (2001) Perilipin ablation results in a lean mouse with aberrant adipocyte lipolysis, enhanced leptin production, and resistance to diet-induced obesity. Proc Natl Acad Sci USA 98:6494–6499

Targett-Adams P, Chambers D, Gledhill S, Hope RG, Coy JF, Girod A, McLauchlan J (2003) Live cell analysis and targeting of the lipid droplet-binding adipocyte differentiation-related protein. J Biol Chem 278:15998–6007

Taskinen MR (2003) Diabetic dyslipidaemia: from basic research to clinical practice. Diabetologia 46:733–749

Trickett JI, Patel DD, Knight BL, Saggerson ED, Gibbons GF, Pease RJ (2001) Characterization of the rodent genes for arylacetamide deacetylase, a putative microsomal lipase, and evidence for transcriptional regulation. J Biol Chem 276:39522–39532

Twisk J, Gillian-Daniel DL, Tebon A, Wang L, Barrett PH, Attie AD (2000) The role of the LDL receptor in apolipoprotein B secretion. J Clin Invest 105:521–532

Wang SM, Hwang RD, Greenberg AS, Yeo HL (2003) Temporal and spatial assembly of lipid droplet-associated proteins in 3T3-L1 preadipocytes. Histochem Cell Biol 120:285–292

Watson RT, Pessin JE (2006) Bridging the GAP between insulin signaling and GLUT4 translocation. Trends Biochem Sci 31:215–222

Watson RT, Pessin JE (2007) GLUT4 translocation: the last 200 nanometers. Cell Signal 19:2209–2217

Welte MA, Gross SP, Postner M, Block SM, Wieschaus EF (1998) Developmental regulation of vesicle transport in *Drosophila* embryos: forces and kinetics. Cell 92:547–557

Wiggins D, Gibbons GF (1992) The lipolysis/esterification cycle of hepatic triacylglycerol. Its role in the secretion of very-low-density lipoprotein and its response to hormones and sulphonylureas. Biochem J 284:457–462

Williams KJ, Tabas I (1995) The respons-to-retention hypothesis in early atherogenesis. Artherioscler Thromb Vasc Biol 15:551–561

Williams KJ, Tabas I (1998) The response-to-retention hypothesis of atherogenesis reinforced. Curr Opin Lipidol 9:471–474

Williams KJ, Brocia RW, Fisher EA (1990) The unstirred water layer as a site of control of apolipoprotein B secretion. J Biol Chem 265:16741–16744

Wolins NE, Rubin B, Brasaemle DL (2001) TIP47 associates with lipid droplets. J Biol Chem 276:5101–5108

Wolins NE, Skinner JR, Schoenfish MJ, Tzekov A, Bensch KG, Bickel PE (2003) Adipocyte protein S3-12 coats nascent lipid droplets. J Biol Chem 278:37713–37721

Wolins NE, Quaynor BK, Skinner JR, Tzekov A, Croce MA, Gropler MC, Varma V, Yao-Borengasser A, Rasouli N, Kern PA, et al. (2006) OXPAT/PAT-1 is a PPAR-induced lipid droplet protein that promotes fatty acid utilization. Diabetes 55:3418–3428

Xu G, Sztalryd C, Lu X, Tansey JT, Gan J, Dorward H, Kimmel AR, Londos C (2005) Post-translational regulation of adipose differentiation-related protein by the ubiquitin/proteasome pathway. J Biol Chem 280:42841–42847

Yamaguchi T, Matsushita S, Motojima K, Hirose F, Osumi T (2006) MLDP, a novel PAT family protein localized to lipid droplets and enriched in the heart, is regulated by peroxisome proliferator-activated receptor alpha. J Biol Chem 281:14232–14240

Yen CL, Monetti M, Burri BJ, Farese RV Jr. (2005) The triacylglycerol synthesis enzyme DGAT1 also catalyzes the synthesis of diacylglycerols, waxes, and retinyl esters. J Lipid Res 46:1502–1511

Yki-Jarvinen H (2002) Ectopic fat accumulation: an important cause of insulin resistance in humans. J R Soc Med 95[Suppl 42]:39–45

Yu YH, Ginsberg HN (2005) Adipocyte signaling and lipid homeostasis: sequelae of insulin-resistant adipose tissue. Circ Res 96:1042–1052

Chapter 2
Oxysterols and Oxysterol-Binding Proteins in Cellular Lipid Metabolism

Vesa M. Olkkonen

This chapter is dedicated to my parents, Raimo and Riitta Olkkonen

Abstract Oxysterols are 27-carbon oxidized derivatives of cholesterol or by-products of cholesterol synthesis with multiple biological activities. The major oxysterols arise either via cholesterol oxidation by cytochrome P450 enzymes or via non-enzymatic autoxidation processes. Oxysterols are present in healthy tissues or plasma at very low levels but are found enriched in pathological structures such as atherosclerotic lesions. Their concentrations in serum are suggested to reflect increased in vivo lipid peroxidation due to oxidative stress. The suggested physiologic functions of oxysterols include transcriptional control of lipid metabolism mediated by liver X receptors and sterol regulatory element binding proteins, as well as modulation of a variety of signaling and differentiation events. The recently characterized family of cytoplasmic oxysterol-binding homologues (ORP proteins) is conserved throughout the eukaryotic kingdom. These proteins are suggested to act as lipid sensors/ transporters with important roles in mediating the impacts of oxysterols and possibly other signaling lipids to the machineries governing cellular lipid metabolism, vesicle transport, and signaling cascades.

2.1 Oxysterols, Their Synthesis and Catabolism

Oxysterols are naturally occurring 27-carbon oxidized derivatives of cholesterol or by-products of cholesterol synthesis with multiple biological activities (Björkhem and Diczfalusy 2002; Javitt 2008; Schroepfer 2000). Kandutsch, Chen, and Heiniger (1978) formulated the so-called oxysterol hypothesis of cholesterol homeostasis. This hypothesis contended that oxysterols, rather than cholesterol itself, mediate the feedback regulation of cholesterol synthesis. However, a wealth of evidence has subsequently accumulated that cholesterol plays a pivotal role in its own feedback regulation (reviewed by Gill et al. 2008). In addition to the proposed role in feedback

V.M. Olkkonen
Department of Molecular Medicine, National Public Health Institute, Biomedicum, P.O. Box 104, 00251 Helsinki, Finland; and Institute of Biomedicine/Anatomy, P.O. Box 63, 00014 University of Helsinki, Finland
e-mail: vesa.olkkonen@thl.fi

C. Ehnholm (ed.), *Cellular Lipid Metabolism*,
DOI 10.1007/978-3-642-00300-4_2,

regulation of cholesterol metabolism, oxysterols have been extensively studied in the context of the oxidative hypothesis of atherosclerosis (Steinberg et al. 1989). Although these sterols are detected in healthy human or animal tissues or plasma at very low concentrations as compared to cholesterol, they are enriched in low-density lipoprotein (LDL) subfractions associated with increased cardiovascular disease risk, macrophage foam cells, and atherosclerotic lesions, evoking the idea that they might play a role in the pathology of atherosclerosis. However, normal physiologic functions of these compounds have more recently emerged. Oxysterols act as ligands of liver X receptors (LXR; Tontonoz and Mangelsdorf 2003) and regulate the intracellular transport and processing of sterol regulatory element binding proteins (SREBP; Goldstein et al. 2006). Thereby these compounds have the potential to act as endogenous regulators of gene expression in lipid metabolism. Furthermore, new intracellular receptors for oxysterols have been identified and novel functions of oxysterols in cell signaling discovered, enhancing the interest in these compounds.

2.1.1 Oxysterols that Arise Through Enzymatic Cholesterol Oxidation

The most common modifications of cholesterol that occur in oxysterols are hydroxyl, keto, hydroperoxy, or epoxy moieties (Fig. 2.1). Oxysterols are markedly more hydrophilic and therefore more mobile within cells than cholesterol, but they do incorporate into the cellular membranes (Lange et al. 1995; Morel et al. 1996; Theunissen et al. 1986). A number of the most abundant oxysterols are generated in cells as products of enzyme-mediated oxidative processes (Fig. 2.1). The quantitatively most important of these are 27-, 24(S)-, and 7α-hydroxycholesterol (OHC; for nomenclature see Table 2.1), which arise in reactions catalyzed by mitochondrial or endoplasmic reticulum cholesterol hydroxylases belonging to the cytochrome P450 family (Luoma 2007; Russell 2000).

27-OHC [also called, more appropriately, 25(R),26-hydroxycholesterol] has an important function in hepatocytes: It serves as an intermediate in the alternative, so-called acid pathway of bile acid biosynthesis (Björkhem and Eggertsen 2001). However, other cell types including macrophages and endothelial cells are also capable of converting cholesterol to 27-OHC, a process that provides one means of removing excess cellular cholesterol (Babiker et al. 1997). This oxysterol is generated by the mitochondrial sterol-27-hydroxylase encoded by *CYP27A1*, genetic defects of which, found in a disease called cerebrotendinous xanthomatosis, result in tissue xanthomas, deposits containing both cholesterol and cholestanol, and an increased risk of premature atherosclerosis (Björkhem et al. 1994; Panzenboeck et al. 2007). Sterol-27-hydroxylase is also capable of converting the primary product 27-OHC further into 3β-hydroxy-5-cholestenoic acid (Fig. 2.1). 27-OHC is a potent suppressor of cholesterol biosynthesis: Synthesis of this oxysterol was

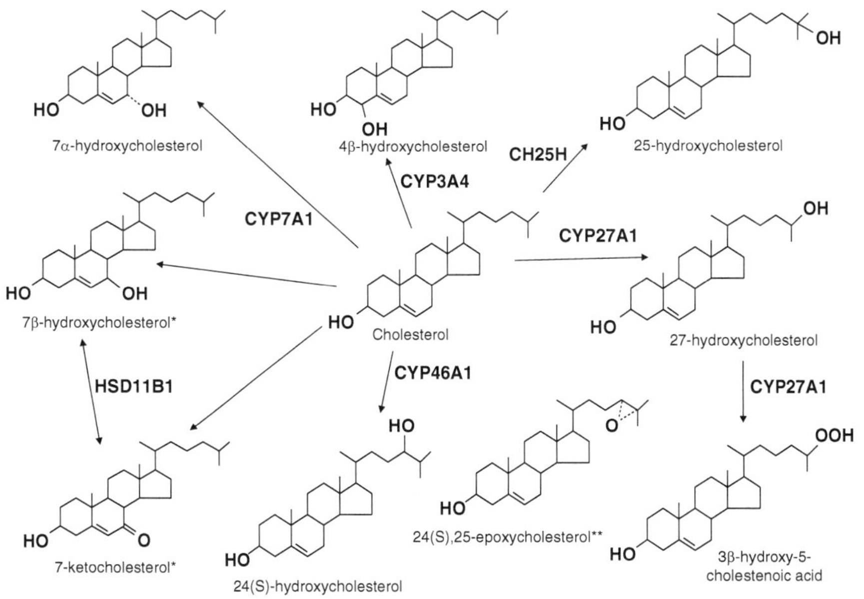

Fig. 2.1 Structure and origin of some common oxysterols. Most of the oxysterols displayed are generated via cholesterol oxidation by enzymes that belong to the cytochrome P450 family (CYP). *CH25H*, cholesterol 25-hydroxylase, is a di-iron enzyme; *HSD11B1* stands for 11β-hydroxysteroid dehydrogenase type 1; *7-ketocholesterol** and *7β-hydroxycholesterol** arise through non-enzymatic oxidation of cholesterol and *24(S),25-epoxycholesterol*** from a shunt of the cholesterol biosynthetic pathway (courtesy of Kati Hyvärinen)

Table 2.1 Nomenclature and abbreviations of the oxysterols discussed in this chapter

Abbreviation	Common name	Systematic name
α-EPOX	5α,6α-Epoxycholesterol	Cholestan-5α,6α-epoxy-3β-ol
β-EPOX	5β,6β-Epoxycholesterol	Cholestan-5β,6β-epoxy-3β-ol
4β-OHC	4β-Hydroxycholesterol	Cholest-5-en-3β,4β-diol
7α-OHC	7α-Hydroxycholesterol	Cholest-5-en-3β,7α-diol
7β-OHC	7β-Hydroxycholesterol	Cholest-5-en-3β,7β-diol
7-KC	7-Ketocholesterol	Cholest-5-en-3β-ol-7-one
	7-Hydroperoxycholesterol	7β-Hydroperoxycholest-5-en-3β-ol
20(R)-OHC	20(*R*)-Hydroxycholesterol	(20*R*) Cholest-5-en-3β,20-diol
20(S)-OHC	20(*S*)-Hydroxycholesterol	(20*S*)-Cholest-5-en-3β,20-diol
22(R)-OHC	22(*R*)-Hydroxycholesterol	(22*R*)-Cholest-5-en-3β,22-diol
24(S)-OHC	24(*S*)-Hydroxycholesterol	(24*S*)-Cholest-5-en-3β,24-diol
24(S),25-EPOX	24(*S*),25-Epoxycholesterol	(24*S*,25)-Epoxycholest-5-en-3β-ol
25-OHC	25-Hydroxycholesterol	Cholest-5-en-3β,25-diol
27-OHC	27-Hydroxycholesterol	(25*R*)-Cholest-5-en-3β,26-diol
	3β-Hydroxy-5-cholestenoic acid	3β-Hydroxy-5-cholesten-25(R) -26-carboxylic acid
		Cholestan-3β,5α,6β-triol

shown to be required for the rapid inactivation of 3-hydroxy-3-methylglutaryl coenzyme A (HMGCoA) reductase upon cellular cholesterol loading (Lange et al. 2008). Furthermore, both 27-OHC and cholestenoic acid can act as ligands of the liver X receptors (see Sect. 2.2.3.1).

Sterol-27-hydroxylase is also able to use also 7- and 8-dehydrocholesterol as its substrates, generating their 27-hydroxymetabolites capable of liganding the LXRs (Wassif et al. 2003). These metabolites were found in vivo in the serum of patients suffering of Smith-Lemli-Opitz syndrome (Yu and Patel 2005) caused by genetic defects of 3β-hydroxysterol-Δ7 reductase (*DHCR7*). CYP27A1 was also found to act in vitro on other intermediates of the cholesterol biosynthetic process, lanosterol, zymosterol, and desmosterol, bringing up the idea that in vivo there may be a large family of previously unstudied 27-hydroxylase products of cholesterol precursor forms, termed oxysteroids (Javitt 2004).

Cholesterol-24-hydroxylase (encoded by *CYP46A1*) is almost exclusively expressed in neurons of the central nervous system (CNS), and the product of the enzyme, 24(S)-OHC, is transferred into the circulation. This process has been suggested to play an important role in the maintenance of the CNS sterol balance (Björkhem et al. 1997; Lütjohann et al. 1996). Increased plasma levels of 24(S)-OHC are associated with Alzheimer's disease (AD) and non-Alzheimer dementia, but the concentration of the sterol falls below control levels at the late stages of AD, most likely due to extensive neuron loss (Bretillon et al. 2000; Kolsch et al. 2003; Lütjohann et al. 2000). Furthermore, changes in plasma 24(S)-OHC levels are found in patients with multiple sclerosis (Leoni et al. 2002). Serum 24(S)-OHC concentration apparently reflects the rate of cholesterol turnover in the brain, and it has been suggested as a new diagnostic indicator of diseases affecting the CNS.

Hepatic cholesterol-7α-hydroxylase (encoded by *CYP7A1*) carries out the initial enzymatic step in the major, so-called neutral pathway of bile acid synthesis (reviewed by Björkhem and Eggertsen 2001). 7α-OHC is thus produced in relatively large quantities in the liver and part of it leaks into the circulation, where it is one of the most abundant oxysterols. 25-OHC has been extensively used for in vitro studies of cellular sterol metabolism, due to its potent inhibitory effect on cholesterol biosynthesis and LDL receptor expression (for a review, see Brown and Jessup 1999), as well as the fact that it is readily available commercially. 25-OHC is present in mammalian tissues or plasma at very low quantities, but has been suggested to have important tissue-specific functions (Chen et al. 2002). 25-OHC can arise via cholesterol autoxidation but it is also generated enzymatically by a di-iron enzyme, cholesterol-25-hydroxylase (CH25H; Lund et al. 1998). Furthermore, also CYP27A1 is capable of generating this oxysterol (Lund et al. 1993). 22(R)-OHC and 20(R)-OHC are formed in steroidogenic tissues as intermediates in the synthesis of steroid hormones (Sugano et al. 1996). A further endogenous enzymatically formed oxysterol important for the control of cellular sterol homeostasis is 24(S),25-epoxycholesterol generated by a shunt of the cholesterol biosynthetic process by the same enzymes that catalyze the synthesis of cholesterol (see Sect. 2.2.3.3; Nelson et al. 1981; Rowe et al. 2003).

2.1.2 Oxysterols Generated via Non-Enzymatic Oxidative Events

Certain oxysterols are generated in non-enzymatic, lipid peroxide or free radical mediated processes, often termed cholesterol autoxidation. It is somewhat unclear to what extent these compounds arise within our body and how much of them are obtained via nutrition. Approximately 1% of the sterol present in the western diet is estimated to be in oxidised form (van de Bovenkamp et al. 1988). The most common oxysterols found in food are 7-ketocholesterol (7-KC), 7α-OHC, 7β-OHC, and cholesterol-5α,6α- and -5β,6β-epoxides (α- and β-EPOX). Sterols modified in the 7-position, as well as some of the side-chain hydroxylated oxysterols, are also suggested to arise in vivo as a result of non-enzymatic oxidation events (for a review, see Brown and Jessup 1999).

2.1.3 Oxysterols in the Circulation

Cholesterol autoxidation that occurs readily during sample preparation has formed a major problem in the analysis of plasma or tissue oxysterol concentrations, resulting in substantial variation in the reported oxysterol concentrations of human plasma (Björkhem 1986; Smith et al. 1981). The most abundant oxysterol in plasma is 27-OHC, the concentration of which is approximately 80 μmol/mol cholesterol. The next in abundancy are 24(S)-OHC, 4β-OHC and 7α-OHC, followed by β-EPOX and cholestan-3β,5α,6β-triol (reviewed by Brown and Jessup 1999). The concentration of 4β-OHC is markedly increased in the serum of patients treated with certain antiepileptic drugs, due to induction of the drug metabolizing enzyme cytochrome P450 CYP3A4, which converts cholesterol to this oxidized derivative (Bodin et al. 2002; Fig. 2.1). 4β-OHC is markedly stable, since it is a poor substrate for the hydroxylases that further modify oxysterols thus routing them to the bile acid biosynthetic pathways. Of the most abundant oxysterols in plasma (apart from 7-KC and the 5,6-epoxides, which are esterified to a lower extent, and 3β-hydroxy-5-cholestanoic acid that is entirely transported in the lipoprotein-free fraction), 70–90% are found as fatty acyl esters associated with lipoproteins (Brown and Jessup 1999; Dzeletovic et al. 1995). Concerning atherosclerosis, it is interesting that increased oxysterol concentrations have been detected in low-density lipoprotein (LDL) subfractions associated with an increased risk of coronary artery disease (CAD). The oxysterol levels in the desialylated LDL subfraction of CAD patients and in circulating LDL immuno-complexes were reported to be 2- to 3-fold higher than in the normally sialylated LDL (Tertov et al. 1992, 1996). Further, oxysterols were found approximately 4-fold enriched in electronegative LDL (LDL^-) as compared to normal LDL, corresponding to approximately 5% of the total amount of sterol in the LDL^- (Chang et al. 1997; Sevanian et al. 1997). The origin of the most abundant oxysterols in human circulation is summarized in Fig. 2.2.

Can oxysterols in human serum be used as an indicator CVD risk, or could they even represent an actual risk factor? Salonen et al. (1997) reported that an

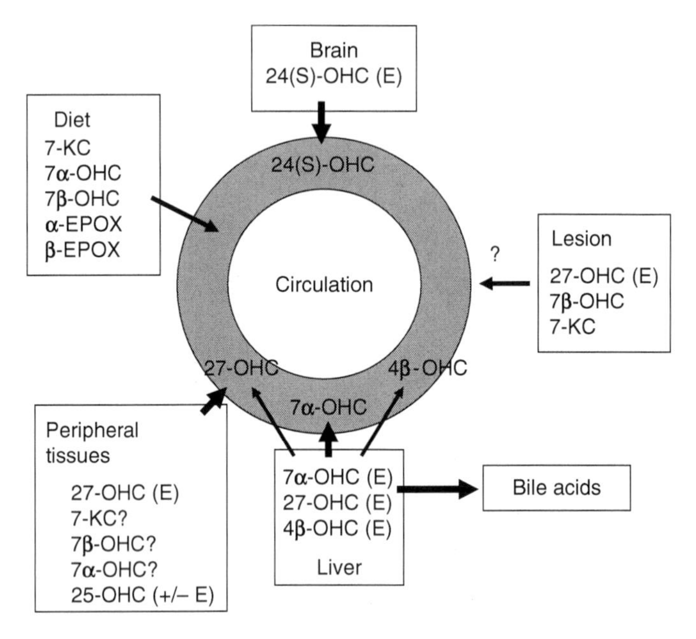

Fig. 2.2 Sources of the most common oxysterols found in human circulation and tissues. Enzymatic origin is indicated with *(E)*. Adapted with permission from Taylor & Francis Healthsciences (Olkkonen and Lehto 2004)

increased plasma concentration of 7β-OHC was the strongest predictor for thickening of carotid artery walls in an analysis of more than 30 variables. Zieden et al. (1999) compared established and possible new risk factors between Swedish and Lithuanian 50-year-old men. The latter had a 4-fold higher risk of coronary death and displayed significantly higher plasma levels of 7β-OHC than the Swedes, while the levels of the established risk factors did not differ significantly between the study groups. Zhou et al. (2000) reported significantly higher plasma concentrations of several oxysterols in a group of patients with angina and 80% stenosis in their coronary arteries than in age- and sex-matched controls. Yasunobu et al. (2001) studied the concentrations of auto-antibodies against oxidized (Ox)LDL and the serum oxysterol content in patients subjected to coronary angiography. When angiographically normal subjects were compared with the stenotic group, both the OxLDL antibodies and the levels of 25-OHC, 27-OHC, and 7β-OHC were significantly higher in the stenotic group. Rimner et al. (2005) reported parallel findings: Plasma 7-KC, β-EPOX, and 7β-OHC were found to be significantly increased in patients with stable coronary artery disease as compared to a control group. Familial combined hyperlipidemia (FCH) is the most common inherited disorder of lipid metabolism, characterized by reduced plasma high-density lipoprotein (HDL) cholesterol, elevated apolipoprotein B concentrations, and a preponderance of small dense LDL subfractions. FCH is associated with increased cardiovascular risk, perhaps due to oxidative damage to lipids accumulating in the plasma of the subjects. Arca et al. (2007)

recently reported increased levels of 7-KC and 7β-OHC in the plasma of FCH patients that persisted after correction for hyperlipidemia and were independent of the presence of clinical atherosclerosis. Furthermore, increased serum 7-KC levels were reported in patients with type 2 diabetes mellitus, associated especially with a group with multiple coronary risk factors (Endo et al. 2008), and elevated oxysterol levels were found in end-stage renal disease patients (Siems et al. 2005). These and a number of other investigations suggest that the serum concentrations of certain oxysterols can be used as indicators of increased in vivo lipid peroxidation and thus of oxidative stress.

2.1.4 Catabolism of Oxysterols

A majority of the oxysterols in plasma are associated with lipoproteins and internalized by cells via various cell surface lipoprotein receptors. The bulk of this internalization occurs in the liver, an organ that excretes sterols as such or in the form of bile acids. Thus, the liver is the major site of oxysterol clearance. In addition to *CYP7A1*, hepatocytes express oxysterol-7α-hydroxylase (encoded by *CYP7B1*), which carries out 7α-hydroxyl modification of both exogenous 27-OHC and that produced by hepatic CYP27A1, thus routing it to the bile acid synthetic pathway (Rose et al. 1997). 27-OHC and 24(S)-OHC are also to some extent 7α-hydroxylated by CYP7A1 (Norlin et al. 2000a, b) in reactions that are, however, inefficient. An additional cytochrome P450 species, CYP39, carries out more efficiently the 7α-hydroxylation of 24(S)-OHC (Li-Hawkins et al. 2000). In the bile acid synthetic pathway, 3β-hydroxy-Δ^5 steroids undergo oxidation and isomerization to produce the corresponding 3-oxo-Δ^4 steroids, followed by saturation of the double bond formed in the A-ring and reduction of the oxo-group into a 3α-hydroxy group. Degradation of the steroid side-chain is initiated by the 27-hydroxylation (see above) either early or late in the process. Synthesis of the primary bile acid cholic acid requires a further 12α-hydroxylation carried out by another P450 enzyme CYP8B1 (for a review, see Björkhem and Eggertsen 2001).

The central non-enzymatically formed oxysterol, 7-ketocholesterol, can be hydroxylated by CYP27A1 to form 27-hydroxy-7-ketocholesterol, which is further metabolized into aqueous-soluble products, apparently bile acids (Brown et al. 2000b; Lyons and Brown 2001b). On the other hand, 7-KC is reduced to 7β-OHC by hepatic 11β-hydroxysteroid dehydrogenase type 1, HSD11B1 (Hult et al. 2004; Schweizer et al. 2004; Fig. 2.1). This enzyme shows species-specific differences in its function: in human and rat it interconverts 7-KC and 7β-OHC, while substrates of the hamster enzyme also include 7α-OHC (Schweizer et al. 2004). The interconversion processes enable rapid hepatic catabolism of 7-KC into aqueous-soluble bile acids, and provide an obvious explanation to the findings that 7-KC added exogenously into the circulation is rapidly metabolized and does not accumulate in the vascular walls (Lyons and Brown 2001a; Lyons et al. 1999). In addition, oxysterols such as 7-KC and 25-OHC are substrates for the steroid/sterol sulfotransferase

SULT2B1b, which modifies them at the 3-position and apparently has a detoxifying function (Fuda et al. 2007; Li et al. 2007).

2.2 Biological Activities of Oxysterols

2.2.1 Effects of Oxysterol Administration on Cells in Vitro

In a number of studies, the effects of OxLDL administration on cultured cells have been at least in part attributed to the biological activities of oxysterols present in these oxidatively modified lipoproteins. The most common approach of generating OxLDL employs Cu^{2+} ions. This method of LDL modification results primarily in the formation of oxysterols derivatised in the 7-position, the most abundant of which is 7-KC (Brown and Jessup 1999).

Treatment of cultured macrophages with OxLDL leads to foam cell formation and a marked intracellular sterol accumulation. In such experiments the amount of oxysterols may account for even 50% of the total sterol content of the cells. Macrophages loaded with OxLDL or ones specifically enriched with 7-KC display impaired ability to efflux cholesterol to apolipoprotein A-I (Gelissen et al. 1996; Kritharides et al. 1995). Why is this? Cholesterol and oxysterols esterified with the oxidized fatty acids of OxLDL tend to accumulate within the lysosomes of the macrophages (Brown et al. 2000a). This type of sterol esters may also in vivo form lipid deposits that are difficult to remove from cells and may facilitate the development of atherosclerotic lesions. Furthermore, the oxysterols present in OxLDL, mainly 7-KC, 7β-OHC and β-EPOX, have cytotoxic properties and induce apoptotic death of several cell types, including those relevant for atherogenesis, monocytes/macrophages, endothelial cells, and smooth muscle cells (Larsson et al. 2006; Lemaire-Ewing et al. 2005; Lordan et al. 2007; Luthra et al. 2007; Ohtsuka et al. 2006; Rimner et al. 2005). The extremely unstable and cytotoxic 7-hydroperoxycholesterol arises as an intermediate in the oxidation reactions targeting the sterol 7-position, and is apparently an important cytotoxic compound in OxLDL (Chisolm et al. 1994). Interestingly, ATP-binding cassette transporter G1 (ABCG1) is capable of protecting macrophages and other cell types from apoptosis induced by oxidized LDL, 7-KC or 7β-OHC, apparently by promoting the efflux of oxysterols to HDL particles (Engel et al. 2007; Terasaka et al. 2007). Other oxysterols such as 25-OHC can also be removed from cells by ABCA1 (Tam et al. 2006, Terasaka et al. 2007).

Several oxysterols have been reported to impair the selective barrier function of endothelial cells (Boissonneault et al. 1991) and to inhibit nitric oxide (NO) synthesis, thus having an adverse effect on vasodilatation (Deckert et al. 1998; Millanvoye-Van Brussel et al. 2004). Oxysterols (mainly 7β-OHC, 25-OHC, 7-KC) induce the expression and secretion of proinflammatory cytokines/chemokines such as interleukins 6 and 8 and monocyte chemotactic protein-1 (Erridge et al. 2007; Knutsen Rydberg et al. 2003; Lemaire-Ewing et al. 2005, 2008; Leonarduzzi et al. 2005; Sung et al. 2008).

Furthermore, oxysterols have the capacity to impact cell differentiation: As examples, 7-KC induces monocyte differentiation into macrophages and further into foam cells (Hayden et al. 2002), 22(S)-OHC, and 20(S)-OHC enhance stem cell differentiation into osteoblasts in vitro and bone healing in vivo (Aghaloo et al. 2007; Richardson et al. 2007) while they inhibit the adipogenic differentiation of the stem cells (Kha et al. 2004; Kim et al. 2007), and 25-OHC modulates the differentiation of Leydig cells (Chen et al. 2002).

The findings based on the use of either OxLDL or pure oxysterol species on cultured cells underscore the potency of oxysterols as effector molecules impacting multiple biologic processes. Many of the effects are obviously harmful in terms of cardiovascular status. One must, however, keep in mind that the adverse effects of oxysterols have often been studied in vitro at pharmacologic concentrations. Furthermore, the studies have in most cases been carried out using single, pure oxysterol species. When applied as mixtures similar to those found in vivo, together with cholesterol or as fatty acyl esters, the adverse cellular effects of oxysterols are markedly alleviated (Biasi et al. 2004; Hietter et al. 1984; Leonarduzzi et al. 2004; Monier et al. 2003). However, certain oxysterol combinations may also have synergistic toxic effects (Larsson et al. 2006).

2.2.2 *Oxysterols in Atherosclerotic Lesions*

The concentrations of oxysterols in atherosclerotic lesions in both humans and in animal models are typically two orders of magnitude higher than in plasma; The lesion oxysterols are almost completely in esterified form. Several studies demonstrate that 27-OHC is the most abundant oxysterol in human lesions. Its concentration correlates with the cholesterol content of the lesions and the severity of atherosclerosis (Brown and Jessup 1999; Carpenter et al. 1993, 1995; Garcia-Cruset et al. 1999, 2001; Vaya et al. 2001). This may reflect activation of the mitochondrial sterol-27-hydroxylase in cells within the lesion as an attempt to solubilize and remove the accumulating sterol excess. Also, 7β-OHC and 7-KC, which arise through non-enzymatic oxidation processes, are abundant in lesions. The portion of these oxysterols is especially high in fatty streaks, perhaps reflecting a strong oxidative activity in lesions at this stage, or selective accumulation of the 7-modified oxysterols in macrophage foam cells, which are the principal component of fatty streaks (Maor et al. 2000; Mattsson-Hulten et al. 1996). The observations on the deposition of oxysterols in lesions, together with the documented cytotoxic and pro-apoptotic effects of these compounds, as well as their impact on macrophage differentiation (see Sect. 2.2.1), have made researchers believe that they may have pathophysiological importance in the development of atherosclerotic lesions: They may inflict endothelial injury, promote formation of the necrotic core of atheroma and vascular calcification, and possibly facilitate plaque rupture (Colles et al. 2001; Guyton et al. 1990; Li et al. 2001; Marchant et al. 1996; Saito et al. 2008; Sevanian et al. 1995).

2.2.3 *Oxysterols as Regulators of Cellular Lipid Metabolism*

2.2.3.1 Oxysterols Act as Liver X Receptor Agonists

The interest of the scientific community towards oxysterols has greatly intensified since the discovery that they act as ligands of the nuclear receptor proteins designated Liver X receptor-α (LXRα) and LXRβ (also known as NR1H3 and NR1H2, respectively; Apfel et al. 1994; Willy et al. 1995). While LXRβ is expressed at relatively even levels in all tissues, LXRα is expressed at high levels in the liver and less abundantly in the adrenal glands, intestine, adipose tissue, macrophages, lung, and kidney (Repa and Mangelsdorf 2000). Oxysterols at physiological concentrations were found to bind to and activate the LXRs in vitro (Forman et al. 1997; Janowski et al. 1996; Lehmann et al. 1997). There has been controversy as to whether naturally occurring oxysterol agonists such as 22(R)-OHC, 24(S),25-EPOX, 24(S)-OHC, 27-OHC, and 3β-hydroxy-5-cholestenoic acid execute the same LXR activating function in vivo as they do in vitro (Björkhem and Diczfalusy 2002). Recent investigations employing genetic manipulation of oxysterol metabolism have shed light on this question, supporting the notion that oxysterols indeed act as endogenous LXR agonists in vivo (Chen et al. 2007).

The LXRs bind to the promoters of specific target genes as heterodimers with the retinoid X receptor (RXR; NR2B1) and impact gene expression via recruitment of transcriptional coactivators or corepressors (Fig. 2.3). The binding is of the permissive type, i.e., ligand binding to either LXR or RXR is sufficient to activate target gene transcription. The sequence elements recognized by the LXRs are direct repeat four (DR4) motifs, termed LXR responsive elements (LXREs; Repa and Mangelsdorf 2000; Tontonoz and Mangelsdorf 2003). Genes regulated by the LXRs are involved in sterol absorption in the intestine, the reverse cholesterol transport process, bile acid synthesis, biliary neutral sterol secretion, hepatic lipogenesis, and synthesis of nascent high-density lipoproteins (Li and Glass 2004; Tontonoz and Mangelsdorf 2003). In addition, new functions have recently assigned to the LXRs: LXRα was reported to silence cholesterol biosynthesis by direct suppression of two genes encoding cholesterogenic enzymes (Wang et al. 2008b). The most extensively studied LXR target genes are those encoding ATP-binding cassette transporters A1 (ABCA1; Costet et al. 2000; Repa et al. 2000b) and G1 (ABCG1; Kennedy et al. 2001; Sabol et al. 2005; Venkateswaran et al. 2000), which mediate the efflux of cholesterol from macrophages to lipid-poor apolipoprotein A1 and to spherical HDL particles, respectively. The lipidation of apolipoprotein A1 by ABCA1 is crucial for the formation of nascent high-density lipoproteins (for a review, see Oram and Vaughan 2006). A recent report suggests that, in addition to control of *ABCA1* transcription, LXRβ/RXR dimers bind directly to the ABCA1 protein, regulating its activity in an oxysterol-dependent manner (Hozoji et al. 2008). In general, the LXRs are regarded antiatherogenic due their central role as regulators of genes involved in the reverse cholesterol transport pathway. Importantly, the stimulation of the LXRs is also reported to inhibit macrophage inflammatory responses, another mechanism that alleviates atherogenesis (Castrillo et al. 2003; Joseph et al. 2003; Zelcer and Tontonoz 2006). Further, Bensinger et al. (2008) reported that ligation of LXR during T-cell activation inhibits mitogen-driven expansion, while loss of LXRβ conferred the cells proliferative advantage. In contrast to the potentially atheroprotective properties of the LXRs, they also activate genes responsible for lipogenesis, such as those encoding sterol regulatory element binding protein 1c (SREBP-1c; Repa et al. 2000a; Yoshikawa et al. 2001) and fatty acyl synthetase (FAS; Joseph et al. 2002). Therefore, LXR stimulation by pharmacologic agonists increases plasma triglycerides representing a cardiovascular risk factor. Development of LXR agonists that selectively activate reverse cholesterol transport in the absence of effects on lipogenesis is therefore an important goal of the pharmaceutical industry.

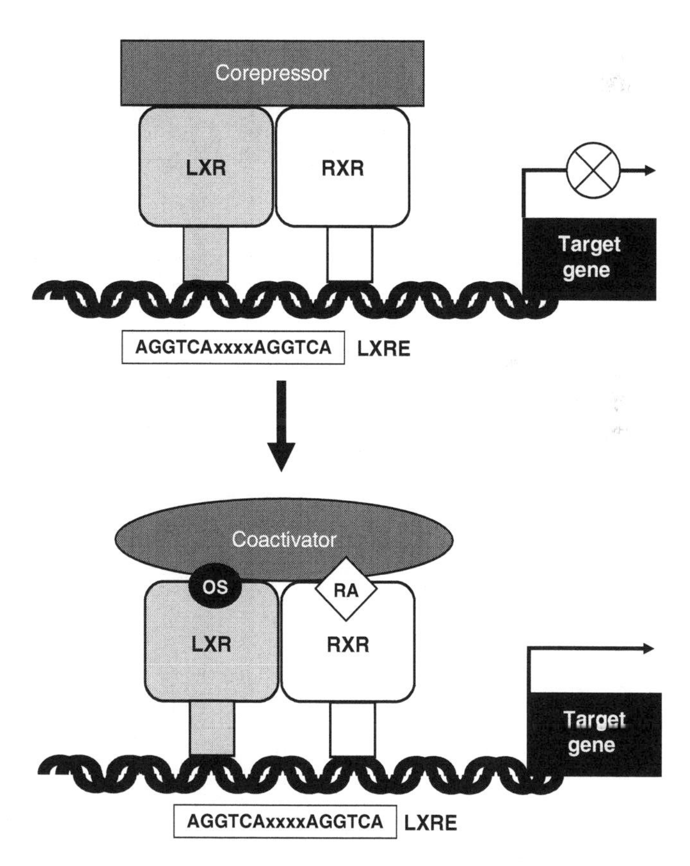

Fig. 2.3 Function of liver X receptors (*LXR*). Within the nucleus, LXR–retinoid X receptor (*RXR*) heterodimers are bound to LXR response elements (*LXRE*) in the promoters of target genes. In the absence of LXR or RXR ligands, the complex recruits corepressors that silence gene expression. When ligands of either LXR (oxysterols, *OS*) or RXR (retinoic acid or its derivatives, *RA*) or both are bound, the co-repressor complexes are exchanged for co-activator complexes and target gene transcription is induced

2.2.3.2 Oxysterols Regulate the Maturation of Sterol Regulatory Element Binding Proteins

Cholesterol biosynthesis and cellular uptake, as well as fatty acid biosynthesis, are controlled by transcription factors named sterol regulatory element binding proteins (SREBPs), and their cholesterol-sensing accessory factor, the SREBP cleavage activating protein (SCAP)(Eberle et al. 2004; Goldstein et al. 2006). Of the three SREBP isoforms produced from two genes, SREBP-1c is particularly abundant in the liver where its expression is regulated by insulin and glucagon, and it plays a major role in controlling hepatic lipogenesis and glucose utilization. SREBP-2, also expressed at relatively high levels in the liver, is predominantly responsible for control of cholesterol metabolism. The third family member, SREBP-1a, is mainly expressed in tissues with a high proliferative capacity, and functions in both cholesterol and triglyceride metabolism. The SREBPs are synthesized as precursors anchored to ER membranes and form complexes with SCAP. When the cellular cholesterol level is low, SREBP-SCAP complexes move to the Golgi complex, where SREBPs undergo a two-step proteolytic processing to release the N-terminal fragment, a basic helix–loop–helix leucine zipper transcription factor. These fragments enter the nucleus and bind to sterol regulatory elements (SRE) in the promoter regions of a number of genes whose products mediate the synthesis of cholesterol and fatty acids. When sterol builds up in cells, SCAP senses cholesterol in the ER membranes and interacts with the Insig (*Ins*ulin-*i*nduced *g*ene) proteins. As a result the SREBP-SCAP complex is retained in the ER (Yabe et al. 2002; Yang et al. 2002). The SREBP machinery is sensitive to both cholesterol and oxysterols, of which 25-OHC is most commonly used in experimental set-ups. While SCAP does not bind 25-OHC (Adams et al. 2004; Brown et al. 2002), the Insig proteins were recently found to directly bind this oxysterol and to mediate the regulatory effect of 25-OHC on SREBP processing. Upon oxysterol binding, the Insigs apparently undergo a conformational change that increases their affinity for SCAP, resulting in an inhibition of ER to Golgi transport of SCAP–SREBP complexes and SREBP activation (Radhakrishnan et al. 2007; Sun et al. 2007).

2.2.3.3 24(S),25-Epoxycholesterol, a Signal for Sterol Biosynthetic Pathway Activity

Unlike most oxysterols, 24(S),25-epoxycholesterol [24(S),25-EPOX] is not derived from preformed cholesterol but is generated as a side product of the cholesterol biosynthetic process by the same enzymes that catalyze the synthesis of cholesterol (Nelson et al. 1981; Rowe et al. 2003). This compound is a potent feedback regulator of cholesterol biosynthesis, suppressing SREBP-2 processing (Janowski et al. 2001) and repressing the cellular HMGCoA reductase activity (Mark et al. 1996). Furthermore, it is the most potent oxysterol activator of the LXRs (Chen et al. 2007; Janowski et al. 1999; Lehmann et al. 1997; Wong et al. 2004), and its role in cellular sterol homeostatic regulation has in the past few years been substantially elucidated. The cellular level of 24(S),25-EPOX can be manipulated with two methods: (i) by using inhibitors of the cholesterol biosynthetic pathway enzyme 2,3-oxidosqualene cyclase (OSC), resulting in elevation of cellular 24(S),25-EPOX levels, or (ii) by overexpressing this enzyme, which reduces the cellular content of 24(S),25-EPOX. With these tools and other approaches the groups of A. Brown and M. Huff have come up with a number of novel and highly important findings: Wong et al. (2004) showed that the treatment of macrophages with statins, inhibitors of HMGCoA reductase and thus of cholesterol synthesis, results in reduced expression of LXR target genes due to concomitant reduction in the synthesis of 24(S),25-EPOX. They also demonstrated that the synthesis of 24(S),25-EPOX parallels that of cholesterol and acts as a signal that fine-tunes the acute control of cellular cholesterol homeostasis and protects cells against accumulation of newly synthesized cholesterol. This occurs via its regulatory impact on both the LXRs which control cholesterol efflux from cells and the SREBP machinery responsible for cellular cholesterol biosynthesis and uptake (Wong et al. 2007a, 2008). Beyea et al. (2007) demonstrated that partial inhibition of OSC in THP-1 macrophages reduced cholesterol synthesis and increased the expression of several LXR target genes, *ABCA1*, *ABCG1*, and *APOE*. Importantly, OSC inhibition did not stimulate lipoprotein lipase (*LPL*) or fatty acid synthase (*FAS*), and the observed induction of the lipogenic transcription factor SREBP-1c was counteracted by a block in its conversion to the active nuclear form. Rowe et al. (2003) suggested that induction of LXRs by partial inhibition of OSC could be used in cardiovascular therapy. However, this idea is undermined by the recent finding that, even though 24(S),25-EPOX facilitates *ABCA1* and *ABCG1* expression and cholesterol efflux from normal macrophages in culture, it strongly inhibits cholesterol efflux from macrophage foam cells, possibly due to inhibition of cholesterol ester hydrolase function (Ouimet et al. 2008).

24(S),25-EPOX also seems to have a specific role in CNS sterol homeostasis: It was shown to be produced by astrocytes in vitro and to be taken up by neurons in which it exerted downstream effects on gene regulation (Wong et al. 2007b). The authors presented the interesting hypothesis that 24(S),25-EPOX acts as a signal from astrocytes that reduces the energetically costly cholesterol biosynthetic capacity of neurons, enabling the neurons to divert resources into other cellular processes.

As a conclusion, the endogenous cellular oxysterols can be regarded as signaling molecules employed as indicators of the cellular sterol status, which, via their effects on the LXR and SREBP machineries of sterol homeostatic control, prevent cellular sterol accumulation and have a potentially atheroprotective impact.

2.2.3.4 27-Hydroxycholesterol, a Selective Estrogen Receptor Modulator

Estrogen receptors α (NR3A1) and β (NR3A2) are nuclear receptors that, in addition to reproductive functions, mediate estrogen regulation of a number of other physiologic processes (Deroo and Korach 2006). One potentially important target of estrogen receptors is the cardiovascular system, where the action of estrogens is believed to be beneficial and to explain in part the fact that women in reproductive age are markedly protected from cardiovascular diseases as compared to men. It was recently reported that 27-OHC, which accumulates in atherosclerotic lesions (see Sect. 2.2.2), antagonized the estrogen-dependent production of nitric oxide by vascular cells, resulting in reduced vasorelaxation of rat aorta, a potentially deleterious effect (Umetani et al. 2007). Furthermore, increasing 27-OHC levels repressed carotid artery re-endothelialization. Interestingly, 27-OHC also had cell type-specific proestrogenic actions (DuSell et al. 2008; Umetani et al. 2007), indicating that it acts as an endogenous selective estrogen receptor modulator (SERM). The above findings suggest that 27-OHC may contribute to the loss of estrogen protection from vascular disease.

2.2.3.5 The Niemann–Pick C1 Protein Binds both Cholesterol and Oxysterols

Niemann–Pick C (NPC) disease is an inborn error of metabolism characterized by accumulation of free cholesterol and sphingolipids within late endocytic compartments of cells (Sturley et al. 2004). The disease is caused by mutations in either of two proteins, NPC1 or NPC2. Of these, NPC1 is a large multi-spanning membrane protein localized in late endosomes. NPC1 has a consensus cholesterol binding motif consisting of five trans-membrane helices, but the mechanism by which NPC1 facilitates cholesterol egress from late endocytic compartments is unknown. Recently, NPC1 was found to bind not only cholesterol but also 25-OHC and most likely also other oxysterols such as 24(S)-OHC and 27-OHC (Infante et al. 2008a). The authors found that cholesterol and 25-OHC bind not to the trans-membrane helix motif but to a luminal loop of NPC1 (Infante et al. 2008b). The relevance of this sterol binding for the function of NPC1 remained unclear in experiments employing cultured fibroblasts, but one can envision that the oxysterol interaction may in vivo play a regulatory role in the function of NPC1 in routing cholesterol, glycolipids, and phospholipids out of late endosomes.

2.2.4 Oxysterols Regulate Hedgehog Signaling

The Hedgehog (Hh) signaling pathway plays a central role in the patterning of multicellular embyos, in post-embryonic development, and in adult tissue homeostasis, including the physiology of stem cells (Varjosalo and Taipale 2007). Corcoran and Scott (2006) reported that cholesterol or certain oxysterols are required for Sonic hedgehog pathway signal transduction and proliferation of medulloblastoma cells. Later, Dwyer et al. (2007) demonstrated convincingly that naturally occurring oxysterols exert their osteoinductive effects (see Sect. 2.2.1) through activation of the Hh signaling pathway. Furthermore, Kim et al. (2007) reported that the inhibition of bone marrow stromal cell differentiation into adipocytes by 20(S)-OHC occurs through a mechanism dependent on the Hh pathway. These findings introduce an

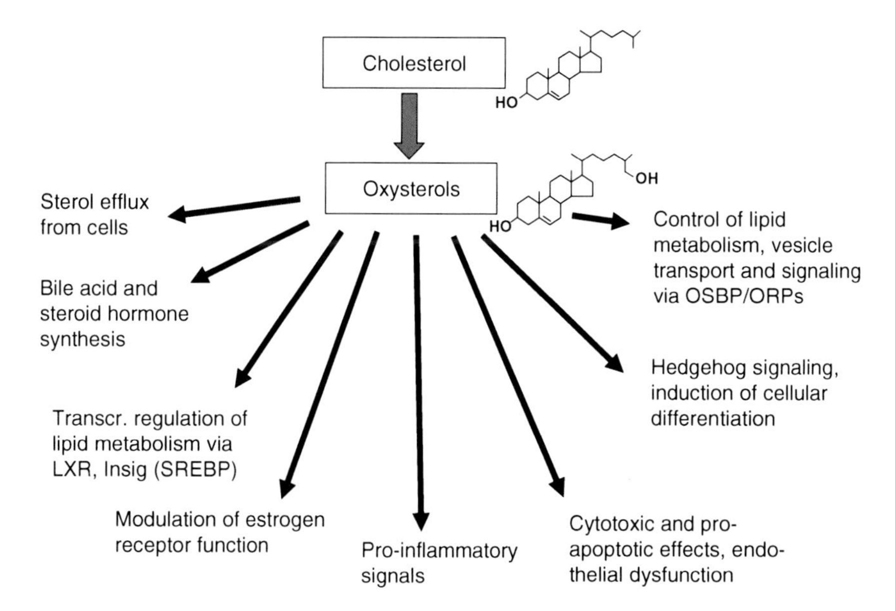

Fig. 2.4 Summary of the most prominent biological activities of oxysterols. The functions are not assigned to oxysterols in general, but each activity is associated with specific oxysterol species (for details and references, see text)

important, novel role of oxysterols as developmental regulators The most prominent biological activities of oxysterols are summarized in Fig. 2.4.

2.3 Cytoplasmic Oxysterol-Binding Proteins

2.3.1 Identification of Oxysterol-Binding Protein-Related Proteins

During early studies of feedback inhibition of cholesterol synthesis, oxysterols such as 25-OHC were found to be more potent than cholesterol in reducing the activity of HMGCoA reductase, a rate-controlling enzyme in cholesterol biosynthesis (Brown and Goldstein 1974; Kandutsch and Chen 1974; Kandutsch et al. 1978). These findings prompted a search for protein factors mediating the effects of oxysterols on cellular lipid metabolism. Protein fractions with oxysterol binding activity were isolated from different sources (Beseme et al. 1986, 1987; Defay et al. 1982; Kandutsch et al. 1977; Kandutsch and Shown 1981; Kandutsch and Thompson 1980). Taylor and co-workers identified a cytosolic oxysterol-binding protein (OSBP) whose sterol binding specificity correlated with the ability of these compounds to suppress the activity HMGCoA reductase (Taylor and Kandutsch 1985; Taylor et al. 1984). OSBP was purified (Dawson et al. 1989b; Taylor et al. 1989) and cDNAs were cloned from rabbit (Dawson et al. 1989a) and human (Levanon et al. 1990). OSBP was found to be a cytoplasmic protein that is translocated from a cytosolic or vesicular compartment to membranes of the Golgi apparatus upon treatment of cells with 25-OHC. Discovery of the SREBPs (see Sect. 2.2.3.2; Hua et al. 1993; Wang et al. 1993; Yokoyama et al. 1993) turned major interest in the field away from OSBP. Furthermore, the LXRs (see Sect. 2.2.3.1) were soon identified as oxysterol-binding transcription factors with central roles in the control of cellular sterol and lipid metabolism (Forman et al. 1997; Janowski et al. 1996; Lehmann et al. 1997). However, work aimed to characterize the function of OSBP continued in the laboratory of N. Ridgway. After some years, novel interest in this protein and its relatives was evoked due to: (i) identification of OSBP-related gene/protein families in eukaryotic organisms from yeast to man, and (ii) functional studies revealing clues for important roles of OSBP homologues in cellular lipid metabolism, vesicle transport, and cell signaling.

Proteins displaying homology with the carboxy-terminal ligand binding domain of OSBP (designated OSBP-related ligand binding domain; ORD) are present in practically all eukaryotic organisms (Lehto and Olkkonen 2003; Yan and Olkkonen 2008). They are called either OSBP-related proteins (ORP) or OSBP-like proteins (OSBPL). In humans (Jaworski et al. 2001; Lehto et al. 2001) and mice (Anniss et al. 2002) the gene family consists of 12 members. Most ORP messages are expressed ubiquitously. However, there are marked quantitative differences in the tissue- and cell type-specific mRNA expression patterns of the family members, and alterations of ORP expression levels occur during cell differentiation processes (Gregorio-King et al. 2001; Johansson et al. 2003; Lehto et al. 2004). The cellular functions of ORPs have mainly been investigated in mammalian cells and in the yeast *Saccharomyces cerevisiae* (Schulz and Prinz 2007; Yan and Olkkonen 2008). However, reports have also been published on ORP family members in *Drosophila melanogaster* (Alphey et al. 1998), *Caenorhabditis elegans* (Sugawara et al. 2001), *Dictyostelium discoideum* (Fukuzawa and Williams 2002), rainbow trout (Ramachandra et al. 2007), the parasitic protist *Cryptosporidium parvum* (Zeng and Zhu 2006), as well as in several plants (Avrova et al. 2004; Li et al. 2008; Skirpan et al. 2006). The presence of the gene family throughout the eukaryotic kingdom provides evidence for a fundamental function of the ORPs that originated early in eukaryotic evolution. The human and *S. cerevisiae* ORP proteins are depicted in Fig. 2.5.

2.3.2 Structure and Ligands of ORPs

The ORPs minimally comprise an OSBP-related ligand-binding domain (ORD), but in mammals a majority of them carry an amino-terminal extension containing a pleckstrin homology (PH) domain that is in several cases known to bind membrane phosphoinositides (Johansson et al. 2005; Lehto et al. 2005; Levine and Munro 1998, 2002). The proteins consisting of an ORD only are here designated "short ORPs", while those carrying a PH domain are called "long ORPs".

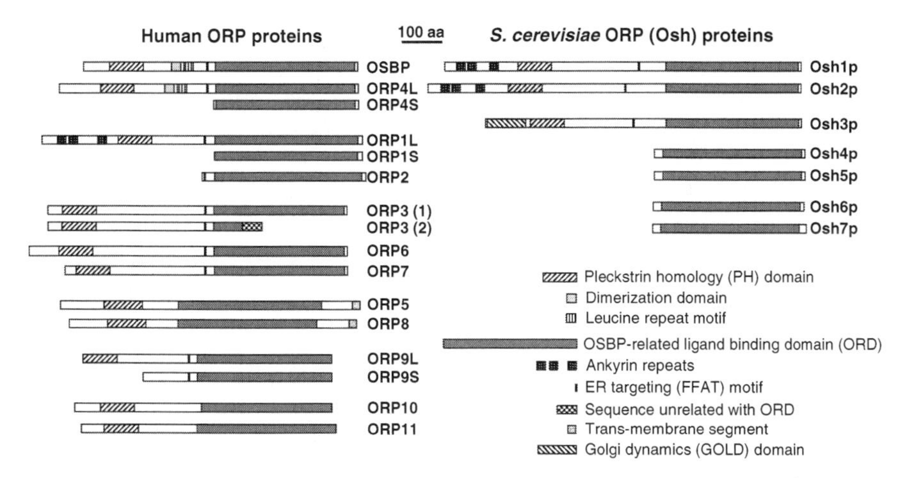

Fig. 2.5 The human and *S. cerevisiase* OSBP-related protein (*ORP*) families. The major structural elements are identified. In humans, splice variation leading to changes in mRNA and protein structure is found for several ORPs (see NCBI database; Collier et al. 2003). Therefore, the amino acid (*aa*) scale given is not exact but indicative; and the proteins depicted do not represent individual splice variants but rather groups formed by several variants

Identification of ligands for the ORDs of the ORPs is crucial for elucidation of the functions of these proteins. The first ORP high-resolution structure, that of a short yeast *S. cerevisiae* ORP, Osh4p/Kes1p, revealed that Osh4p is a sterol-binding protein (Im et al. 2005). It was crystallized in complex with five different sterols, and has a sterol binding pocket formed by 19 β-strands in an antiparallel arrangement. The sheet bends to an almost complete roll that is, in the presence of bound ligand, closed by a lid containing an amphipathic α-helix connected by a flexible linker. Sterols bind within the pocket oriented with the 3β-hydroxyl group at the bottom of the hydrophobic binding tunnel. The sterol side-chain interacts with the lid, stabilizing its closed conformation. Importantly, many of the interactions of the bound sterol are mediated via water molecules within the pocket, giving the ligand interaction substantial flexibility. This provides a plausible explanation to the ability of the pocket to accommodate structurally different sterols and possibly also other types of lipid ligands. The structure of Osh4p suggested that this protein and its homologues might act as sterol transporters or mediators of sterol signals (Im et al. 2005).

Of the mammalian ORPs, OSBP, ORP4 (also designated OSBP2), ORP1, and ORP2 have been shown to bind oxysterols (Moreira et al. 2001; Suchanek et al. 2007; Taylor et al. 1984; Wang et al. 2002; Yan et al. 2007a). Moreover, ORP8 was reported to show affinity for 25-OHC (Yan et al. 2008). Based on evidence obtained through the use of UV cross-linkable sterol derivatives in live cells, Suchanek et al. (2007) suggested that a majority of the other human ORPs may bind sterols. However, this data must be interpreted with caution since it is extremely difficult to verify the specificity of the live cell cross-linking signals. The structure of Osh4p

(Im et al. 2005) was used as template for molecular homology modeling of mammalian ORP2 (Suchanek et al. 2007) and OSBP (Wang et al. 2008a). In the former study, analysis of site-specific mutants designed based on the model suggested that ORP2 has a sterol-binding pocket similar to that of Osh4p. Interestingly, Wang et al. (2008a) showed evidence that the intact ORD of OSBP is not required for binding of 25-OHC or cholesterol, which is also known to be a ligand of OSBP (Wang et al. 2005c). Specific and high-affinity sterol binding was also detected with a 51-amino-acid (aa) protein fragment that corresponds to the amino-terminal end of the ORD and comprises part of the modeled lid region (Fig. 2.6). One interpretation of the data is that the 51-aa segment forms the actual sterol binding site, and the β-barrel structure merely protects the bound sterol. The authors also found that a glycine/alanine-rich region at the amino-terminal end of OSBP, together with the PH domain, controls the binding of cholesterol but not of 25-OHC, suggesting interactions of the amino-terminal PH domain region and the ORD, most likely regulated by conformational changes triggered by ligand interactions. At the carboxy-terminal end of the ORD of OSBP and several other ORPs, there is a predicted

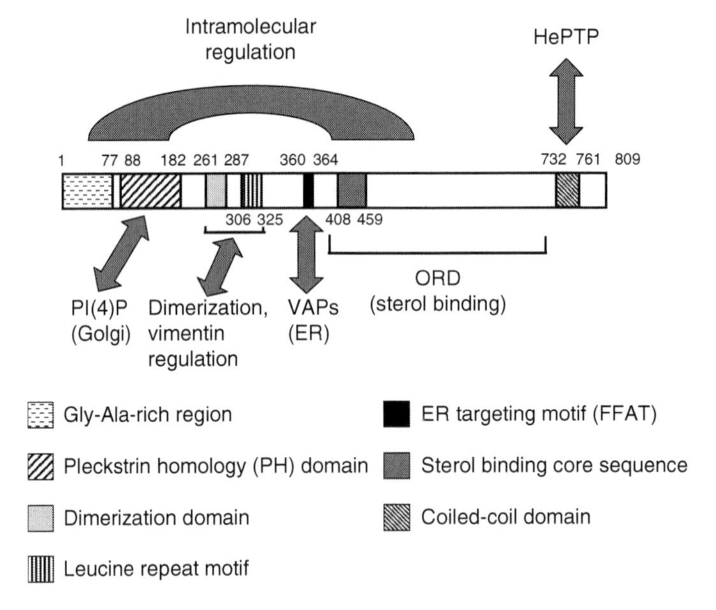

Fig. 2.6 Structural elements important for OSBP function. The domains and sequence motifs are identified at the bottom. The pleckstrin homology (PH) domain specifies Golgi targeting mediated by PI(4)P and the ER targeting motif (FFAT) interacts with the VAP proteins, functions relevant for the control of sphingomyelin synthesis. OSBP forms homodimers and heterodimers with ORP4L, the latter process being connected with regulation of vimentin intermediate filament organization by ORP4L. A coiled-coil domain near the carboxy-terminus of OSBP interacts with HePTP, a tyrosine phosphatase that regulates the activity of extracellular signal regulated kinases, ERK. The Gly/Ala-rich region at the amino-terminus works with the PH domain to control cholesterol binding by the ORD, while occupancy of the ORD by oxysterol ligands regulates the Golgi-targeting function of the PH domain

coiled-coil forming region. Wang et al. (2008a) demonstrated that this region is required for the interaction of OSBP with a phosphatase acting on the ERK. The coiled-coil motif may play an important role in the protein–protein interactions of other ORPs as well (see Sect. 2.3.7).

In addition to sterol ligands, the ORDs of ORP1, ORP2, ORP9, and ORP10 have been suggested to show affinity for phosphoinositides (PIPs; Fairn and McMaster 2005a, b; Hynynen et al. 2005), but it is unclear whether these interactions involve a pocket such as that found in yeast Osh4p. Most likely, they represent binding of positively charged amino acid clusters on the protein surface to negatively charged membrane phosphoinositides (see Im et al. 2005). The data by Raychaudhuri et al. (2006) demonstrates that PIPs facilitate the transfer of sterols between membranes by Osh4p, suggesting that the PIP interaction plays an important regulatory role in the sterol sensor/transporter function of the short ORPs.

2.3.3 Subcellular Distribution of ORPs

ORPs are in principle cytosolic proteins with the capacity to associate peripherally with cellular membranes. To obtain clues of ORP function, one of the avenues of research has been the analysis of their subcellular localization and of the determinants that specify the membrane targeting of the proteins. The Golgi targeting of OSBP (Ridgway et al. 1992) is specified by a pleckstrin homology (PH) domain in the amino-terminal part of the protein (Fig. 2.6; Lagace et al. 1997; Levine and Munro 1998). PH domains are also present in the N-terminal region of ten "long" mammalian OSBP homologues. The amino-terminal PH domain-containing extensions present in these ORPs contain predominant targeting information: (i) ORP9, the Golgi complex (Wyles and Ridgway 2004), (ii) ORP1L, late endosomes (Johansson et al. 2003, 2005), and (iii) ORP3,6, and 7, plasma membrane (Lehto et al. 2004). Binding of 25-OHC to the carboxy-terminal domain of OSBP has been suggested to induce a conformational change that unmasks the PH domain, thus inducing a shift of the protein to Golgi membranes. The interaction of the OSBP PH domain with phosphatidylinositol-4-phosphate [PI(4)P] is crucial for targeting of this protein to the Golgi complex and essential for its function (Lagace et al. 1997; Levine and Munro 1998, 2002). Similarly, the PH domain of *S. cerevisiae* Osh1p displays Golgi-targeting specificity and interacts with PI(4)P (Roy and Levine 2004). The Golgi localization of the OSBP and Osh1p PH domains also depends on ADP-ribosylation factors, ARFs (Levine and Munro 2002; Roy and Levine 2004), small GTPases essential for transport vesicle formation (Kahn et al. 2005).

The long variant of mammalian ORP1, ORP1L (Johansson et al. 2003), and two of the long ORPs in *S. cerevisiae*, Osh1p and Osh2p (Beh et al. 2001; Schmalix and Bandlow 1994), have at their very amino-terminus a region containing ankyrin repeats (ANK), motifs typically involved in protein–protein interactions (Fig. 2.5). The ANK region of ORP1L interacts with the GTP-bound active form of the late endosomal (LE) small GTPase Rab7 and plays an important role in targeting of

ORP1L to these compartments. The ANK repeat region of yeast Osh1p was reported to target the protein to the nucleus–vacuole junction (NVJ; Levine and Munro 2001). Since Osh1p was shown to interact with the NVJ protein component Nvj1p (Kvam and Goldfarb 2004), this protein is most likely recognized by the Osh1p ANK region. Findings by Johansson et al. (2003) suggest that the ORP1L ANK region and PH domain synergize in targeting the protein to LE. The ANK repeat region is most likely used to achieve specific membrane targeting of select ORP family members via protein–protein interactions.

Eight of the mammalian ORPs (OSBP, ORP1, 2, 3, 4, 6, 7, 9) possess a sequence motif denoted FFAT (two phenylalanines in an acidic tract) with the consensus sequence EFFDAxE, in the region between the PH domain and the ORD (Fig. 2.5). This motif binds to VAMP-associated proteins (VAP), trans-membrane proteins of the endoplasmic reticulum (ER), conferring the proteins the ability to associate with ER membranes (Kaiser et al. 2005; Loewen et al. 2003). Furthermore, ORP5 and ORP8 have a carboxy-terminal trans-membrane segment that, at least in the case of ORP8, specifies ER targeting (Yan et al. 2008).

The VAPs act as ER docking receptors for several factors involved in the regulation of lipid metabolism, including the ORPs, Goodpasture antigen binding protein (GBPB)/ceramide transporter (CERT; Hanada et al. 2003), the retinal degeneration B (rdgB)/Nir proteins (Amarilio et al. 2005; Lev 2004), and *S. cerevisiae* Opi1p, a transcriptional repressor of inositol synthesis (Loewen et al. 2004). In the case of Opi1p, the association with a yeast VAP homologue, Scs2p, is regulated by the phospholipid composition of the ER: phosphatidic acid (PA) facilitates Opi1p binding to Scs2p in the ER. Addition of inositol that results in phosphatidylinositol synthesis consumes ER PA, leading to the detachment and nuclear translocation of Opi1p. One can envision that also the ORPs could bind to VAP proteins dependent on the presence/absence of a lipid signal, such as ER phospholipid composition or occupancy of the ORD by a ligand. These signals could induce shuttling of ORPs from the ER to another compartment, specified by the targeting determinants in the amino-terminal PH domain region. After executing a given function at the non-ER target organelle, the ORP could undergo a conformational change and return to the ER. This functional cycle could involve transfer of the bound ORD ligand between the two membranes, but binding of the ligand could also serve a signaling function.

Junctions at which ER membranes are closely juxtaposed with other organelles (ER junctions) have been implicated in several crucial cellular processes such as inter-compartmental lipid transport, store-operated Ca^{2+} entry, excitation–contraction coupling in striated muscle, and coupling of Ca^{2+} transport between the ER and mitochondria (reviewed by Levine 2004). *S. cerevisiae* Osh1p localizes, in addition to the Golgi complex, to the nucleus–vacuole junction, an ER junction characteristic of yeast, and Osh2p as well as Osh3p seem to target ER–plasma membrane junctions (Kvam and Goldfarb 2004; Levine and Munro 2001; Loewen et al. 2003). In a scheme arising from these findings, ORPs interact with the two (ER, non-ER) membrane compartments simultaneously via the FFAT motif and the amino-terminal PH domain region. The model predicts a role of long ORPs in the generation of ER junctions with other organelles or the regulation of some aspects of their function, possibly lipid transport through these membrane contact sites (Levine 2004; Olkkonen and Levine 2004; Peretti et al. 2008).

2.3.4 *Function of OSBP in Lipid Metabolism*

Of the mammalian ORPs, the founder member of the family, OSBP, has been characterized most extensively. It is translocated from a cytosolic or vesicular/ER localization to the Golgi complex upon addition of its high-affinity ligand 25-OHC (Ridgway et al. 1992). OSBP overexpression in Chinese hamster ovary (CHO) cells results in increased cholesterol biosynthesis and a reduction of cholesterol esterification (Lagace et al. 1997). Manipulations of the cellular sterol status were reported impact the Golgi localization of OSBP, suggesting that the Golgi sterol content or the flux of cholesterol through the Golgi complex are sensed by OSBP (Mohammadi et al. 2001; Ridgway et al. 1998; Storey et al. 1998). Furthermore, OSBP undergoes cholesterol-sensitive phosphorylation of specific serine residues, dephosphorylation accompanying the Golgi association of the protein. 25-OHC does not affect the phosphorylation status of OSBP, nor does phosphorylation affect the binding of 25-OHC by OSBP, indicating that the subcellular localization of the protein may be regulated by phosphorylation only in the absence of a "master" 25-OHC signal (Ridgway et al. 1998). The mechanisms by which OSBP affects cholesterol metabolism, and whether these impacts are oxysterol-dependent, remain unclear. A study employing siRNA-mediated silencing of OSBP demonstrated that the endogenous OSBP in HeLa cells plays no role in the suppression of cholesterol biosynthesis by 25-OHC (Nishimura et al. 2005). Instead, the effect of 25-OHC on cellular cholesterol homeostasis is apparently mediated through a direct interaction with the Insigs, proteins that retain SREBP-SCAP complexes in the ER when sterols are abundant (see Sect. 2.2.3.2; Radhakrishnan et al. 2007; Sun et al. 2007).

OSBP overexpression in CHO cells in the presence of 25-OHC was reported to enhance the synthesis of sphingomyelin (Lagace et al. 1999). Mutant OSBP with an amino acid substitution in the PH domain displayed enhanced association with ER membranes and was found to arrest a fluorescent ceramide analogue in the ER, indicating that the function of OSBP may involve transport of ceramide from the ER to Golgi sites where sphingomyelin synthase is located (Wyles et al. 2002). Solid evidence for this hypothesis was provided by the finding that the Golgi translocation and activation of ceramide transport protein, CERT (Hanada et al. 2003), are abolished when OSBP expression is knocked down through RNA interference (Perry and Ridgway 2006). The authors proposed that OSBP acts as a sterol sensor whose function is to integrate sphingomyelin biosynthesis with the cellular sterol status. Consistently, Peretti et al. (2008) reported new evidence that OSBP, together with VAPs, CERT, and the FFAT-motif containing phosphatidylcholine/phosphatidylinositol transfer protein Nir2, co-ordinates lipid transport between the ER and the Golgi apparatus and thus the Golgi lipid composition as well as the functional properties of this organelle. This finding may be connected with the recent observation that OSBP has the capacity to modulate the intracellular trafficking and processing of amyloid precursor protein (Zerbinatti et al. 2008).

We recently showed that adenoviral overexpression of rabbit OSBP in mouse liver leads to an increase of plasma very-low-density lipoprotein (VLDL) and liver tissue triglycerides (TG; Yan et al. 2007b). The increase of plasma TG was attributed to an enhancement of hepatic TG secretion. Investigation of the underlying mechanism revealed up-regulation of SREBP-1c expression and increase of the active nuclear form of this lipogenic transcription factor in the OSBP-transduced liver. Furthermore, we found evidence that siRNA-mediated silencing of OSBP in cultured hepatocytes attenuated the insulin induction of SREBP-1c and fatty acid synthetase (FAS), as well as TG synthesis. OSBP overexpression was also found to inhibit phosphorylation of the extracellular signal-regulated kinases (ERK). As changes in ERK activity were reported to impact the stability of nuclear SREBP-1c (Botolin et al. 2006), this observation provides a putative mechanistic explanation to the OSBP overexpression phenotype. The above findings suggest a new role of OSBP as a regulator of TG metabolism and its control by insulin signaling cascades.

Bowden and Ridgway (2008) demonstrated that reducing the cellular amount of OSBP by RNA interference increased the cellular content and cholesterol efflux activity of ABCA1. No change was detected in the *ABCA1* mRNA level; and OSBP silencing was shown to increase the half-life of the ABCA1 protein, the effect depending on an intact OSBP sterol-binding domain. Even though the underlying mechanism remained unclear, this finding underscores the multiple sterol-dependent regulatory functions of OSBP.

2.3.5 Evidence for the Involvement of Mammalian OSBP Homologues in Lipid Metabolism

The physiologic functions of the OSBP homologues in mammals are as yet poorly understood. The closest relative of OSBP, ORP4/OSBP2, exists as two major variants (Wang et al. 2002): ORP4L (long) and ORP4S (short). Like OSBP, ORP4 was shown to bind the oxysterols 25-OHC and 7-KC (Moreira et al. 2001; Wang et al. 2002). Both ORP4S and ORP4L were reported to localize on vimentin intermediate filaments in CHO cells (Wyles et al. 2007). Unlike OSBP, the subcellular localization of the ORP4 variants was not affected by treatment of cells with 25-OHC. Interestingly, overexpression of ORP4S or ORP4L mutants in leucine repeat motif induced abnormal bundling or aggregation of the vimentin filaments and significant inhibition of the esterification of LDL-derived cholesterol, suggesting a functional role of this protein in cholesterol transport to the ER (Wang et al. 2002; Wyles et al. 2007). The authors envisioned that ORP4 could use vimentin filaments as a scaffold or tracks for transport of cholesterol or regulatory oxysterols between endocytic compartments and the ER. They also provided evidence that ORP4L heterodimerizes with OSBP.

Interestingly, even though OSBP does not localize on vimentin filaments, OSBP leucine repeat mutants caused a collapse of vimentin filaments similar to corresponding ORP4L mutants, probably reflecting dimerization and functional interplay of OSBP with ORP4L (Wyles et al. 2007). These findings have important implications, considering that each cell type expresses simultaneously a large number of different ORPs, which could form homo- and heterodimers with distinct functional properties.

There is increasing evidence for functions of the closely related ORP1 and ORP2 in cellular lipid metabolism. Overexpression of the long ORP1 variant, ORP1L, or its ankyrin repeat region leads to enhanced recruitment of microtubule-dependent dynein/dynactin motor complexes on late endosomes (LE) and clustering of the endosomes in the juxtanuclear region (see Sect. 2.3.3). Since the ORD of ORP1L binds both PIPs (Fairn and McMaster 2005a) and sterols (Suchanek et al. 2007), and also the PH domain interacts with PIPs (Johansson et al. 2005), ORP1L can be envisioned to act as a lipid sensor that, in complex with Rab7 and its other effector protein RILP (Johansson et al. 2007), modulates the motility and/or distribution of LE according to lipid cues. Our latest work demonstrates that macrophage ORP1L overexpression in LDL receptor-deficient mice increases the size of atherosclerotic lesions (Yan et al. 2007a). The transgenic macrophages were shown to display a defect in cholesterol efflux to spherical HDL and reduced expression of ABCG1 and apolipoprotein E (apoE), as well as increased expression of phospholipid transfer protein (PLTP). These changes in gene expression in response to ORP1L overexpression provide putative explanations to the observed enhancement of atherogenesis and provide compelling evidence for a functional role of ORP1L in macrophage sterol metabolism. However, the relationship between ORP1L function in LE motility and the latter findings remains so far poorly understood.

Overexpression of ORP2, a short human ORP, in CHO or HeLa cells results in an up-regulation of cellular cholesterol efflux (Hynynen et al. 2005; Laitinen et al. 2002). Furthermore, the transport of newly synthesized cholesterol from the ER to the cell surface was enhanced by overexpressed ORP2, as determined by an assay based on cyclodextrin extraction of plasma membrane cholesterol (Hynynen et al. 2005). The HeLa cells expressing ORP2 also showed, probably as a homeostatic response to cholesterol loss, up-regulation of LDL receptor expression and LDL uptake, as well as increased HMGCoA reductase activity. The results were consistent with enhancement of intracellular cholesterol transport by ORP2, or to impaired ability of the cellular plasma membrane to sequester cholesterol. The mechanism underlying the findings is so far unclear, but since we know that ORP2 binds several sterols, we find it possible that ORP2 could (at least in the overexpression situation) transport sterols between subcellular membrane compartments. Interestingly, we also observed in the ORP2 expressing CHO cells a defect in neutral lipid (both triglyceride and cholesterol ester) storage, associated with altered phospholipid fatty acid composition especially under conditions of lipoprotein starvation (Käkelä et al. 2005). Together with our recent unpublished findings, this suggests that a central physiologic function of endogenous cellular ORP2 may involve neutral lipid metabolism.

2.3.6 Functional Interplay of ORPs with the Transcriptional Regulators of Lipid Metabolism

One is tempted to speculate that ORPs could modulate the availability of oxysterol ligands to the LXRs or the Insigs. Overexpression of ORP1L was found to mildly enhance the LXR-mediated transactivation of a reporter gene, dependent on the presence of LXR agonist, either 22(R)OHC or a synthetic non-sterol agonist (Johansson et al. 2003). However, the mechanism underlying this effect remained unclear. We recently demonstrated (Yan et al. 2007a) that transgenic macrophages overexpressing ORP1L display a defect in cholesterol efflux to spherical HDL and reduced expression of ABCG1 and apoE, as well as increased expression of PLTP (see Sect. 2.3.5). The affected genes are subject to transcriptional regulation by the LXRs. Moreover, ORP1L overexpression in cultured mouse macrophages attenuated the response of the *ABCG1* mRNA to the LXR agonist 22(R)OHC, which was also shown to be a ligand of ORP1L. One possible interpretation of the results is that ORP1L modulates the LXR ligand interactions, thereby affecting the expression of LXR target genes and the development of atherosclerosis. However, we find it equally probable that other, more indirect mechanisms may account for the observed phenotypic effect. The strongest evidence for functional interplay between ORPs and the LXRs is found in our recent study (Yan et al. 2008) demonstrating that silencing of ORP8 expression in THP-1 macrophages induces the transcription of *ABCA1* and, as a plausible consequence, cholesterol efflux to apolipoprotein A-I. This effect was reproduced using a luciferase reporter driven by the *ABCA1* promoter. ORP8 silencing synergized with a synthetic LXR agonist and was significantly suppressed when a mutant *ABCA1* promoter devoid of a functional LXR response (DR4) element was used. Interestingly, ORP8 was found to be abundant in the macrophages of human coronary artery lesions, bringing up the possibility that ORP8 might play a role in the development of atherosclerosis.

Our recent finding that hepatic OSBP overexpression enhances lipogenesis putatively via up-regulation of SREBP-1c (see Sect. 2.3.4; Yan et al. 2007b) suggests a functional link between the ORPs and the SREBPs. However, it is unclear if the underlying mechanism involves modulation of Insig function. A change in ERK activity and a resulting change in the stability of nuclear SREBP-1c (Botolin et al. 2006) provide another putative explanation for the findings. The functional connections between the ORP and the LXR and SREBP systems of lipid homeostatic control are an attractive topic for future investigations, relevant for the mechanisms underlying the development of dyslipidemias and atherosclerosis.

2.3.7 Function of Yeast Osh Proteins in Sterol Metabolism

Baker's yeast *S. cerevisiae* is used extensively as a model organism in molecular cell biology due to its amenability to a spectrum of powerful genetic approaches. Sterol homeostasis in *S. cerevisiae* shares a number of similarities with that in mammalian cells (Henneberry and Sturley 2005). The predominant sterol in yeast is ergosterol, the structure of which differs only slightly from cholesterol. A majority of the enzymes responsible for the synthesis of ergosterol are localized in the ER but, just like cholesterol in mammalian cells (Ikonen 2008), ergosterol concentrates at the plasma membrane (Zinser et al. 1991). Under anaerobic conditions yeast sterol biosynthesis is inhibited and the cells are able to take up sterol from the growth medium. Sterol uptake can also be achieved in specific genetic set-ups (Schulz and Prinz 2007). Sterol transport between the yeast ER and plasma membrane is reported to occur via non-vesicular mechanisms (Baumann et al. 2005; Li and Prinz 2004). *S. cerevisiae* lacks homologues of the putative sterol carriers found in mammalian cells, but has, however, seven ORPs named Osh1–7p. Three of them (Osh1–3p) belong to the category of long ORPs and four (Osh4–7p) are of the short subtype (Beh et al. 2001; Schmalix and Bandlow 1994).

Even before the Osh4p structure was solved, several studies provided evidence for the involvement of yeast Osh proteins in sterol metabolism. Jiang et al. (1994) investigated strains mutant for *OSH1*, *OSH4/KES1*, and *OSH5/HES1*. In the double or triple mutants they discovered pleiotropic sterol-related phenotypes, including tryptophan transport defects and nystatin resistance, as well as mild reductions of membrane ergosterol levels. Beh et al. (2001) determined the phenotypic effects of all 127 permutations of *OSH* deletion alleles. The results demonstrated that the individual *OSH* genes are not essential, but deletion of all seven is lethal, suggesting that the genes have a shared function essential for viability. The viable combinations of *OSH* deletions displayed distinct sterol-related defects and depletion of all seven proteins resulted in cellular sterol accumulation, providing evidence for a disturbance of sterol homeostatic control. Elimination of *OSH* function resulted in a redistribution of ergosterol from the plasma membrane to intracellular locations, vacuolar fragmentation, and cellular accumulation of lipid droplets (Beh and Rine 2004). Furthermore, the authors reported disturbances of endocytosis, cell budding, and cell wall deposition in cells with *OSH* defects. These findings suggest that the function of *S. cerevisiae* Osh proteins involves the subcellular sterol distribution, the other phenotypic effects possibly being secondary to this. Alternatively, disturbance of cellular membrane dynamics by Osh depletion could lead to an abnormal sterol distribution. In support of the former hypothesis, Raychaudhuri et al. (2006) presented evidence for function of the Osh proteins (Osh3–5p) in sterol transport from the yeast plasma membrane to the esterification compartment, the ER. Consistent with the in vivo findings, the authors showed that Osh4p is capable of extracting sterols from donor liposomes and transferring them to acceptor vesicles in vitro. The sterol transfer was more rapid between membranes that contain PIPs, suggesting that interaction of ORPs with the negatively charged PIP headgroups on membrane surfaces facilitates the sterol transport function. Consistently, the authors found that the transport of sterol from the plasma membrane to the ER is slowed down in mutant strains with defects in PI(4)P or PI(4,5)P_2 synthesis. Therefore, also the interactions of several short mammalian ORPs with PIPs (Fairn and McMaster 2005a, b; Hynynen et al. 2005) may play an important role in the ability of these proteins to extract sterol from cellular membranes. It is unclear whether the yeast ORPs investigated by Raychaudhuri et al. (2006) act as true sterol transporters – lack of all seven Osh proteins reduced sterol transport to the ER only 5-fold. According to an alternative hypothesis, the ORPs affect sterol transport indirectly by affecting the ability of the cellular plasma membrane to sequester these lipids (Sullivan et al. 2006).

Work by the group of H. Yang shows that the association of *S. cerevisiae* Osh6p and Osh7p with cellular membranes is regulated by the AAA family ATPase Vps4p (Wang et al. 2005a, b). The carboxy-terminal coiled-coil domain of Osh7p was shown to determine the interaction with Vps4p. Deletion of *VPS4* or *OSH6–OSH7* double deletion resulted in a defect of sterol ester synthesis, and Osh7p overexpression partially replenished sterol esterification in the *vps4Δ* strain. One interpretation of these findings is that Osh6p and Osh7p act as sterol transfer proteins, and that Vps4p catalyzes their dissociation from membranes as an essential part of their functional cycle. Alternatively, Os6p and Osh7p could regulate the activity of Vps4p. Since Vps4p dissociates the ESCRT III complex responsible for sorting of cargo proteins to the multivesicular body (Babst et al. 2002a, b), Osh6p and Osh7p could via Vps4p mediate lipid signals to the machinery regulating endosomal sorting/membrane trafficking.

2.3.8 Osh4p Regulates Secretory Vesicle Transport

The best known example of ORP involvement in intracellular vesicle transport was described in *S. cerevisiae*. Fang et al. (1996) demonstrated that deletion of *OSH4/KES1* leads to by-pass of the temperature-sensitivity of mutants in *SEC14*, a gene encoding a phosphatidylinositol transfer protein (PITP; Sec14p) essential for secretory vesicle biogenesis (Bankaitis et al. 2005). This suggested that yeast Osh4p acts as a negative regulator of Golgi secretory function. Sec14p is thought to maintain a membrane composition permissive to Gcs1p, a GTPase activating protein (GAP) for Arf1, a small GTPase with a central role in transport vesicle formation (Yanagisawa et al. 2002). Also, Osh4p was suggested to exert its effect via regulation of Arf1 activity (Li et al. 2002). Positive membrane curvature plays a distinct role in the Arf GTPase cycle and in transport vesicle formation (Antonny et al. 2005). In addition, Osh4p was shown to contain a specific type of amphipathic helix that targets membranes with a positive curvature (Bigay et al. 2005), consistent with a functional interplay of Osh4p and Arf1p. Intriguingly, a recent study (Fairn et al. 2007) shows that Osh4p reduces both the cellular content of PI(4)P and its availability for recognition by other proteins, which include components involved in transport vesicle formation (D'Angelo et al. 2008). PI(4)P and Arf act in concert to recruit transport factors on Golgi membranes; For example, the PI(4)P adaptor proteins (FAPPs) essential for Golgi to plasma membrane vesicle transport bind both Arf and PI(4)P (Godi et al. 2004). Therefore, the observation of Fairn and co-workers may provide one explanation to the connection of Osh4p function with Arf. As a conclusion, the mechanism by which Osh4p impacts post-Golgi vesicle transport is not completely understood, but recent findings have brought us close to the solution of this dilemma. Analysis of site-specific *osh4* mutants revealed no clear correlation between sterol binding capacity and the ability to inhibit Golgi-derived vesicular transport (Im et al. 2005). Therefore, the relationship between the functions of Osh4p in post-Golgi membrane trafficking and in sterol transport remains unclear.

Interestingly, a recent study demonstrated that the Osh proteins, including Osh4p, play important roles in yeast cell polarization by maintaining the proper subcellular localization of septins, the Rho GTPases Cdc42p and Rho1p, and the Rab GTPase Sec4p (Kozminski et al. 2006). The mechanisms underlying this observation are as yet poorly understood. However, since a functional secretory pathway is essential for polarized growth, the impact of ORPs on cell polarization may involve a lipid-dependent action in secretory vesicle transport.

2.3.9 Mammalian ORPs and Intracellular Vesicle Transport

Even though the most studied case of ORP function in vesicle transport is *S. cerevisiae* Osh4p/Kes1p (see Sect. 2.3.8), there are several reports suggesting the involvement mammalian ORPs in vesicle transport along the secretory or the endocytic pathways of membrane trafficking. Most of these studies employed overexpression of an intact ORP or of a PH domain separated from its context. ORP1L overexpression was shown to interfere with the delivery of endocytosed cargo to late endosomes/lysosomes (Johansson et al. 2005) and excess ORP2 was reported to interfere with endocytotic uptake of markers from the cell surface (Hynynen et al. 2005). Furthermore, the overexpression of human ORP2 in CHO cells and in *S. cerevisiae* was reported to interfere with Golgi vesicle transport (Laitinen et al. 2002; Xu et al. 2001). Similarly, Levine and Munro (1998) showed that overexpression of the OSBP PH domain interferes with Golgi secretory function.

The mechanisms underlying the cellular effects of ORP overexpression are unclear and may not faithfully reflect the physiologic function of the proteins for the following reasons:

1. Overexpression of certain ORPs is known to cause distortion of the structure of the ER and Golgi compartments (Lehto et al. 2005; Wyles and Ridgway 2004), which could lead to artefactual disturbances in membrane trafficking.
2. Association of an excessive amount of an ORP with membrane phosphoinositides may disturb vesicle transport through interference with the trafficking machinery known to interact with these negatively charged phospholipids (see Fairn et al. 2007).

Johansson et al. (2007) employed ORP1L knock-down with siRNA to show that the protein is required for the clustering of late endocytic compartments in the pericentriolar region. However, the actual late endocytic trafficking functions were not assessed in this study. Studies employing gene silencing are now necessary to make reliable observations on ORP function in intracellular membrane trafficking.

2.3.10 ORPs – Integrating Lipid Cues with Cell Signaling Cascades

The first convincing evidence for the involvement of ORPs in cell signalling was provided by the study of Sugawara et al. (2001), who identified a *Caenorhabditis elegans* ORP, designated BRAM-interacting protein, BIP, as a modulator of transforming growth factor (TGF)-β signalling. The authors carried out a two-hybrid screen using bone morphogenetic protein (BMP) receptor associated protein (BRAM) as a bait, and identified a *Xenopus* ORP (BIP) as a BRAM binding partner. Thereafter they isolated the *C. elegans* homologue of this cDNA and showed that it interacts with the *C. elegans* BRAM homologues BRA-1 and BRA-2. They further demonstrated that inhibition of BIP expression by RNA interference produces a Sma phenotype characteristic of disturbance of the *C. elegans* TGF-β pathway that regulates body length. The documented interaction of the ORP (BIP) with the BRA proteins evidences for a direct role as a modulator of the signaling pathway.

Increasing evidence is accumulating for the involvement of ORPs in signaling events in mammals. Wang et al. (2005c) identified OSBP as a sterol-sensing scaffolding factor that regulates the dephosphorylation and thereby the activity of extracellular signal-regulated kinases (ERK), key components of the mitogen activated protein kinase (MAPK) signaling pathways. OSBP was shown to bind both cholesterol and 25-OHC; and the cholesterol-bound state scaffolds a protein phosphatase complex (PP2A serine/threonine phosphatase and HePTP, a PTPPBS family tyrosine phosphatase) which dephosphorylates and thereby inactivates ERK. Reduction of the cellular cholesterol content or addition of 25-OHC dissociates the phosphatase complex, leading to hyperphosphorylation of the ERKs. A second signaling function of OSBP was reported by Romeo and Kazlauskas (2008), who found evidence that up-regulation of profilin-1 (an actin-binding protein implicated in endothelial dysfunction and atherosclerosis) by 7-KC is mediated by OSBP. The signal transduction route involves interaction of OSBP-7-KC complex with the tyrosine kinase JAK-2, which phosphorylates Tyr394 on OSBP. This leads to the activation of STAT3, which is responsible for the induction of profilin. An interesting possibility brought up by these findings is that also other members of the ORP family could have lipid-specific scaffolding functions in signaling pathways. Indeed, Lessman et al. (2007) demonstrated that ORP9 contains a phosphoinositide-dependent kinase-2 (PDK-2) phosphorylation site, the phosphorylation of which is dependent on PKC-β or mTOR in bone marrow-derived mast cells or HEK293 cells, respectively. The authors also provided RNA interference and immunoprecipitation evidence that ORP9 interacts with these kinases to negatively regulate phosphorylation of the PKD-2 site in Akt/protein kinase B, a major controller of cell survival, cell cycle progression, and glucose metabolism (Hanada et al. 2004).

Interestingly, ORP3 and ORP7 were recently found to interact with R-Ras, a small GTPase that regulates cell adhesion and migration (Goldfinger et al. 2007; Kinbara et al. 2003), implying a functional role of these ORPs in Ras signaling. Consistent with the above findings, Lehto et al. (2008) demonstrated that ORP3 controls cell adhesion and spreading, organization of the actin cytoskeleton, β1-integrin activity, and macrophage phagocytic function, cellular processes also subject to regulation by R-Ras. ORP3 is present at highest levels in leukocytes (T-cells, B-cells, monocytes, macrophages) and the epithelia of several tissues; and abnormally high expression is detected in certain forms of leukemia and solid tumors, suggesting that the protein may modify the signaling processes and adhesion properties of cells in a manner that facilitates malignant growth.

Even though the role of ORP interactions with lipidous ligands has not been addressed in all of the above studies, it is plausible that liganding of both the PH domain and the ORD is crucial for the signaling functions outlined above. One can thus envision that, in analogy with the sterol-dependent function of OSBP in ERK and JAK-2/STAT3 signaling, the other ORPs may also relay essential lipid cues to cellular signaling cascades.

2.4 Future Perspectives

New roles of oxysterols as potent regulators not only of cellular and body lipid metabolism but also of other vital cellular processes, such as signaling cascades, differentiation, and apoptosis, have emerged at an increasing pace. In addition to the suggested roles of oxysterols as potentially harmful substances that accumulate in pathophysiologic states, these compounds have at their normal concentrations at least the potential to function as signaling molecules that maintain cellular and body lipid homeostasis and determine cell fate. The fact that oxysterols act as ligands of the LXRs, transcription factors with potentially antiatherogenic effects, has prompted intense efforts towards the development of synthetic agonists that would selectively induce cardioprotective LXR target genes. In analogy, the discovery of oxysterol interactions with new receptor proteins involved in lipid metabolism, such as the Insig proteins, the estrogen receptors, NPC1 and the OSBP-related proteins, brings up the possibility of developing new oxysterol-related pharmacologic compounds that modify the function of these target proteins, with beneficial effects on lipid metabolism and atherogenesis. Furthermore, modulation of the expression levels or activity of the cytochrome P450 enzymes responsible for generating oxysterols, or inhibition of 2,3-oxidosqualene cyclase could be feasible therapeutic approaches.

Discovery of oxysterols as regulators of Hedgehog signaling uncovered a new role of oxysterols as regulators of embryonic development. Together with the increasing number of reports on the involvement of ORP proteins in cell signaling, this finding has opened new perspectives. We can expect a plethora of novel connections of oxysterols, oxysteroids, and their receptor proteins with differentiation and developmental processes.

The ORPs constitute a highly interesting protein family present ubiquitously in eukaryotic cells, thus representing machinery of fundamental importance in cell physiology. The functions suggested for ORPs in different organisms are summarized in Table 2.2. The bulk of the literature suggests that binding of sterols or other lipids within the ligand binding domain of ORPs serves a regulatory function, but in certain

Table 2.2 A summary of the suggested ORP functions

Organism	Protein	Suggested function	References
Mammals	OSBP	Sterol-dependent regulation of ERK, JAK-2/STAT3 and sphingomyelin synthesis; modulation of hepatic lipogenesis, ABCA1 stability and processing of the amyloid precursor protein	Wang et al. (2005c), Perry and Ridgway (2006), Yan et al. (2007b), Bowden and Ridgway (2008), Romeo and Kazlauskas (2008), Zerbinatti et al. (2008)
	ORP1L	Motility and distribution of late endosomes; macrophage lipid metabolism	Johansson et al. (2003, 2005, 2007), Yan et al. (2007a)
	ORP1S	Vesicle transport from Golgi	Xu et al. (2001)
	ORP2	Sterol transport; neutral lipid metabolism; vesicle transport from Golgi	Xu et al. (2001), Laitinen et al. (2002), Hynynen et al. (2005), Käkelä et al. (2005)
	ORP3	Regulation of cell–cell and cell–matrix adhesion	Lehto et al. (2008)
	ORP4	Vimentin-dependent sterol transport and/or signalling	Wang et al. (2002), Wyles et al. (2007)
	ORP8	Regulation of *ABCA1* expression and cholesterol efflux	Yan et al. (2008)
	ORP9	Regulation of Akt phosphorylation	Lessmann et al. (2007)
Rainbow trout	OORP-T	Utilization of yolk lipids in oocyte development	Ramachandra et al. (2007)
Drosophila melanogaster	OSBP-Dm (CG6708)	Cell cycle control	Alphey et al. (1998)
Caenorhabditis elegans	BIP (obr-3)	Modulation of TGF-β signalling	Sugawara et al. (2001)
Dictyostelium discoideum	OSBPa	Regulation of slug-fruiting body switch	Fukuzawa and Williams (2002)
Cryptosporidium parvum	CpORP1	Lipid transport across the parasitophorous vacuole membrane	Zeng and Zhu (2006)
Solanum tuberosum	StOBP1	Function in a non-specific defence pathway	Avrova et al. (2004)
Petunia inflata	PiORP1	Pollen development	Skirpan et al. (2006)
Glycine max	GmOSBP	Reactions to stress response and cotyledon senescence	Li et al. (2008)
Saccharomyces cerevisiae	Osh1p	Post-synthetic sterol regulation; piecemeal microautophagy of the nucleus; cell polarity establishment	Jiang et al. (1994), Beh et al. (2001), Kvam and Goldfarb (2004), Kozminski et al. (2006)
	Osh2p	Sterol metabolism; cell polarity establishment	Daum et al. (1999), Beh et al. (2001), Kozminski et al. (2006)

	Osh3p	Sterol transport; sphingolipid metabolism; regulation of nuclear fusion during mating and pseudohyphal growth	Park et al. (2002), Yano et al. (2004), Hur et al. (2006), Raychaudhuri et al. (2006)
	Osh4p/Kes1p	Sterol transport and metabolism; post-Golgi vesicle transport; cell polarity establishment	Jiang et al. (1994), Fang et al. (1996), Li et al. (2002), Kozminski et al. (2006), Raychaudhuri et al. (2006)
	Osh5p	Sterol transport and metabolism	Jiang et al. (1994), Raychaudhuri et al. (2006)
	Osh6p	Sterol transport; regulation of Vps4p function; cell polarity establishment	Wang et al. (2005a, b), Kozminski et al. (2006)
	Osh7p	Sterol transport; regulation of Vps4p function	Wang et al. (2005b)
Candida albicans	Osh3	Regulation of pseudohyphal growth	Hur et al. (2006)

cases there is also evidence for a sterol transport function. The initial findings on ORP function (*S. cerevisiae* Osh4p/Kes1p) involved intracellular vesicle transport. However, more recent studies in various organisms have connected a majority of ORP proteins with cellular lipid metabolism or cell signaling events. The interest in this protein family is constantly expanding and the ground-breaking studies published during the past few years have paved the way for creating new functional hypotheses. Testing these hypotheses in cultured cell set-ups and in animal models will in near future enable major progress in our understanding of the physiologic role of ORPs. Most likely, many central functions of the ORPs turn out to involve regulation of cellular and body lipid metabolism. However, lipid homeostasis must be integrated with a number of other regimes. Numerous new functional connections of the ORPs with the control of intracellular vesicle transport, cell differentiation, proliferation, polarity, adhesion, migration, and survival/death, will no doubt be discovered. Furthermore, it will be of major importance to elucidate the role of the ORP proteins and genetic variation within the *OSBPL* genes in human diseases such as cancer, dyslipidemias, and atherosclerosis.

Acknowledgements Work in the author's group is supported by the Academy of Finland (grant 121457), the Sigrid Juselius Foundation, the Finnish Foundation for Cardiovascular Research, the Finnish Cultural Foundation, the Magnus Ehrnrooth Foundation, and the European Union FP7 (LipidomicNet, agreement no. 202272).

References

Adams CM, Reitz J, De Brabander JK, Feramisco JD, Li L, Brown MS, Goldstein JL (2004) Cholesterol and 25-hydroxycholesterol inhibit activation of SREBPs by different mechanisms, both involving SCAP and Insigs. J Biol Chem 279:52772–52780

Aghaloo TL, Amantea CM, Cowan CM, Richardson JA, Wu BM, Parhami F, Tetradis S (2007) Oxysterols enhance osteoblast differentiation in vitro and bone healing in vivo. J Orthop Res 25:1488–1497

Alphey L, Jimenez J, Glover D (1998) A *Drosophila* homologue of oxysterol binding protein (OSBP) - implications for the role of OSBP. Biochim Biophys Acta. 1395:159–164

Amarilio R, Ramachandran S, Sabanay H, Lev S (2005) Differential regulation of endoplasmic reticulum structure through VAP-Nir protein interaction. J Biol Chem 280:5934–5944

Anniss AM, Apostolopoulos J, Dworkin S, Purton LE, Sparrow RL (2002) An oxysterol-binding protein family identified in the mouse. DNA Cell Biol 21:571–580

Antonny B, Bigay J, Casella JF, Drin G, Mesmin B, Gounon P (2005) Membrane curvature and the control of GTP hydrolysis in Arf1 during COPI vesicle formation. Biochem Soc Trans 33:619–622

Apfel R, Benbrook D, Lernhardt E, Ortiz MA, Salbert G, Pfahl M (1994) A novel orphan receptor specific for a subset of thyroid hormone-responsive elements and its interaction with the retinoid/thyroid hormone receptor subfamily. Mol Cell Biol 14:7025–7035

Arca M, Natoli S, Micheletta F, Riggi S, Di Angelantonio E, Montali A, Antonini TM, Antonini R, Diczfalusy U, Iuliano L (2007) Increased plasma levels of oxysterols, in vivo markers of oxidative stress, in patients with familial combined hyperlipidemia: reduction during atorvastatin and fenofibrate therapy. Free Radic Biol Med 42:698–705

Avrova AO, Nawsheen T, Rokka V-M, Heilbronn J, Campbell E, Hein I, Gilroy EM, Cardle L, Bradshaw JE, Stewart HE, Fakim YJ, Loake G, Birch PRJ (2004) Potato oxysterol binding protein and cathepsin B are rapidly up-regulated in independent defence pathways that distinguish *R* gene-mediated and field resistences to *Phytophthora infestans*. Mol Plant Pathol 5:45–46

Babiker A, Andersson O, Lund E, Xiu RJ, Deeb S, Reshef A, Leitersdorf E, Diczfalusy U, Björkhem I (1997) Elimination of cholesterol in macrophages and endothelial cells by the sterol 27-hydroxylase mechanism. Comparison with high density lipoprotein-mediated reverse cholesterol transport. J Biol Chem 272:26253–26261

Babst M, Katzmann DJ, Estepa-Sabal EJ, Meerloo T, Emr SD (2002a) Escrt-III: an endosome-associated heterooligomeric protein complex required for mvb sorting. Dev Cell 3:271–282

Babst M, Katzmann DJ, Snyder WB, Wendland B, Emr SD (2002b) Endosome-associated complex, ESCRT-II, recruits transport machinery for protein sorting at the multivesicular body. Dev Cell 3:283–289

Bankaitis VA, Phillips S, Yanagisawa L, Li X, Routt S, Xie Z (2005) Phosphatidylinositol transfer protein function in the yeast Saccharomyces cerevisiae. Adv Enzyme Regul 45:155–170

Baumann NA, Sullivan DP, Ohvo-Rekilä H, Simonot C, Pottekat A, Klaassen Z, Beh CT, Menon AK (2005) Transport of newly synthesized sterol to the sterol-enriched plasma membrane occurs via nonvesicular equilibration. Biochemistry 44:5816–5826

Beh CT, Rine J (2004) A role for yeast oxysterol-binding protein homologs in endocytosis and in the maintenance of intracellular sterol-lipid distribution. J Cell Sci 117:2983–2996

Beh CT, Cool L, Phillips J, Rine J (2001) Overlapping functions of the yeast oxysterol-binding protein homologues. Genetics 157:1117–1140

Bensinger SJ, Bradley MN, Joseph SB, Zelcer N, Janssen EM, Hausner MA, Shih R, Parks JS, Edwards PA, Jamieson BD, Tontonoz P (2008) LXR signaling couples sterol metabolism to proliferation in the acquired immune response. Cell 134:97–111

Beseme F, Astruc ME, Defay R, Descomps B, Crastes de Paulet A (1986) Characterization of oxysterol-binding protein in rat embryo fibroblasts and variations as a function of the cell cycle. Biochim Biophys Acta 886:96–108

Beseme F, Astruc ME, Defay R, Crastes de Paulet A (1987) Rat liver cytosol oxysterol-binding protein. Characterization and comparison with the HTC cell protein. FEBS Lett 210:97–103

Beyea MM, Heslop CL, Sawyez CG, Edwards JY, Markle JG, Hegele RA, Huff MW (2007) Selective up-regulation of LXR-regulated genes ABCA1, ABCG, and APOE in macrophages through increased endogenous synthesis of 24(S),25-epoxycholesterol. J Biol Chem 282:5207–5216

Biasi F, Leonarduzzi G, Vizio B, Zanetti D, Sevanian A, Sottero B, Verde V, Zingaro B, Chiarpotto E, Poli G (2004) Oxysterol mixtures prevent proapoptotic effects of 7-ketocholesterol in macrophages: implications for proatherogenic gene modulation. FASEB J 18:693–695

Bigay J, Casella JF, Drin G, Mesmin B, Antonny B (2005) ArfGAP1 responds to membrane curvature through the folding of a lipid packing sensor motif. EMBO J 24:2244–2253

Björkhem I (1986) Assay of unesterified 7-oxocholesterol in human serum by isotope dilution-mass spectrometry. Anal Biochem 154:497–501

Björkhem I, Diczfalusy U (2002) Oxysterols: friends, foes, or just fellow passengers. Arterioscler Thromb Vasc Biol 22:734–742

Björkhem I, Eggertsen G (2001) Genes involved in initial steps of bile acid synthesis. Curr Opin Lipidol 12:97–103

Björkhem I, Andersson O, Diczfalusy U, Sevastik B, Xiu RJ, Duan C, Lund E (1994) Atherosclerosis and sterol 27-hydroxylase: evidence for a role of this enzyme in elimination of cholesterol from human macrophages. Proc Natl Acad Sci USA 91:8592–8596

Björkhem I, Lütjohann D, Breuer O, Sakinis A, Wennmalm A (1997) Importance of a novel oxidative mechanism for elimination of brain cholesterol. Turnover of cholesterol and 24(S)-hydroxycholesterol in rat brain as measured with 18O2 techniques in vivo and in vitro. J Biol Chem 272:30178–30184

Bodin K, Andersson U, Rystedt E, Ellis E, Norlin M, Pikuleva I, Eggertsen G, Björkhem I, Diczfalusy U (2002) Metabolism of 4 beta -hydroxycholesterol in humans. J Biol Chem 277:31534–31540

Boissonneault GA, Hennig B, Ouyang CM (1991) Oxysterols, cholesterol biosynthesis, and vascular endothelial cell monolayer barrier function. Proc Soc Exp Biol Med 196:338–343

Botolin D, Wang Y, Christian B, Jump DB (2006) Docosahexaneoic acid (22:6,n-3) regulates rat hepatocyte SREBP-1 nuclear abundance by Erk- and 26S proteasome-dependent pathways. J Lipid Res 47:181–192

Bowden K, Ridgway ND (2008) Oxysterol binding protein (OSBP) negatively regulates ATP binding cassette transporter A1 (ABCA1) protein stability. J Biol Chem 283:18210–18217

Bretillon L, Siden A, Wahlund LO, Lutjohann D, Minthon L, Crisby M, Hillert J, Groth CG, Diczfalusy U, Björkhem I (2000) Plasma levels of 24S-hydroxycholesterol in patients with neurological diseases. Neurosci Lett 293:87–90

Brown MS, Goldstein JL (1974) Suppression of 3-hydroxy-3-methylglutaryl coenzyme A reductase activity and inhibition of growth of human fibroblasts by 7-ketocholesterol. J Biol Chem 249:7306–7314

Brown AJ, Jessup W (1999) Oxysterols and atherosclerosis. Atherosclerosis 142:1–28

Brown AJ, Mander EL, Gelissen IC, Kritharides L, Dean RT, Jessup W (2000a) Cholesterol and oxysterol metabolism and subcellular distribution in macrophage foam cells. Accumulation of oxidized esters in lysosomes. J Lipid Res 41:226–237

Brown AJ, Watts GF, Burnett JR, Dean RT, Jessup W (2000b) Sterol 27-hydroxylase acts on 7-ketocholesterol in human atherosclerotic lesions and macrophages in culture. J Biol Chem 275:27627–27633

Brown AJ, Sun L, Feramisco JD, Brown MS, Goldstein JL (2002) Cholesterol addition to ER membranes alters conformation of SCAP, the SREBP escort protein that regulates cholesterol metabolism. Mol Cell 10:237–245

Carpenter KL, Taylor SE, Ballantine JA, Fussell B, Halliwell B, Mitchinson MJ (1993) Lipids and oxidised lipids in human atheroma and normal aorta. Biochim Biophys Acta 1167:121–130

Carpenter KL, Taylor SE, van der Veen C, Williamson BK, Ballantine JA, Mitchinson MJ (1995) Lipids and oxidised lipids in human atherosclerotic lesions at different stages of development. Biochim Biophys Acta 1256:141–150

Castrillo A, Joseph SB, Vaidya SA, Haberland M, Fogelman AM, Cheng G, Tontonoz P (2003) Crosstalk between LXR and toll-like receptor signaling mediates bacterial and viral antagonism of cholesterol metabolism. Mol Cell 12:805–816

Chang YH, Abdalla DS, Sevanian A (1997) Characterization of cholesterol oxidation products formed by oxidative modification of low density lipoprotein. Free Radic Biol Med 23:202–214

Chen JJ, Lukyanenko Y, Hutson JC (2002) 25-hydroxycholesterol is produced by testicular macrophages during the early postnatal period and influences differentiation of Leydig cells in vitro. Biol Reprod 66:1336–1341

Chen W, Chen G, Head DL, Mangelsdorf DJ, Russell DW (2007) Enzymatic reduction of oxysterols impairs LXR signaling in cultured cells and the livers of mice. Cell Metab 5:73–79

Chisolm GM, Ma G, Irwin KC, Martin LL, Gunderson KG, Linberg LF, Morel DW, DiCorleto PE (1994) 7 beta-hydroperoxycholest-5-en-3 beta-ol, a component of human atherosclerotic lesions, is the primary cytotoxin of oxidized human low density lipoprotein. Proc Natl Acad Sci USA 91:11452–11456

Colles SM, Maxson JM, Carlson SG, Chisolm GM (2001) Oxidized LDL-induced injury and apoptosis in atherosclerosis. Potential roles for oxysterols. Trends Cardiovasc Med 11:131–138

Collier FM, Gregorio-King CC, Apostolopoulos J, Walder K, Kirkland MA (2003) ORP3 splice variants and their expression in human tissues and hematopoietic cells. DNA Cell Biol 22:1–9

Corcoran RB, Scott MP (2006) Oxysterols stimulate Sonic hedgehog signal transduction and proliferation of medulloblastoma cells. Proc Natl Acad Sci USA 103:8408–8413

Costet P, Luo Y, Wang N, Tall AR (2000) Sterol-dependent transactivation of the ABC1 promoter by the liver X receptor/retinoid X receptor. J Biol Chem 275:28240–28245

D'Angelo G, Vicinauza M, Di Campli A, De Matteis MA (2008) The multiple roles of PtdIns(4) P - not just the precursor of PtdIns(4,5)P2. J Cell Sci 121:1955–1963

Daum G, Tuller G, Nemec T, Hrastnik C, Balliano G, Cattel L, Milla P, Rocco F, Conzelmann A, Vionnet C, Kelly DE, Kelly S, Schweizer E, Schuller HJ, Hojad U, Greiner E, Finger K (1999) Systematic analysis of yeast strains with possible defects in lipid metabolism. Yeast 15:601–614
Dawson PA, Ridgway ND, Slaughter CA, Brown MS, Goldstein JL (1989a) cDNA cloning and expression of oxysterol-binding protein, an oligomer with a potential leucine zipper. J Biol Chem 264:16798–16803
Dawson PA, Van der Westhuyzen DR, Goldstein JL, Brown MS (1989b) Purification of oxysterol binding protein from hamster liver cytosol. J Biol Chem 264:9046–9052
Deckert V, Brunet A, Lantoine F, Lizard G, Millanvoye-van Brussel E, Monier S, Lagrost L, David-Dufilho M, Gambert P, Devynck MA (1998) Inhibition by cholesterol oxides of NO release from human vascular endothelial cells. Arterioscler Thromb Vasc Biol 18:1054–1060
Defay RE, Astruc ME, Roussillon S, Descomps B, Crastes De Paulet A (1982) A specific hydroxysterol binding protein in human lymphocyte cytosol. Biochimie 64:331–339
Deroo BJ, Korach KS (2006) Estrogen receptors and human disease. J Clin Invest 116:561–570
DuSell CD, Umetani M, Shaul PW, Mangelsdorf DJ, McDonnell DP (2008) 27-hydroxycholesterol is an endogenous selective estrogen receptor modulator. Mol Endocrinol 22:65–77
Dwyer JR, Sever N, Carlson M, Nelson SF, Beachy PA, Parhami F (2007) Oxysterols are novel activators of the hedgehog signaling pathway in pluripotent mesenchymal cells. J Biol Chem 282:8959–8968
Dzeletovic S, Breuer O, Lund E, Diczfalusy U (1995) Determination of cholesterol oxidation products in human plasma by isotope dilution-mass spectrometry. Anal Biochem 225:73–80
Eberle D, Hegarty B, Bossard P, Ferre P, Foufelle F (2004) SREBP transcription factors: master regulators of lipid homeostasis. Biochimie 86:839–848
Endo K, Oyama T, Saiki A, Ban N, Ohira M, Koide N, Murano T, Watanabe H, Nishii M, Miura M, Sekine K, Miyashita Y, Shirai K (2008) Determination of serum 7-ketocholesterol concentrations and their relationships with coronary multiple risks in diabetes mellitus. Diabetes Res Clin Pract 80:63–68
Engel T, Kannenberg F, Fobker M, Nofer J-R, Bode G, Lueken A, Assmann G, Seedorf U (2007) Expression of ATP-binding cassette-transporter ABCG1 prevents cell death by transporting cytotoxic 7β-hydroxycholesterol. FEBS Lett 581:1673–1680
Erridge C, Webb DJ, Spickett CM (2007) 25-Hydroxycholesterol, 7beta-hydroxycholesterol and 7-ketocholesterol upregulate interleukin-8 expression independently of Toll-like receptor 1, 2, 4 or 6 signalling in human macrophages. Free Radic Res 41:260–266
Fairn GD, McMaster CR (2005a) Identification and assessment of the role of a nominal phospholipid binding region of ORP1S (oxysterol-binding-protein-related protein 1 short) in the regulation of vesicular transport. Biochem J 387:889–896
Fairn GD, McMaster CR (2005b) The roles of the human lipid-binding proteins ORP9S and ORP10S in vesicular transport. Biochem Cell Biol 83:631–636
Fairn GD, Curwin AJ, Stefan CJ, McMaster CR (2007) The oxysterol binding protein Kes1p regulates Golgi apparatus phosphatidylinositol-4-phosphate function. Proc Natl Acad Sci USA 104:15352–15357
Fang M, Kearns BG, Gedvilaite A, Kagiwada S, Kearns M, Fung MK, Bankaitis VA (1996) Kes1p shares homology with human oxysterol binding protein and participates in a novel regulatory pathway for yeast Golgi-derived transport vesicle biogenesis. EMBO J 15:6447–6459
Forman BM, Ruan B, Chen J, Schroepfer GJ Jr, Evans RM (1997) The orphan nuclear receptor LXRalpha is positively and negatively regulated by distinct products of mevalonate metabolism. Proc Natl Acad Sci USA 94:10588–10593
Fuda H, Javitt NB, Mitamura K, Ikegawa S, Strott CA (2007) Oxysterols are substrates for cholesterol sulfotransferase. J Lipid Res 48:1343–1352
Fukuzawa M, Williams JG (2002) OSBPa, a predicted oxysterol binding protein of Dictyostelium, is required for regulated entry into culmination. FEBS Lett 527:37–42
Garcia-Cruset S, Carpenter KL, Guardiola F, Mitchinson MJ (1999) Oxysterols in cap and core of human advanced atherosclerotic lesions. Free Radic Res 30:341–50
Garcia-Cruset S, Carpenter KL, Guardiola F, Stein BK, Mitchinson MJ (2001) Oxysterol profiles of normal human arteries, fatty streaks and advanced lesions. Free Radic Res 35:31–41

Gelissen IC, Brown AJ, Mander EL, Kritharides L, Dean RT, Jessup W (1996) Sterol efflux is impaired from macrophage foam cells selectively enriched with 7-ketocholesterol. J Biol Chem 271:17852–17860
Gill S, Chow R, Brown AJ (2008) Sterol regulators of cholesterol homeostasis and beyond: the oxysterol hypothesis revisited and revised. Progr Lipid Res 47:391–404
Godi A, Di Campli A, Konstantakopoulos A, Di Tullio G, Alessi DR, Kular GS, Daniele T, Marra P, Lucocq JM, De Matteis MA (2004) FAPPs control Golgi-to-cell-surface membrane traffic by binding to ARF and PtdIns(4)P. Nat Cell Biol 6:393–404
Goldfinger LE, Ptak C, Jeffery ED, Shabanowitz J, Han J, Haling JR, Sherman NE, Fox JW, Hunt DF, Ginsberg MH (2007) An experimentally derived database of candidate Ras-interacting proteins. J Proteome Res 6:1806–1811
Goldstein JL, DeBose-Boyd RA, Brown MS (2006) Protein sensors for membrane sterols. Cell 124:35–46
Gregorio-King CC, Collier GR, McMillan JS, Waugh CM, McLeod JL, Collier FM, Kirkland MA (2001) ORP-3, a human oxysterol-binding protein gene differentially expressed in hematopoietic cells. Blood 98:2279–2281
Guyton JR, Black BL, Seidel CL (1990) Focal toxicity of oxysterols in vascular smooth muscle cell culture. A model of the atherosclerotic core region. Am J Pathol 137:425–434
Hanada K, Kumagai K, Yasuda S, Miura Y, Kawano M, Fukasawa M, Nishijima M (2003) Molecular machinery for non-vesicular trafficking of ceramide. Nature 426:803–809
Hanada M, Feng J, Hemmings BA (2004) Structure, regulation and function of PKB/AKT-a major therapeutic target. Biochim Biophys Acta 1697:3–16
Hayden JM, Brachova L, Higgins K, Obermiller L, Sevanian A, Khandrika S, Reaven PD (2002) Induction of monocyte differentiation and foam cell formation in vitro by 7-ketocholesterol. J Lipid Res 43:26–35
Henneberry AL, Sturley SL (2005) Sterol homeostasis in the budding yeast, *Saccharomyces cerevisiae*. Semin Cell Dev Biol 16:155–161
Hietter H, Trifilieff E, Richert L, Beck JP, Luu B, Ourisson G (1984) Antagonist action of cholesterol towards the toxicity of hydroxysterols on cultured hepatoma cells. Biochem Biophys Res Commun 120:657–664
Hojozi M, Munehira Y, Ikeda Y, Makishima M, Matsuo M, Kioka N, Ueda K (2008) Direct interaction of nuclear receptor LXRβ with ABCA1 modulates cholesterol efflux. J Biol Chem 283:30057–30063
Hua X, Yokoyama C, Wu J, Briggs MR, Brown MS, Goldstein JL, Wang X (1993) SREBP-2, a second basic-helix-loop-helix-leucine zipper protein that stimulates transcription by binding to a sterol regulatory element. Proc Natl Acad Sci USA 90:11603–11607
Hult M, Elleby B, Shafqat N, Svensson S, Rane A, Jornvall H, Abrahmsen L, Oppermann U (2004) Human and rodent type 1 11beta-hydroxysteroid dehydrogenases are 7beta-hydroxycholesterol dehydrogenases involved in oxysterol metabolism. Cell Mol Life Sci 61:992–999
Hur HS, Ryu JH, Kim KH, Kim J (2006) Characterization of Osh3, an oxysterol-binding protein, in filamentous growth of *Saccharomyces cerevisiae* and *Candida albicans*. J Microbiol 44:523–529
Hynynen R, Laitinen S, Käkelä R, Tanhuanpää K, Lusa S, Ehnholm C, Somerharju P, Ikonen E, Olkkonen VM (2005) Overexpression of OSBP-related protein 2 (ORP2) induces changes in cellular cholesterol metabolism and enhances endocytosis. Biochem J 390:273–283
Ikonen E (2008) Cellular cholesterol trafficking and compartmentalization. Nat Rev Mol Cell Biol 9:125–138
Im YJ, Raychaudhuri S, Prinz WA, Hurley JH (2005) Structural mechanism for sterol sensing and transport by OSBP-related proteins. Nature 437:154–158
Infante RE, Abi-Mosleh L, Radhakrishnan A, Dale JD, Brown MS, Goldstein JL (2008a) Purified NPC1 protein. I. Binding of cholesterol and oxysterols to a 1278-amino acid membrane protein. J Biol Chem 283:1052–1063
Infante RE, Radhakrishnan A, Abi-Mosleh L, Kinch LN, Wang ML, Grishin NV, Goldstein JL, Brown MS (2008b) Purified NPC1 protein: II. Localization of sterol binding to a 240-amino acid soluble luminal loop. J Biol Chem 283:1064–10675

Janowski BA, Willy PJ, Devi TR, Falck JR, Mangelsdorf DJ (1996) An oxysterol signalling pathway mediated by the nuclear receptor LXR alpha. Nature 383:728–731

Janowski BA, Grogan MJ, Jones SA, Wisely GB, Kliewer SA, Corey EJ, Mangelsdorf DJ (1999) Structural requirements of ligands for the oxysterol liver X receptors LXRalpha and LXRbeta. Proc Natl Acad Sci USA 96:266–271

Janowski BA, Shan B, Russell DW (2001) The hypocholesterolemic agent LY295427 reverses suppression of sterol regulatory element-binding protein processing mediated by oxysterols. J Biol Chem 276:45408–45416

Javitt NB (2004) Oxysteroids: a new class of steroids with autocrine and paracrine functions. Trends Endocrinol Metab 15:393–397

Javitt NB (2008) Oxysterols: novel biologic roles for the 21st century. Steroids 73:149–157

Jaworski CJ, Moreira E, Li A, Lee R, Rodriguez IR (2001) A family of 12 human genes containing oxysterol-binding domains. Genomics 78:185–196

Jiang B, Brown JL, Sheraton J, Fortin N, Bussey H (1994) A new family of yeast genes implicated in ergosterol synthesis is related to the human oxysterol binding protein. Yeast 10:341–353

Johansson M, Bocher V, Lehto M, Chinetti G, Kuismanen E, Ehnholm C, Staels B, Olkkonen VM (2003) The two variants of oxysterol binding protein-related protein-1 display different tissue expression patterns, have different intracellular localization, and are functionally distinct. Mol Biol Cell 14:903–915

Johansson M, Lehto M, Tanhuanpää K, Cover TL, Olkkonen VM (2005) The oxysterol-binding protein homologue ORP1L interacts with Rab7 and alters functional properties of late endocytic compartments. Mol Biol Cell 16:5480–5492

Johansson M, Rocha N, Zwart W, Jordens I, Janssen L, Kuijl C, Olkkonen VM, Neefjes J (2007) Activation of endosomal dynein motors by stepwise assembly of Rab7-RILP-p150Glued, ORP1L, and the receptor betalll spectrin. J Cell Biol 176:459–471

Joseph SB, Laffitte BA, Patel PH, Watson MA, Matsukuma KE, Walczak R, Collins JL, Osborne TF, Tontonoz P (2002) Direct and indirect mechanisms for regulation of fatty acid synthase gene expression by liver X receptors. J Biol Chem 277:11019–11025

Joseph SB, Castrillo A, Laffitte BA, Mangelsdorf DJ, Tontonoz P (2003) Reciprocal regulation of inflammation and lipid metabolism by liver X receptors. Nat Med 9:213–219

Kahn RA, Volpicelli-Daley L, Bowzard B, Shrivastava-Ranjan P, Li Y, Zhou C, Cunningham L (2005) Arf family GTPases: roles in membrane traffic and microtubule dynamics. Biochem Soc Trans 33:1269–1272

Kaiser SE, Brickner JH, Reilein AR, Fenn TD, Walter P, Brunger AT (2005) Structural basis of FFAT motif-mediated ER targeting. Structure 13:1035–1045

Käkelä R, Tanhuanpää K, Laitinen S, Somerharju P, Olkkonen VM (2005) Overexpression of OSBP-related protein 2 (ORP2) in CHO cells induces alterations of phospholipid species composition. Biochem Cell Biol 83:677–683

Kandutsch AA, Chen HW (1974) Inhibition of sterol synthesis in cultured mouse cells by cholesterol derivatives oxygenated in the side chain. J Biol Chem 249:6057–6061

Kandutsch AA, Shown EP (1981) Assay of oxysterol-binding protein in a mouse fibroblast, cell-free system. Dissociation constant and other properties of the system. J Biol Chem 256:13068–13073

Kandutsch AA, Thompson EB (1980) Cytosolic proteins that bind oxygenated sterols. Cellular distribution, specificity, and some properties. J Biol Chem 255:10813–10821

Kandutsch AA, Chen HW, Shown EP (1977) Binding of 25-hydroxycholesterol and cholesterol to different cytoplasmic proteins. Proc Natl Acad Sci USA 74:2500–2503

Kandutsch AA, Chen HW, Heiniger HJ (1978) Biological activity of some oxygenated sterols. Science 201:498–501

Kennedy MA, Venkateswaran A, Tarr PT, Xenarios I, Kudoh J, Shimizu N, Edwards PA (2001) Characterization of the human ABCG1 gene: liver X receptor activates an internal promoter that produces a novel transcript encoding an alternative form of the protein. J Biol Chem 276:39438–39447

Kha HT, Basseri B, Shouhed D, Richardson J, Tetradis S, Hahn TJ, Parhami F (2004) Oxysterols regulate differentiation of mesenchymal stem cells: pro-bone and anti-fat. J Bone Miner Res 19:830–840
Kim WK, Meliton V, Amantea CM, Hahn TJ, Parhami F (2007) 20(S)-hydroxycholesterol inhibits PPARgamma expression and adipogenic differentiation of bone marrow stromal cells through a hedgehog-dependent mechanism. J Bone Miner Res 22:1711–1719
Kinbara K, Goldfinger LE, Hansen M, Chou FL, Ginsberg MH (2003) Ras GTPases: integrins' friends or foes. Nat Rev Mol Cell Biol 4:767–776
Knutsen Rydberg E, Salomonsson L, Hulten LM, Noren K, Bondjers G, Wiklund O, Bjornheden T, Ohlsson BG (2003) Hypoxia increases 25-hydroxycholesterol-induced interleukin-8 protein secretion in human macrophages. Atherosclerosis 170:245–252
Kolsch H, Lutjohann D, von Bergmann K, Heun R (2003) The role of 24S-hydroxycholesterol in Alzheimer's disease. J Nutr Health Aging 7:37–41
Kozminski KG, Alfaro G, Dighe S, Beh CT (2006) Homologues of oxysterol-binding proteins affect Cdc42p- and Rho1p-mediated cell polarization in *Saccharomyces cerevisiae*. Traffic 7:1–19
Kritharides L, Jessup W, Mander EL, Dean RT (1995) Apolipoprotein A-I-mediated efflux of sterols from oxidized LDL-loaded macrophages. Arterioscler Thromb Vasc Biol 15:276–289
Kvam E, Goldfarb DS (2004) Nvj1p is the outer-nuclear-membrane receptor for oxysterol-binding protein homolog Osh1p in *Saccharomyces cerevisiae*. J Cell Sci 117:4959–4968
Lagace TA, Byers DM, Cook HW, Ridgway ND (1997) Altered regulation of cholesterol and cholesteryl ester synthesis in Chinese-hamster ovary cells overexpressing the oxysterol-binding protein is dependent on the pleckstrin homology domain. Biochem J 326:205–213
Lagace TA, Byers DM, Cook HW, Ridgway ND (1999) Chinese hamster ovary cells overexpressing the oxysterol binding protein (OSBP) display enhanced synthesis of sphingomyelin in response to 25-hydroxycholesterol. J Lipid Res 40:109–116
Laitinen S, Lehto M, Lehtonen S, Hyvärinen K, Heino S, Lehtonen E, Ehnholm C, Ikonen E, Olkkonen VM (2002) ORP2, a homolog of oxysterol binding protein, regulates cellular cholesterol metabolism. J Lipid Res 43:245–255
Lange Y, Ye J, Strebel F (1995) Movement of 25-hydroxycholesterol from the plasma membrane to the rough endoplasmic reticulum in cultured hepatoma cells. J Lipid Res 36:1092–1097
Lange Y, Ory DS, Ye J, Lanier MH, Hsu FF, Steck TL (2008) Effectors of rapid homeostatic responses of endoplasmic reticulum cholesterol and 3-hydroxy-3-methylglutaryl-CoA reductase. J Biol Chem 283:1445–1455
Larsson DA, Baird S, Nyhalah JD, Yuan XM, Li W (2006) Oxysterol mixtures, in atheroma-relevant proportions, display synergistic and proapoptotic effects. Free Radic Biol Med 41:902–910
Lehmann JM, Kliewer SA, Moore LB, Smith-Oliver TA, Oliver BB, Su JL, Sundseth SS, Winegar DA, Blanchard DE, Spencer TA, Willson TM (1997) Activation of the nuclear receptor LXR by oxysterols defines a new hormone response pathway. J Biol Chem 272:3137–3140
Lehto M, Olkkonen VM (2003) The OSBP-related proteins: a novel protein family involved in vesicle transport, cellular lipid metabolism, and cell signalling. Biochim Biophys Acta 1631:1–11
Lehto M, Laitinen S, Chinetti G, Johansson M, Ehnholm C, Staels B, Ikonen E, Olkkonen VM (2001) The OSBP-related protein family in humans. J Lipid Res 42:1203–1213
Lehto M, Tienari J, Lehtonen S, Lehtonen E, Olkkonen VM (2004) Subfamily III of mammalian oxysterol-binding protein (OSBP) homologues: the expression and intracellular localization of ORP3, ORP6, and ORP7. Cell Tissue Res 315:39–57
Lehto M, Hynynen R, Karjalainen K, Kuismanen E, Hyvarinen K, Olkkonen VM (2005) Targeting of OSBP-related protein 3 (ORP3) to endoplasmic reticulum and plasma membrane is controlled by multiple determinants. Exp Cell Res 310:445–462
Lehto M, Mäyränpää MI, Pellinen T, Po I, Lehtonen S, Kovanen PT, Groop P-H, Ivaska J, Olkkonen VM (2008) The R-Ras interaction partner ORP3 regulates cell adhesion. J Cell Sci 121:695–705

Lemaire-Ewing S, Prunet C, Montange T, Vejux A, Berthier A, Bessede G, Corcos L, Gambert P, Neel D, Lizard G (2005) Comparison of the cytotoxic, pro-oxidant and pro-inflammatory characteristics of different oxysterols. Cell Biol Toxicol 21:97–114

Lemaire-Ewing S, Berthier A, Royer MC, Logette E, Corcos L, Bouchot A, Monier S, Prunet C, Raveneau M, Rebe C, Desrumaux C, Lizard G, Neel D (2008) 7β-Hydroxycholesterol and 25-hydroxycholesterol-induced interleukin-8 secretion involves a calcium-dependent activation of c-fos via the ERK1/2 signaling pathway in THP-1 cells: oxysterols-induced IL-8 secretion is calcium-dependent. Cell Biol Toxicol (in press), DOI: PMID 18317936

Leonarduzzi G, Biasi F, Chiarpotto E, Poli G (2004) Trojan horse-like behavior of a biologically representative mixture of oxysterols. Mol Aspects Med 25:155–167

Leonarduzzi G, Gamba P, Sottero B, Kadl A, Robbesyn F, Calogero RA, Biasi F, Chiarpotto E, Leitinger N, Sevanian A, Poli G (2005) Oxysterol-induced up-regulation of MCP-1 expression and synthesis in macrophage cells. Free Radic Biol Med 39:1152–1161

Leoni V, Masterman T, Diczfalusy U, De Luca G, Hillert J, Björkhem I (2002) Changes in human plasma levels of the brain specific oxysterol 24S-hydroxycholesterol during progression of multiple sclerosis. Neurosci Lett 331:163–166

Lessmann E, Ngo M, Leitges M, Minguet S, Ridgway ND, Huber M (2007) Oxysterol-binding protein-related protein (ORP) 9 is a PDK-2 substrate and regulates Akt phosphorylation. Cell Signal 19:384–392

Lev S (2004) The role of the Nir/rdgB protein family in membrane trafficking and cytoskeleton remodeling. Exp Cell Res 297:1–10

Levanon D, Hsieh CL, Francke U, Dawson PA, Ridgway ND, Brown MS, Goldstein JL (1990) cDNA cloning of human oxysterol-binding protein and localization of the gene to human chromosome 11 and mouse chromosome 19. Genomics 7:65–74

Levine T (2004) Short-range intracellular trafficking of small molecules across endoplasmic reticulum junctions. Trends Cell Biol 14:483–490

Levine TP, Munro S (1998) The pleckstrin homology domain of oxysterol-binding protein recognises a determinant specific to Golgi membranes. Curr Biol 8:729–739

Levine TP, Munro S (2001) Dual targeting of Osh1p, a yeast homologue of oxysterol-binding protein, to both the Golgi and the nucleus-vacuole junction. Mol Biol Cell 12:1633–1644

Levine TP, Munro S (2002) Targeting of Golgi-specific pleckstrin homology domains involves both PtdIns 4-kinase-dependent and -independent components. Curr Biol 12:695–704

Li AC, Glass CK (2004) PPAR- and LXR-dependent pathways controlling lipid metabolism and the development of atherosclerosis. J Lipid Res 45:2161–2173

Li DY, Inoue H, Takahashi M, Kojima T, Shiraiwa M, Takahara H (2008) Molecular characterization of a novel salt-inducible gene for an OSBP (oxysterol-binding protein)-homologue from soybean. Gene 407:12–20

Li W, Dalen H, Eaton JW, Yuan XM (2001) Apoptotic death of inflammatory cells in human atheroma. Arterioscler Thromb Vasc Biol 21:1124–1130

Li X, Rivas MP, Fang M, Marchena J, Mehrotra B, Chaudhary A, Feng L, Prestwich GD, Bankaitis VA (2002) Analysis of oxysterol binding protein homologue Kes1p function in regulation of Sec14p-dependent protein transport from the yeast Golgi complex. J Cell Biol 157:63–77

Li X, Pandak WM, Erickson SK, Ma Y, Yin L, Hylemon P, Ren S (2007) Biosynthesis of the regulatory oxysterol, 5-cholesten-3beta,25-diol 3-sulfate, in hepatocytes. J Lipid Res 48:2587–2596

Li Y, Prinz WA (2004) ATP-binding cassette (ABC) transporters mediate nonvesicular, raft-modulated sterol movement from the plasma membrane to the endoplasmic reticulum. J Biol Chem 279:45226–45234

Li-Hawkins J, Lund EG, Turley SD, Russell DW (2000) Disruption of the oxysterol 7alpha-hydroxylase gene in mice. J Biol Chem 275:16536–16542

Loewen CJ, Roy A, Levine TP (2003) A conserved ER targeting motif in three families of lipid binding proteins and in Opi1p binds VAP. EMBO J 22:2025–2035

Loewen CJ, Gaspar ML, Jesch SA, Delon C, Ktistakis NT, Henry SA, Levine TP (2004) Phospholipid metabolism regulated by a transcription factor sensing phosphatidic acid. Science 304:1644–1647

Lordan S, O'Callaghan YC, O'Brien NM (2007) Death-signaling pathways in human myeloid cells by oxLDL and its cytotoxic components 7beta-hydroxycholesterol and cholesterol-5beta,6beta-epoxide. J Biochem Mol Toxicol 21:362–372

Lund E, Björkhem I, Furster C, Wikvall K (1993) 24-, 25- and 27-hydroxylation of cholesterol by a purified preparation of 27-hydroxylase from pig liver. Biochim Biophys Acta 1166:177–182

Lund EG, Kerr TA, Sakai J, Li WP, Russell DW (1998) cDNA cloning of mouse and human cholesterol 25-hydroxylases, polytopic membrane proteins that synthesize a potent oxysterol regulator of lipid metabolism. J Biol Chem 273:34316–34327

Luoma PV (2007) Cytochrome P450 - physiological key factor against cholesterol accumulation and the atherosclerotic vascular process. Ann Med 39:359–370

Luthra S, Dong J, Gramajo AL, Chwa M, Kim DW, Neekhra A, Kuppermann BD, Kenney MC (2007) 7-Ketocholesterol activates caspases-3/7, -8, and -12 in human microvascular endothelial cells in vitro. Microvasc Res 75:343–350

Lütjohann D, Breuer O, Ahlborg G, Nennesmo I, Siden A, Diczfalusy U, Björkhem I (1996) Cholesterol homeostasis in human brain: evidence for an age-dependent flux of 24S-hydroxycholesterol from the brain into the circulation. Proc Natl Acad Sci USA 93:9799–804

Lütjohann D, Papassotiropoulos A, Björkhem I, Locatelli S, Bagli M, Oehring RD, Schlegel U, Jessen F, Rao ML, von Bergmann K, Heun R (2000) Plasma 24S-hydroxycholesterol (cerebrosterol) is increased in Alzheimer and vascular demented patients. J Lipid Res 41:195–198

Lyons MA, Brown AJ (2001a) 7-Ketocholesterol delivered to mice in chylomicron remnant-like particles is rapidly metabolised, excreted and does not accumulate in aorta. Biochim Biophys Acta 1530:209–218

Lyons MA, Brown AJ (2001b) Metabolism of an oxysterol, 7-ketocholesterol, by sterol 27-hydroxylase in HepG2 cells. Lipids 36:701–711

Lyons MA, Samman S, Gatto L, Brown AJ (1999) Rapid hepatic metabolism of 7-ketocholesterol in vivo: implications for dietary oxysterols. J Lipid Res 40:1846–1857

Maor I, Kaplan M, Hayek T, Vaya J, Hoffman A, Aviram M (2000) Oxidized monocyte-derived macrophages in aortic atherosclerotic lesion from apolipoprotein E-deficient mice and from human carotid artery contain lipid peroxides and oxysterols. Biochem Biophys Res Commun 269:775–780

Marchant CE, Van der Veen C, Law NS, Hardwick SJ, Carpenter KL, Mitchinson MJ (1996) Oxidation of low-density lipoprotein by human monocyte-macrophages results in toxicity to the oxidising culture. Free Radic Res 24:333–342

Mark M, Muller P, Maier R, Eisele B (1996) Effects of a novel 2,3-oxidosqualene cyclase inhibitor on the regulation of cholesterol biosynthesis in HepG2 cells. J Lipid Res 37:148–158

Mattsson-Hulten L, Lindmark H, Diczfalusy U, Björkhem I, Ottosson M, Liu Y, Bondjers G, Wiklund O (1996) Oxysterols present in atherosclerotic tissue decrease the expression of lipoprotein lipase messenger RNA in human monocyte-derived macrophages. J Clin Invest 97:461–468

Millanvoye-Van Brussel E, Topal G, Brunet A, Do Pham T, Deckert V, Rendu F, David-Dufilho M (2004) Lysophosphatidylcholine and 7-oxocholesterol modulate Ca2+ signals and inhibit the phosphorylation of endothelial NO synthase and cytosolic phospholipase A2. Biochem J 380:533–539

Mohammadi A, Perry RJ, Storey MK, Cook HW, Byers DM, Ridgway ND (2001) Golgi localization and phosphorylation of oxysterol binding protein in Niemann-Pick C and U18666A-treated cells. J Lipid Res 42:1062–1071

Monier S, Samadi M, Prunet C, Denance M, Laubriet A, Athias A, Berthier A, Steinmetz E, Jurgens G, Negre-Salvayre A, Bessede G, Lemaire-Ewing S, Neel D, Gambert P, Lizard G (2003) Impairment of the cytotoxic and oxidative activities of 7 beta-hydroxycholesterol and 7-ketocholesterol by esterification with oleate. Biochem Biophys Res Commun 303:814–824

Moreira EF, Jaworski C, Li A, Rodriguez IR (2001) Molecular and biochemical characterization of a novel oxysterol-binding protein (OSBP2) highly expressed in retina. J Biol Chem 276:18570–18578

Morel DW, Edgerton ME, Warner GE, Johnson WJ, Phillips MC, Rothblat GH (1996) Comparison of the intracellular metabolism and trafficking of 25-hydroxycholesterol and cholesterol in macrophages. J Lipid Res 37:2041–2051

Nelson JA, Steckbeck SR, Spencer TA (1981) Biosynthesis of 24,25-epoxycholesterol from squalene 2,3;22,23-dioxide. J Biol Chem 256:1067–1068

Nishimura T, Inoue T, Shibata N, Sekine A, Takabe W, Noguchi N, Arai H (2005) Inhibition of cholesterol biosynthesis by 25-hydroxycholesterol is independent of OSBP. Genes Cells 10:793–801

Norlin M, Andersson U, Björkhem I, Wikvall K (2000a) Oxysterol 7 alpha-hydroxylase activity by cholesterol 7 alpha-hydroxylase (CYP7A). J Biol Chem 275:34046–34053

Norlin M, Toll A, Björkhem I, Wikvall K (2000b) 24-hydroxycholesterol is a substrate for hepatic cholesterol 7alpha-hydroxylase (CYP7A). J Lipid Res 41:1629–1639

Ohtsuka M, Miyashita Y, Shirai K (2006) Lipids deposited in human atheromatous lesions induce apoptosis of human vascular smooth muscle cells. J Atheroscler Thromb 13:256–262

Olkkonen VM, Lehto M (2004) Oxysterols and oxysterol binding proteins: role in lipid metabolism and atherosclerosis. Ann Med 36:562–572

Olkkonen VM, Levine TP (2004) Oxysterol binding proteins: in more than one place at one time. Biochem Cell Biol 82:87–98

Ouimet M, Wang M-D, Cadotte N, Ho K, Marcel YL (2008) Epoxycholesterol impairs cholesterol ester hydrolysis in macrophage foam cells, resulting in decreased cholesterol efflux. Arterioscler Thromb Vasc Biol 28:1144–1150

Oram JF, Vaughan AM (2006) ATP-Binding cassette cholesterol transporters and cardiovascular disease. Circ Res 99:1031–1043

Panzenboeck U, Andersson U, Hansson M, Sattler W, Meaney S, Björkhem I (2007) On the mechanism of cerebral accumulation of cholestanol in patients with cerebrotendinous xanthomatosis. J Lipid Res 48:1167–1174

Park YU, Hwang O, Kim J (2002) Two-hybrid cloning and characterization of OSH3, a yeast oxysterol-binding protein homolog. Biochem Biophys Res Commun 293:733–740

Peretti D, Dahan N, Shimoni E, Hirschberg K, Lev S (2008) Coordinated lipid transfer between the ebnoplasmic reticulum and the Golgi complex requires the VAP proteins and is essential for Golgi-mediated transport. Mol Biol Cell 19:3871–3884

Perry RJ, Ridgway ND (2006) Oxysterol-binding protein and vesicle-associated membrane protein-associated protein are required for sterol-dependent activation of the ceramide transport protein. Mol Biol Cell 17:2604–2616

Radhakrishnan A, Ikeda Y, Kwon HJ, Brown MS, Goldstein JL (2007) Sterol-regulated transport of SREBPs from endoplasmic reticulum to Golgi: oxysterols block transport by binding to Insig. Proc Natl Acad Sci USA 104:6511–6518

Ramachandra RK, Lankford SE, Weber GM, Rexroad CE, 3rd, Yao J (2007) Identification of OORP-T, a novel oocyte-specific gene encoding a protein with a conserved oxysterol binding protein domain in rainbow trout. Mol Reprod Dev 74:502–511

Raychaudhuri S, Im YJ, Hurley JH, Prinz WA (2006) Nonvesicular sterol movement from plasma membrane to ER requires oxysterol-binding protein-related proteins and phosphoinositides. J Cell Biol 173:107–119

Repa JJ, Mangelsdorf DJ (2000) The role of orphan nuclear receptors in the regulation of cholesterol homeostasis. Annu Rev Cell Dev Biol 16:459–481

Repa JJ, Liang G, Ou J, Bashmakov Y, Lobaccaro JM, Shimomura I, Shan B, Brown MS, Goldstein JL, Mangelsdorf DJ (2000a) Regulation of mouse sterol regulatory element-binding protein-1c gene (SREBP-1c) by oxysterol receptors, LXRalpha and LXRbeta. Genes Dev 14:2819–2830

Repa JJ, Turley SD, Lobaccaro JA, Medina J, Li L, Lustig K, Shan B, Heyman RA, Dietschy JM, Mangelsdorf DJ (2000b) Regulation of absorption and ABC1-mediated efflux of cholesterol by RXR heterodimers. Science 289:1524–1529

Richardson JA, Amantea CM, Kianmahd B, Tetradis S, Lieberman JR, Hahn TJ, Parhami F (2007) Oxysterol-induced osteoblastic differentiation of pluripotent mesenchymal cells is mediated through a PKC- and PKA-dependent pathway. J Cell Biochem 100:1131–1145

Ridgway ND, Dawson PA, Ho YK, Brown MS, Goldstein JL (1992) Translocation of oxysterol binding protein to Golgi apparatus triggered by ligand binding. J Cell Biol 116:307–319

Ridgway ND, Lagace TA, Cook HW, Byers DM (1998) Differential effects of sphingomyelin hydrolysis and cholesterol transport on oxysterol-binding protein phosphorylation and Golgi localization. J Biol Chem 273:31621–31628

Rimner A, Al Makdessi S, Sweidan H, Wischhusen J, Rabenstein B, Shatat K, Mayer P, Spyridopoulos I (2005) Relevance and mechanism of oxysterol stereospecifity in coronary artery disease. Free Radic Biol Med 38:535–544

Romeo GR, Kazlauskas A (2008) Oxysterol and diabetes activate STAT3, and control endothelial expression of profilin-1 via OSBP1. J Biol Chem 283:9595–9605

Rose KA, Stapleton G, Dott K, Kieny MP, Best R, Schwarz M, Russell DW, Björkhem I, Seckl J, Lathe R (1997) Cyp7b, a novel brain cytochrome P450, catalyzes the synthesis of neurosteroids 7alpha-hydroxy dehydroepiandrosterone and 7alpha-hydroxy pregnenolone. Proc Natl Acad Sci USA 94:4925–4930

Rowe AH, Argmann CA, Edwards JY, Sawyez CG, Morand OH, Hegele RA, Huff MW (2003) Enhanced synthesis of the oxysterol 24(S),25-epoxycholesterol in macrophages by inhibitors of 2,3-oxidosqualene:lanosterol cyclase: a novel mechanism for the attenuation of foam cell formation. Circ Res 93:717–725

Roy A, Levine TP (2004) Multiple pools of phosphatidylinositol 4-phosphate detected using the pleckstrin homology domain of Osh2p. J Biol Chem 279:44683–44689

Russell DW (2000) Oxysterol biosynthetic enzymes. Biochim Biophys Acta 1529:126–135

Sabol SL, Brewer HB, Jr., Santamarina-Fojo S (2005) The human ABCG1 gene: identification of LXR response elements that modulate expression in macrophages and liver. J Lipid Res 46:2151–2167

Saito E, Wachi H, Saro F, Seyama Y (2008) 7-ketocholesterol, a major oxysterol, promotes Pi-induced vascular calcification in cultured smooth muscle cells. J Atheroscler Thromb 15:130–137

Salonen JT, Nyyssonen K, Salonen R, Porkkala-Sarataho E, Tuomainen TP, Diczfalusy U, Björkhem I (1997) Lipoprotein oxidation and progression of carotid atherosclerosis. Circulation 95:840–845

Schmalix WA, Bandlow W (1994) SWH1 from yeast encodes a candidate nuclear factor containing ankyrin repeats and showing homology to mammalian oxysterol-binding protein. Biochim Biophys Acta 1219:205–210

Schroepfer GJ Jr (2000) Oxysterols: modulators of cholesterol metabolism and other processes. Physiol Rev 80:361–554

Schulz TA, Prinz WA (2007) Sterol transport in yeast and the oxysterol binding protein homologue (OSH) family. Biochim Biophys Acta 1771:769–780

Schweizer RA, Zurcher M, Balazs Z, Dick B, Odermatt A (2004) Rapid hepatic metabolism of 7-ketocholesterol by 11beta-hydroxysteroid dehydrogenase type 1: species-specific differences between the rat, human, and hamster enzyme. J Biol Chem 279:18415–18424

Sevanian A, Hodis HN, Hwang J, McLeod LL, Peterson H (1995) Characterization of endothelial cell injury by cholesterol oxidation products found in oxidized LDL. J Lipid Res 36:1971–1986

Sevanian A, Bittolo-Bon G, Cazzolato G, Hodis H, Hwang J, Zamburlini A, Maiorino M, Ursini F (1997) LDL- is a lipid hydroperoxide-enriched circulating lipoprotein. J Lipid Res 38:419–428

Siems W, Quast S, Peter D, Augustin W, Carluccio F, Grune T, Sevanian A, Hampl H, Wiswedel I (2005) Oxysterols are increased in plasma of end-stage renal disease patients. Kidney Blood Press Res 28:302–306

Skirpan AL, Dowd PE, Sijacic P, Jaworski CJ, Gilroy S, Kao TH (2006) Identification and characterization of PiORP1, a Petunia oxysterol-binding-protein related protein involved in receptor-kinase mediated signaling in pollen, and analysis of the ORP gene family in *Arabidopsis*. Plant Mol Biol 61:553–565

Smith LL, Teng JI, Lin YY, Seitz PK, McGehee MF (1981) Sterol metabolism XLVII. Oxidized cholesterol esters in human tissues. J Steroid Biochem 14:889–900

Steinberg D, Parthasarathy S, Carew TE, Khoo JC, Witztum JL (1989) Beyond cholesterol. Modifications of low-density lipoprotein that increase its atherogenicity. N Engl J Med 320:915–924

Storey MK, Byers DM, Cook HW, Ridgway ND (1998) Cholesterol regulates oxysterol binding protein (OSBP) phosphorylation and Golgi localization in Chinese hamster ovary cells: correlation with stimulation of sphingomyelin synthesis by 25-hydroxycholesterol. Biochem J 336:247–256

Sturley SL, Patterson MC, Balch W, Liscum L (2004) The pathophysiology and mechanisms of NP-C disease. Biochim Biophys Acta 1685:83–87

Suchanek M, Hynynen R, Wohlfahrt G, Lehto M, Johansson M, Saarinen H, Radzikowska A, Thiele C, Olkkonen VM (2007) The mammalian OSBP-related proteins (ORP) bind 25-hydroxycholesterol in an evolutionarily conserved pocket. Biochem J 405:473–480

Sugano S, Miura R, Morishima N (1996) Identification of intermediates in the conversion of cholesterol to pregnenolone with a reconstituted cytochrome p-450scc system: accumulation of the intermediate modulated by the adrenodoxin level. J Biochem 120:780–787

Sugawara K, Morita K, Ueno N, Shibuya H (2001) BIP, a BRAM-interacting protein involved in TGF-beta signalling, regulates body length in *Caenorhabditis elegans*. Genes Cells 6:599–606

Sullivan DP, Ohvo-Rekila H, Baumann NA, Beh CT, Menon AK (2006) Sterol trafficking between the endoplasmic reticulum and plasma membrane in yeast. Biochem Soc Trans 34:356–358

Sun LP, Seemann J, Goldstein JL, Brown MS (2007) Sterol-regulated transport of SREBPs from endoplasmic reticulum to Golgi: Insig renders sorting signal in Scap inaccessible to COPII proteins. Proc Natl Acad Sci USA 104:6519–6526

Sung SC, Kim K, Lee KA, Choi KH, Kim SM, Son YH, Moon YS, Eo SK, Rhim BY (2008) 7-ketocholesterol upregulates interleukin-6 via mechanisms that are distinct from those of tumor necrosis factor-alpha, in vascular smooth muscle cells. J Vasc Res 46:36–44

Tam SP, Mok L, Chimini G, Vasa M, Deeley RG (2006) ABCA1 mediates high-affinity uptake of 25-hydroxycholesterol by membrane vesicles and rapid efflux of oxysterol by intact cells. Am J Physiol Cell Physiol 291:C490–C502

Taylor FR, Kandutsch AA (1985) Oxysterol binding protein. Chem Phys Lipids 38:187–194

Taylor FR, Saucier SE, Shown EP, Parish EJ, Kandutsch AA (1984) Correlation between oxysterol binding to a cytosolic binding protein and potency in the repression of hydroxymethylglutaryl coenzyme A reductase. J Biol Chem 259:12382–12387

Taylor FR, Shown EP, Thompson EB, Kandutsch AA (1989) Purification, subunit structure, and DNA binding properties of the mouse oxysterol receptor. J Biol Chem 264:18433–18439

Terasaka N, Wang N, Yvan-Charvet L, Tall AR (2007) High-density lipoprotein protects macrophages from oxidized low-density lipoprotein-induced apoptosis by promoting efflux of 7-ketocholesterol via ABCG1. Proc Natl Acad Sci USA 104:15093–15098

Tertov VV, Sobenin IA, Gabbasov ZA, Popov EG, Jaakkola O, Solakivi T, Nikkari T, Smirnov VN, Orekhov AN (1992) Multiple-modified desialylated low density lipoproteins that cause intracellular lipid accumulation. Isolation, fractionation and characterization. Lab Invest 67:665–675

Tertov VV, Sobenin IA, Orekhov AN, Jaakkola O, Solakivi T, Nikkari T (1996) Characteristics of low density lipoprotein isolated from circulating immune complexes. Atherosclerosis 122:191–199

Theunissen JJ, Jackson RL, Kempen HJ, Demel RA (1986) Membrane properties of oxysterols. Interfacial orientation, influence on membrane permeability and redistribution between membranes. Biochim Biophys Acta 860:66–74

Tontonoz P, Mangelsdorf DJ (2003) Liver X receptor signaling pathways in cardiovascular disease. Mol Endocrinol 17:985–993

Umetani M, Domoto H, Gormley AK, Yuhanna IS, Cummins CL, Javitt NB, Korach KS, Shaul PW, Mangelsdorf DJ (2007) 27-Hydroxycholesterol is an endogenous SERM that inhibits the cardiovascular effects of estrogen. Nat Med 13:1185–1192

van de Bovenkamp P, Kosmeijer-Schuil TG, Katan MB (1988) Quantification of oxysterols in Dutch foods: egg products and mixed diets. Lipids 23:1079–1085

Varjosalo M, Taipale J (2007) Hedgehog signaling. J Cell Sci 120:3–6

Wang C, JeBailey L, Ridgway ND (2002) Oxysterol-binding-protein (OSBP)-related protein 4 binds 25-hydroxycholesterol and interacts with vimentin intermediate filaments. Biochem J 361:461–472

Wang P, Duan W, Munn AL, Yang H (2005a) Molecular characterization of Osh6p, an oxysterol binding protein homolog in the yeast Saccharomyces cerevisiae. FEBS J 272:4703–4715

Wang P, Zhang Y, Li H, Chieu HK, Munn AL, Yang H (2005b) AAA ATPases regulate membrane association of yeast oxysterol binding proteins and sterol metabolism. EMBO J 24:2989–2999

Wang PY, Weng J, Anderson RG (2005c) OSBP is a cholesterol-regulated scaffolding protein in control of ERK 1/2 activation. Science 307:1472–1476

Wang PY, Weng J, Lee S, Anderson RG (2008a) N-terminus controls sterol binding while c-terminus regulates scaffolding function of OSBP. J Biol Chem 283:8034–8045

Wang X, Briggs MR, Hua X, Yokoyama C, Goldstein JL, Brown MS (1993) Nuclear protein that binds sterol regulatory element of low density lipoprotein receptor promoter. II. Purification and characterization. J Biol Chem 268:14497–14504

Wang Y, Rogers P, Su C, Varga G, Stayrook KR, Burris TP (2008b) Regulation of cholesterologenesis by the oxysterol receptor, LXRα. J Biol Chem 283:26332–26339

Wassif CA, Yu J, Cui J, Porter FD, Javitt NB (2003) 27-Hydroxylation of 7- and 8-dehydrocholesterol in Smith-Lemli-Opitz syndrome: a novel metabolic pathway. Steroids 68:497–502

Vaya J, Aviram M, Mahmood S, Hayek T, Grenadir E, Hoffman A, Milo S (2001) Selective distribution of oxysterols in atherosclerotic lesions and human plasma lipoproteins. Free Radic Res 34:485–497

Venkateswaran A, Repa JJ, Lobaccaro JM, Bronson A, Mangelsdorf DJ, Edwards PA (2000) Human white/murine ABC8 mRNA levels are highly induced in lipid-loaded macrophages. A transcriptional role for specific oxysterols. J Biol Chem 275:14700–14707

Willy PJ, Umesono K, Ong ES, Evans RM, Heyman RA, Mangelsdorf DJ (1995) LXR, a nuclear receptor that defines a distinct retinoid response pathway. Genes Dev 9:1033–1045

Wong J, Quinn CM, Brown AJ (2004) Statins inhibit synthesis of an oxysterol ligand for the liver x receptor in human macrophages with consequences for cholesterol flux. Arterioscler Thromb Vasc Biol 24:2365–2371

Wong J, Quinn CM, Brown AJ (2007a) Synthesis of the oxysterol, 24(S), 25-epoxycholesterol, parallels cholesterol production and may protect against cellular accumulation of newly-synthesized cholesterol. Lipids Health Dis 6:10

Wong J, Quinn CM, Guillemin G, Brown AJ (2007b) Primary human astrocytes produce 24(S),25-epoxycholesterol with implications for brain cholesterol homeostasis. J Neurochem 103:1764–1773

Wong J, Quinn CM, Gelissen IC, Brown AJ (2008) Endogenous 24(S),25-epoxycholesterol fine-tunes acute control of cellular cholesterol homeostasis. J Biol Chem 283:700–707

Wyles JP, Ridgway ND (2004) VAMP-associated protein-A regulates partitioning of oxysterol-binding protein-related protein-9 between the endoplasmic reticulum and Golgi apparatus. Exp Cell Res 297:533–547

Wyles JP, McMaster CR, Ridgway ND (2002) Vesicle-associated membrane protein-associated protein-A (VAP-A) interacts with the oxysterol-binding protein to modify export from the endoplasmic reticulum. J Biol Chem 277:29908–29918

Wyles JP, Perry RJ, Ridgway ND (2007) Characterization of the sterol-binding domain of oxysterol-binding protein (OSBP)-related protein 4 reveals a novel role in vimentin organization. Exp Cell Res 313:1426–1437

Xu Y, Liu Y, Ridgway ND, McMaster CR (2001) Novel members of the human oxysterol-binding protein family bind phospholipids and regulate vesicle transport. J Biol Chem 276:18407–18414

Yabe D, Brown MS, Goldstein JL (2002) Insig-2, a second endoplasmic reticulum protein that binds SCAP and blocks export of sterol regulatory element-binding proteins. Proc Natl Acad Sci USA 99:12753–12758

Yan D, Olkkonen VM (2008) Characteristics of oxysterol binding proteins. Int Rev Cytol 265:253–285

Yan D, Jauhiainen M, Hildebrand RB, van Dijk KW, Van Berkel TJC, Ehnholm C, Van Eck M, Olkkonen VM (2007a) Expression of human OSBP-related protein 1L in macrophages enhances atherosclerotic lesion development in LDL receptor-deficient mice. Arterioscler Thromb Vasc Biol 27:1618–1624

Yan D, Lehto M, Rasilainen L, Metso J, Ehnholm C, Ylä-Herttuala S, Jauhiainen M, Olkkonen VM (2007b) Oxysterol binding protein induces upregulation of SREBP-1c and enhances hepatic lipogenesis. Arterioscler Thromb Vasc Biol 27:1108–1114

Yan D, Mäyränpää MI, Wong J, Perttilä J, Lehto M, Jauhiainen M, Kovanen PT, Ehnholm C, Brown AJ, Olkkonen VM (2008) OSBP-related protein 8 (ORP8) suppresses ABCA1 expression and cholesterol efflux from macrophages. J Biol Chem 283:332–340

Yanagisawa LL, Marchena J, Xie Z, Li X, Poon PP, Singer RA, Johnston GC, Randazzo PA, Bankaitis VA (2002) Activity of specific lipid-regulated ADP ribosylation factor-GTPase-activating proteins is required for Sec14p-dependent Golgi secretory function in yeast. Mol Biol Cell 13:2193–206

Yang T, Espenshade PJ, Wright ME, Yabe D, Gong Y, Aebersold R, Goldstein JL, Brown MS (2002) Crucial step in cholesterol homeostasis: sterols promote binding of SCAP to INSIG-1, a membrane protein that facilitates retention of SREBPs in ER. Cell 110:489–500

Yano T, Inukai M, Isono F (2004) Deletion of OSH3 gene confers resistance against ISP-1 in Saccharomyces cerevisiae. Biochem Biophys Res Commun 315:228–234

Yasunobu Y, Hayashi K, Shingu T, Yamagata T, Kajiyama G, Kambe M (2001) Coronary atherosclerosis and oxidative stress as reflected by autoantibodies against oxidized low-density lipoprotein and oxysterols. Atherosclerosis 155:445–453

Yokoyama C, Wang X, Briggs MR, Admon A, Wu J, Hua X, Goldstein JL, Brown MS (1993) SREBP-1, a basic-helix-loop-helix-leucine zipper protein that controls transcription of the low density lipoprotein receptor gene. Cell 75:187–197

Yoshikawa T, Shimano H, Amemiya-Kudo M, Yahagi N, Hasty AH, Matsuzaka T, Okazaki H, Tamura Y, Iizuka Y, Ohashi K, Osuga J, Harada K, Gotoda T, Kimura S, Ishibashi S, Yamada N (2001) Identification of liver X receptor-retinoid X receptor as an activator of the sterol regulatory element-binding protein 1c gene promoter. Mol Cell Biol 21:2991–3000

Zelcer N, Tontonoz P (2006) Liver X receptors as integrators of metabolic and inflammatory signaling. J Clin Invest 116:607–614

Zeng B, Zhu G (2006) Two distinct oxysterol binding protein-related proteins in the parasitic protist Cryptosporidium parvum (Apicomplexa). Biochem Biophys Res Commun 346:591–399

Zerbinatti CV, Cordy JM, Chen C-D, Guillily M, Suon S, Ray WJ, Seabrook GR, Abraham CR, Wolozin B (2008) Oxysterol-binding protein-1 (OSBP1) modulates processing and trafficking of the amyloid precursor protein. Mol Neurodegener 3:5

Zhou Q, Wasowicz E, Handler B, Fleischer L, Kummerow FA (2000) An excess concentration of oxysterols in the plasma is cytotoxic to cultured endothelial cells. Atherosclerosis 149:191–197

Zieden B, Kaminskas A, Kristenson M, Kucinskiene Z, Vessby B, Olsson AG, Diczfalusy U (1999) Increased plasma 7 beta-hydroxycholesterol concentrations in a population with a high risk for cardiovascular disease. Arterioscler Thromb Vasc Biol 19:967–971

Zinser E, Sperka-Gottlieb CD, Fasch EV, Kohlwein SD, Paltauf F, Daum G (1991) Phospholipid synthesis and lipid composition of subcellular membranes in the unicellular eukaryote *Saccharomyces cerevisiae*. J Bacteriol 173:2026–2034

Chapter 3
Cellular Lipid Traffic and Lipid Transporters: Regulation of Efflux and HDL Formation

Yves L. Marcel, Mireille Ouimet, and Ming-Dong Wang

Abstract Cellular cholesterol homeostasis integrates multiple pathways of cellular cholesterol movement and the transcriptional regulation of genes implicated in cholesterol metabolism as a function of cholesterol uptake and export. The uptake of lipoprotein-derived cholesterol and the export of cholesterol by lipoprotein secretion and/or efflux to exogenous acceptors are highly regulated processes dependent on intracellular cholesterol transport. Exogenous cholesterol derived from lipoprotein internalization is delivered to LE and lysosomes, where cholesteryl esters are hydrolyzed to generate free cholesterol. Cholesterol from the LE must be re-distributed to the ER, where together with newly synthesized cholesterol, it can be esterified by ACAT and stored in LD. LD-associated cholesterol undergoes multiple cycles of hydrolysis and esterification as a function cellular requirements and can be mobilized for efflux, a process dependent on the presence of exogenous acceptors and on upstream events that remain to be fully elucidated. ABCA1, which is highly regulated by transcriptional and posttranscriptional mechanisms, is the major transporter for phospholipid and cholesterol efflux and, along with mediators of cellular lipid transport, regulates HDL formation.

3.1 Introduction

Cellular export of lipids includes the secretion of lipoproteins, the shedding of membranes and the regulated lipid efflux pathway, which occurs through secondary interactions between the core constituents of the secreted lipoproteins, apolipoproteins and multiple cell types in peripheral tissues.

Apolipoprotein A-I (apoA-I)-containing lipoproteins, the major class of circulating HDL particles, are secreted by hepatocytes and intestinal cells, whereas apoE-containing lipoproteins are synthesized and secreted by a number of cells, including

Y.L. Marcel, M. Ouimet, and M.-D. Wang
Lipoprotein and Atherosclerosis Research Group, University of Ottawa Heart Institute, 40 Ruskin Street, Ottawa, K1Y 4W7 Ontario, Canada
e-mail: ylmarcel@ottawaheart.ca

C. Ehnholm (ed.), *Cellular Lipid Metabolism*,
DOI 10.1007/978-3-642-00300-4_3, © Springer-Verlag Berlin Heidelberg 2009

macrophages, hepatocytes and neuronal cells. Several lipid transporters are known to participate in cellular lipid traffic and eventually in the secretory and efflux pathways, including members of the ATP-binding cassette (ABC) family, ABCA1, ABCA2, ABCA7, ABCG1, ABCG4, as well as the scavenger receptor-BI (SR-BI).

Newly secreted apoA-I-containing lipoproteins released into the circulation by hepatocytes and intestinal epithelial cells, are heterogeneous and not fully lipidated. These nascent lipoproteins acquire additional lipids at the cell surface of hepatocytes and peripheral cells through ABCA1- and ABCG1-mediated pathways. Further modification of the secreted HDL particles by esterification of cholesterol and interactions with postprandial remnant lipoproteins eventually generates the mature circulating HDL_3 and HDL_2 particles, which can be selectively taken up via SR-BI or CETP by hepatocytes. These lipoproteins are further remodeled by hepatic lipase, endothelial lipase, cholesteryl ester transfer protein and phospholipid transfer protein, culminating in the release of preβ-apoA-I, a partially phospholipidated apoA-I that is highly active in mediating efflux via ABCA1. However, many potentially important details are lacking in the well-studied HDL metabolism pathway. How much cholesterol does the liver contribute to the secreted HDL and how much is contributed by extrahepatic tissues under normal physiological conditions? Is nascent hepatic HDL similar to the HDL formed by lipid efflux to preβ-apoA-I in peripheral cells?

In this review, we compare the specificity of cells involved in lipid export and secretion and focus on differences that are relevant to the physiological assembly of HDL particles. The lipid efflux pathway depends not only on the activity of transporters that mediate lipid export to extracellular acceptors but also on upstream cellular trafficking events required for lipid mobilization. We summarize current understanding of how cellular cholesterol trafficking regulates lipid efflux.

3.2 Regulation of apoA-I Synthesis, Lipidation and Secretion in Hepatocytes: Genesis of apoA-I-Containing Lipoproteins and HDL

The liver synthesizes about 70% of plasma apoA-I, with the remainder being produced by the intestine (Eisenberg 1984; Tall 1990). ApoA-I is part of the apoA-I/apoCIII/apoA-IV gene cluster, which is controlled by a common enhancer upstream of the apoCIII gene. Important transcription factors regulating enhancer activity include SP1, HNF-4 and other nuclear receptors (Zannis et al. 2001). Hepatic expression of apoA-I is regulated by nutrients and hormones, in particular estradiol (Jin et al. 1998; Hargrove et al. 1999; Lamon-Fava et al. 1999; Mooradian et al. 2006). Early studies demonstrated generation of a spectrum of poorly lipidated apoA-I and nascent HDL by hepatocytes (Dixon and Ginsberg 1992) and HepG2 cells (Chisholm et al. 2002). Elevated expression of ABCA1 in hepatocytes (Wellington et al. 2002) first suggested a role for hepatic ABCA1 activity in the lipidation of HDL. Subsequently, it was shown that ABCA1 mediates the transfer of both phospholipids and cholesterol to apoA-I. In primary murine hepatocytes infected with an adenovirus for human apoA-I, endogenously synthesized human apoA-I forms a population of apoA-I-containing lipoproteins (LpA-I) similar to those formed by incubation with exogenous apoA-I (Kiss et al. 2003). Using this model, our studies indicate that ABCA1 is responsible for more than 60% of endogenous apoA-I phospholipidation, and the remainder occurs independently of hepatic ABCA1.

Importantly, ABCA1 inactivation in hepatocytes does not affect apoA-I secretion and similarly sized particles are formed, although fewer large particles accumulate. Therefore, apoA-I secretion is not controlled by lipid availability, and the type of particles generated appears to be dictated by apoA-I structure (i.e. its ability to form dimers and tetramers (Kiss et al. 2003), and by ABCA1 oligomerization (Denis et al. 2004a, b). Investigation of the early steps of apoA-I lipidation as it traffics from the endoplasmic reticulum (ER) to the Golgi shows that phospholipidation in the ER is independent of ABCA1 activity (Maric et al. 2005). The bulk of lipidation of apoA-I, however, occurs in the Golgi and at the plasma membrane (Maric et al. 2005), consistent with the cellular distribution of ABCA1 (Neufeld et al. 2001, 2004). In conclusion, apoA-I can be secreted in a lipidated form by hepatocytes, but the observation that a large proportion of apoA-I particles are poorly lipidated suggests that lipid supply and/or ABCA1 expression may be limiting.

3.3 Cell Specificity of ABCA1 Expression and HDL Formation in Vivo: Insight from Genetically Modified Mice

Targeted *Abca1* inactivation in mice demonstrated that ABCA1 activity in the liver, intestines and macrophages could contribute 80%, 30% and 20%, respectively, of the apoA-I lipidation necessary to maintain plasma HDL levels (Haghpassand et al. 2001; Timmins et al. 2005; Brunham et al. 2006). Furthermore, hepatic and intestinal ABCA1 activities are additive and mice deficient in both have only 10% of wild-type HDL levels (Brunham et al. 2006).

Overexpression of hepatic ABCA1 in mice by adenoviral infection or transgenesis (Vaisman et al. 2001 Basso 2003; Joyce et al. 2006) has also shown that hepatic ABCA1 activity is limiting for HDL lipidation. Both approaches increased plasma apoA-I and HDL cholesterol, but since apoA-I lipidation and secretion were not assessed, it is unclear whether the elevated apoA-I-associated cholesterol was due to a decrease in apoA-I catabolism, increased hepatic de novo cholesterol synthesis or both (Joyce et al. 2006). Wellington and colleagues infected mice with increasing doses of ad-ABCA1 (Wellington et al. 2003), which elevated HDL cholesterol to a plateau at relatively low ABCA1 protein levels (1.2-fold compared to control). At higher doses of ad-ABCA1, non-HDL cholesterol increased. Similar results have been obtained by others using an ABCA1 transgene driven by the apoE promoter (Joyce et al. 2006). Mice expressing a BAC ABCA1 transgene demonstrated a pattern of tissue specific expression similar to the endogenous mouse gene

(Cavelier et al. 2001; Singaraja et al. 2001). BAC transgenic mice with low BAC copy number had increased HDL, which increased with ABCA1 expression. Of note, the latter BAC ABCA1 transgenic line showed a 65% increase in HDL cholesterol (Singaraja et al. 2001). Therefore, ABCA1 driven by its natural promoter and ubiquitously expressed increases HDL-cholesterol.

These studies have established that ABCA1 activity is limiting for the lipidation of apoA-I and the maintenance of HDL levels. Together, hepatic and intestinal ABCA1 activities account for the near totality of HDL production, which indicates that apoA-I lipidation by peripheral cells is negligible with the exception of macrophage ABCA1, which normally exports the excess cholesterol accumulating in these cells.

3.4 Transcriptional and Posttranscriptional Regulation of ABCA1

ABCA1 expression is highly regulated, both transcriptionally and posttranscriptionally (Schmitz and Langmann 2005). Transcription of ABCA1 is markedly regulated by oxysterols (Lawn et al. 1999) ligands for LXRα and LXRβ, and by retinoic acid, acting via RXR (Costet et al. 2000; Schwartz et al. 2000). Treatment of cells with oxysterols and/or retinoic acid induces ABCA1 expression and the combined effect is synergistic (Repa and Mangelsdorf 1999; Costet et al. 2000; Schwartz et al. 2000). The activation of ABCA1 transcription by the LXRα/RXR and LXRβ/RXR heterodimers may be modulated the transcriptional factor SP1, which binds to the proximal ABCA1 promotor, upstream of the LXR/RXR responsive element, and physically interacts with LXR/RXR receptors bound to the promoter (Thymiakou et al. 2007). Two major ABCA1 transcripts, p-ABCA1 and l-ABCA1, have been identified that are mainly expressed in peripheral tissues and the liver, respectively; these transcripts contain different corresponding promoter regions (Tamehiro et al. 2007). Compactin, a member of the class of cholesterol-lowering agents, HMG-CoA reductase inhibitors (statins), increases the liver-type transcript and decreases the peripheral-type transcript. The stimulation of ABCA1 transcription by statins in rat hepatoma McARH7777 cells was associated with the binding of SREBP-2 to the novel liver-type promoter. The same two transcripts were also dominant in human and mouse livers, whereas the intestine contains only the peripheral-type transcript (Tamehiro et al. 2007). This study further documents the complex tissue-specific regulation of ABCA1 expression and offers an explanation for the conflicting results of statin treatment on HDL formation and cholesterol efflux from macrophages and SREBP-2 effects on vascular cells (Wong et al. 2004; Zeng et al. 2004). In addition, the human ABCA1 promoter contains an activator protein (AP)2-binding site between positions –368 and –147, upstream of LXR-binding site, which negatively regulates ABCA1 transcription in CHO cells. HepG2 cells lack AP2α expression, an additional complexity of the tissue-specific regulation of ABCA1 expression (Iwamoto et al. 2007).

The significant discordance between ABCA1 mRNA and protein expression is consistent with a high level of posttranscriptional regulation (Wellington et al. 2002). Under basal conditions in cultured cells, ABCA1 protein is highly unstable, with a half-life of 1–2 h (Oram et al. 2000; Wang and Oram 2002; Wang et al. 2003). Binding of apoA-I to ABCA1-expressing cells reduces ABCA1 degradation by preventing its proteolysis by calpain (Wang et al. 2003) and thiol proteases (Arakawa and Yokoyama 2002). Pathological conditions such as free cholesterol loading (i.e. ER stress) can also induce ABCA1 degradation by the ubiquitin–proteasome-mediated pathway (Feng and Tabas 2002). Whereas the binding of calpain to the phosporylated PEST sequence of ABCA1 targets the latter to the degradative pathway, binding of apoA-I to ABCA1 causes the dephosphorylation of the PEST sequence that is followed by a decrease of ABCA1 degradation by calpain, resulting in increased ABCA1 at the cell surface (Martinez et al. 2003). Similarly, ABCA1 with a deleted PEST domain remains at the cell surface, does not traffic to late endosomes (LE) and consequently effluxes cholesterol from the plasma membrane but not from LE (Chen et al. 2005).

ABCA1 stability is also modulated by PKC activity. ApoA-I activates PKCα by PC-PLC-mediated generation of diacylglycerol initiated by the removal of cellular sphingomyelin (Ito et al. 2002) and activated PKCα subsequently phosphorylates and stabilizes ABCA1 (Yamauchi et al. 2003). However, the phosphorylated amino acid residues are yet to be defined. Suppression was also reported for cholesterol efflux, but not for phospholipids, to apoA-I in human monocyte-derived macrophages and human THP-1 macrophages treated with a panPKC inhibitor (Kiss et al. 2005). These results indicate that the regulation of ABCA1-mediated efflux by PKC activity may be more complicated than previously anticipated. Indeed, unsaturated fatty acids that are elevated in diabetes increase ABCA1 phosphorylation but destabilize ABCA1 through a PKCδ pathway, perhaps contributing to the abnormal HDL metabolism in patients with diabetes (Wang and Oram 2005, 2007).

The incubation of apolipoproteins with ABCA1-expressing cells activates the protein–tyrosine kinase, JAK2, which in turn initiates a process that enhances apolipoprotein interactions with ABCA1 to positively regulate lipid efflux. However, tyrosine kinase-mediated phosphorylation of ABCA1 is undetectable, suggesting that the target of JAK2 is not ABCA1 but rather an associated protein, perhaps autophosphorylation of JAK2 itself (Tang et al. 2004). An active JAK2 is required for apolipoprotein binding to ABCA1 and for lipid efflux, yet apoA-I cellular binding stabilizes ABCA1 independently of JAK2 activity (Tang et al. 2006). This has been interpreted as evidence that apolipoproteins coordinate the activity of ABCA1 through several distinct pathways that are likely to involve numerous and possibly novel molecules. Probable candidates include the syntrophins and their associated proteins as constituents of the ABCA1 protein complex (see below).

Whereas efflux to apoA-I is impaired in NPC1-deficient macrophages, it is increased by up to 8-fold in NPC1-deficient hepatocytes (Wang et al. 2007a). The increased efflux correlates with a marked increase in ABCA1 protein in NPC1-deficient hepatocytes, which contrasts to the observed reduction of ABCA1 in NPC1-deficient macrophages. Elevated ABCA1 expression is largely posttranslational, since ABCA1 mRNA is only slightly increased in NPC1–/– hepatocytes. Moreover, LXRα mRNA in these cells is comparable to wild-type hepatocytes, whereas multiple LXRα target genes are reduced (possibly as a result of mitochondrial dysfunction). Interestingly, the lysosomal enzyme, cathepsin D (CTSD; which is a positive modulator of ABCA1; Haidar et al. 2006), is markedly increased at both mRNA and protein levels by NPC1 inactivation in hepatocytes but not in macrophages (Wang et al. 2007a). Although PPARγ and CREB mRNA levels are upregulated by NPC1 inactivation, the stimulation of hepatocytes with a PPARγ agonist or cAMP does not increase ABCA1 expression. In contrast, administration of the PPARγ agonist, rosiglitazone, and the CREB inducer, cAMP, markedly increase ABCA1 expression in murine macrophages. Therefore, ABCA1 expression is apparently subject to unique regulation in hepatocytes, where NPC1 activity modulates cathepsin D and ABCA1 expression (Wang et al. 2007a).

3.5 Cellular Traffic of ABCA1

3.5.1 Syntrophin and the Regulation of Lipid Efflux Activity

Neufeld and colleagues demonstrated a complex intracellular trafficking route for ABCA1, which was suggested to play a role in modulating lipid efflux and cellular cholesterol homeostasis (Neufeld et al. 2001): ABCA1-GFP was shown to reside on the cell surface and in subsets of early endosomes (EE), LE and lysosomes; and time-lapse microscopy revealed dynamic interactions between these compartments. Subsequent studies established that ABCA1 in LE plays a role in lipid efflux from that organelle and that apoA-I shuttles between the plasma membrane (PM) and LE (Neufeld et al. 2004). LE traffic is defective in Tangier disease cells and LE accumulates high levels of cholesterol, sphingolipids and NPC1 protein (Neufeld et al. 2004). The addition of apoA-I reduces NPC1 content in LE of normal but not Tangier cells (Neufeld et al. 2004). Noteworthy is the impaired secretory vesicular transport from the Golgi to PM in Tangier cells, which shows that ABCA1-mediated efflux of cholesterol also modulates secretory activity and the movement of membranes from the Golgi (Zha et al. 2003). However, the phospholipid flippase activity of ABCA1 may also contribute to this process (Roosbeek et al. 2004). The PEST sequence regulates ABCA1 internalization from the PM and ABCA1-δPEST display defective internalization and trafficking to LE, which in turn impairs cholesterol efflux from LE (specifically labeled with acetylated-LDL in SRA expressing cells) but not cholesterol efflux from the PM (labeled with LDL; Chen et al. 2005). However the significance of this pathway for efflux to apoA-I was recently challenged (Denis et al. 2008; Faulkner et al. 2008; see below).

The complex intracellular trafficking of ABCA1 suggests the involvement of additional accessory proteins. A yeast two-hybrid library identified α1-syntrophin and Lin7 as ABCA1-interacting proteins and α1-syntrophin expression was found to increase the half-life of ABCA1 and enhance cholesterol efflux (Munehira, Ohnishi et al. 2004). A putative PDZ-binding domain at the C terminal region of ABCA1 has been shown to bind the β2-syntrophin/utrophin complex (Buechler, Boettcher et al. 2002) and the β1-syntrophin/utrophin complex (Okuhira, Fitzgerald et al. 2005). The syntrophin family comprises five isoforms (α1, β1, β2, γ1, γ2), each with the same molecular organization including two tandem pleckstrin homology (PH) domains, a PDZ domain and a C-terminal syntrophin unique (SU) domain (Ahn et al. 1996). The C-terminal region, including the PH2 and SU domains, interacts with the dystrophin family of proteins, dystrophin, utrophin, and dystrobrevin (Albrecht and Froehner 2002). Thus, the syntrophins link the dystrophin family of proteins to other proteins that bind to their PDZ domains, providing a scaffold at the cell surface Fig. 3.1.

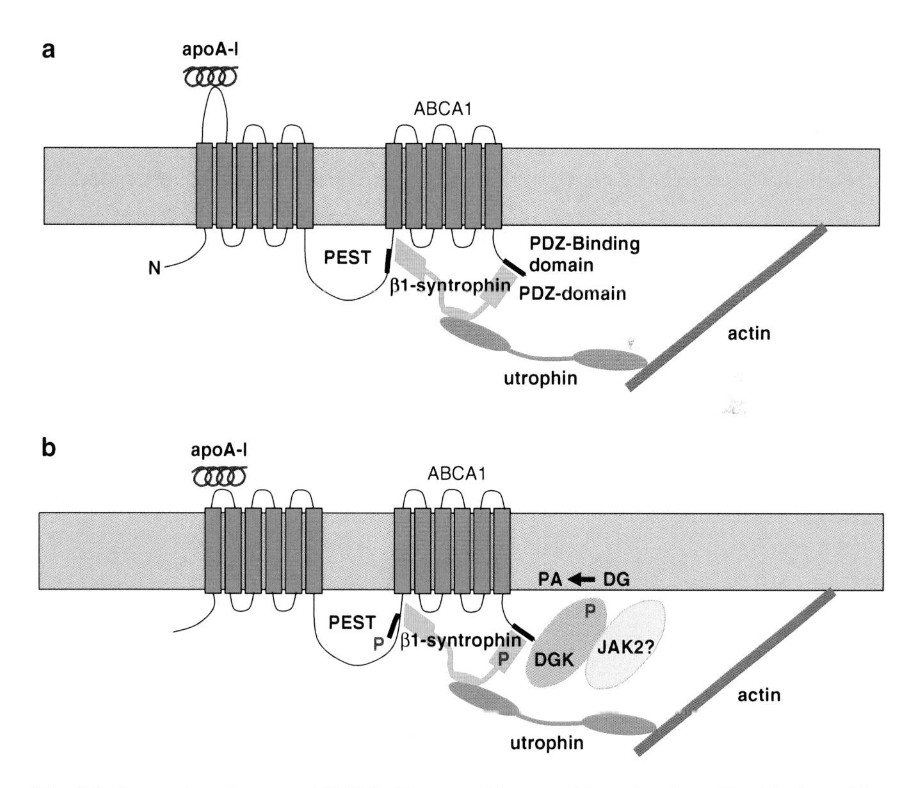

Fig. 3.1 Interactions between ABCA1, β1-syntrophin, utrophin and actin. **a** The binding of β1-syntrophin PDZ domain to ABCA1 PTZ-binding domain at its C-terminus. β1-Syntrophin binds to utrophin, which anchors the complex to actin and retains ABCA1 at the cell surface (adapted from Buechler et al. 2002; Okuhira et al. 2005). **b** The putative regulation of the β1-syntrophin–ABCA1 complex. β1-Syntrophin binds to a diglyceride kinase (*DGK*), which is proposed to phosphorylate β1-syntrophin and a residue adjacent to the PEST sequence on ABCA1. JAK2 is a protein-tyrosine kinase, which is activated upon binding of apoA-I to ABCA1 and phosphorylates an unknown target that activates the ABCA1 complex (Tang et al. 2004), independently of JAK2 activity (Tang et al. 2006)

The tissue-specific expression of syntrophins contributes to their specificity: the α1 isoform is expressed highly in heart, brain and muscle, β1 in the liver and β2 is highest in the intestines. Freeman and colleagues have shown that β1-syntrophin expression regulates ABCA1 levels and cholesterol efflux. Inhibition of β1-syntrophin expression accelerates the degradation of newly synthesized ABCA1, and co-expression of β1-syntrophin and ABCA1 results in the formation of cell surface ABCA1 clusters, which protects it from degradation and increases its concentration at the cell surface and efflux activity (Okuhira et al. 2005). The authors also identified utrophin and β-dystrobrevin as ABCA1-interacting proteins. Since utrophin is an adaptor protein that binds both to syntrophin and actin, it can provide an important bridge between ABCA1 at the plasma membrane and the cytoskeleton, providing motility and controlling its endocytosis and traffic. Another conserved C-terminal motif (VFVNFA) in ABCA1 between residues 41–46, next to the PDZ-binding domain (first four residues) binds a yet unidentified protein (Fitzgerald et al. 2004). Mutation of this motif eliminated both apoA-I binding and efflux without alteration of ABCA1 trafficking. Therefore, β1-syntrophin and utrophin form complexes with ABCA1 that stabilize this transporter, link it to the cytoskeleton and contribute to the regulation of its activity. Interestingly, we have found significantly decreased β1-syntrophin (but not utrophin) mRNA expression in a transcriptome analysis of monocyte-derived macrophages from low HDL subjects with efflux defects despite normal ABCA1 sequence (Kiss et al. 2007; Sarov-Blat et al. 2007). The enhanced ABCA1 protein level in NPC1-deficient hepatocytes, which is mainly regulated by a posttranslational mechanism, is also associated with increased β1-syntrophin expression (Marcel et al., unpublished data). These observations could be compatible with regulation of ABCA1 stability by syntrophins. The mechanism by which syntrophin stabilizes ABCA1 and increases cholesterol efflux remains to be elucidated.

The exocytosis of neurosecretory granules is controlled by a system that presents striking analogies with the mobilization of ABCA1 (Ort et al. 2001). The islet cell autoantigen (ICA) 512, a tyrosine kinase receptor, is regulated by insulin; its cytoplasmic tail binds to β2-syntrophin, which connects it to utrophin and actin. Its tail also has a PEST sequence and a υ-calpain cleavage site adjacent to the PDZ-binding motif. Binding of β2-syntrophin shields ICA512 from degradation and arrests the secretory granules on the cytomatrix. Dephosphorylation of β2-syntrophin weakens its binding to ICA512, which is then degraded, triggering the release of the secretory granules to facilitate their exocytosis (such as the release of insulin by islet β-cells; Ort et al. 2001). The subcellular localization of diacylglycerol kinase (DGK)ζ in skeletal muscle, which converts DAG into PA, follows a similar mechanism: DGKζ, GDP-bound Rac1 and syntrophin form a cytosolic complex, which upon PKC-mediated phosphorylation of a MARCKS domain of DGKζ, induces translocation of the complex to the plasma membrane, where Rac1 is activated by GTP hydrolysis and Rac1 dissociates from DGKζ and syntrophins (Hogan et al. 2001; Abramovici et al. 2003; Yakubchyk et al. 2005). One can thus propose: (1) a similar model may apply to ABCA1, (2) syntrophins contribute to its retention in subcellular reticulated structures, large intracellular puncta or close to the cell surface (Okuhira et al. 2005 and (3) syntrophins are recruited to these sites in response to increased cellular cholesterol load, vesicular cholesterol traffic and/or apoA-I binding.

Tamehiro et al. recently identified an ABCA1-binding protein that is a subunit of the serine palmitoyltransferase enzyme, SPTLC1, which negatively regulates ABCA1 stability and cholesterol efflux (Tamehiro et al. 2008). Pharmacologic inhibition of SPTLC1 with myriocin (which resulted in the disruption of the SPTLC1–ABCA1 complex) and siRNA knockdown of SPTLC1 expression both stimulated ABCA1 efflux by nearly 60%. In 293 cells, SPTLC1 inhibition of ABCA1 activity led to the blockade of ABCA1 exit from the ER. In contrast, myriocin treatment of macrophages increased the level of cell surface ABCA1. These results indicate that the physical interaction of ABCA1 and SPTLC1 results in intracellular retention of ABCA1 protein, in opposition to the effect of syntrophins (Tamehiro et al. 2008).

3.5.2 *Sorting of ABCA1 Between Golgi, Plasma Membrane and LE-Lysosomes: Contribution of Sortilin*

The short half-life of ABCA1 requires ongoing synthesis and selective trafficking to replenish the active pools. Little is known about ABCA1 secretion and sorting from the Golgi. ABCA1 activity affects or is affected by the transport of cholesterol and PL, particularly sphingolipids. Secretory vesicular transport, including both raft-rich (defined by GPI-anchored proteins) and raft-poor (defined by VSVG) vesicles, is accelerated by lipid efflux to apoA-I, an effect not observed in cells with inactive ABCA1 (Zha et al. 2003). This suggests that ABCA1 activity in the Golgi alters membrane composition and stimulates vesicular movement. Recent observations indicate that the transport of prosaposin by LRP1 and GULP regulates ABCA1 in the LE compartment (Kiss et al. 2006). Prosaposin and acid sphingomyelinase (ASM) are first transported from the Golgi to lysosomes by mannose-6-phosphate receptors (M6PR) and sortilin (Ni and Morales 2006). Sortilin is an alternative receptor capable of sorting soluble lysosomal enzymes from the Golgi to lysosomes. CTSD expression increases prosaposin and saposins and upregulates ABCA1, whereas inhibition of CTSD decreases ABCA1 levels (Haidar et al. 2006). These observations suggest that the ABCA1 pathway, which is linked to the metabolism and transport of sphingolipids through LE and lysosomes, is contingent on the sorting and transport of both sphingolipids and lysosomal enzymes from the Golgi. Preliminary data suggest that sortilin but not M6PR is regulated by NPC1 (unpublished results) and may contribute to the sorting of ABCA1 and ASM, either directly or indirectly via effects on sphingolipids and cholesterol Fig. 3.2.

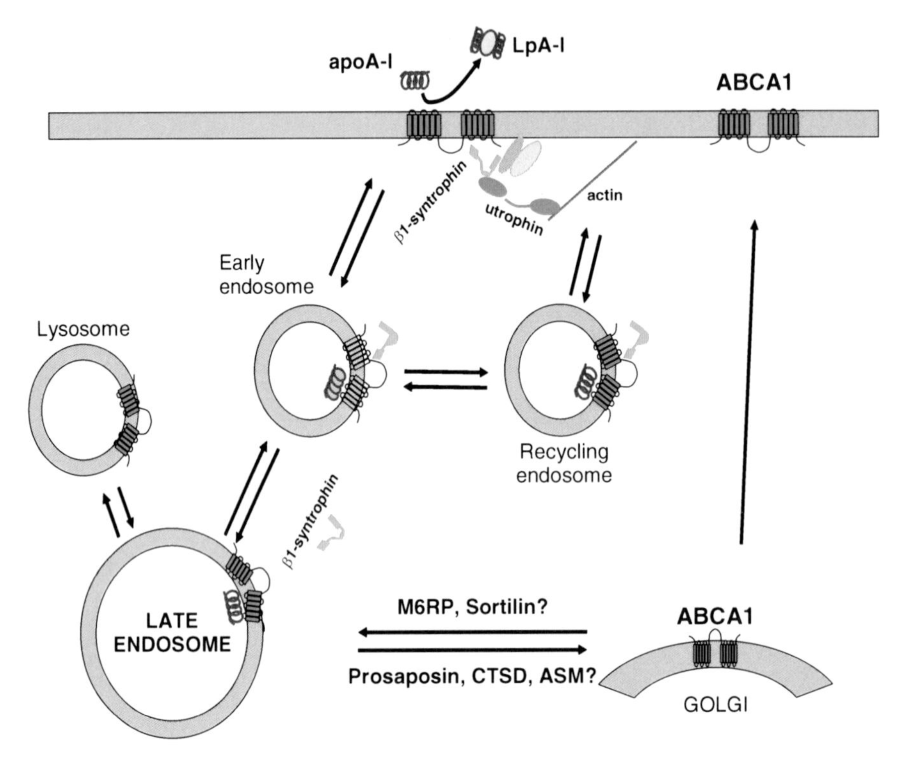

Fig. 3.2 ABCA1 cellular traffic. Newly synthesized ABCA1 is delivered at the cell surface by the secretory pathway; and ABCA1 activity also modulates secretory activity and the movement of membranes from the Golgi (Zha et al. 2003). It is proposed that the ABCA1–β1-syntrophin complex illustrated in Fig. 3.1 may be released from its interaction with the utrophin–actin scaffold at the plasma membrane to enter the endosomal pathway as described by Neufeld and colleagues (2001, 2004). P6RP and sortilin have been proposed to contribute to traffic between Golgi and LE. This traffic may also include prosaposin, CTSD and ASM, which contribute to the regulation of ABCA1 activity through unknown mechanisms

3.6 Integrated Models of Lipid Efflux and Lipoprotein Assembly: Nascent HDL Formation

As discussed above, ABCA1 presentation at the cell surface of hepatocytes, intestinal cells, macrophages and other cell types is essential for the formation and maturation of HDL. Lipid free and/or lipid poor apoA-I are the seed molecules in the initiation of HDL formation, first intracellularly during secretion in hepatocytes and epithelial intestinal cells and second by interaction with ABCA1 at cell surface. Relevant to this process, several questions remain:

1. How does apoA-I interact with cell surface ABCA1 protein and what are the consequences?

2. Does endocytosis of the apoA-I/ABCA1 complex contribute to HDL formation?

3.6.1 Interaction of apoA-I with Cell Surface ABCA1

Earlier observations showed that optimal cholesterol efflux in macrophages requires binding of the C-terminal domain of apoA-I to a cell surface-binding site and the subsequent translocation of intracellular cholesterol to an efflux-competent pool (Burgess et al. 1999). Vedhachalam and colleagues, investigating the mechanisms of cholesterol and phospholipid efflux in macrophages, showed by cross-linking that only about 10% of cell surface bound apoA-I directly interacting with ABCA1 and the major band migrated as monomeric apoA-I. This study indicated that two apoA-I cell surface-binding sites as well as the apoA-I C-terminal domain are required for the acquisition of lipids. ABCA1 activity thus creates two types of high affinity apoA-I-binding sites at the cell surface. The low capacity site formed by direct interaction of apoA-I and ABCA1 has a regulatory role and contributes to the genesis of a second site, which has a very high binding capacity for apoA-I C-terminal-dependent binding and lipidation (Vedhachalam et al. 2007a).

In further studies, a three step model for apoA-I binding and lipidation was formulated (Vedhachalam et al. 2007b). First, the binding of apoA-I to ABCA1 at the cell surface generates a net phospholipid translocation to the plasma membrane exofacial leaflet, causing the asymmetric packing of the two leaflets and inducing a strain in the membrane. Second, the membrane strain is released by membrane bending and the creation of exovesiculated lipid domains (protrusion curvature). Third, this lipid domain allows apoA-I molecules to insert and remove phospholipids and cholesterol by C-terminal dependent binding, which results in the release of discoidal particles containing two, three or four molecules of apoA-I (Vedhachalam et al. 2007b). Here, other questions arise:

1. Does the apoA-I that first interacts with ABCA1 immediately acquire phospholipids upon interaction or does this occur while the membrane is bending?
2. Does it migrate laterally to the protrusion curvature and acquire phospholipids and cholesterol?
3. If this is the case, does this model reconcile with the two-step lipidation theory (Fielding et al. 2000)?
4. Given that the last and productive step in apoA-I binding and lipidation is dependent on an unstable phospholipid bilayer and does not involve ABCA1 interaction, should it be dependent on the oligomerization of ABCA1 (Denis, Haidar et al. 2004a)?

ApoA-I normally self-associates to form oligomers. Its three-dimensional structure is that of a helical bundle structure in which the N- and C-termini are in close proximity (Silva et al. 2005) and their interaction mediates the self-association of lipid-free apoA-I (Silva et al. 2005; Zhu and Atkinson 2007). This suggests that oligomerization during efflux may be largely dependent on apoA-I itself and occurs when apoA-I concentration increases at the site of its recruitment for efflux. Kinetic studies of the formation of apoA-I lipidated particles by ABCA1 in J774 macrophages and HEK-293 cells showed concurrent formation of particles with two, three and four apoA-I molecules with no evidence of a precursor–product relationship, therefore compatible with a model of nascent apoA-I oligomers being lipidated at the plasma membrane.

3.6.2 Contribution of Retroendocytosis

The contribution of retroendocytosis of ABCA1 and apoA-I to the ABCA1-mediated lipidation of apoA-I was initially proposed based on the observation that ABCA1 traffic is triggered by the addition of apoA-I (Neufeld et al. 2001, 2004). This issue has recently been revisited. Cell surface biotinylation experiments revealed that two-thirds of apoA-I was bound at the cell surface and one-third in intracellular compartments in an ABCA1 and apoA-I C-terminal domain-dependent manner. Quantification of the turnover of apoA-I at these sites demonstrated a four-fold faster dissociation of apoA-I from the internal compartment compared to the plasma membrane, suggesting a greater contribution of the internal ABCA1 compartment to apoA-I lipidation (Hassan et al. 2008). This conclusion was not consistent with the studies of the other two groups. Faulkner and colleagues determined the functional significance of ABCA1-mediated endocytosis of bacterially expressed apoA-I in which cysteines were introduced at strategic positions to allow either fluorescent labeling or ^{35}S-cysteine labeling in murine macrophage cell line RAW264.7. The majority of internalized ^{35}S-apoA-I was re-secreted as a degraded protein. Furthermore, specific inhibitors were able to uncouple apoA-I lipidation from apoA-I internalization (Faulkner et al. 2008), casting a doubt on the significance of the endocytic pathway in lipid efflux or HDL formation. Denis and colleagues, who also analyzed apoA-I internalization and lipidation as a function of ABCA1 expression by confocal microscopy and efflux assays, concluded that the majority of HDL formation occurs at the cell surface, whereas the ABCA1-dependent internalization of apoA-I leads to its targeting to lysosomes and degradation (Denis et al. 2008). The apparent contradictions concerning the role of apoA-I endocytosis in ABCA1-mediated lipidation may be related to the duration of apoA-I interaction with the cells or to the cholesterol loading status and cell types. Hassan et al. (2008) have suggested that the half-time for apoA-I and ABCA1 interaction, leading to the resecretion of lipidated apoA-I was shorter than that which culminated in lysosomal targeting and degradation (Denis et al. 2008).

Further studies will be required to identify and validate the endocytic pathways that lead to apoA-I lipidation rather than to lysosomal targeting and degradation. It is still unclear whether the targeting of apoA-I and ABCA1 to lysosomes and their degradation is a function of the length of physical interactions or possibly a function of cellular cholesterol levels. Cells with excess cholesterol should increase ABCA1 presentation at the cell surface in order to bind apoA-I for its productive lipidation, whereas cells with low cholesterol levels should express little cell surface ABCA1 and target apoA-I that binds to ABCA1 for degradation in lysosomes.

3.7 Complementarities of ABCA1, ABCG1 and SR-BI in Lipid Efflux and HDL Formation and Their Combined Role in Reverse Cholesterol Transport in Vivo

Many lipid transporters have been identified that display tissue specific expression and regulation and exhibit different functions. For example, ABCA1 and ABCG1/ABCG4 exist in almost all tissues, while ABCG5/ABCG8 is exclusively expressed on the apical (bile canalicular) surface of hepatocytes where it mediates cholesterol excretion via bile. ABCA7 is highly homologous to ABCA1 and mediates cellular cholesterol and phospholipid release to apolipoproteins when transfected in vitro. Intriguingly, in murine fibroblasts, ABCA7 expression is lowered by increased cellular cholesterol whereas ABCA1, in turn, is upregulated. However, silencing ABCA7 does not reduce efflux to apoA-I but decreases phagocytic activity (Iwamoto et al. 2006), suggesting a role distinct from that of ABCA1. SR-BI is an HDL receptor and a multifunctional cholesterol transporter (Trigatti et al. 2000), which mediates selective uptake of HDL cholesteryl esters in hepatocytes and adrenergic cells. It also contributes to bidirectional cholesterol flux to and from lipoproteins in various cell types, including macrophages.

3.7.1 HDL Genesis in Various Types of Cells

The predominant pathway for efflux to apoA-I and the formation of pre-β-migrating HDL is an ABCA1-dependent mechanism. It is still unclear whether ABCA1 alone operates in all cell types for mature HDL formation when the incubation time is extended, or when HDL is remodeled in the circulation. Also unclear are the conditions and factors that modulate the maturation of pre-β-HDL to α-HDL. When human fibroblasts, HepG2 or Caco2 cells are treated with LXR and RXR agonists (22-OH, 9-cis RA) or when macrophages are activated with cAMP and then incubated with apoA-I for 24 h, α-migrating HDL are formed (Krimbou et al. 2005; Duong et al. 2006). In contrast, pre-β-migrating HDL particles are observed after 16 h incubation of non-stimulated cells, including human hepatoma cells (HepG2), human intestinal transformed cells (Caco2) and CHO cells stably expressing full-length human apoA-I (Chau et al. 2006). Similarly, HEK293 cells overexpressing ABCA1, but not expressing ABCG1, SR-BI, LCAT, PLTP or apoM, form pre-β-HDL particles of various sizes, but not α-HDL (Mulya et al. 2007). This study, designed to eliminate the contribution of the known HDL-modifying factors, showed that overexpression of ABCA1 alone did not generate α-migrating HDL. Most importantly, the minimally lipidated pre-β-HDL particles formed by ABCA1 in these cells had reduced ability to again interact with ABCA1. This observation suggested that ABCA1 could depend on other transporters for further lipidation of the various pre-β-HDL particles formed. ABCG1 is the best candidate for this process. ABCG1 is a half transporter, which may homodimerize or heterodimerize with ABCG4 to form a full transporter that functions at the cell surface (Cserepes et al. 2004). The cholesterol efflux activity of ABCG1 and ABCG4 to both smaller HDL_3 and larger HDL_2, but not to lipid free apoA-I, was first described in 293 cells transfected with both transporters (Wang et al. 2004). The complementarities between ABCA1 and either the ABCG1/ABCG4 heterodimer or potentially the homodimers of ABCG1 or ABCG4 were extended and demonstrated in several studies (Gelissen et al. 2006; Vaughan and Oram 2006), whereas SR-BI activity could not complement ABCA1-mediated efflux (Lorenzi et al. 2008). However, ABCG1–/– mice display normal levels of normal plasma HDL, indicating that a compensatory mechanism may exist or that ABCG1 is redundant for HDL formation. Further investigations are needed to clarify the physiological complementarities of these lipid transporters. We also need to identify the factors controlling the transformation of pre-β- to α-HDL, particularly in hepatocytes, which account for up to 80% of total circulating HDL, but which mainly form pre-β-HDL in vitro (Kiss et al. 2003).

3.7.2 Cholesterol Efflux to apoA-I in Macrophages

Lipid efflux to apoA-I is solely dependent on ABCA1 in cholesterol loaded macrophages, but to a lesser extent in unloaded macrophages (Wang et al. 2007b). ABCG1 inactivation does not affect circulating HDL, however *Abcg1–/–* mice challenged with a high fat diet show significant cellular accumulation of neutral lipids and phospholipids in hepatocytes, macrophages and other cells (Kennedy et al. 2005), pointing to a very different role for ABCG1 in cellular lipid transport. ABCG1 might also provide an additional mechanism for cholesterol removal under stress conditions. Of interest, macrophages of patients with type 2 diabetes patients have no detectable ABCG1 but normal levels of ABCA1 (Mauldin et al. 2008). ABCG1 can efflux cholesterol to lipoprotein particles arising from ABCA1-mediated efflux to apoA-I, suggesting that the two transporters act in concert to maximize the efflux of excess cholesterol from macrophages (Gelissen et al. 2006; Vaughan and Oram 2006), although this process and its functional significance are not fully resolved (Cavelier et al. 2006; Jessup et al. 2006). Adorni et al. (2007) quantified the relative contribution of different transporters to cholesterol efflux to human sera in loaded or unloaded macrophages, both in vitro and in vivo. In normal unloaded macrophages, ABCA1 ablation has no effect on total efflux whereas in cholesterol loaded macrophages ABCA1 accounted for 37% of the net efflux, ABCG1 for 21% and SR-BI for 9%, with the remainder attributable to diffusional efflux, consistent with similar results for the relative contributions of these transporters to in vivo reverse cholesterol transport (Adorni et al. 2007).

3.7.3 In Vivo Cholesterol Efflux from Macrophages and Reverse Cholesterol Transport

Ablation of macrophage *Abca1* (Wang et al. 2007c) and ABCG1 (Wang et al. 2007b) significantly decrease in vivo reverse cholesterol transport in wild-type mice. In contrast, total body ablation of *Abca1* does not affect hepatobiliary metabolism, a result linked to the low circulating HDL and plasma cholesterol levels in these mice (Groen et al. 2001). Interestingly, experiments in mice with either *Abca1* ablation, liver specific *Abca1* overexpression or total *Abca1* expression showed that hepatic ABCA1 levels contributed to HDL catabolism and HDL-associated CE selective uptake in hepatocytes (Singaraja et al. 2006), suggesting that ABCA1 in hepatocytes might also be involved in reverse cholesterol transport. Although the contribution of SR-BI to efflux from cholesterol loaded macrophages may be minimal since SR-BI expression in these cells is almost completely suppressed (Yu et al. 2004; Wang et al. 2007b), bone marrow transplantation experiments have demonstrated that macrophage SR-BI deficiency leads to increased atherosclerotic lesion size (Zhang et al. 2003). Increased liver specific expression of SR-BI was also shown to significantly increase reverse cholesterol transport from peripheral tissues (Zhang et al. 2005), further corroborating the critical role of liver SR-BI in HDL uptake.

Macrophage ABCG1 is a mediator of cholesterol efflux to HDL in vitro and a significant contributor to in vivo reverse cholesterol transport from cholesterol loaded macrophages, but its effects on atherosclerosis in mouse models are complex. Studies of bone marrow transfer of *Abcg1–/–* macrophages to *Ldlr–/–* or *Apoe–/–* mice by several investigators showed unexpectedly that arterial atherosclerotic lesions were not increased, but rather decreased (Baldan et al. 2006; Ranalletta et al. 2006). In contrast, Out et al. (2007) reported that whole body *Abcg1* deficiency conferred a significant 1.9-fold increase in atherosclerotic lesion size in mice fed with atherogenic diet for 12 weeks as compared to wild-type mice (Out et al. 2007). Further experiments by Yvan-Charvet et al. (2008) showed that bone marrow transfer from mice with combined *Abcg1* and ABCA1 deficiency to *Ldlr–/–* mice fed a high fat diet resulted in a 2.5-fold increase in atherosclerosis as compared to control or single *Abca1–/–* bone marrow recipients. The isolated *Abca1–/–* and *Abca1–/– Abcg1–/–* macrophages also had impaired efflux to apoA-I and HDL and increased susceptibility to apoptosis compared to *Abcg1–/–* macrophages (Yvan-Charvet et al. 2007). A follow up study by Out et al. (2008) demonstrated that *Ldlr–/–* mice transplanted with bone marrow from *Abca1–/–Abcg1–/–*- mice and fed a Western diet for 6 weeks developed atherosclerotic lesions smaller than those of *Ldlr–/–* mice receiving ABCA1–/– macrophages, but larger than those receiving *Abcg1–/–* macrophages. A striking observation was the massive accumulation of foam cells in mice receiving *Abca1–/– Abcg1–/–* bone marrow, despite their very low plasma cholesterol (Yvan-Charvet et al. 2007; Out et al. 2008), suggesting a purely cellular phenotype. Interestingly, the *Abca1–/– Abcg1–/–* macrophages displayed marked lipid accumulation, both in vitro and in vivo, but were not recruited to the arterial wall (Out et al. 2008), presumably because the low circulating cholesterol level in these mice did not create the arterial wall inflammation required for monocyte recruitment. Together these experiments suggest that the activation of ABCA1 and ABCG1 has both additive and distinct effects on lipid accumulation that affect macrophage susceptibility to apoptosis.

3.8 Cellular Lipid Traffic Through the Late Endosomes

3.8.1 Egress of Cholesterol from LE

Cellular lipids originating from either exogenous lipoproteins internalized via endocytosis or phagocytosis of cell debris, as well as de novo synthesized lipids, need to be sorted and directed to the proper cellular organelle. Circulating plasma lipoproteins (i.e. LDL) are taken up by the endocytic route, which fuses with early endosomes and rapidly traffics to sorting endosomes, where some constituents are recycled to the cell surface. The recycling endosome membranes are rich in cholesterol and henceforth in phospholipids with cholesterol affinity (Ikonen 2008), which account for the high proportion of LDL-derived cholesterol that is transferred to this compartment (Wang et al. 2007c). Receptor-bound LDL particles are then delivered to the late endosome/lysosome compartments. The hydrolysis of cholesteryl ester in the LDL particle is initiated in early endosomes, but mainly occurs in late endosomes/lysosomes (Brown and Goldstein 1986; Sugii et al. 2003). Normally, cholesterol released from the late endosomes transfers to other membranes, including recycling endosomes and the plasma membrane for efflux, ER for esterification, LD for storage and to mitochondria. This transport is assumed to involve the segregation of endosomal membranes, but its rapidity has so far prevented the identification of the primary intermediate membranes. Molecular transfer by cholesterol binding proteins also takes place, but it is unclear whether it mediates the bulk transport. Sterol-carrier protein 2 (SCP-2) and its homologue SCP-x, were first considered but appeared to traffic toward peroxisomes and to be involved in the oxidation of branched fatty acids (Gallegos et al. 2001). Furthermore, overexpression of SCP-2 in hepatocytes minimally altered cholesterol traffic to mitochondria (Ren et al. 2004). MLN64 is a well characterized cholesterol binding protein. Its N-terminal domain binds to late endosomes, leaving the C-terminal cholesterol binding domain START exposed to the cytosolic side (Alpy and Tomasetto 2006) where it participates in actin-mediated dynamics of late endosomes (Holtta-Vuori et al. 2005). MLN64 presumably transports cholesterol to cytosolic acceptors or membranes. Cellular overexpression of MLN64 contributes to the mobilization of lysosomal cholesterol and its transfer to mitochondria in a START domain-dependent fashion (Zhang et al. 2002), supporting a role for MLN64 in lipid transport from the LE. However, targeted mutation of the MLN64 cholesterol-binding domain, the StAR-related lipid transfer (START) domain, caused only moderate alterations in cellular sterol metabolism

(Kishida et al. 2004), perhaps reflecting the functional redundancy of the 15 murine proteins containing a START domain (Alpy and Tomasetto 2006). The START member, StARD1, appears essential for cholesterol transfer to the inner mitochondrial membrane in steroidogenic tissues for the regulated synthesis of steroid hormone (Alpy and Tomasetto 2005). StarD1 overexpression in hepatocytes increased bile acid biosynthesis by 5.7-fold, as compared to a 1.2-fold increase with overexpression of MLN64 (Ren et al. 2004). The members of StarD4 subfamily of START proteins (StarD4, StarD5, StarD6) are newly identified cytosolic cholesterol binding proteins of the START-domain protein family. Like StarD1, StarD4 only binds cholesterol (Rodriguez-Agudo et al. 2008), and its expression in mouse liver is negatively regulated by cholesterol loading and positively regulated by SREBP-2 (Soccio et al. 2002). StarD4 transfection in COS-1 cells significantly stimulated LXRE reporter activity as did transfection with StarD1, MLN64 and STARD5 (Soccio et al. 2005). In contrast, StarD5, which binds both cholesterol and 25-hydroxycholesterol, is regulated by ER stress (Soccio et al. 2005). StarD5 is highly expressed in macrophages (Rodriguez-Agudo et al. 2006) and further upregulated in free cholesterol loaded macrophages (Soccio et al. 2002, 2005).

3.8.2 Regulation of Cholesterol Traffic in LE

Cholesterol export from LE is regulated by the sequential and independent actions of late endosome luminal soluble NPC2 and membrane-bound NPC1 (Liscum and Sturley 2004), as demonstrated by the arrest of cholesterol transport at that stage by functional mutations in the genes encoding for these two proteins. NPC2 protein has a unique cholesterol binding pocket, whereas the 240-amino-acid soluble luminal loop of the late endosome membrane bound NPC1 can bind cholesterol and oxysterols and is regulated by cellular cholesterol levels (Infante et al. 2008a, b). In addition to cholesterol, other lipids, particularly sphingolipids, accumulate in the late endosomes, where together with lyso-bis phosphatidic acid (LBPA), which induces membrane invagination, they form luminal membranes that are typical of the multi-vesicular bodies (MVB; Matsuo et al. 2004). These densely packed membrane structures account for the high cholesterol storage capacity of LE and MVB (Russell et al. 2006). Glycosphingolipid storage disorders show that the storage of sphingolipids in LE contributes to a secondary accumulation of cholesterol (Pagano 2003), suggesting that the enrichment of LE in both sphingolipids and cholesterol impairs endosomal traffic. Incubating cells with exogenous sphingolipids results in their intracellular accumulation, which in turn increases LDL internalization by the LDL receptor and consequently cholesterol uptake (Puri et al. 2003). This suggests a model in which sphingolipids in endocytic compartments stimulate cholesterol retention in membranes, thus serving as "a molecular trap". Hence, overload and storage of either sphingolipids or cholesterol can impair late endosomal functions Fig. 3.3.

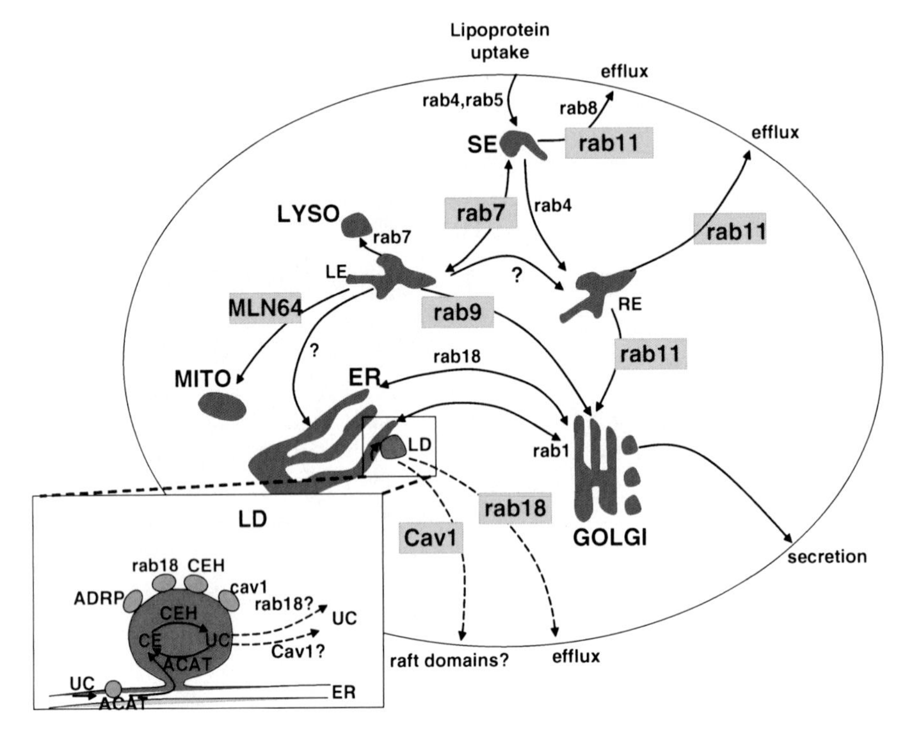

Fig. 3.3 Cellular cholesterol transport pathways. Various Rab GTPases have been shown to regulate the vesicular traffic of cholesterol-rich membrane and cargo from the sorting endosome (*SE*), the recycling endosome (*RE*), the late endosome (*LE*) to the Golgi and the endoplasmic reticulum (*ER*) and the lipid droplet (*LD*). *Highlighted* Rab proteins have been shown to enhance transport of cholesterol to the plasma membrane. LD cholesterol undergoes cycles of esterification by ACAT and hydrolysis by cholesteryl ester hydrolase that allows its export from the cell by a number of putative pathways, including the secretory pathway, vesicular transport via rab18, caveolar transport and others

The exact mechanism by which cholesterol and sphingomyelin are mobilized from the LE is not known, but several lines of evidence show that cholesterol traffic can be rescued by Rab-dependent membrane transport. Overexpression of Rab7 (which mediates cargo progression from early to late endosomes) and Rab9 (which characterizes LE to lysosome conversion) but not Rab11 (which is associated with recycling endosomes that recycle to the plasma membrane (Zerial and McBride 2001), reduced LE cholesterol accumulation. This was accompanied by an increase in cholesteryl ester formation and by normalization of glycosphingolipid transport (Choudhury et al. 2002; Narita et al. 2005). Interestingly, TIP47, a cargo adaptor for MPR, was reported to traffic from LE to the Golgi together with MPR (Diaz and Pfeffer 1998; Carroll et al. 2001). TIP47 binds to a cytosolic domain of MPR and activated (GTP-bound) Rab9 (Carroll et al. 2001). Upon reaching the Golgi, TIP47 may be released to its stable cytosolic form and/or transferred to the lipid droplet (LD). In cells exposed to fatty acids and accumulating triacylglycerides, TIP47 association with LD increases (Wolins et al. 2001) and stabilizes the forming LD (Wolins et al. 2006). Interestingly, TIP47 was shown to be an inhibitor of retinyl ester hydrolysis in keratinocytes (Gao and Simon 2006), suggesting that it could also regulate the mobilization of cholesteryl esters from LD.

In summary, traffic of lipids and specially cholesterol out of LE clearly depends on functional NPC1 and NPC2 proteins, on molecular transporters such as MLN64 and on membrane-mediated transport that are either inhibited by accumulation of sphingolipids and/or cholesterol or stimulated by vesicular mechanisms that depend on specific Rab GTPases. Exit of cholesterol and sphingolipids from LE allows transport to a various cellular sites, including movement of excess cholesterol to the ER for storage in lipid droplets and efflux to exogenous acceptors.

3.9 Cholesterol Traffic Through the Lipid Droplet

All cells can assemble lipid droplets, which are dynamic organelles that serve as the sites of accumulation and storage of excess fatty acids and cholesterol in the form of neutral triacylglycerides and cholesteryl esters. Newly synthesized cholesterol as well as lipoprotein-derived cholesterol can be incorporated into lipid droplets in the ER where the ER-resident protein acyl-CoA:cholesterol acyl transferase (ACAT) catalyzes the esterification of excess cholesterol for storage in LD. In this organelle, cycles of hydrolysis and re-esterification control cholesterol availability for cell membranes and for efflux to cholesterol acceptors. The LD is recognized as a metabolically active but atypical intracellular organelle (Brasaemle 2007) constituted of a hydrophobic core of triacylglycerides (TG) and cholesteryl esters (CE) which are surrounded by a phospholipid monolayer coated with specific proteins. Several proteomic studies of lipid droplet proteins have identified signature coat proteins, including the PAT family of proteins [perilipin, adipophilin/adipocyte differentiation-related protein (ADRP), TIP47, S3-12, OXPAT, LSD1, LSD2]; for a review, see Brasaemle (2007). Many of these proteins are exchangeable and stable either in lipid bound form or free in the cytosol, a property which may be structurally linked to plasma exchangeable apolipoproteins. TIP47 for example contains 11-mer repeats which can form amphipathic α-helices organized in a domain similar to the N-terminal α-helical bundle of apoE, which could explain its stability in both lipid bound and soluble forms (Wolins 2006). In keeping with the postulated ER origin of the LD, several enzymes of the neutral lipid synthetic pathways that reside in the ER have also been found to be associated with this organelle (Coleman and Lee 2004), including ACAT1 (Lin et al. 1999). SNARE proteins, including SNAP23, syntaxin-5, VAMP4, NSF and α-SNAP, are associated with LD and represent an essential mechanism to mediate their fusion and growth (Bostrom et al. 2007).

3.9.1 Regulation of Cholesterol Traffic in the Adipocyte LD

The cholesterol content and nature of LD differ between cells, reflecting the specialized nature of lipid storage. The adipocyte LD is primarily enriched in TG, but its cholesterol store is unique because of its high free cholesterol localized to the surface of the LD and very low cholesteryl ester content. It is remarkable that this LD-associated free cholesterol store represents up to 30% of adipocyte cellular cholesterol (Prattes et al. 2000). Since the adipocyte synthesizes very little cholesterol (Angel 1970), cellular cholesterol homeostasis is maintained through a balance between cholesterol export and import. Early studies demonstrated cholesterol efflux to HDL by adipocytes (Barbaras et al. 1986, 1987), but it is not known whether this efflux is mediated by ABCG1 or SR-BI. Insulin and angiotensin II induce the translocation of SR-BI from intracellular sites to the plasma membrane in adipocytes, a process associated with increased HDL cholesterol uptake and increased adipocyte cholesterol content (Tondu et al. 2005; Yvan-Charvet et al. 2007). In these adipogenic conditions, selective uptake of cholesteryl esters by SR-BI is stimulated, leading to cholesterol accumulation. In human adipocytes, SR-BI, CETP and LRP1 make similar contributions to selective uptake of HDL-CE (Vassiliou and McPherson 2004). Upon β-adrenergic stimulation of lipolysis, cholesterol efflux to HDL increases, likely related to liberation of free cholesterol on the surface of the LD. Under these conditions, SR-BI expression is unaffected (Verghese et al. 2007). There are no data on ABCG1-mediated efflux to HDL in adipocytes; however ablation of ABCG1 in mice results in lower adipose tissue mass and decreased adipocyte size (Buchmann et al. 2007) but no changes in plasma HDL cholesterol (Kennedy et al. 2005).

Cholesterol ester transfer protein (CETP) is expressed in adipocytes (Jiang et al. 1991) and in preadipocytes (Gauthier et al. 1999), where its expression is regulated by cellular cholesterol (Radeau et al. 1995) and where it mediates selective uptake of HDL cholesteryl esters (Benoist et al. 1997). Adipocyte-specific overexpression of CETP in mice results in decreased HDL plasma levels (Zhou et al. 2006), likely due to both secretion of CETP into plasma and selective uptake of HDL-derived cholesterol.

3.9.2 Regulation of Cholesterol Traffic in the Macrophage LD

Primary monocyte-derived macrophages and macrophage cell lines primarily accumulate CE in lipid droplets upon incubation with acetylated or aggregated LDL, whereas they accumulate TG upon exposure to TG-rich lipoproteins, such as VLDL and remnant lipoproteins. This raises the question as to whether neutral lipid storage occurs in separate pure CE and TG droplets or rather in mixed CE and TG droplets. The incubation of acetylated LDL-loaded macrophages with the acyl-CoA synthetase inhibitor triacsin D, which depletes cells of accumulated TG, has been used to vary the cellular TG/CE ratio. This results in a progressive change in lipid droplet anisotropy, indicating that TG and CE originally accumulated in mixed lipid droplets (Lada et al. 2002). Additionally, this study showed that the CE in mixed droplets in the liquid or isotropic state is effluxed more effectively than that in the liquid crystalline or anisotropic droplets resulting from the TG depletion. Interestingly, independently of the nature of the neutral lipid accumulating in the macrophage LD, lipoprotein loading leads to increased adipophilin (human ortholog of ADRP) expression in differentiated THP-1 cells concomitant with decreased perilipin content (Persson et al. 2007). This is reminiscent of the adipocyte LD maturation program, whereby as the lipid droplets acquire more lipids and enlarge, the composition of the coat proteins changes from ADRP to exclusively perilipin in the mature adipocyte (Brasaemle 2007).

3.9.3 Regulation of Cholesterol Traffic in the Hepatocyte LD

The hepatocyte LD participates in high throughput and turnover of triglycerides and cholesteryl esters, the latter either transiting through the ER secretory pathway or trafficking to the plasma membrane and other organelles. The LD of a human hepatic cell line, HuH7, apparently includes adipophilin, acyl-CoA synthetase 3 and 17β-hydroxy steroid desaturase 11 as its major coat proteins and contains significantly more CE than triglycerides (Fujimoto 2004). However, the latter observation may have been a result of de-differentiation of the cell line as observed for HepG2 hepatoma cells, where the impairment of triglyceride mobilization for LD storage impairs VLDL secretion (Wu et al. 1996; Tsai et al. 2007). Hepatocytes accumulate LD in both the cytosol and in the lumen of the ER, which arise from a microsomal fraction and function to provide lipids for VLDL assembly. Proteomic and lipid analyses of murine ER luminal LD showed a unique composition with the absence of coat proteins found on cytosolic lipid droplets (ADRP, TIP-47). Rather, luminal LD are characterized by the presence of proteins not found in cytosolic LD, including ER luminal carboxylesterase Ces3, commonly referred to as triacylglycerol hydrolase (TGH), the microsomal triglyceride transfer protein (MTP) and apoE (Wang et al. 2007). Intraluminal LD isolated from liver homogenates of mice fasted overnight was shown to contain significantly more TG than CE. The composition of hepatic LD, their enrichment in TG or CE, can thus be manipulated as a function of culture conditions. However, exceptions to this observation can occur. Notably, incubation of HepG2 in lipoprotein-deficient serum (LPDS)-containing medium supplemented or not with oleic acid does not appreciably change the CE/TG ratio (10%) and the CE/TC ratio (27%) (Borradaile et al. 2002).

The LD of hepatocytes as well as fibroblasts and adipocytes are associated with specific Rab GTPases, including Rab5c, Rab7, Rab11 and Rab18, supporting the dynamic nature of LD. Of these, Rab18 exclusively labels the LD surface (Martin et al. 2005; Ozeki et al. 2005). This seemingly LD-specific Rab18 is particularly interesting because its localization to the LD in adipocytes is dynamically regulated by lipolytic stimulation (Martin et al. 2005). In response to lipolytic stimuli, Rab18 is recruited to LD and, intriguingly, Rab18-labeled LD are frequently found in association with ER (Ozeki et al. 2005) as well as with non-identified tubulovesicular elements (Martin et al. 2005). It is tempting to speculate that Rab18 functions in the displacement of lipids from metabolically active LD to supply lipids as required elsewhere in the cell. Preliminary results indicate that Rab18 overexpression in macrophages stimulates cholesterol efflux, suggesting a role for Rab18 in the mobilization of cholesterol from lipid droplets (R.S. Kiss and Y.L. Marcel, unpublished data). In hepatocytes, Rab18 localizes to LD displacing ADRP and promoting the close apposition of LD to ER-derived membranes (Ozeki et al. 2005).

3.10 Caveolin and Cellular Cholesterol Transport

The present discussion is limited to the contribution of caveolins to lipid traffic through LD and to the efflux pathway. The facts that caveolin-1 associates with cholesterol in a 1:1 stoichiometry (Murata et al. 1995) and is the major cellular protein associated with photoactivatible-cholesterol (Thiele et al. 2000) represent the most compelling arguments for the structural association of caveolin with cholesterol. However, the role of caveolin as an essential cholesterol carrier remains controversial. Caveolin-1 associates with LD in a regulated and transient manner (Martin and Parton 2005). Caveolin-1 and -2 traffic to LD upon addition of fatty acids and leave the LD as lipids are mobilized (Pol et al. 2004). Liver regeneration is strikingly dependent on caveolin-1, because the formation of LD is essential to this regenerative process (Fernandez et al. 2006). The function of caveolin-1 in LD regulation is only emerging. Clear evidence that caveolin is involved in the maintenance of LD integrity and cholesterol homeostasis is provided by the effect of the truncated mutant, caveolin-3DGV, which blocks LD motility and prevents re-distribution of cholesterol to the cell surface, resulting in the accumulation of neutral lipids in LD and perturbation of cell surface lipid rafts (Pol et al. 2004). Additionally, this mutant has been associated with increased free cholesterol retention in late endosomes that impairs their motility, similarly to the well-characterized phenotype of NPC1–/– cells (Pol et al. 2001). Recent work in our laboratory reveals that NPC1–/– hepatocytes themselves display a 3-fold increase in caveolin-1 levels compared to the wild type, which parallels the increase in cholesterol storage in these cells (Wang et al., unpublished data). Interestingly, in these studies caveolin-1 and cholesterol co-localize in perinuclear regions of hepatocytes and accumulate in the same subcellular density fractions, but in hepatocytes cultured without oleate, caveolin-1 and cholesterol do not co-localize with LD.

Plasma membrane caveolae have been reported to mediate the efflux of cellular free cholesterol, including newly synthesized cholesterol (Fielding and Fielding 1995). Caveolin-1 expression was also shown to increase the efflux of newly synthesized cholesterol in hepatocytes (Fu et al. 2004), an effect which we also observed in HepG2 cells transfected with adeno caveolin-1 (Marcel, unpublished data). Moreover, the Cav^{DGV} mutant has been shown to cause a reduction in both cholesterol efflux and cholesterol biosynthesis (Pol et al. 2001). However, rat hepatoma McA-RH7777 cells stably transfected with caveolin-1, which formed more caveolae and synthesized higher levels of cholesterol, showed no change in cholesterol efflux to either HDL or apoA-I (A. McKenzie, P. Links et al., unpublished data). In conclusion, there is consensus that caveolin-1 plays a role in the transport of newly synthesized cholesterol to the cell surface for efflux but the contribution of caveolin-1 to the transport of LD-derived cholesterol remains to be further investigated.

3.11 Mobilization of LD Lipids for Efflux

3.11.1 The LD is the Major Source of Cholesterol for Efflux

In macrophages, cholesteryl esters derived from endocytosed LDL or modified LDL are delivered to LE/lysosomes, where they are hydrolyzed. Excess lipoprotein-derived cholesterol is esterified by the ER-resident protein, ACAT, and stored as CE in cytoplasmic LD. Cholesterol in the LD undergoes constitutive cycles of esterification–hydrolysis (Brown et al. 1979; McGookey and Anderson 1983), whereby unesterified cholesterol released from the LD by CE hydrolysis can be effluxed to a cholesterol acceptor or re-esterified by ACAT. Although the LD represents the major source of cholesterol available to the efflux pathway, efflux from the LD has been ignored because of the emphasis placed on the mobilization of LE cholesterol. An early study concluded that acetylated LDL-derived cholesterol deposited in late endosomes/lysosomes was the preferential source of cholesterol for ABCA1-mediated efflux in murine macrophages, whereas this pathway was defective in NPC1 mutant macrophages (Chen et al. 2001). In fibroblasts, ABCA1 traffics to LE, and its overexpression stimulates efflux and corrects the Tangier phenotype (Neufeld et al. 2004). This was interpreted as evidence that the pool of ABCA1 in LE was important for efflux and could normalize the elevated cholesterol levels characteristic of Tangier cells. In parallel, deletion of the ABCA1 PEST sequence was shown to impair the internalization of cell surface ABCA1 and its transport to the LE, resulting in decreased efflux of acetylated LDL-derived cholesterol to apoA-I (Chen et al. 2005). However, with the exception of acid lipase and NPC1 deficiency, cholesterol accumulates as cholesteryl esters in LD, and not in LE in normal cells, including macrophages (Brown and Goldstein 1986). Upon addition of apoA-I, cellular LD CE stores are rapidly reduced. Thus, the maintenance of normal cellular cholesterol homeostasis requires mechanisms for the rapid mobilization and re-distribution of cholesterol from LD to efflux competent compartments. The routes of cholesterol traffic from LD to cholesterol acceptors, such as HDL and apoA-I, are not well established. It is unclear whether this process primarily occurs via vesicular or non-vesicular transport, or both. Potential contributors in this process include Rab18 and caveolin-1, as discussed above. Whether LD-associated CE hydrolysis in macrophages is a constitutive process that results in efflux when cholesterol acceptors are present, or whether this process can be triggered by a signaling cascade initiated by the binding of cholesterol acceptors to lipid transporters at the plasma membrane is an exciting prospect to be explored.

3.11.2 Hydrolysis and Mobilization of LD Cholesteryl Esters for Efflux

Although mechanisms for cholesterol traffic from LD to an efflux compartment remain to be fully elucidated, it is becoming clear that LD-associated CE hydrolysis can be a limiting factor in cholesterol efflux. In macrophage foam cells, addition of a LXR ligand, epoxycholesterol, markedly enhanced ABCA1 and ABCG1 expression but failed to increase cholesterol efflux to HDL and/or apoA-I (Ouimet et al. 2008). Epoxycholesterol impaired the mobilization of cholesteryl esters from LD in cholesterol-loaded macrophages, indicating that hydrolysis of CE in LD is a prerequisite for cholesterol efflux. Dysregulation of hydrolysis decreases efflux, even when essential cholesterol transporters are overexpressed and cholesterol acceptors are present in excess. Additional evidence is provided by studies demonstrating that overexpression of CE hydrolases markedly increases cholesterol efflux from macrophages, whereas impaired CE hydrolase activity contributes to lipid accumulation, characteristic of foam cells in atherosclerotic lesions.

Neutral cholesteryl ester hydrolase belongs to the multigene family of carboxylesterases that are involved in drug metabolism, detoxification and these enzymes hydrolyze ester, thioester and amide bonds (for a review, see Dolinsky et al. 2004)). The murine carboxylesterase 3 (Ces3), commonly referred to as triglyceride hydrolase (TGH), is expressed predominantly in the liver with lower levels in adipose tissue, macrophages, intestine, heart and kidney; and it hydrolyzes triglycerides and to a lesser extend cholesteryl esters. The human homolog of TGH, CES1, cloned from human peripheral blood macrophages, THP-1 macrophages (Ghosh 2000) and from human livers (Zhao et al. 2005), was shown to have CE hydrolase activity and has been called neutral CEH. Interestingly, nCEH overexpression in THP1 cells reduces CE accumulation and enhances ABCA1- and ABCG1-mediated cholesterol efflux (Zhao et al. 2007b), while macrophage-specific CEH expression in vivo reduces atherosclerosis in *Ldlr(–/–)/Ceh* transgenic mice fed an atherogenic diet (Zhao et al. 2007a). Enhanced reverse cholesterol transport from CEH overexpressing macrophages was characteristic of these *Ldlr(–/–)/Ceh* transgenic mice, demonstrating the physiological importance of the CE hydrolase and its limiting activity in murine macrophages (Zhao et al. 2007a).

Hormone sensitive lipase (HSL), a triacylglycerol lipase, also has a rate-limiting cholesteryl ester hydrolase activity in adipose tissue (Kraemer and Shen 2006) and contributes significantly to CE hydrolysis in steroidogenic tissues (Ali et al. 2005). HSL overexpression was shown to increase CE hydrolysis in macrophage foam cells and lowered total cholesterol cellular content, although the effect of enhanced CE hydrolysis on lipid efflux in this study was not documented (Escary et al. 1998). Macrophage-specific transgenic expression of rat HSL and apoA-IV in mice challenged with an atherogenic diet has also been shown to reduce the size of aortic lesions as compared to wild-type C57BL/6J littermates (Choy et al. 2003). Taken together, these studies support the view that enhanced LD CE hydrolysis is antiatherogenic.

3.11.3 Is ABCA1 Involved in the Mobilization and Traffic of LD Cholesterol for Efflux?

The ABCA1 transporter functions both in cellular cholesterol trafficking and in plasma membrane cholesterol efflux. ABCA1 expression re-distributes cholesterol in the plasma membrane, resulting in expansion of non-raft domains with which apoA-I preferentially associates (Zha et al. 2001; Landry et al. 2006). In addition, secretory vesicular transport from the Golgi is enhanced by stimulation of cholesterol efflux by apoA-I in normal cells, but not in Tangier cells with deficient cholesterol efflux due to inactive ABCA1 (Zha et al. 2003). Structural and functional abnormalities in caveolar processing and in the *trans*-Golgi secretory network have been reported in cells lacking functional ABCA1, which indicates that the activity of this transporter affects the vesicular budding between the Golgi and the plasma membrane (Orsó et al. 2000). Together, these findings indicate that ABCA1 functions to promote vesicular trafficking of membranes in the secretory pathway and contributes to the flow of lipids from intracellular storage sites to the plasma membrane via the Golgi (Zha et al. 2003). These observations may link ABCA1 activity with the mobilization of lipids from LD to the ER and subsequent trafficking via the Golgi to the plasma membrane for lipid efflux. It is tempting to speculate that Rab18, which tethers the LD to the ER, may be involved in regulating the delivery of LD-associated cholesterol to the ER for its subsequent ensuing trafficking through the Golgi to the plasma membrane for efflux.

3.12 Conclusions

Cellular lipid traffic and homeostasis integrates the regulation of cellular lipid metabolism with cellular lipid uptake and with cellular lipid export. We emphasize here that lipid and particularly cholesterol traffic are intimately linked to lipid secretion and efflux. The coordinated regulation of cholesterol efflux pathways starts with de novo synthesis of cholesterol and lipoprotein uptake and with their cellular traffic by vesicular and molecular transport mechanisms, which together with extracellular lipid acceptors control the transporters involved in cellular export.

References

Abramovici H, Hogan AB, Obagi C, Topham MK, Gee SH (2003) Diacylglycerol kinase-zeta localization in skeletal muscle is regulated by phosphorylation and interaction with syntrophins. Mol Biol Cell 14:4499–4511

Adorni MP, et al (2007) The roles of different pathways in the release of cholesterol from macrophages. J Lipid Res 48:2453–2462

Ahn AH, Freener CA, Gussoni E, Yoshida M, Ozawa E, Kunkel LM (1996) The three human syntrophin genes are expressed in diverse tissues, have distinct chromosomal locations, and each bind to dystrophin and its relatives. J Biol Chem 271:2724–2730

Albrecht DE, Froehner SC (2002) Syntrophins and dystrobrevins: defining the dystrophin scaffold at synapses. Neurosignals 11:123–129

Ali YB, Carriere F, Verger R, Petry S, Muller G, Abousalham A (2005) Continuous monitoring of cholesterol oleate hydrolysis by hormone-sensitive lipase and other cholesterol esterases. J Lipid Res 46:994–1000

Alpy F, Tomasetto C (2005) Give lipids a START: the StAR-related lipid transfer. (START) domain in mammals. J Cell Sci 118:2791–2801

Alpy F, Tomasetto C (2006) MLN64 and MENTHO, two mediators of endosomal cholesterol transport. Biochem Soc Trans 34:343–345

Angel A (1970) Studies on the compartmentation of lipid in adipose cells. I. Subcellular distribution, composition, and transport of newly synthesized lipid: liposomes. J Lipid Res 11:420–432

Arakawa R, Yokoyama S (2002) Helical apolipoproteins stabilize ATP-binding cassette transporter A1 by protecting it from thiol protease-mediated degradation. J Biol Chem 277:22426–22429

Baldan A, et al (2006) Deletion of the transmembrane transporter ABCG1 results in progressive pulmonary lipidosis. J Biol Chem 281:29401–29410

Barbaras R, Grimaldi P, Negrel R, Ailhaud G (1986) Characterization of high-density lipoprotein binding and cholesterol efflux in cultured mouse adipose cells. Biochim Biophys Acta 888:143–156

Barbaras R, Puchois P, Fruchart JC, Ailhaud G (1987) Cholesterol efflux from cultured adipose cells is mediated by LpAI particles but not by LpAI:AII particles. Biochem Biophys Res Commun 142:63–69

Basso F, et al (2003) Role of the hepatic ABCA1 transporter in modulating intrahepatic cholesterol and plasma HDL cholesterol concentrations. J Lipid Res 44:296–302

Benoist F, Lau P, McDonnell M, Doelle H, Milne R, McPherson R (1997) Cholesteryl ester transfer protein mediates selective uptake of high density lipoprotein cholesteryl esters by human adipose tissue. J Biol Chem 272:23572–23577

Borradaile NM, de Dreu LE, Barrett PH, Huff MW (2002) Inhibition of hepatocyte apoB secretion by naringenin: enhanced rapid intracellular degradation independent of reduced microsomal cholesteryl esters. J Lipid Res 43:1544–1554

Bostrom P, et al (2007) SNARE proteins mediate fusion between cytosolic lipid droplets and are implicated in insulin sensitivity. Nat Cell Biol 9:1286–1293

Brasaemle DL (2007) Thematic review series: adipocyte biology. The perilipin family of structural lipid droplet proteins: stabilization of lipid droplets and control of lipolysis. J Lipid Res 48:2547–2559

Brown MS, Goldstein JL (1986) A receptor-mediated pathway for cholesterol homeostasis. Science 232:34–47

Brown MS, Goldstein JL, Krieger M, Ho YK, Anderson RG (1979) Reversible accumulation of cholesteryl esters in macrophages incubated with acetylated lipoproteins. J Cell Biol 82:597–613

Brunham LR, et al (2006) Intestinal ABCA1 directly contributes to HDL biogenesis in vivo. J Clin Invest 116:1052–1062

Buchmann J, et al (2007) Ablation of the cholesterol transporter adenosine triphosphate-binding cassette transporter G1 reduces adipose cell size and protects against diet-induced obesity. Endocrinology 148:1561–1573

Buechler C, Boettcher A, Bared SM, Probst MC, Schmitz G (2002) The carboxyterminus of the ATP-binding cassette transporter A1 interacts with a beta2-syntrophin/utrophin complex. Biochem Biophys Res Commun 293:759–765

Burgess JW, et al (1999) Deletion of the C-terminal domain of apolipoprotein A-I impairs cell surface binding and lipid efflux in macrophage. Biochemistry 38:14524–14533

Carroll KS, Hanna J, Simon I, Krise J, Barbero P, Pfeffer SR (2001) Role of Rab9 GTPase in facilitating receptor recruitment by TIP47. Science 292:1373–1376

Cavelier C, Lorenzi I, Rohrer L, Von Eckardstein A (2006) Lipid efflux by the ATP-binding cassette transporters ABCA1 and ABCG1. Biochim Biophys Acta 1761:655–666

Cavelier LB, Qiu Y, Bielicki JK, Afzal V, Cheng JF, Rubin EM (2001) Regulation and activity of the human ABCA1 gene in transgenic mice. J Biol Chem 276:18046–18051

Chau P, Nakamura Y, Fielding CJ, Fielding PE (2006) Mechanism of prebeta-HDL formation and activation. Biochemistry 45:3981–3987

Chen W, et al (2001) Preferential ATP-binding cassette transporter A1-mediated cholesterol efflux from late endosomes/lysosomes. J Biol Chem 276:43564–43569

Chen W, Wang N, Tall AR (2005) A PEST deletion mutant of ABCA1 shows impaired internalization and defective cholesterol efflux from late endosomes. J Biol Chem 280:29277–29281

Chisholm JW, Burleson ER, Shelness GS, Parks JS (2002) ApoA-I secretion from HepG2 cells: evidence for the secretion of both lipid-poor apoA-I and intracellularly assembled nascent HDL. J Lipid Res 43:36–44

Choudhury A, et al (2002) Rab proteins mediate Golgi transport of caveola-internalized glycosphingolipids and correct lipid trafficking in Niemann-Pick C cells. J Clin Invest 109:1541–1550

Choy HA, Wang XP, Schotz MC (2003) Reduced atherosclerosis in hormone-sensitive lipase transgenic mice overexpressing cholesterol acceptors. Biochim Biophys Acta 1634:76–85

Coleman RA, Lee DP (2004) Enzymes of triacylglycerol synthesis and their regulation. Prog Lipid Res 43:134–176

Costet P, Luo Y, Wang N, Tall AR (2000) Sterol-dependent transactivation of the human ABC1 promoter by LXR/RXR. J Biol Chem 275:28240–28245

Cserepes J, et al (2004) Functional expression and characterization of the human ABCG1 and ABCG4 proteins: indications for heterodimerization. Biochem Biophys Res Commun 320:860–867

Denis M, Haidar B, Marcil M, Bouvier M, Krimbou L, Genest J (2004a) Characterization of oligomeric human ATP binding cassette transporter A1. Potential implications for determining the structure of nascent high density lipoprotein particles. J Biol Chem 279:41529–41536

Denis M, Haidar B, Marcil M, Bouvier M, Krimbou L, Genest J, Jr (2004b) Molecular and cellular physiology of apolipoprotein A-I lipidation by the ATP-binding cassette transporter A1. (ABCA1). J Biol Chem 279:7384–7394

Denis M, Landry YD, Zha X (2008) ATP-binding Cassette A1-mediated Lipidation of Apolipoprotein A-I Occurs at the Plasma Membrane and Not in the Endocytic Compartments. J Biol Chem 283:16178–16186

Diaz E, Pfeffer SR (1998) TIP47: a cargo selection device for mannose 6-phosphate receptor trafficking. Cell 93:433–443

Dixon JL, Ginsberg HN (1992) Hepatic synthesis of lipoproteins and apolipoproteins. Semin Liver Dis 12:364–372

Dolinsky VW, Gilham D, Alam M, Vance DE, Lehner R (2004) Triacylglycerol hydrolase: role in intracellular lipid metabolism. Cell Mol Life Sci 61:1633–1651

Duong PT, Collins HL, Nickel M, Lund-Katz S, Rothblat GH, Phillips MC (2006) Characterization of nascent HDL particles and microparticles formed by ABCA1-mediated efflux of cellular lipids to apoA-I. J Lipid Res 47:832–843

Eisenberg S (1984) High density lipoprotein metabolism (review; 398 refs). J Lipid Res 25:1017–1058

Escary JL, Choy HA, Reue K, Schotz MC (1998) Hormone-sensitive lipase overexpression increases cholesteryl ester hydrolysis in macrophage foam cells. Arterioscler Thromb Vasc Biol 18:991–998

Faulkner LE, et al (2008) An analysis of the role of a retroendocytosis pathway in ATP-binding cassette transporter (ABCA1)-mediated cholesterol efflux from macrophages. J Lipid Res 49:1322–1332

Feng B, Tabas I (2002) ABCA1-mediated cholesterol efflux is defective in free cholesterol-loaded macrophages. Mechanism involves enhanced ABCA1 degradation in a process requiring full NPC1 activity. J Biol Chem 277:43271–43280

Fernandez MA, et al (2006) Caveolin-1 is essential for liver regeneration. Science 313:1628–1632

Fielding PE, Fielding CJ (1995) Plasma membrane caveolae mediate the efflux of cellular free cholesterol. Biochemistry 34:14288–14292

Fielding PE, Nagao K, Hakamata H, Chimini G, Fielding CJ (2000) A two-step mechanism for free cholesterol and phospholipid efflux from human vascular cells to apolipoprotein A-1. Biochemistry 39:14113–14120

Fitzgerald ML, Okuhira K, Short GF, III, Manning JJ, Bell SA, Freeman MW (2004) ATP-binding cassette transporter A1 contains a novel C-terminal VFVNFA motif that is required for its cholesterol efflux and ApoA-I binding activities. J Biol Chem 279:48477–48485

Fu Y, Hoang A, Escher G, Parton RG, Krozowski Z, Sviridov D (2004) Expression of caveolin-1 enhances cholesterol efflux in hepatic cells. J Biol Chem 279:14140–14146

Fujimoto Y, et al (2004) Identification of major proteins in the lipid droplet-enriched fraction isolated from the human hepatocyte cell line HuH7. Biochim Biophys Acta 1644:47–59

Gallegos AM, et al (2001) Gene structure, intracellular localization, and functional roles of sterol carrier protein-2. Prog Lipid Res 40:498–563

Gao JG, Simon M (2006) Molecular screening for GS2 lipase regulators: inhibition of keratinocyte retinylester hydrolysis by TIP47. J Invest Dermatol 126:2087–2095

Gauthier B, Robb M, McPherson R (1999) Cholesteryl ester transfer protein gene expression during differentiation of human preadipocytes to adipocytes in primary culture. Atherosclerosis 142:301–307

Gelissen IC, et al (2006) ABCA1 and ABCG1 synergize to mediate cholesterol export to apoA-I. Arterioscler Thromb Vasc Biol 26:534–540

Ghosh S (2000) Cholesteryl ester hydrolase in human monocyte/macrophage: cloning, sequencing, and expression of full-length cDNA. Physiol Genomics 2:1–8

Groen AK, Bloks VW, Bandsma RH, Ottenhoff R, Chimini G, Kuipers F (2001) Hepatobiliary cholesterol transport is not impaired in Abca1-null mice lacking HDL. J Clin Invest 108:843–850

Haghpassand M, Bourassa PA, Francone OL, Aiello RJ (2001) Monocyte/macrophage expression of ABCA1 has minimal contribution to plasma HDL levels. J Clin Invest 108:1315–1320

Haidar B, et al (2006) Cathepsin D, a lysosomal protease, regulates ABCA1-mediated lipid efflux. J Biol Chem 281:39971–39981

Hargrove GM, Junco A, Wong NC (1999) Hormonal regulation of apolipoprotein AI. J Mol Endocrinol 22:103–111

Hassan HH, et al (2008) Quantitative analysis of ABCA1-dependent compartmentalization and trafficking of apolipoprotein A-I: implications for determining cellular kinetics of nascent high density lipoprotein biogenesis. J Biol Chem 283:11164–11175

Hogan A, et al (2001) Interaction of gamma 1-syntrophin with diacylglycerol kinase-zeta. Regulation of nuclear localization by PDZ interactions. J Biol Chem 276:26526–26533

Holtta-Vuori M, Alpy F, Tanhuanpaa K, Jokitalo E, Mutka AL, Ikonen E (2005) MLN64 is involved in actin-mediated dynamics of late endocytic organelles. Mol Biol Cell 16:3873–3886

Ikonen E (2008) Cellular cholesterol trafficking and compartmentalization. Nat Rev Mol Cell Biol 9:125–138

Infante RE, Abi-Mosleh L, Radhakrishnan A, Dale JD, Brown MS, Goldstein JL (2008a) Purified NPC1 protein. I. Binding of cholesterol and oxysterols to a 1278-amino acid membrane protein. J Biol Chem 283:1052–1063

Infante RE, et al (2008b) Purified NPC1 protein: II. Localization of sterol binding to a 240-amino acid soluble luminal loop. J Biol Chem 283:1064–1075

Ito J, Nagayasu Y, Ueno S, Yokoyama S (2002) Apolipoprotein-mediated cellular lipid release requires replenishment of sphingomyelin in a phosphatidylcholine-specific phospholipase C-dependent manner. J Biol Chem 277:44709–44714

Iwamoto N, Abe-Dohmae S, Sato R, Yokoyama S (2006) ABCA7 expression is regulated by cellular cholesterol through the SREBP2 pathway and associated with phagocytosis. J Lipid Res 47:1915–1927

Iwamoto N, et al (2007) ATP-binding cassette transporter A1 gene transcription is downregulated by activator protein 2alpha. Doxazosin inhibits activator protein 2alpha and increases high-density lipoprotein biogenesis independent of alpha1-adrenoceptor blockade. Circ Res 101:156–165

Jessup W, Gelissen IC, Gaus K, Kritharides L (2006) Roles of ATP binding cassette transporters A1 and G1, scavenger receptor BI and membrane lipid domains in cholesterol export from macrophages. Curr Opin Lipidol 17:247–257

Jiang XC, et al (1991) Mammalian adipose tissue and muscle are major sources of lipid transfer protein mRNA. J Biol Chem 266:4631–4639

Jin FY, Kamanna VS, Kashyap ML (1998) Estradiol stimulates apolipoprotein A-I- but not A-II-containing particle synthesis and secretion by stimulating mRNA transcription rate in Hep G2 cells. Arterioscler Thromb Vasc Biol 18:999–1006

Joyce CW, et al (2006) ABCA1 overexpression in the liver of LDLr-KO mice leads to accumulation of pro-atherogenic lipoproteins and enhanced atherosclerosis. J Biol Chem 281:33053–33065

Kennedy MA, et al (2005) ABCG1 has a critical role in mediating cholesterol efflux to HDL and preventing cellular lipid accumulation. Cell Metab 1:121–131

Kishida T, et al (2004) Targeted mutation of the MLN64 START domain causes only modest alterations in cellular sterol metabolism. J Biol Chem 279:19276–19285

Kiss RS, et al (2003) The lipidation by hepatocytes of human apolipoprotein A-I occurs by both ABCA1-dependent and -independent pathways. J Biol Chem 278:10119–10127

Kiss RS, Maric J, Marcel YL (2005) Lipid efflux in human and mouse macrophagic cells: evidence for differential regulation of phospholipid and cholesterol efflux. J Lipid Res 46:1877–1887

Kiss RS, et al (2006) The lipoprotein receptor-related protein-1. (LRP) adapter protein GULP mediates trafficking of the LRP ligand prosaposin, leading to sphingolipid and free cholesterol accumulation in late endosomes and impaired efflux. J Biol Chem 281:12081–12092

Kiss RS, et al (2007) Genetic etiology of isolated low HDL syndrome: incidence and heterogeneity of efflux defects. Arterioscler Thromb Vasc Biol 27:1139–1145

Kraemer FB, Shen WJ (2006) Hormone-sensitive lipase knockouts. Nutr Metab. (Lond) 3:12

Krimbou L, et al (2005) Biogenesis and speciation of nascent apoA-I-containing particles in various cell lines. J Lipid Res 46:1668–1677

Lada AT, Willingham MC, St Clair RW (2002) Triglyceride depletion in THP-1 cells alters cholesteryl ester physical state and cholesterol efflux. J Lipid Res 43:618–628

Lamon-Fava S, Ordovas JM, Schaefer EJ (1999) Estrogen increases apolipoprotein. (Apo) A-I secretion in Hep G2 cells by modulating transcription of the apo A-I gene promoter. Arterioscler Thromb Vasc Biol 19:2960–2965

Landry YD, Denis M, Nandi S, Bell S, Vaughan AM, Zha X (2006) ATP-binding cassette transporter A1 expression disrupts raft membrane microdomains through its ATPase-related functions. J Biol Chem 281:36091–36101

Lawn RM, et al (1999) The Tangier disease gene product ABC1 controls the cellular apolipoprotein-mediated lipid removal pathway. J Clin Invest 104:R25–R31
Lin S, Cheng D, Liu MS, Chen J, Chang TY (1999) Human Acyl-CoA:Cholesterol acyltransferase-1 in the endoplasmic reticulum contains seven transmembrane domains. J Biol Chem 274:23276–23285
Liscum L, Sturley SL (2004) Intracellular trafficking of Niemann-Pick C proteins 1 and 2: obligate components of subcellular lipid transport. Biochim Biophys Acta 1685:22–27
Lorenzi I, Von Eckardstein A, Radosavljevic S, Rohrer L (2008) Lipidation of apolipoprotein A-I by ATP-binding cassette transporter (ABC) A1 generates an interaction partner for ABCG1 but not for scavenger receptor BI. Biochim Biophys Acta 1781:306–313
Maric J, Kiss RS, Franklin V, Marcel YL (2005) Intracellular lipidation of newly synthesized apolipoprotein A-I in primary murine hepatocytes. J Biol Chem 280:39942–39949
Martin S, Parton RG (2005) Caveolin, cholesterol, and lipid bodies. Semin Cell Dev Biol 16:163–174
Martin S, Driessen K, Nixon SJ, Zerial M, Parton RG (2005) Regulated localization of Rab18 to lipid droplets: effects of lipolytic stimulation and inhibition of lipid droplet catabolism. J Biol Chem 280:42325–42335
Martinez L, Agerholm-Larsen B, Wang N, Chen W, Tall AR (2003) Phosporylation of a pest sequence in ABCA1 promotes calpain degradation and is reversed by APOA-I. J Biol Chem 278:37368–37374
Matsuo H, et al (2004) Role of LBPA and Alix in multivesicular liposome formation and endosome organization. Science 303:531–534
Mauldin JP, et al (2008) Reduced expression of ATP-binding cassette transporter G1 increases cholesterol accumulation in macrophages of patients with type 2 diabetes mellitus. Circulation 117:2785–2792
McGookey DJ, Anderson RG (1983) Morphological characterization of the cholesteryl ester cycle in cultured mouse macrophage foam cells. J Cell Biol 97:1156–1168
Mooradian AD, Haas MJ, Wong NC (2006) The effect of select nutrients on serum high-density lipoprotein cholesterol and apolipoprotein A-I levels. Endocr Rev 27:2–16
Mulya A, Lee JY, Gebre AK, Thomas MJ, Colvin PL, Parks JS (2007) Minimal lipidation of pre-beta HDL by ABCA1 results in reduced ability to interact with ABCA1. Arterioscler Thromb Vasc Biol 27:1828–1836
Munehira Y, et al (2004) Alpha1-syntrophin modulates turnover of ABCA1. J Biol Chem 279:15091–15095
Murata M, Peranen J, Schreiner R, Wieland F, Kurzchalia TV, Simons K (1995) VIP21/caveolin is a cholesterol-binding protein. Proc Natl Acad Sci USA 92:10339–10343
Narita K, et al (2005) Protein transduction of Rab9 in Niemann-Pick C cells reduces cholesterol storage. FASEB J 19:1558–1560
Neufeld EB, et al (2001) Cellular localization and trafficking of the human ABCA1 transporter. J Biol Chem 276:27584–27590
Neufeld EB, et al (2004) The ABCA1 transporter modulates late endocytic trafficking: insights from the correction of the genetic defect in Tangier disease. J Biol Chem 279:15571–15578
Ni X, Morales CR (2006) The lysosomal trafficking of acid sphingomyelinase is mediated by sortilin and mannose 6-phosphate receptor. Traffic 7:889–902
Okuhira K, et al (2005) Purification of ATP-binding cassette transporter A1 and associated binding proteins reveals the importance of beta1-syntrophin in cholesterol efflux. J Biol Chem 280:39653–39664
Oram JF, Lawn RM, Garvin MR, Wade DP (2000) ABCA1 is the cAMP-inducible apolipoprotein receptor that mediates cholesterol secretion from macrophages. J Biol Chem 275: 34508–34511
Orsó E, et al (2000) Transport of lipids from Golgi to plasma membrane is defective in Tangier disease patients and *Abc1*-deficient mice. Nature Genet 24:192–196
Ort T, et al (2001) Dephosphorylation of beta2-syntrophin and Ca2+/mu-calpain-mediated cleavage of ICA512 upon stimulation of insulin secretion. EMBO J 20:4013–4023

Ouimet M, Wang MD, Cadotte N, Ho K, Marcel YL (2008) Epoxycholesterol impairs cholesteryl ester hydrolysis in macrophage foam cells, resulting in decreased cholesterol efflux. Arterioscler Thromb Vasc Biol 28:1144-1150

Out R, et al (2007) Total body ABCG1 expression protects against early atherosclerotic lesion development in mice. Arterioscler Thromb Vasc Biol 27:594–599

Out R, et al (2008) Combined deletion of macrophage ABCA1 and ABCG1 leads to massive lipid accumulation in tissue macrophages and distinct atherosclerosis at relatively low plasma cholesterol levels. Arterioscler Thromb Vasc Biol 28:258–264

Ozeki S, Cheng J, Tauchi-Sato K, Hatano N, Taniguchi H, Fujimoto T (2005) Rab18 localizes to lipid droplets and induces their close apposition to the endoplasmic reticulum-derived membrane. J Cell Sci 118:2601–2611

Pagano RE (2003) Endocytic trafficking of glycosphingolipids in sphingolipid storage diseases. Philos Trans R Soc Lond B Biol Sci 358:885–891

Persson J, Degerman E, Nilsson J, Lindholm MW (2007) Perilipin and adipophilin expression in lipid loaded macrophages. Biochem Biophys Res Commun 363:1020–1026

Pol A, Luetterforst R, Lindsay M, Heino S, Ikonen E, Parton RG (2001) A caveolin dominant negative mutant associates with lipid bodies and induces intracellular cholesterol imbalance. J Cell Biol 152:1057–1070

Pol A, et al (2004) Dynamic and regulated association of caveolin with lipid bodies: modulation of lipid body motility and function by a dominant negative mutant. Mol Biol Cell 15:99–110

Prattes S, et al (2000) Intracellular distribution and mobilization of unesterified cholesterol in adipocytes: triglyceride droplets are surrounded by cholesterol-rich ER-like surface layer structures. J Cell Sci 113:2977–2989

Puri V, Jefferson JR, Singh RD, Wheatley CL, Marks DL, Pagano RE (2003) Sphingolipid storage induces accumulation of intracellular cholesterol by stimulating SREBP-1 cleavage. J Biol Chem 278:20961–20970

Radeau T, Lau P, Robb M, McDonnell M, Ailhaud G, McPherson R (1995) Cholesteryl ester transfer protein (CETP) mRNA abundance in human adipose tissue: Relationship to cell size and membrane cholesterol content. J Lipid Res 36:2552–2561

Ranalletta M, Wang N, Han S, Yvan-Charvet L, Welch C, Tall AR (2006) Decreased atherosclerosis in low-density lipoprotein receptor knockout mice transplanted with Abcg1-/- bone marrow. Arterioscler Thromb Vasc Biol 26:2308–2315

Ren S, et al (2004) Effect of increasing the expression of cholesterol transporters (StAR, MLN64, and SCP-2) on bile acid synthesis. J Lipid Res 45:2123–2131

Repa JJ, Mangelsdorf DJ (1999) Nuclear receptor regulation of cholesterol and bile acid metabolism. Curr Opin Biotechnol 10:557–563

Rodriguez-Agudo D, et al (2006) Localization of StarD5 cholesterol binding protein. J Lipid Res 47:1168–1175

Rodriguez-Agudo D, et al (2008) Intracellular cholesterol transporter StarD4 binds free cholesterol and increases cholesterol ester formation. J Lipid Res 49:1409–1419

Roosbeek S, et al (2004) Phosphorylation by protein kinase CK2 modulates the activity of the ATP binding cassette A1 transporter. J Biol Chem 279:37779–37788

Russell MR, Nickerson DP, Odorizzi G (2006) Molecular mechanisms of late endosome morphology, identity and sorting. Curr Opin Cell Biol 18:422–428

Sarov-Blat L, et al (2007) Predominance of a proinflammatory phenotype in monocyte-derived macrophages from subjects with low plasma HDL-cholesterol. Arterioscler Thromb Vasc Biol 27:1115–1122

Schmitz G, Langmann T (2005) Transcriptional regulatory networks in lipid metabolism control ABCA1 expression. Biochim Biophys Acta 1735:1–19

Schwartz K, Lawn RM, Wade DP (2000) ABC1 Gene Expression and ApoA-I-Mediated Cholesterol Efflux Are Regulated by LXR. Biochem Biophys Res Commun 274:794–802

Silva RA, Hilliard GM, Fang J, Macha S, Davidson WS (2005) A three-dimensional molecular model of lipid-free apolipoprotein A-I determined by cross-linking/mass spectrometry and sequence threading. Biochemistry 44:2759–2769

Singaraja RR, et al (2001) Human *ABCA1* BAC transgenic mice show increased high density lipoprotein cholesterol and apoAI-dependent efflux stimulated by an internal promoter containing liver X receptor response elements in intron 1. J Biol Chem 276: 33969–33979

Singaraja RR, et al (2006) Hepatic ATP-binding cassette transporter A1 is a key molecule in high-density lipoprotein cholesteryl ester metabolism in mice. Arterioscler Thromb Vasc Biol 26:1821–1827

Soccio RE, Adams RM, Romanowski MJ, Sehayek E, Burley SK, Breslow JL (2002) The cholesterol-regulated StarD4 gene encodes a StAR-related lipid transfer protein with two closely related homologues, StarD5 and StarD6. Proc Natl Acad Sci USA 99:6943–6948

Soccio RE, Adams RM, Maxwell KN, Breslow JL (2005) Differential gene regulation of StarD4 and StarD5 cholesterol transfer proteins. Activation of StarD4 by sterol regulatory element-binding protein-2 and StarD5 by endoplasmic reticulum stress. J Biol Chem 280:19410–19418

Sugii S, Reid PC, Ohgami N, Du H, Chang TY (2003) Distinct endosomal compartments in early trafficking of low density lipoprotein-derived cholesterol. J Biol Chem 278:27180–27189

Tall AR (1990) Plasma high density lipoproteins. Metabolism and relationship to atherogenesis. J Clin Invest 86:379–384

Tamehiro N, et al (2007) Sterol regulatory element-binding protein-2- and liver X receptor-driven dual promoter regulation of hepatic ABC transporter A1 gene expression: mechanism underlying the unique response to cellular cholesterol status. J Biol Chem 282:21090–21099

Tamehiro N, et al (2008) SPTLC1 binds ABCA1 to negatively regulate trafficking and cholesterol efflux activity of the transporter. Biochemistry 47:6138–6147

Tang C, Vaughan AM, Oram JF (2004) Janus kinase 2 modulates the apolipoprotein interactions with ABCA1 required for removing cellular cholesterol. J Biol Chem 279:7622–7628

Tang C, Vaughan AM, Anantharamaiah GM, Oram JF (2006) Janus kinase 2 modulates the lipid-removing but not protein-stabilizing interactions of amphipathic helices with ABCA1. J Lipid Res 47:107–114

Thiele C, Hannah MJ, Fahrenholz F, Huttner WB (2000) Cholesterol binds to synaptophysin and is required for biogenesis of synaptic vesicles. Nat Cell Biol 2:42–49

Thymiakou E, Zannis VI, Kardassis D (2007) Physical and functional interactions between liver X receptor/retinoid X receptor and Sp1 modulate the transcriptional induction of the human ATP binding cassette transporter A1 gene by oxysterols and retinoids. Biochemistry 46:11473–11483

Timmins JM, et al (2005) Targeted inactivation of hepatic Abca1 causes profound hypoalphalipoproteinemia and kidney hypercatabolism of apoA-I. J Clin Invest 115:1333–1342

Tondu AL, et al (2005) Insulin and angiotensin II induce the translocation of scavenger receptor class B, type I from intracellular sites to the plasma membrane of adipocytes. J Biol Chem 280:33536–33540

Trigatti B, Rigotti A, Krieger M (2000) The role of the high-density lipoprotein receptor SR-BI in cholesterol metabolism. Curr Opin Lipidol 11:123–131

Tsai J, Qiu W, Kohen-Avramoglu R, Adeli K (2007) MEK-ERK inhibition corrects the defect in VLDL assembly in HepG2 cells: potential role of ERK in VLDL-ApoB100 particle assembly. Arterioscler Thromb Vasc Biol 27:211–218

Vaisman BL, et al (2001) ABCA1 overexpression leads to hyperalphalipoproteinemia and increased biliary cholesterol excretion in transgenic mice. J Clin Invest 108:303–309

Vassiliou G, McPherson R (2004) Role of cholesteryl ester transfer protein in selective uptake of high density lipoprotein cholesteryl esters by adipocytes. J Lipid Res 45:1683–1693

Vaughan AM, Oram JF (2006) ABCA1 and ABCG1 or ABCG4 act sequentially to remove cellular cholesterol and generate cholesterol-rich HDL. J Lipid Res 47:2433–2443

Vedhachalam C, Ghering AB, Davidson WS, Lund-Katz S, Rothblat GH, Phillips MC (2007a) ABCA1-induced cell surface binding sites for ApoA-I. Arterioscler Thromb Vasc Biol 27:1603–1609

Vedhachalam C, et al (2007b) Mechanism of ATP-binding cassette transporter A1-mediated cellular lipid efflux to apolipoprotein A-I and formation of high density lipoprotein particles. J Biol Chem 282:25123–25130

Verghese PB, Arrese EL, Soulages JL (2007) Stimulation of lipolysis enhances the rate of cholesterol efflux to HDL in adipocytes. Mol Cell Biochem 302:241–248

Wang H, Gilham D, Lehner R (2007a) Proteomic and lipid characterization of apolipoprotein B-free luminal lipid droplets from mouse liver microsomes: implications for very low density lipoprotein assembly. J Biol Chem 282:33218–33226

Wang MD, et al (2007b) Differential regulation of ATP binding cassette protein A1 expression and ApoA-I lipidation by Niemann–Pick type C1 in murine hepatocytes and macrophages. J Biol Chem 282:22525–22533

Wang MD, Kiss RS, Franklin V, McBride HM, Whitman SC, Marcel YL (2007c) Different cellular traffic of LDL-cholesterol and acetylated LDL-cholesterol leads to distinct reverse cholesterol transport pathways. J Lipid Res 48:633–645

Wang N, et al (2003) A PEST sequence in ABCA1 regulates degradation by calpain protease and stabilization of ABCA1 by apoA-I. J Clin Invest 111:99–107

Wang N, Lan D, Chen W, Matsuura F, Tall AR (2004) ATP-binding cassette transporters G1 and G4 mediate cellular cholesterol efflux to high-density lipoproteins. Proc Natl Acad Sci USA 101:9774–9779

Wang X, et al (2007d) Macrophage ABCA1 and ABCG1, but not SR-BI, promote macrophage reverse cholesterol transport in vivo. J Clin Invest 117:2216–2224

Wang Y, Oram JF (2002) Unsaturated fatty acids inhibit cholesterol efflux from macrophages by increasing degradation of ATP-binding cassette transporter A1. J Biol Chem 277:5692–5697

Wang Y, Oram JF (2005) Unsaturated fatty acids phosphorylate and destabilize ABCA1 through a phospholipase D2 pathway. J Biol Chem 280:35896–35903

Wang Y, Oram JF (2007) Unsaturated fatty acids phosphorylate and destabilize ABCA1 through a protein kinase C delta pathway. J Lipid Res 48:1062–1068

Wellington CL, et al (2002) ABCA1 mRNA and protein distribution patterns predict multiple different roles and levels of regulation. Lab Invest 82:273-283

Wellington CL, et al (2003) Alterations of plasma lipids in mice via adenovirual mediated hepatic overexpression of human ABCA1. J Lipid Res 44:1470–1480

Wolins NE, Rubin B, Brasaemle DL (2001) TIP47 associates with lipid droplets. J Biol Chem 276:5101–5108

Wolins NE, Brasaemle DL, Bickel PE (2006) A proposed model of fat packaging by exchangeable lipid droplet proteins. FEBS Lett 580:5484–5491

Wong J, Quinn CM, Brown AJ (2004) Statins inhibit synthesis of an oxysterol ligand for the liver x receptor in human macrophages with consequences for cholesterol flux. Arterioscler Thromb Vasc Biol 24:2365–2371

Wu XJ, Shang AM, Jiang HS, Ginsberg HN (1996) Low rates of apoB secretion from HepG2 cells result from reduced delivery of newly synthesized triglyceride to a "secretion- coupled" pool. J Lipid Res 37:1198–1206

Yakubchyk Y, et al (2005) Regulation of neurite outgrowth in N1E-115 cells through PDZ-mediated recruitment of diacylglycerol kinase zeta. Mol Cell Biol 25:7289–7302

Yamauchi Y, Hayashi M, Abe-Dohmae S, Yokoyama S (2003) ApoA-I activates PKCalpha signaling to phosphorylate and stabilize ABCA1 for the HDL assembly. J Biol Chem 278:47890–47897

Yu L, Cao G, Repa J, Stangl H (2004) Sterol regulation of scavenger receptor class B type I in macrophages. J Lipid Res 45:889–899

Yvan-Charvet L, et al (2007a) In vivo evidence for a role of adipose tissue SR-BI in the nutritional and hormonal regulation of adiposity and cholesterol homeostasis. Arterioscler Thromb Vasc Biol 27:1340–1345

Yvan-Charvet L, et al (2007b) Combined deficiency of ABCA1 and ABCG1 promotes foam cell accumulation and accelerates atherosclerosis in mice. J Clin Invest 117:3900–3908

Zannis VI, Kan HY, Kritis A, Zanni EE, Kardassis D (2001) Transcriptional regulatory mechanisms of the human apolipoprotein genes *in vitro* and *in vivo*. Curr Opin Lipidol 12:181–207

Zeng L, et al (2004) Sterol-responsive element-binding protein (SREBP) 2 down-regulates ATP-binding cassette transporter A1 in vascular endothelial cells: a novel role of SREBP in regulating cholesterol metabolism. J Biol Chem 279:48801–48807

Zerial M, McBride H (2001) Rab proteins as membrane organizers. Nat Rev Mol Cell Biol 2:107–117

Zha X, Genest J Jr, McPherson R (2001) Endocytosis is enhanced in Tangier fibroblasts: possible role of ATP-binding cassette protein A1 in endosomal vesicular transport. J Biol Chem 276:39476–39483

Zha X, Gauthier A, Genest J, McPherson R (2003) Secretory vesicular transport from the Golgi is altered during ATP-binding cassette protein A1 (ABCA1)-mediated cholesterol efflux. J Biol Chem 278:10002–10005

Zhang M, et al (2002) MLN64 mediates mobilization of lysosomal cholesterol to steroidogenic mitochondria. J Biol Chem 277:33300–33310

Zhang W, et al (2003) Inactivation of macrophage scavenger receptor class B type I promotes atherosclerotic lesion development in apolipoprotein E-deficient mice. Circulation 108:2258–2263

Zhang Y, Da Silva JR, Reilly M, Billheimer JT, Rothblat GH, Rader DJ (2005) Hepatic expression of scavenger receptor class B type I (SR-BI) is a positive regulator of macrophage reverse cholesterol transport in vivo. J Clin Invest 115:2870–2874

Zhao B, Natarajan R, Ghosh S (2005) Human liver cholesteryl ester hydrolase: cloning, molecular characterization, and role in cellular cholesterol homeostasis. Physiol Genomics 23:304–310

Zhao B, Song J, Chow WN, St Clair RW, Rudel LL, Ghosh S (2007a) Macrophage-specific transgenic expression of cholesteryl ester hydrolase significantly reduces atherosclerosis and lesion necrosis in Ldlr mice. J Clin Invest 117:2983–2992

Zhao B, Song J, St Clair R, Ghosh S (2007b) Stable over-expression of human macrophage cholesteryl ester hydrolase (CEH) results in enhanced free cholesterol efflux from human THP1-macrophages. Am J Physiol Cell Physiol 292:C405–C412

Zhou H, et al (2006) Adipose tissue-specific CETP expression in mice: impact on plasma lipoprotein metabolism. J Lipid Res 47:2011–2019

Zhu HL, Atkinson D (2007) Conformation and lipid binding of a C-terminal (198-243) peptide of human apolipoprotein A-I. Biochemistry 46:1624–1634

Chapter 4
Bile Acids and Their Role in Cholesterol Homeostasis

Nora Bijl, Astrid van der Velde, and Albert K. Groen

Abstract Bile acids are synthesized from cholesterol and have long been thought to be just a degradation product with an additional function in food digestion. During the past decade many new functions of bile acids emerged and, instead of functioning at the interphase of the outside world and the body, bile acids turned out to be extremely important signal transduction molecules which play an important role in balancing flux through diverse metabolic pathways. In this chapter we focus on the function of bile acids in regulation of cholesterol homeostasis at both the cellular and organismal level.

4.1 Introduction

The word cholesterol has a negative connotation due to its association with cardiovascular disease. However, cholesterol in itself is a molecule of undisputed biological importance and has a variety of functions in higher eukaryotes. Cholesterol is first of all an essential structural component of cell membranes. Owing to its bipolar structure, it is located inside the lipid bilayer, generates a semi-permeable barrier between cellular compartments and regulates membrane fluidity. As a component of HDL and LDL, it travels through the blood as part of a system that regulates the distribution of cholesterol in various tissues. Cholesterol is also a precursor for steroid hormones and, although this process is quantitatively not very important, it has a major physiological impact. The six different steroid hormones in humans function as lipophylic signalling molecules during metabolism, growth and reproduction. The human body is capable of producing the daily need of cholesterol and therefore does not need cholesterol from food as an additional source. However, it does take up cholesterol from food highly efficiently and, when the food supply of cholesterol is high, the excess cholesterol can be stored for short-term buffering as

N. Bijl, A.V.D. Velde, and A.K. Groen
Department of Medical Biochemistry (Room K1-106), Academic Medical Centre,
University of Amsterdam, Meibergdreef 9, 1105 AZ, Amsterdam, The Netherlands
e-mail: a.k.groen@amc.uva.nl

C. Ehnholm (ed.), *Cellular Lipid Metabolism*,
DOI 10.1007/978-3-642-00300-4_4,

cholesterol esters in the liver. Cholesterol in itself cannot be degraded but high levels of cholesterol provide a negative feedback signal that stops de novo synthesis, preventing cholesterol overload (Engelking et al. 2005). The biosynthesis of bile represents the prime pathway of cholesterol catabolism. Approximately 90% of the cholesterol that is taken up by food or that is produced de novo is eventually converted into bile acids. In this manner superfluous cholesterol can be eliminated from the body. Cholesterol can also be directly excreted via a pathway involving direct transintestinal excretion (Kruit et al. 2005; van der Velde et al. 2007) although the contribution of this pathway differs strongly between different species.

Under normal conditions, depending on the *species*, about 50% of cholesterol is absorbed (Bhattacharyya and Eggen 1980; Crouse and Grundy 1978; Wang et al. 2001). Bile acids play a role in cholesterol absorption by emulsifying lipids and allowing them to travel from the aqueous luminal milieu to the brush border membrane of enterocytes. Here, cholesterol is taken up by specific receptors as discussed in this chapter. There appears to be a relationship between circulating levels of bile acids and cholesterol (Bays and Goldberg 2007). This association is now the subject of extensive research and it is clear that bile acids do not only serve as physiological detergents in the intestine. By acting as a nuclear hormone receptor activator, they regulate the expression of important genes in homeostasis of lipid, glucose and cholesterol as well as their own synthesis (Scotti et al. 2007; Thomas et al. 2008; Zimber and Gespach 2008). Furthermore, bile acids have been described to regulate energy homeostasis at least in mice (Houten et al. 2006).

A vision emerges of a complex interplay between bile acids and cholesterol, whereby bile acids control cholesterol homeostasis by regulation of synthesis, catabolism and uptake, and where the supply of cholesterol is needed for de novo bile acid synthesis. This interplay involves both physical interactions and control at the level of gene transcription. In this chapter we discuss the homeostasis of bile acids and cholesterol, how these two are related and how they influence each other.

4.2 Bile Acid Synthesis

Bile is formed by the liver and consists of bile acids, cholesterol, phospholipids and waste products. After synthesis, bile acids are transported across the canalicular membrane into bile canaliculi and, in species such as mice and humans, are stored in the gallbladder (Hofmann 1990; Hofmann and Hagey 2008). When food is ingested they stream into the small intestine, where they help with the digestion. At the end of the ileum, they are re-absorbed by active transport and returned to the liver. This enterohepatic cycling of bile acids is highly efficient and can take place two to three times during a meal. However, about 5% escapes re-absorption in the intestine and is lost in the faeces (Hofmann 1990; Hofmann and Hagey 2008). This loss is compensated for by neosynthesis from cholesterol in the liver. The biosynthesis involves a variety of enzymes in the endoplasmic reticulum, mitochondria, cytosol and peroxisomes.

During the process of bile acid synthesis, the conformation of the cholesterol molecule changes from *trans* to *cis*. As a consequence, all hydrophilic groups move to one side of the molecule, making it strongly amphiphatic and providing it with the ability to form micelles (for reviews, see Hofmann 1999; Hofmann and Hagey 2008).

The most abundant human bile acids are the primary bile acids cholic acid (CA) and chenodeoxycholic acid (CDCA). After synthesis the amino acids glycine or taurine can be added to carbon 24 via an amide bond; and formally we should than speak in terms of bile salts. This final conjugation takes place before secretion into bile and provides the bile with stronger detergent properties to facilitate lipid and vitamin absorption in the intestine. Mice have a different bile acid pool with mostly muricholic acid conjugated to taurine. These bile acids can then be modified by the intestinal flora, giving rise to secondary bile acids, lithocholic acid and deoxycholic acid. These secondary bile acids are highly toxic, especially lithocholic acid (Hofmann and Hagey 2008; Russell 2003).

According to the current concept, two different routes exist for bile acid synthesis: the "classic or neutral" pathway and the "alternative or acidic" pathway. The classic route is also called the neutral route because the intermediates in this pathway are neutral up to the last steps of the synthesis route. This pathway consists of various steps in which first the sterol nucleus is hydroxylated by the microsomal enzyme cholesterol 7a-hydroxylase (CYP7A1). This is also the step that is highly regulated, as discussed in the next paragraph. Other modifications of the sterol nucleus include saturation of the double bond, epimerization of the 3β-hydroxyl group and hydroxylations. This is followed by shortening of the side-chain to 3 C-atoms and, finally, carboxylation of the last C-atom of the side-chain (for an extensive description of BA biosynthesis, see the reviews by Chiang 2004; Russell 2003). The alternative pathway starts with side-chain modification by the enzyme sterol 27 hydroxylase (CYP27A1). In subsequent steps in the pathway the steroid ring structure of the formed oxysterols are modified by 7α-hydroxylation. This step is not catalysed by CYP7A1, but by CYP7B1. The enzyme CYP7B1 is structurally similar to CYP7A1, but has different and broader substrate specificity (Norlin and Wikvall 2007). The relative contribution of these pathways to total bile acid synthesis varies between species and with various physiological and pathological conditions. The neutral pathway is considered quantitatively the most important because its contribution to total bile salt synthesis is ~90% in humans and ~75% in mice (Chiang 2004; Russell 2003). Both CA and CDCA are formed by this pathway in roughly equal amounts in humans.

4.2.1 Regulation of Synthesis by Nuclear Receptors

Maintaining a balance between bile acid synthesis, secretion and intestinal re-absorption is vital since every aspect of their homeostasis is linked to various important physiological processes. Under normal conditions, accumulation of bile acids in hepatocytes is avoided through a tight control of uptake, synthesis and secretion; and this control is organized by a series of feedback and feed-forward autoregulatory processes. These mechanisms involve the participation of a series of nuclear receptors which function as ligand inducible transcription factors.

4.2.2 Oxysterol Feed-Forward Regulation of Bile Synthesis

In mammals, cholesterol homeostasis is maintained by the control of uptake, de novo synthesis, storage as cholesterol esters and catabolism. When the cholesterol supply from food is high, a feed-forward pathway is activated that leads to the catabolic elimination of cholesterol as bile acids. The rate of bile acid synthesis parallels the activity of CYP7A1, which is the probably the main rate-controlling enzyme of the bile acid biosynthetic pathway (Russell 2003). Activation of CYP7A1 expression is mediated by LXRα, a nuclear receptor that binds oxysterols formed during the de novo synthesis of cholesterol (Chen et al. 2007). LXR belongs to the family of nuclear receptors; it is expressed in liver, spleen, adipose tissue, lung and pituitary and requires heterodimerization with RXR to become functionally active (Goodwin et al. 2008). The generation of LXR knockout mice provided insight in the action of LXR. LXR knockout mice accumulate large amounts of cholesterol in the liver on a cholesterol-rich diet, but cannot respond to this with up-regulation of CYP7A1 (Peet et al. 1998a, b). Furthermore, these mice have changes in various genes such as SREBP1 and 2, HMGCR, HMGCS and SCD, suggesting a wide range of functions for this nuclear receptor (Peet et al. 1998b; Quinet et al. 2006). LXR is highly conserved between humans and rodents. However, LXR has much less effect on hamster and human CYP7A1, which lack a LXR binding motif. It therefore seems that rat and mouse are unique in the ability to convert excess cholesterol to bile acids by activation of LXR and subsequent stimulation of CYP7A1 (Peet et al. 1998a).

4.2.3 Bile Acid Feedback Regulation of Bile Synthesis

Initial studies showed that interruption of the enterohepatic circulation of bile acids by biliary diversion or treatment with bile acid binding resins increases the rate of bile acid synthesis and the activity of CYP7A1 by about 3- to 4-fold (Dueland et al. 1991; Gustafsson 1978). Conversely, expansion of the bile acid pool by intraduodenal infusion of bile acids suppresses CYP7A1 expression and reduces the rate of bile acid synthesis (Nagano et al. 2004). No direct bile acid binding site was detected in the promoter of CYP7A1 and the mechanism behind the feedback regulation was unknown for a long time. This changed in 1999, when it was discovered that bile acids are the endogenous ligands for FXR (NR1H4; Makishima et al. 1999). It was known by that time that oxysterols positively induce bile synthesis via LXR and it was postulated that the feedback mechanism also involves the activation of a nuclear receptor. The nuclear receptor FXR was a candidate because: (i) it is specifically expressed in tissues where bile acids function (such as the liver, intestine, kidney),

(ii) it is evolutionarily related to LXR and (iii) it also functions as a heterodimer with the retinoid X receptor (RXR). FXR belongs to a family of transcription factors [the nuclear receptor (NR) superfamily] that is involved in diverse physiological functions such as reproduction, development and metabolism (Kuipers et al. 2007; Rader 2007). Various studies have now shown that FXR regulates a network of genes involved in synthesis, metabolism and transport of bile acids.

The suppression of CYP7A1 promoter activity through the activated FXR-RXR complex is mediated by an indirect mechanism involving interaction with other transcription proteins. Binding of bile acids to the FXR-RXR complex induces the transcription of SHP (small heterodimer partner). SHP is a receptor that binds to and inhibits a third receptor, the liver receptor homologue 1 (LRH-1 or NR5A2). LRH-1 is an orphan receptor that positively regulates CYP7A1 by binding to BARE-II in the CYP7A1 promoter (del Castillo-Olivares and Gil 2000; Lee and Moore 2002). The interaction between SHP and LRH-1 blocks the transcriptional activity of LRH-1; and CYP7A1 expression is stopped leading to a drastic decrease in bile acid output (Goodwin et al. 2008). Studies in FXR knockout mice show that indeed these mice no longer react to bile acids by down-regulation of CYP7A1 (Kuipers et al. 2007; Lambert et al. 2003).

Bile acids also regulate the expression of other genes via FXR. Cyp8b1, the enzyme that controls the ratio in which the primary bile acid species cholate and chenodeoxycholate are formed, seems to be under the same negative feedback control as CYP7A1 (Sinal et al. 2000; not reproduced by Kok et al. 2003; Box 4.1).

4.2.4 FGF-Regulated Feedback of Bile Synthesis

Although most studies initially focused at the liver to unravel the mechanism underlying regulation of bile acid synthesis, it was clear that this could not be the only organ involved and another pathway must exist. This idea originated from the observation

Box 4.1

FXR is called the bile acid sensor and plays a major role in regulation of bile acid homeostasis. After binding to DNA as a heterodimer with the retinoid X receptor, FXR controls the synthesis, conjugation, secretion, detoxification, excretion and uptake of bile acids. In the liver, FXR controls bile acid biosynthesis (CYP7A1; sterol 12a-hydroxylase, CYP8B1), simusoidal uptake (NTCP), and canalicular secretion (BSEP). In the intestine, FXR controls almost all genes involved in bile acid detoxification. In the enterocyte, FXR controls bile acid absorption (ASBT), intracellular trafficking (IBABP) and basolateral efflux (OSTa, OSTb). The development of specific FXR agonists (GW4064, fexaramine, AGN34, 6a-ethyl-chenodeoxycholicacid) and the generation of FXR −/− mice have provided powerful tools to study the pathways that are in part controlled by FXR (Hubbert et al. 2007).

that, in rodents, blocking the flow of bile acids into the intestine by bile duct ligation increased the expression of CYP7A1 and activity in the liver (Dueland et al. 1991; Gustafsson 1978). Hepatic concentrations of bile acids increase under these conditions but the expected down-regulation of CYP7A1 does not occur. Furthermore, studies in rat showed that the intraduodenal administration of taurocholic acid inhibited CYP7A1 expression in the liver, whereas direct intravenous or portal administration did not (Nagano et al. 2004; Pandak et al. 1991). It was therefore suggested as early as 1991 that the intestine must be involved by secreting a factor in response to bile acids, which either changes bile acid composition or in itself signals back to the liver. It is now clear that this factor exists and that it is a member of the fibroblast growth factor family, namely FGF19, or the mouse orthologue FGF15 (Box 4.2).

The first evidence that FGF19 was involved came from studies that aimed to find FXR target genes. Treatment of human hepatocytes with the FXR agonist GW4064 strongly up-regulated the mRNA expression of FGF19 (Inagaki et al. 2005). In the search for a ligand, FGF19 was reported to bind to the FGFR4 in vitro (Xie et al. 1999). The expression of this receptor is found mainly in liver, in large hepatocytes adjacent to the central vein and in smaller hepatocytes throughout the liver; and it is the sole FGFR expressed significantly in mature liver hepatocytes. FGFR4 null mice exhibit depleted gallbladders, an elevated bile acid pool, reduced activity of

Box 4.2

FGFs constitute a large family of growth factors that influence a wide variety of biological processes, such as angiogenesis, embryogenesis, differentiation (Galzie et al. 1997; Goldfarb 1996). FGFs induce their biological effects by binding to and activating FGFRs (for a review, see Schlessinger et al. 2000). This occurs by dimerization of the *trans*-membrane receptors upon binding of the FGF, followed by autophosphorylation of a number of tyrosine residues and recruitment of downstream effectors, such as the FGF receptor substrate protein 2 alpha (FRS2a). The FRS2a is a membrane-linked docking protein that contains myristyl anchors, phosphotyrosine binding domains (PTB) and multiple tyrosine phosphorylation sites at its C-terminus. These are docking sites for Grb2 and Shp2 linking the FGFR with the Ras and MAPK cascade (Wiedlocha and Sorensen 2004). Within the family of human FGFs, there are seven phylogenetic subfamilies based on amino acid sequence identities (Bottcher and Niehrs 2005; Wiedlocha and Sorensen 2004). In general, FGFs within the same family tend to share functional activity, but this does not hold true for the subfamily consisting of FGF19, FGF21 and FGF23. The core sequences of these three FGFs are very diverse and so are their functions. The mouse orthologue of human FGF19 is FGF15; and although they share only 53% total amino acid identity, their role in regulation of bile synthesis is very similar.

JNK and elevated excretion of bile acids, due to elevated levels of Cyp7a1 (Yu et al. 2000). Conversely, transgenic mice expressing a constitutively active FGFR4 have increased JNK activity, decreased CYP7A1 expression and a reduced bile acid pool size (Yu et al. 2005).

The data from FGFR4 null mice, together with the observation that FGF19 is a target gene for FXR and that it specifically binds to FGFR4 in vitro, prompted researchers to test the ability of FGF19 to down-regulate CYP7A1 expression. Indeed it was found to decrease transcription of CYP7A1 in a dose-dependent manner in human hepatocytes. However, no expression was found in human or mouse liver (Inagaki et al. 2005). The solution came from studies in mice, where it was shown that FGF15 is predominantly expressed in the small intestine following administration of GW4064 or cholic acid. FGF15 mRNA is highly expressed in the ileum and only at low levels in other enterohepatic organs. FGFR4 shows the reverse expression pattern, with high expression in the liver and with little or no expression in the intestine (Inagaki et al. 2005).

It is now clear that FGF15/19 is predominantly responsible for the feedback inhibition triggered by bile acids in the intestine. In the intestine, bile acids activate the nuclear transcription factor FXR, which in turn activates the transcription of FGF15/19. This results in the secretion of FGF15/19 to either the lymph or blood from where it reaches the liver to provide a feedback signal for bile synthesis (Kim et al. 2007; Fig. 4.1).

Activation of FGFR4 by FGF in vivo is dependent on the presence of beta-klotho. The function of this protein was initially unknown and beta-klotho null mice appeared normal upon examination (Ito et al. 2005). However, they had pronounced alterations in bile acid metabolism similar to that observed in FGFR4 null mice or FGF15 null mice, including increased expression of CYP7A1 and bile excretion in faeces (Ito et al. 2005). Beta-klotho was found to be expressed in liver, pancreas and fat (Ito et al. 2000). Beta-klotho is a membrane protein that contains two regions in the extracellular domain with homology to those in family 1 glycosidases, which hydrolyse glycosidic bonds (Ito et al. 2002). The function of beta-klotho remains elusive, although it has been suggested that it regulates the concentration of cofactors for FGFR4 (glycosaminoglycans). The reliance on beta-klotho adds another level of selectivity to the signalling capacity of FGF15/19.

4.2.5 Other Pathways

The nuclear receptor regulation of bile acid synthesis involving FXR, SHP, LXR and LRH-1 explains a major part of the regulatory responses in bile acid biosynthesis. However, experiments in SHP knockout mice have indicated the existence of SHP-independent mechanisms for the suppression of CYP7A1. SHP knockout mice increase the synthesis and accumulation of bile acids and produce more cholic acid compared to their controls (Wang et al. 2003). These increases are caused as expected by a decreased down-regulation of CYP7A1 gene expression (Boulias et al. 2005). However, this effect is not as dramatic as in FXR knockout mice, suggesting

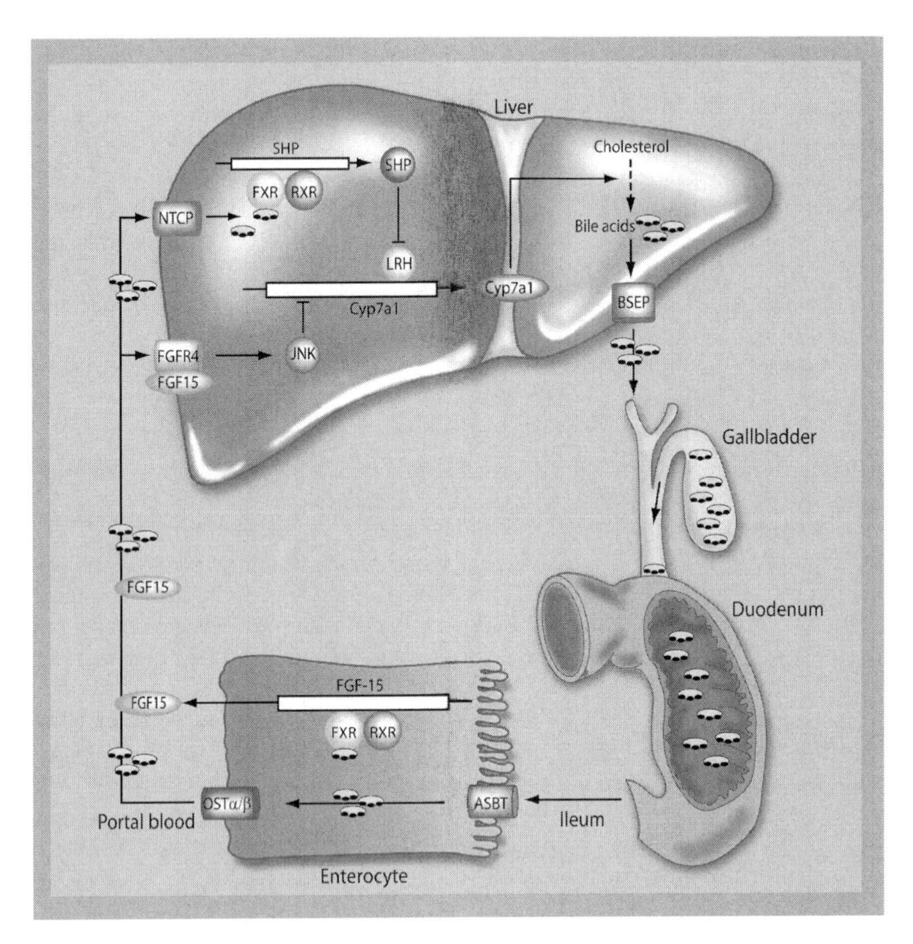

Fig. 4.1 Feedback regulation of bile synthesis is mediated by pathways involving both FGF15 and bile acids. Cholesterol in the liver is converted to bile acids through 7a-hydroxylation by the enzyme *Cyp7a1*. The produced bile acids are secreted via *BSEP* and stored in the gall bladder. After ingestion of a meal, they flow into the duodenum where they help with food digestion. In the distal ileum they are reabsorbed via *ASBT*. In the enterocyte, the bile acids activate *FXR* which leads to induction of the *FGF-15* gene. Both bile acids and FGF-15 are secreted to either the lymph or blood from where they reach the liver. FGF-15 binds the *FGFR4* and this result in suppression of Cyp7a1 expression through a *JNK*-dependent signaling cascade. Bile acids are taken up from the portal blood via *NTCP*. Activation of FXR in the liver by bile acids leads to induction of the *SHP* gene, which can directly suppress the *Cyp7a1* gene through an interaction with LRH-1

the existence of SHP independent mechanisms for the down-regulation of CYP7A1 (Watanabe et al. 2004). SHP knockout mice fail to repress CYP7A1 in response to the FXR agonist GW4064 as expected (Inagaki et al. 2005; Wang et al. 2003). Remarkably, these mice remain responsive to bile acid feeding by down-regulation of CYP7A1 (Wang et al. 2003). This pathway may be secondary to liver injury from high bile acid levels. Bile acids are known to induce inflammatory cytokines,

TNFα and IL-1β, in hepatic macrophages (Kupffer cells; Li et al. 2008). These cytokines may cross the sinusoid and inhibit CYP7A1 gene expression in hepatocytes via activation of JNK. Bile acids can also signal via JNK by the protein kinase C pathway. This may involve phosphorylation and inactivation of transcription factors such as HNF4α, which is a crucial factor stimulating the expression of CYP7a1 (Li et al. 2008). In addition, the pregnane X receptor (PXR) and the vitamin D receptor (VDR; Jiang et al. 2006) may suppress CYP7A1 by SHP-independent mechanisms (Chiang 2004).

4.3 Regulation of the Enterohepatic Circulation

During their function as emulsifiers of dietary lipids and fat-soluble vitamins, bile acids cycle through several organs that are part of the enterohepatic circulation, including the liver, the bile ducts, the gallbladder and the intestine; and they return to the liver via the portal vein. During this journey they have to pass different organs; and to do this they need to cross cell membranes various times. The majority of bile acids are conjugated to taurine or glycine amino acids and exist in the form of membrane impermeable anions. Specialized transporters located in the membranes of organs in the enterohepatic cycle allow the proper transport of bile acids (for reviews, see Meier and Stieger 2000; Oude Elferink et al. 2006). These transporters influence bile acid concentrations in different compartments of the intestinal tract and it is not surprising that bile acids have a regulatory role in the expression of these transporters. The following paragraphs discuss the various bile acid transporters and how they are regulated.

4.3.1 Liver

Bile acids returning from the intestine via portal blood are taken up by the hepatocytes, transported through the hepatocyte and re-secreted at the other side to continue cycling between the liver and the intestine. The large pool of bile acids that are fluxed in this way through the hepatocytes with a high transport rate provide the main force for bile flow. This flux through the hepatocytes occurs against a steep concentration gradient and therefore requires distinct active transport systems expressed in a polarized fashion. Large pores (fenestrae) allow bile acids to enter the space of Disse. The process of bile acid extraction is efficient: 75–90% from the first pass of portal blood (for reviews, see Meier and Stieger 2000; Oude Elferink et al. 2006). Sodium-dependent and sodium-independent transport pathways have been identified to play a key role in hepatic uptake of bile acids from sinusoidal blood. The sodium-dependent process is represented by the sodium taurocholate cotransporter polypeptide (Ntcp, Slc10a1; for a review, see Hagenbuch and Dawson 2004), the substrate specificity of which is essentially limited to conjugated bile acids and certain sulfated steroids. NTCP accounts for more than 80% of conjugated (i.e., taurocholate and glycocholate) but less than 50% of unconjugated (i.e., cholate) bile acid uptake. In contrast, the sodium-independent pathway is represented by different members of the superfamily of organic anion-transporting polypeptides (OATP/SLCO; for a review, see Hagenbuch and Meier 2004). In human liver, the highest expressions are found for OATP1B1 (SLCO1B1) and its 80% sequence homologue OATP1B3 (SLCO1B3), both of which are predominantly if not exclusively expressed in the liver (Stieger et al. 2007).

The expression of NTCP and OATP1B1 (SLC2A1) is repressed by high levels of bile acids, as can be observed in cholestasis. Under these conditions, FXR induces SHP; and this blocks the stimulating effect of retinoic acid receptor and RXR heterodimer on the NTCP promoter (in rats; Denson et al. 2001). Similarly, activation of SHP leads to repression of hepatocyte nuclear factor 1, which is the major transcriptional activator of OATP1B1. In mice this regulatory mechanism is different or at least another pathway exists, since the repression in gene expression under high bile acid conditions still takes place in SHP –/– mice. It is of interest to note that cholestasis leads to activation of OATP1B3. It is speculated that the up-regulation of this transporter might constitute an escape mechanism promoting the hepatocellular clearance of xenobiotics during cholestasis.

After uptake of bile acids at the basolateral membrane, they are transported to the canalicular membrane. The movement is still not very well understood and might be mediated by vesicle transport or by transport proteins. Evidence for the latter comes from studies demonstrating rapid cytosolic diffusion of fluorescent derivates of bile acid before their canalicular secretion (Bahar and Stolz 1999). At the canalicular membrane the bile acids are effluxed against a steep concentration gradient into the bile. Whereas the influx at the basolateral membrane is primarily driven by a net influx of Na^+, the efflux ability of the canalicular pump depends on the availability of ATP. Canalicular transport is mediated by the bile acid export pump (BSEP or ABCB11). BSEP has a broad specificity for bile acids and pumps conjugated bile acids such as taurine- or glycine-conjugated cholate, chenodeoxycholate and deoxycholate into bile (Meier and Stieger 2000; Stieger et al. 2007). A minor bile acid export pump is the ABC transporter MRP2 which mediates the export of bilirubin conjugates and a wide variety of organic substrates, such as glutathione, glucuronide and sulfate-conjugated drugs. MRP2 also mediates the transport of divalent bile acids such as sulfated tauro- and glycolithocholate. MRP2 is the main driving force for bile acid-independent bile flow through canalicular excretion of reduced glutathione (Meier and Stieger 2000; Oude Elferink et al. 2006). MDR3 (ABCB4) was shown to function as an ATP-dependent phospholipid flippase, translocating phosphatidylcholine from the inner to the outer leaflet of the canalicular membrane (Oude Elferink et al. 2006). Canalicular phospholipids are then solubilized by canalicular bile acids to form mixed micelles, thereby protecting cholangiocytes from the detergent properties of bile acids. The expression of BSEP, MRP2 and MDR3 is regulated by FXR and leads to an increase in bile efflux when intracellular bile acid levels rise (Kuipers et al. 2007).

4.3.2 Intestine

Apical sodium dependent bile acid transporter (ASBT) is the ileal counterpart of the hepatic NTCP. It efficiently transports conjugated and unconjugated bile acids with a preference for the taurine and glycine conjugates over the unconjugated form (Dawson and Oelkers 1995). The essential role of ASBT in intestinal bile acid absorption is evident from studies in ASBT null mice that show intestinal bile acid malabsorption and consequent disturbance of the enterohepatic circulation. Recent studies showed that ASBT is under control of 25-hydroxycholesterol and the presence of this sterol inhibits ASBT activity in human intestinal epithelial cells (Alrefai et al. 2005). The negative feedback is mediated by the induction of SHP via FXR activation. SHP in its turn represses LRH-1-dependent gene activation.

After uptake by the enterocyte, bile acids are bound to the intestinal bile acid binding protein (IBABP). IBABP is a small soluble protein, expressed exclusively in the terminal ileum. I-BABP gene is positively expressed by bile acids via FXR activation (Nakahara et al. 2005). The concerted action of FXR activation (positive regulation of bile binding and negative regulation of uptake) helps to reduce the cytotoxic effects of bile acids in the ileum.

Although the proteins responsible for uptake and intracellular transport of bile acids have been known for many years, the mechanism of transport across the basolateral membrane was unknown until recently. Two proteins have been identified that function as a heterodimer to transport bile acids: Ostα and Ostβ (Dawson et al. 2005). The regulation of expression seems to be under positive control of bile acids via activation of FXR (Dawson et al. 2005). The discovery of Osta/b also further supported the central role for FGF15 signalling in control of bile synthesis. Osta/b null mice show increased expression of FGF15 and down-regulation of CYP7A1 in the liver (Dawson et al. 2008). This is likely due to an increase in enterocyte bile acid concentration. Conversely, mice lacking the bile acid uptake transporter ASBT show decreased FGF expression and an increase in bile acid synthesis (Jung et al. 2007).

4.4 Cholesterol in the Enterohepatic Circulation

Body cholesterol derives from two sources, i.e., de novo biosynthesis and diet. Since, in humans, total body sterol output almost always exceeds dietary intake, continuous synthesis is essential. The liver is the predominant organ for cholesterol synthesis but the intestine also plays an important role, particularly in rodents (Dietschy and Turley 2002). Cholesterol is synthesized from two-carbon acetyl-CoA moieties. The rate-limiting enzyme in the synthetic pathway is HMG-CoA reductase, a highly regulated enzyme that catalyses the conversion of HMG-CoA into mevalonate. Cholesterol itself regulates feedback inhibition of HMG-CoA reductase activity, as accumulation of (oxy)sterols in the endoplasmic reticulum (ER) membrane triggers HMG-CoA reductase to bind to Insig proteins, which leads to ubiquitination and degradation of HMG-CoA reductase (Gong et al.

2006). In addition, cholesterol regulates the gene expression of HMG-CoA reductase indirectly by blocking the activation of the transcription factor, sterol regulatory element-binding protein 2 (SREBP2). Under low-cholesterol conditions, SREBP2 in the ER is escorted by the SREBP cleavage activating protein (SCAP) to the Golgi. In the Golgi, SREBP2 is cleaved to generate its transcriptionally active form, which activates transcription of the HMG-CoA reductase encoding gene (Brown and Goldstein 1999). Conversely, when sterols accumulate in the ER membrane, cholesterol binding to the sterol-sensing domain of SCAP causes a conformation change, which induces binding of SCAP to the ER anchor protein Insig, preventing the exit of SCAP-SREBP2 complexes to the Golgi and thereby preventing activation of SREBP2. This effect is transduced by oxysterols which bind to Insigs (Gong et al. 2006; Radhakrishnan et al. 2007), causing Insigs to bind to SCAP. Mutational analysis of the six transmembrane helices of Insigs reveals that the third and fourth are important when Insigs are binding to oxysterols and Scap (Gong et al. 2006). The interaction of oxysterols with Insigs finally explains the long-known ability of oxysterols to inhibit cholesterol synthesis in animal cells.

4.4.1 Cholesterol Absorption in the Intestine

In an aqueous milieu the solubility of unesterified cholesterol is about 1 μM. Depending on the amount of cholesterol taken in via food, the concentration in the intestinal lumen is at least three orders of magnitude higher. Cholesterol in food stuffs is mostly present in an oily phase, whereas cholesterol coming into the intestine via the bile is present either as a mixed micelle together with bile acid and phospholipids or as a vesicle with phospholipids only (Hernell et al. 1990). Until relatively recently, cholesterol was generally assumed to be absorbed via passive diffusion. In model systems the sterol can flip-flop rapidly through lipid bilayers, so in principle no proteins seemed necessary to assist in cholesterol absorption. However, this situation changed drastically with the discovery of the cholesterol absorption inhibitor ezetimibe (Rosenblum et al. 1998). Ezetimibe and analogues comprise a new class of sterol absorption inhibitors that reduce diet-induced hypercholesterolemia. Using a bioinfomatics approach, the NPC1L1 protein was identified as a putative cholesterol transporter in intestinal cells and target for ezetimibe (Altmann et al. 2004). Indeed the NPC1L1 knockout mice show a 69% reduction in cholesterol absorption which cannot be further reduced by ezetimibe treatment (Altmann et al. 2004). The NPC1L1 protein contains 13 putative transmembrane domains: the third to seventh transmembrane helices are thought to constitute a sterol sensing domain also present in NPC1, SCAP and HMG-CoA reductase (Alrefai et al. 2007; Altmann et al. 2004). Recent studies show that interaction of this domain with cholesterol induces a conformational change in the protein; this in turn induces endocytosis, taking a cholesterol-rich domain with it into the cell (Ge et al. 2008). Thus cholesterol regulates its own absorption in the intestine via a feed-forward mechanism. The mechanism does not account for down-regulation in the presence of a cholesterol overload. Perhaps an excess of cholesterol in the plasma membrane may hamper the formation of endocytotic vesicles. The form in which cholesterol is transported to the brush border membrane of the enterocytes has not been studied. It has always been assumed that micelization via bile acids is an essential step. In principle, with the discovery of the NPC1L1-mediated mechanism, uptake via diffusion could be possible. Yet, a number of studies have shown bile acids to be essential for cholesterol absorption. Almost no cholesterol is taken up in the absence of bile acids (Wang 2007; Wang and Lee 2008). Whether the bile acids are just necessary for donating the cholesterol to the membrane or whether they accelerate NPC1L1-mediated cholesterol uptake directly has not yet been studied.

In addition to NPC1L1 several other proteins have been suggested to play a role in cholesterol absorption. In in vitro experiments the scavenger receptors SR-B1 and CD 36 have been shown to mediate the uptake of unesterified cholesterol (Knopfel et al. 2007). Aminopeptidase N has also been suggested to be actively involved in cholesterol absorption (Kramer et al. 2000, 2005). Evidence for a substantial contribution of these three proteins to cholesterol uptake in vivo is, however, lacking.

4.4.2 Intestinal Cholesterol Secretion

In addition to protein mediated uptake also a mechanism for protein mediated sterol efflux has been identified. In 2000 Berge et al. reported that mutations in the ABC transporter heterodimer ABCG5/ABCG8 were responsible for the excessive accumulation of plant sterols in patients with the disease sitosterolemia (Berge et al. 2000). Subsequently, the same group showed these proteins to be present also on the canalicular membrane of hepatocytes where they mediate transport of cholesterol and plant sterols into the bile (Yu et al. 2002a). Overexpression of ABCG5/ABCG8 in transgenic mice leads to an increase in biliary cholesterol secretion and a reduced intestinal absorption of dietary cholesterol, providing strong evidence for ABCG5/ABCG8 being involved in hepatocellular secretion and intestinal efflux of cholesterol (Yu et al. 2002b). The identification of these proteins has been a major step forward in the elucidation of the mechanism of biliary lipid secretion. Until the discovery of ABCG5 and ABCG8, biliary cholesterol secretion was supposed to be a largely passive process driven by the transport of bile acids and phospholipids. Because of the above-mentioned rapid flip-flop of cholesterol across membranes, no transporter was deemed necessary. Small explained this enigma by assuming that the heterodimer Abcg5/Abcg8 is a "liftase" instead of a floppase (Small 2003). Assuming that cholesterol easily flops through the membrane, the proteins lift cholesterol out of the plane of the membrane so that it is more accessible, reducing the activation energy for uptake in bile acid/phospholipids mixed micelles. Recently it was shown that this hypothesis may not hold true. Abcg8 knockout mice showed a 50% decrease in the cholesterol content of the canalicular membrane, which is more compatible with floppase activity of the Abcg5/g8 heterodimer (Kosters et al. 2006). Yet, in vitro experiments with cells overexpressing the proteins also demonstrated an absolute requirement for the presence of bile acid micelles, suggesting that Abcg5/g8 donate cholesterol directly to bile acid micelles (Vrins et al. 2007). Taken together these results demonstrate restricted diffusion of cholesterol through the canalicular membrane. This may also explain why a protein such as NPC1L1 is required for cholesterol transport into the enterocyte. Restricted diffusion of cholesterol across the brush border membrane necessitates the presence of proteins to facilitate cholesterol import into the enterocyte.

The functional coupling of ABCG5/G8 activity to bile acids not only affects the kinetics of biliary cholesterol secretion but also influences plant sterol absorption in the intestine. It is generally assumed that efflux of plant sterols is a primary function of intestinal ABCG5/G8. In patients with defects in bile acid synthesis one might therefore expect to find accumulation of plant sterol in the body. This has indeed been observed. Whether intestinal ABCG5 and G8 also play an important role in cholesterol secretion is not yet clear. The group of Hobbs constructed mice overexpressing human ABCG5/G8 in liver and intestine (Yu et al. 2002b). These mice showed strongly increased faecal neutral sterol output. However, when these mice were crossbred with Abcb4 knockout mice that have both abrogated biliary phospholipid and cholesterol secretion, faecal sterol secretion normalized despite the overexpression of intestinal ABCG5/G8, indicating a minor role for the intestinal proteins in cholesterol homeostasis (Langheim et al. 2005). Experiments with tissue-specific knockout models are required to substantiate these findings.

4.4.3 Novel Pathways for Cholesterol Excretion

It is generally accepted that the only important route for cholesterol to leave the body is the above-described hepatobiliary excretion followed by intestinal passage into the faeces. Probably because the design of this pathway seems so logical and its dynamics have been investigated in many species there has been very little research on alternative pathways. Yet very early work has hinted at the existence of non-hepatobiliary pathways for cholesterol excretion. As early as 1927 Sperry concluded from studies with bile-diverted dogs that these animals continue to excrete cholesterol and he concluded that "that under some conditions the cholesterol of the faeces comes from neither food nor the bile which may be secretion through the intestinal wall, desquamated epithelium, or bacteria" (Sperry 1927). The data from Sperry's work have been largely ignored and it took almost 50 years before Pertsemlidis et al. (1973) confirmed the data of Sperry, also in studies with dogs. Likewise, faecal sterols of non-dietary origin are present in the faeces of patients with biliary obstruction (Cheng and Stanley 1959b) or rats with long-term bile diversion. A major drawback in such studies is the lack of biliary components in the enterohepatic cycle under these conditions. Particularly, the absence of bile acids compromises cholesterol absorption, and consequentlys affect intestinal cholesterol synthesis, as well as lipid absorption, with unknown side-effects. This is probably the reason that these and similar studies have gone largely unnoticed in the literature. With time, experimental set-ups improved and, in the early 1980s, Miettinen et al. (1981) investigated the origins of faecal neutral steroids in normal rats, using an isotopic balance method developed in their laboratory (Miettinen 1970; Miettinen et al. 1990) and the isotopic steady-state balance procedure (Chevallier 1967; Wilson 1964), they established that the specific activity of faecal cholesterol was consistently lower than that of plasma cholesterol or the faecal bile acids. An observation consistent with earlier reports (Chevallier 1960; Danielsson 1960; Peng et al. 1974), this result indicated that a considerable portion of the faecal neutral steroids was derived from cholesterol not in equilibrium with the rapidly exchangeable pool of body cholesterol. The study of Miettinen et al. (1981) showed that approximately 40–50% of faecal neutral sterols in rats fed a sterol-free diet arise from a source of non-exchanging cholesterol. They concluded that these sterols have at least two origins: fur-licking and sterols originating directly from the intestine. Under conditions that prevent fur-licking, either by acetone-washing of the animals or by physical restraint, the contribution of non-exchanging cholesterol to total faecal neutral sterol output was still approximately 33%. A similar phenomenon has been shown also to occur in humans. In 1967 Simmonds et al. performed elegant intestinal perfusion studies in humans and observed significant direct secretion of cholesterol in the small intestine. Until recently the origin of this cholesterol has remained an enigma. The development of transgenic mouse models that have abrogated biliary lipid secretion has made it possible to study the role of the intestine in cholesterol excretion, isolated from the biliary system. A case in point is the Abcb4 knockout mouse. In this model biliary phospholipid and cholesterol secretion is completely absent. Yet bile acid secretion and bile flow are not affected. Neutral sterol excretion is completely normal in these mice proving that an alternative pathway for cholesterol output is present or can be activated. Kruit et al. demonstrated that intravenously administered radiolabelled cholesterol finds its way to the faeces demonstrating a direct pathway for cholesterol from the blood compartment to the intestinal lumen (Kruit et al. 2005). Activation of LXR target genes with the agonist T0901317 stimulated the pathway indicating active transport. To obtain more insight in the underlying mechanism cholesterol secretion was studied in an isolated intestinal perfusion set-up. By cannulating intestinal segments in situ, keeping the blood supply intact, the intestinal lumen can be perfused with buffers to which cholesterol acceptors can be added. Measurement of cholesterol in the perfusate allows assessment of cholesterol transport across the intestinal wall. Surprisingly, cholesterol secretion was highest in the proximal 10 cm of mouse small intestine encompassing duodenum and the first part of jejunum (van der Velde et al. 2007). Output gradually decreased to reach very low levels in the colon (Fig. 4.2). The presence of bile acids and particularly phospholipid was necessary to induce the process. The molecular mechanism by which cholesterol is transported from blood through intestinal wall is still incompletely understood. Surprisingly, Abcg5/g8 seem not to be involved because the rate of transintestinal

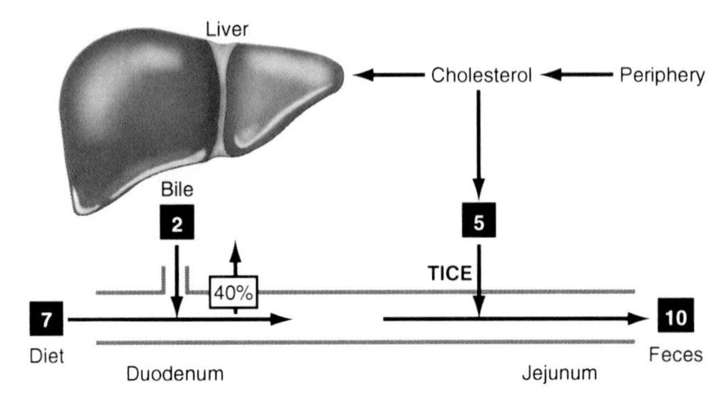

Fig. 4.2 A simplified scheme illustrating the different sources of cholesterol in the intestine. In wild-type mice that have free access to normal chow diet, the cholesterol input into the intestine from both diet and bile (per 100 g bodyweight) is approximately 9 μmol/day. Approximately 40% of this cholesterol is reabsorbed from the intestine and this should result in an output in the faeces of 4–5 μmol/day. Twice as much (10 μmol/day) is found, indicating that a significant amount (5 μmol/day) of the cholesterol output in faeces cannot be explained by input via diet and bile alone. Intestinal perfusion studies demonstrated that direct transintestinal cholesterol excretion (*TICE*) exists and that this output is highest in the proximal 10 cm of mouse small intestine. TICE gradually decreases to reach very low levels in the colon

cholesterol secretion (TICE) was unaltered in Abcg8 knockout mice (van der Velde et al. 2007). Also Srb1 does not mediate TICE, in contrast the rate of TICE was 2-fold higher in Srb1 knockout mice. In addition, feeding mice ezetimibe failed to influence TICE, indicating that also NPC1L1 plays no role in this process. Yet, in addition to bile acids and phospholipids TICE could be influenced via diet and an agonist of the nuclear receptor PPAR-δ (Vrins, unpublished data). High-fat diet doubled neutral sterol output in wildtype mice and also doubled the capacity of TICE. Interestingly, a diet high in cholesterol failed to exert a similar activity (van der Velde et al. 2008). The effects of high-fat diet could be mimicked by giving the mice GW742X, an agonist of the nuclear receptor PPAR-δ, suggesting that the effects are mediated through this nuclear receptor.

Which lipoprotein serves as donor of cholesterol for TICE is not yet clear. Literature data obtained with Abca1 null mice that have no HDL suggest that this lipoprotein may not be important because these mice have completely unaltered neutral sterol excretion. Triglyceride-rich lipoproteins could be involved. In a recent report, Brown et al. (2008) demonstrated increased TICE in mice in which the enzyme acylCoA; cholesterol acyltransferase-2 was knocked down by treating the mice with antisense oligonucleotide. Surprisingly, although cholesterol-esterifying capacity in the liver was almost gone, these mice responded not by increasing biliary cholesterol secretion but instead by enhancing output of triglyceride-rich VLDL. In elegant experiments, Brown et al, (2008) isolated radiolabelled particles in an isolated liver perfusion set-up of both wild-type and liver Acat-2 knockdown mice. Subsequently the particles were infused in a cross-over design in both wild-type and Acat-2 knockdown recipient mice and the fate of the radiolabelled cholesterol was determined. No change in biliary output was observed yet labelling of lumen and cells of the proximal 10 cm intestine increased when mice were infused with isolated pefusate of Acat-2 knockdown mice, suggesting that the substrate for TICE was increased in these samples (Brown et al. 2008). Clearly more work is required to isolate the active component. As indicated above, the heterodimer Abcg5/g8 does not contribute to TICE as measured in isolated intestinal perfusion studies. Additional factors may be required in the perfusate to induce Abcg5/g8 activity. Recent in vivo studies by van der Veen et al. determined the contribution of TICE in mice that were treated with the LXR agonist T0901317. This compound has been shown to increase neutral faecal sterol excretion 2- to 3-fold and also to directly stimulate faecal secretion of macrophage-derived cholesterol. Van der Veen et al. determined whole-body cholesterol fluxes by using elaborate stable isotope methodology. TICE strongly increased after treating wild-type mice with T0901317. In addition they demonstrated significantly decreased TICE in Abcg5 knockout mice. Taken together these results indicate a role for Abcg5/g8 in TICE in vivo. Use of similar methodology may be employed to quantify TICE in humans. The pathway does exist in man as well. Faecal sterols of non-dietary origin are present in the faeces of patients with biliary obstruction (Cheng and Stanley 1959a); and Simmonds et al. (1967) demonstrated direct intestinal cholesterol secretion in intestinal perfusion studies in man. However, the contribution of TICE to faecal sterol excretion is probably lower in humans than in mice. On average, humans secrete about 1 g/day of neutral sterols (Grundy 1983; Grundy and Ahrens 1969) Dietary cholesterol intake is about 400 mg (Samuel and McNamara 1983) and biliary cholesterol secretion amounts to 1000 mg (Hernell et al. 1990; Phillips 1960). Cholesterol absorption has been estimated to be about 50% (Grundy 1983; Grundy and Ahrens 1969; Miettinen 1970; Miettinen et al. 1990). Hence, the average TICE in humans can be estimated to be around 300 mg/day for 70 kg body weight, which is about one-third of the amount secreted into bile. The much higher biliary cholesterol secretion in man is probably the reason for the lower contribution of TICE to sterol excretion compared to mice.

4.5 Role of the Enterohepatic Cycle in the Control of Cholesterol Homeostasis

Secretion of bile is generally thought to be necessary for adequate digestion and handling of lipids in the food. Bile acids are the primary component; phospholipids are assumed to be added to prevent detergent bile acid action in the biliary tree. Since there seems to be no other pathway for cholesterol excretion from the body, the biliary route seems to be designed to accomplish this function. The identification of TICE forces re-evaluation of these paradigms. Particularly in mice, TICE is the predominant pathway for removing cholesterol from the body. In addition, the simple function of bile acids as emulsifiers of dietary lipids has been challenged. The past decade revealed multiple functions of bile acids in the fine transcriptional control of lipid metabolism and even energy homeostasis (Scotti et al. 2007; Thomas et al. 2008; Zimber and Gespach 2008). When bile acids only serve to help in digesting food, one would expect biliary secretion to strongly decrease during prolonged starvation. In mice the opposite has been observed. Bile formation increases in mice starved for up to 48 h (Kok et al. 2003; Scotti et al. 2007; Thomas et al. 2008; Zimber and Gespach 2008). This is mainly due to an increase in bile acid secretion; yet bile acid synthesis progressively decreases, indicating that the animal increases the rate of energetically costly enterohepatic cycling during prolonged starvation. Apparently, the enterohepatic cycle serves an important role in maintaining lipid homeostasis. It is not clear which biliary component is most important in this homeostatic mechanism. It will be interesting to carry out prolonged starvation in an animal model with a (partially) disrupted enterohepatic cycle.

4.6 Concluding Remarks

During the past two decades, progress in the field of bile acid and cholesterol research has been enormous. The role of both steroids in controlling intricate transcription networks has emerged. Particularly, bile acids have lost their role as relatively non-specific detergents, to become important connectors of metabolic pathways. The importance of both bile acid and cholesterol for lipid homeostasis in mammals is exemplified by the extremely complex networks involved in regulation of expression and activity of the key enzymes in the pathways, i.e., HMG-CoA reductase and 7-α-hydroxylase. The liver is the most important site at which the activity of these enzymes is regulated but intensive cross-talk between liver and intestine plays a major role.

References

Alrefai WA, Sarwar Z, Tyagi S, Saksena S, Dudeja PK, Gill RK (2005) Cholesterol modulates human intestinal sodium-dependent bile acid transporter. Am J Physiol Gastrointest Liver Physiol 288:G978–G985

Alrefai WA, Annaba F, Sarwar Z, Dwivedi A, Saksena S, Singla A, Dudeja PK, Gill RK (2007) Modulation of human Niemann–Pick C1-like 1 gene expression by sterol: role of sterol regulatory element binding protein 2. Am J Physiol Gastrointest Liver Physiol 292:G369–G376

Altmann SW, Davis HR Jr, Zhu LJ, Yao X, Hoos LM, Tetzloff G, Iyer SP, Maguire M, Golovko A, Zeng M, Wang L, Murgolo N, Graziano MP (2004) Niemann-Pick C1 Like 1 protein is critical for intestinal cholesterol absorption. Science 303:1201–1204

Bahar RJ, Stolz A (1999) Bile acid transport. Gastroenterol Clin North Am 28:27–58

Bays HE, Goldberg RB (2007) The 'forgotten' bile acid sequestrants: is now a good time to remember? Am J Ther 14:567–580

Berge KE, Tian H, Graf GA, Yu L, Grishin NV, Schultz J, Kwiterovich P, Shan B, Barnes R, Hobbs HH (2000) Accumulation of dietary cholesterol in sitosterolemia caused by mutations in adjacent ABC transporters. Science 290:1771–1775

Bhattacharyya AK, Eggen DA (1980) Cholesterol absorption and turnover in rhesus monkeys as measured by two methods. J Lipid Res 21:518–524

Bottcher RT, Niehrs C (2005) Fibroblast growth factor signaling during early vertebrate development. Endocr Rev 26:63–77

Boulias K, Katrakili N, Bamberg K, Underhill P, Greenfield A, Talianidis I (2005) Regulation of hepatic metabolic pathways by the orphan nuclear receptor SHP. EMBO J 24:2624–2633

Brown JM, Bell TA III, Alger HM, Sawyer JK, Smith TL, Kelley K, Shah R, Wilson MD, Davis MA, Lee RG, Graham MJ, Crooke RM, Rudel LL (2008) Targeted depletion of hepatic ACAT2-driven cholesterol esterification reveals a non-biliary route for fecal neutral sterol loss. J Biol Chem 283:10522–10534

Brown MS, Goldstein JL (1999) A proteolytic pathway that controls the cholesterol content of membranes, cells, and blood, Proc Natl Acad Sci USA 96:11041–11048

Chen W, Chen G, Head DL, Mangelsdorf DJ, Russell DW (2007) Enzymatic reduction of oxysterols impairs LXR signaling in cultured cells and the livers of mice. Cell Metab 5:73–79

Chevallier F (1960) Study of the origin of fecal sterols in the rat by means of radioactive indicators. 1. Demonstration of the secretion of sterols into the intestinal contents. Bull Soc Chim Biol 42:623–632

Chevallier F (1967) Dynamics of cholesterol in rats, studied by the isotopic equilibrium method. Adv Lipid Res 5:209–239

Chiang JY (2004) Regulation of bile acid synthesis: pathways, nuclear receptors, and mechanisms. J Hepatol 40:539–551

Crouse JR, Grundy SM (1978) Evaluation of a continuous isotope feeding method for measurement of cholesterol absorption in man. J Lipid Res 19:967–971

Danielsson H (1960) On the origin of the neural fecal sterols and their relation to cholesterol metabolism in the rat. Acta Physiol Scand 48:364–372

Dawson PA, Oelkers P (1995) Bile acid transporters. Curr Opin Lipidol 6:109–114

Dawson PA, Hubbert M, Haywood J, Craddock AL, Zerangue N, Christian WV, Ballatori N (2005) The heteromeric organic solute transporter alpha-beta, Ostalpha-Ostbeta, is an ileal basolateral bile acid transporter, J Biol Chem 280:6960–6968

del Castillo-Olivares A, Gil G (2000) Role of FXR and FTF in bile acid-mediated suppression of cholesterol 7alpha-hydroxylase transcription. Nucleic Acids Res 28:3587–3593

Denson LA, Sturm E, Echevarria W, Zimmerman TL, Makishima M, Mangelsdorf DJ, Karpen SJ (2001) The orphan nuclear receptor, shp, mediates bile acid-induced inhibition of the rat bile acid transporter, ntcp. Gastroenterology 121:140–147

Dietschy JM, Turley SD (2002) Control of cholesterol turnover in the mouse. J Biol Chem 277:3801–3804

Dueland S, Reichen J, Everson GT, Davis RA (1991) Regulation of cholesterol and bile acid homoeostasis in bile-obstructed rats. Biochem J 280:373–377

Engelking LJ, Liang G, Hammer RE, Takaishi K, Kuriyama H, Evers BM, Li WP, Horton JD, Goldstein JL, Brown MS (2005) Schoenheimer effect explained – feedback regulation of cholesterol synthesis in mice mediated by Insig proteins. J Clin Invest 115:2489–2498

Galzie Z, Kinsella AR, Smith JA (1997) Fibroblast growth factors and their receptors. Biochem Cell Biol 75:669–685

Ge L, Wang J, Qi W, Miao HH, Cao J, Qu YX, Li BL, Song BL (2008) The cholesterol absorption inhibitor ezetimibe acts by blocking the sterol-induced internalization of NPC1L1. Cell Metab 7:508–519

Goldfarb M (1996) Functions of fibroblast growth factors in vertebrate development. Cytokine Growth Factor Rev 7:311–325

Gong Y, Lee JN, Lee PC, Goldstein JL, Brown MS, Ye J (2006) Sterol-regulated ubiquitination and degradation of Insig-1 creates a convergent mechanism for feedback control of cholesterol synthesis and uptake. Cell Metab 3:15–24

Goodwin BJ, Zuercher WJ, Collins JL (2008) Recent advances in liver X receptor biology and chemistry. Curr Top Med Chem 8:781–791

Grundy SM (1983) Absorption and metabolism of dietary cholesterol. Annu Rev Nutr 3:71–96

Grundy SM, Ahrens EH Jr (1969) Measurements of cholesterol turnover, synthesis, and absorption in man, carried out by isotope kinetic and sterol balance methods. J Lipid Res 10:91–107

Gustafsson J (1978) Effect of biliary obstruction on 26-hydroxylation of C27-steroids in bile acid synthesis J Lipid Res 19:237–243

Hagenbuch B, Dawson P (2004) The sodium bile salt cotransport family SLC10. Pflugers Arch 447:566–570

Hagenbuch B, Meier PJ (2004) Organic anion transporting polypeptides of the OATP/ SLC21 family: phylogenetic classification as OATP/ SLCO superfamily, new nomenclature and molecular/functional properties. Pflugers Arch 447:653–665

Hernell O, Staggers JE, Carey MC (1990) Physical-chemical behavior of dietary and biliary lipids during intestinal digestion and absorption. 2. Phase analysis and aggregation states of luminal lipids during duodenal fat digestion in healthy adult human beings. Biochemistry 29:2041–2056

Hofmann AF (1990) Bile acid secretion, bile flow and biliary lipid secretion in humans. Hepatology 12:17S–25S

Hofmann AF (1999) The continuing importance of bile acids in liver and intestinal disease. Arch Intern Med 159:2647–2658

Hofmann AF, Hagey LR (2008) Bile acids: chemistry, pathochemistry, biology, pathobiology, and therapeutics. Cell Mol Life Sci 65:2461–2483

Houten SM, Watanabe M, Auwerx J (2006) Endocrine functions of bile acids. EMBO J 25:1419–1425

Inagaki T, Choi M, Moschetta A, Peng L, Cummins CL, McDonald JG, Luo G, Jones SA, Goodwin B, Richardson JA, Gerard RD, Repa JJ, Mangelsdorf DJ, Kliewer SA (2005) Fibroblast growth factor 15 functions as an enterohepatic signal to regulate bile acid homeostasis. Cell Metab 2:217–225

Ito S, Fujimori T, Furuya A, Satoh J, Nabeshima Y, Nabeshima Y (2005) Impaired negative feedback suppression of bile acid synthesis in mice lacking betaKlotho. J Clin Invest 115:2202–2208

Ito S, Kinoshita S, Shiraishi N, Nakagawa S, Sekine S, Fujimori T, Nabeshima YI (2000) Molecular cloning and expression analyses of mouse betaklotho, which encodes a novel Klotho family protein. Mech Dev 98:115–119

Ito S, Fujimori T, Hayashizaki Y, Nabeshima Y (2002) Identification of a novel mouse membrane-bound family 1 glycosidase-like protein, which carries an atypical active site structure. Biochim Biophys Acta 1576:341–345

Jelinek DF, Andersson S, Slaughter CA, Russell DW (1990) Cloning and regulation of cholesterol 7 alpha-hydroxylase, the rate-limiting enzyme in bile acid biosynthesis, J Biol Chem 265:8190–8197

Jiang W, Miyamoto T, Kakizawa T, Nishio SI, Oiwa A, Takeda T, Suzuki S, Hashizume K (2006) Inhibition of LXRalpha signaling by vitamin D receptor: possible role of VDR in bile acid synthesis. Biochem Biophys Res Commun 351:176–184

Jung D, Inagaki T, Gerard RD, Dawson PA, Kliewer SA, Mangelsdorf DJ, Moschetta A (2007) FXR agonists and FGF15 reduce fecal bile acid excretion in a mouse model of bile acid malabsorption. J Lipid Res 48:2693–2700

Kim I, Ahn SH, Inagaki T, Choi M, Ito S, Guo GL, Kliewer SA, Gonzalez FJ (2007) Differential regulation of bile acid homeostasis by the farnesoid X receptor in liver and intestine. J Lipid Res 48:2664–2672

Knopfel M, Davies JP, Duong PT, Kvaerno L, Carreira EM, Phillips MC, Ioannou YA, Hauser H (2007) Multiple plasma membrane receptors but not NPC1L1 mediate high-affinity, ezetimibe-sensitive cholesterol uptake into the intestinal brush border membrane. Biochim Biophys Acta 1771:1140–1147

Kok T, Hulzebos CV, Wolters H, Havinga R, Agellon LB, Stellaard F, Shan B, Schwarz M, Kuipers F (2003) Enterohepatic circulation of bile salts in farnesoid X receptor-deficient mice: efficient intestinal bile salt absorption in the absence of ileal bile acid-binding protein. J Biol Chem 278:41930–41937

Kosters A, Karpen SJ (2008) Bile acid transporters in health and disease. Xenobiotica 38:1043–1071

Kosters A, Kunne C, Looije N, Patel SB, Oude Elferink RP, Groen AK (2006) The mechanism of ABCG5/ABCG8 in biliary cholesterol secretion in mice, J Lipid Res 47:1959–1966

Kramer W, Glombik H, Petry S, Heuer H, Schafer H, Wendler W, Corsiero D, Girbig F, Weyland C (2000) Identification of binding proteins for cholesterol absorption inhibitors as components of the intestinal cholesterol transporter. FEBS Lett 487:293–297

Kramer W, Girbig F, Corsiero D, Pfenninger A, Frick W, Jahne G, Rhein M, Wendler W, Lottspeich F, Hochleitner EO, Orso E, Schmitz G (2005) Aminopeptidase N (CD13) is a molecular target of the cholesterol absorption inhibitor ezetimibe in the enterocyte brush border membrane. J Biol Chem 280:1306–1320

Kruit JK, Plosch T, Havinga R, Boverhof R, Groot PH, Groen AK, Kuipers F (2005) Increased fecal neutral sterol loss upon liver X receptor activation is independent of biliary sterol secretion in mice. Gastroenterology 128:147–156

Kuipers F, Stroeve JH, Caron S, Staels B (2007) Bile acids, farnesoid X receptor, atherosclerosis and metabolic control. Curr Opin Lipidol 18:289–297

Lambert G, Amar MJ, Guo G, Brewer HB Jr, Gonzalez FJ, Sinal CJ (2003) The farnesoid X-receptor is an essential regulator of cholesterol homeostasis. J Biol Chem 278:2563–2570

Langheim S, Yu L, Von Bergmann K, Lutjohann D, Xu F, Hobbs HH, Cohen JC (2005) ABCG5 and ABCG8 require MDR2 for secretion of cholesterol into bile. J Lipid Res 46:1732–1738

Lee YK, Moore DD (2002) Dual mechanisms for repression of the monomeric orphan receptor liver receptor homologous protein-1 by the orphan small heterodimer partner. J Biol Chem 277:2463–2467

Li T, Ma H, Chiang JY (2008) TGFbeta1, TNFalpha, and insulin signaling crosstalk in regulation of the rat cholesterol 7alpha-hydroxylase gene expression, J Lipid Res 49:1981–1989

Makishima M, Okamoto AY, Repa JJ, Tu H, Learned RM, Luk A, Hull MV, Lustig KD, Mangelsdorf DJ, Shan B (1999) Identification of a nuclear receptor for bile acids. Science 284:1362–1365

Meier PJ, Stieger B (2000) Molecular mechanisms in bile formation. News Physiol Sci 15:89–93

Miettinen TA (1970) Detection of changes in human cholesterol metabolism. Ann Clin Res 2:300–320

Miettinen TA, Proia A, McNamara DJ (1981) Origins of fecal neutral steroids in rats. J Lipid Res 22:485–495

Miettinen TA, Tilvis RS, Kesaniemi YA (1990) Serum plant sterols and cholesterol precursors reflect cholesterol absorption and synthesis in volunteers of a randomly selected male population. Am J Epidemiol 131:20–31

Nagano M, Kuroki S, Mizuta A, Furukawa M, Noshiro M, Chijiiwa K, Tanaka M (2004) Regulation of bile acid synthesis under reconstructed enterohepatic circulation in rats. Steroids 69:701–719

Nakahara M, Furuya N, Takagaki K, Sugaya T, Hirota K, Fukamizu A, Kanda T, Fujii H, Sato R (2005) Ileal bile acid-binding protein, functionally associated with the farnesoid X receptor or the ileal bile acid transporter, regulates bile acid activity in the small intestine. J Biol Chem 280:42283–42289

Norlin M, Wikvall K (2007) The enzymes in the conversion of cholesterol into bile acids. Curr Mol Med 7:199–218

Oude Elferink RP, Paulusma CC, Groen AK (2006) Hepatocanalicular transport defects: pathophysiologic mechanisms of rare diseases. Gastroenterology 130:908–925

Pandak WM, Schwarz C, Hylemon PB, Mallonee D, Valerie K, Heuman DM, Fisher RA, Redford K, Vlahcevic ZR (2001) Effects of CYP7A1 overexpression on cholesterol and bile acid homeostasis. Am J Physiol Gastrointest Liver Physiol 281:G878–G889

Peet DJ, Janowski BA, Mangelsdorf DJ (1998a) The LXRs: a new class of oxysterol receptors. Curr Opin Genet Dev 8:571–575

Peet DJ, Turley SD, Ma W, Janowski BA, Lobaccaro JM, Hammer RE, Mangelsdorf DJ (1998b) Cholesterol and bile acid metabolism are impaired in mice lacking the nuclear oxysterol receptor LXR alpha. Cell 93:693–704

Peng SK, Ho KJ, Taylor CB (1974) The role of the intestinal mucosa in cholesterol metabolism: its relation to plasma and luminal cholesterol. Exp Mol Pathol 21:138–153

Pertsemlidis D, Kirchman EH, Ahrens EH Jr (1973) Regulation of cholesterol metabolism in the dog I Effects of complete bile diversion and of cholesterol feeding on absorption, synthesis, accumulation, and excretion rates measured during life. J Clin Invest 52:2353–2367

Phillips GB (1960) The lipid composition of human bile. Biochim Biophys Acta 41:361–363
Quinet EM, Savio DA, Halpern AR, Chen L, Schuster GU, Gustafsson JA, Basso MD, Nambi P (2006) Liver X receptor (LXR)-beta regulation in LXRalpha-deficient mice: implications for therapeutic targeting. Mol Pharmacol 70:1340–1349
Rader DJ (2007) Liver X receptor and farnesoid X receptor as therapeutic targets. Am J Cardiol 100:n15–n19
Radhakrishnan A, Ikeda Y, Kwon HJ, Brown MS, Goldstein JL (2007) Sterol-regulated transport of SREBPs from endoplasmic reticulum to Golgi: oxysterols block transport by binding to Insig. Proc Natl Acad Sci USA 104:6511–6518
Rosenblum SB, Huynh T, Afonso A, Davis HR Jr, Yumibe N, Clader JW, Burnett DA (1998) Discovery of 1-(4-fluorophenyl)-(3R)-[3-(4-fluorophenyl)-(3S)-hydroxypropyl]-(4S)-(4 -hydroxyphenyl)-2-azetidinone (SCH 58235): a designed, potent, orally active inhibitor of cholesterol absorption. J Med Chem 41:973–980
Russell DW (2003) The enzymes, regulation, and genetics of bile acid synthesis. Annu Rev Biochem 72:137–174
Samuel P, McNamara DJ (1983) Differential absorption of exogenous and endogenous cholesterol in man. J Lipid Res 24:265–276
Schlessinger J, Plotnikov AN, Ibrahimi OA, Eliseenkova AV, Yeh BK, Yayon A, Linhardt RJ, Mohammadi M (2000) Crystal structure of a ternary FGF-FGFR-heparin complex reveals a dual role for heparin in FGFR binding and dimerization. Mol Cell 6:743–750
Scotti E, Gilardi F, Godio C, Gers E, Krneta J, Mitro N, De FE, Caruso D, Crestani M (2007) Bile acids and their signaling pathways: eclectic regulators of diverse cellular functions. Cell Mol Life Sci 64:2477–2491
Simmonds WJ, Hofmann AF, Theodor E (1967) Absorption of cholesterol from a micellar solution: intestinal perfusion studies in man. J Clin Invest 46:874–890
Sinal CJ, Tohkin M, Miyata M, Ward JM, Lambert G, Gonzalez FJ (2000) Targeted disruption of the nuclear receptor FXR/BAR impairs bile acid and lipid homeostasis. Cell 102:731–744
Small DM (2003) Role of ABC transporters in secretion of cholesterol from liver into bile. Proc Natl Acad Sci USA 100:4–6
Sperry WM (1927) Lipid excretion IV A study of the relationship of the bile to the fecal lipids with special reference to certain problems of sterol metabolism. J Biol Chem 1:351–378
Stieger B, Meier Y, Meier PJ (2007) The bile salt export pump. Pflugers Arch 453:611–620
Thomas C, Auwerx J, Schoonjans K (2008) Bile acids and the membrane bile acid receptor TGR5 – connecting nutrition and metabolism. Thyroid 18:167–174
van der Velde AE, Vrins CL, van den O K, Kunne C, Oude Elferink RP, Kuipers F, Groen AK (2007) Direct intestinal cholesterol secretion contributes significantly to total fecal neutral sterol excretion in mice. Gastroenterology 133:967–975
van der Velde AE, Vrins CL, van den O K, Seemann I, Oude Elferink RP, Kuipers F, Groen AK (2008) Regulation of direct transintestinal cholesterol excretion in mice. Am J Physiol Gastrointest Liver Physiol 295:G203–G208
Vrins C, Vink E, Vandenberghe KE, Frijters R, Seppen J, Groen AK (2007) The sterol transporting heterodimer ABCG5/ABCG8 requires bile salts to mediate cholesterol efflux. FEBS Lett 581:4616–4620
Wang DQ (2007) Regulation of intestinal cholesterol absorption. Annu Rev Physiol 69:221–248
Wang DQ, Lee SP (2008) Physical chemistry of intestinal absorption of biliary cholesterol in mice. Hepatology 48:177–185
Wang DQ, Paigen B, Carey MC (2001) Genetic factors at the enterocyte level account for variations in intestinal cholesterol absorption efficiency among inbred strains of mice. J Lipid Res 42:1820–1830
Wang L, Han Y, Kim CS, Lee YK, Moore DD (2003) Resistance of SHP-null mice to bile acid-induced liver damage. J Biol Chem 278:44475–44481
Watanabe M, Houten SM, Wang L, Moschetta A, Mangelsdorf DJ, Heyman RA, Moore DD, Auwerx J (2004) Bile acids lower triglyceride levels via a pathway involving FXR, SHP, and SREBP-1c. J Clin Invest 113:1408–1418

Wiedlocha A, Sorensen V (2004) Signaling, internalization, and intracellular activity of fibroblast growth factor. Curr Top Microbiol Immunol 286:45–79

Wilson JD (1964) The quantification of cholesterol excretion and degradation in the isotopic steady state in the rat: the influence of dietary cholesterol. J Lipid Res 5:409–417

Xie MH, Holcomb I, Deuel B, Dowd P, Huang A, Vagts A, Foster J, Liang J, Brush J, Gu Q, Hillan K, Goddard A, Gurney AL (1999) FGF-19, a novel fibroblast growth factor with unique specificity for FGFR4. Cytokine 11:729–735

Yu C, Wang F, Kan M, Jin C, Jones RB, Weinstein M, Deng CX, McKeehan WL (2000) Elevated cholesterol metabolism and bile acid synthesis in mice lacking membrane tyrosine kinase receptor FGFR4. J Biol Chem 275:15482–15489

Yu C, Wang F, Jin C, Huang X, McKeehan WL (2005) Independent repression of bile acid synthesis and activation of c-Jun N-terminal kinase (JNK) by activated hepatocyte fibroblast growth factor receptor 4 (FGFR4) and bile acids. J Biol Chem 280:17707–17714

Yu L, Hammer RE, Li-Hawkins J, Von Bergmann K, Lutjohann D, Cohen JC, Hobbs HH (2002a) Disruption of Abcg5 and Abcg8 in mice reveals their crucial role in biliary cholesterol secretion. Proc Natl Acad Sci USA 99:16237–16242

Yu L, Li-Hawkins J, Hammer RE, Berge KE, Horton JD, Cohen JC, Hobbs HH (2002b) Overexpression of ABCG5 and ABCG8 promotes biliary cholesterol secretion and reduces fractional absorption of dietary cholesterol. J Clin Invest 110:671–680

Zimber A, Gespach C (2008) Bile acids and derivatives, their nuclear receptors FXR, PXR and ligands: role in health and disease and their therapeutic potential. Anticancer Agents Med Chem 8:540–563

Chapter 5
Cholesterol Trafficking in the Brain

Dieter Lütjohann, Tim Vanmierlo, and Monique Mulder

Abstract After it became clear that aberrations in cerebral cholesterol metabolism could lead to severe neurological diseases the interest in the regulation of brain cholesterol homeostasis increased. In particular when evidence was obtained for an important role of cholesterol in the still largely unknown molecular mechanisms underlying Alzheimer's Disease. Many proteins involved in peripheral cholesterol metabolism are also present in the brain. Yet, brain cholesterol metabolism is very different from that in the remainder of the body. The present chapter first addresses the overall cholesterol turnover in the brain; where cholesterol is synthesized, where it resides and how it is secreted from the brain. Subsequently, the focus is on mechanisms related to intercellular cholesterol trafficking between astrocytes and neurons.

5.1 Introduction

Brain cholesterol metabolism must be strictly regulated for optimal brain functioning. This is evident from the fact that disturbances herein can lead to severe neurological diseases, such as Smith–Lemli Opitz syndrome (Bjorkhem et al. 2001), Niemann–Pick type C1 (Wiegand et al. 2003) and cerebrotendinous xhantomatosis (Moghadasian et al. (2002). Disturbed brain cholesterol metabolism may also play a role in the development and progression of AD (Marx 2001; Puglielli et al. 2004).

Many proteins involved in cholesterol transport are expressed within the brain and lipoprotein like particles are thought to mediate the intercellular trafficking of

D. Lütjohann
Department of Clinical Pharmacology, Zentrallabor, Bonn University, Sigmund-Freud-Strasse 25, 53105 Bonn, Germany
e-mail: dieter.luetjohann@ukb.uni-bonn.de

T. Vanmierlo and M. Mulder
Department of Internal Medicine, Division of Pharmacology, Vascular and Metabolic Diseases, Erasmus Medical Center, 3015 CE Rotterdam, The Netherlands
e-mail: m.t.mulder@erasmusmc.nl

C. Ehnholm (ed.) *Cellular Lipid Metabolism*,
DOI 10.1007/978-3-642-00300-4_5,

cholesterol. Yet, brain cholesterol metabolism is very different from that in the remainder of the body. Cholesterol in the body can be derived from the diet or from de novo synthesis. The brain in contrast fully relies on de novo synthesis (Dietschy and Turley 2001) and cannot retrieve cholesterol from the circulation (Bjorkhem and Meaney 2004). A major site of cholesterol synthesis in the brain is the astrocytes, which secrete it in the form of apoE-containing HDL-like lipoproteins. Neurons can synthesize cholesterol, but after birth and full differentiation of astrocytes they are thought to shut down their cholesterol synthesis and rely on astrocytes for their cholesterol supply. How neurons regulate their cholesterol supply remains to be determined in detail. Evidence has been obtained indicating a role for the brain-specific cholesterol metabolite, 24(*S*)-hydroxycholesterol.

5.2 Cholesterol Turnover in the Brain

The brain is the most cholesterol-rich organ in the body. It contains about 25% of all free cholesterol present in the body, while it represents only 2% of the total body weight. Essentially all cholesterol in the brain is unesterifed. Surprisingly relatively little is still known about the maintenance of cholesterol homeostasis and about its specific roles within the central nervous system (Snipes and Suter U1997).

Almost all of the cholesterol present in the brain is synthesized locally. In spite of the high cholesterol content, cholesterol turnover in the brain was found to be much slower than that in the rest of the body (20- to 80-fold lower than in the liver; Spady and Dietschy 1983) and its concentration is kept remarkably stable. This may explain why, until about 15 years ago, little attention had been paid to cholesterol metabolism in the brain. This changed when it turned out that cerebral cholesterol homeostasis is much more dynamic than initially thought. It is generally assumed that, under non-diseased conditions, cholesterol from plasma lipoproteins does not enter the brain, since it cannot cross the endothelial cells of the BBB. The cerebral capillary endothelium is the anatomical substrate of the BBB, isolating the brain neuropil from the systemic circulation. The cerebral endothelium lining the blood vessel lumen consists of a single layer of cells joined together by tight intercellular junctions. This layer of cells is supported by a basement membrane, which is the laminar structure formed by the fusion of the endothelial and glial vascular basement membrane. The end feet of astrocytes make up a discontinuous sheath at the abluminal surface of the basement membrane (Fig. 5.1). This has been confirmed by studies in mice, sheep and rabbits with use of either labeled lipoproteins or labeled cholesterol (for a review, see Dietschy and Turley 2001). For example, apoE-deficient mice display dramatically increased plasma cholesterol levels, but no alterations in brain cholesterol levels (Lomnitski et al. 1999). However, a recent study with guinea-pigs suggested that minor amounts (~1%) of cholesterol might cross the BBB (Lutjohann et al. 2004).

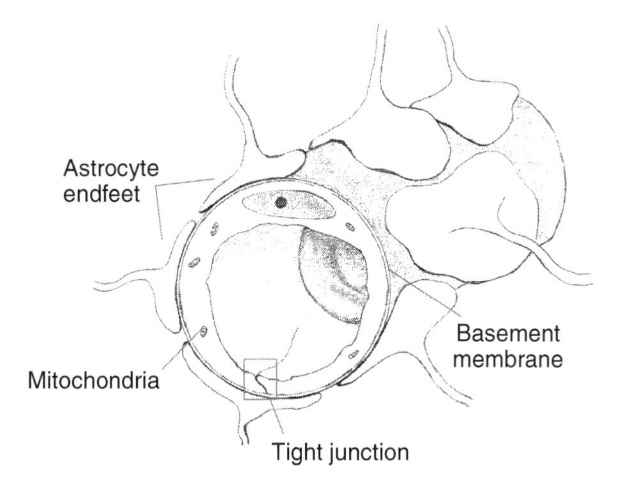

Fig. 5.1 Schematic representation of the blood–brain barrier. The blood–brain barrier consists of endothelial cells lining the vessel wall, connected via tight junctions. These are surrounded by a thick basement membrane which is covered by astrocytic endfeet

Cholesterol, or better its steroid ring structure, cannot be degraded in the human body and high concentrations of free cholesterol can lead to the formation of crystals which are toxic to cells, in particular neurons (Lemaire-Ewing et al. 2005; Travert et al. 2006). Therefore, excess cholesterol is secreted from the brain into the circulation (Brown and Goldstein 1997) and finally released from the body via the liver where it is converted into bile acids (Lutjohann et al. 1996). Within the brain, cholesterol can be modified into its major brain metabolite 24(*S*)-hydroxycholesterol by the enzyme CYP46A1. Based on experiments with mice deficient for the enzyme CYP46A1, it was calculated that about 64% is being secreted in the form of this polar cholesterol metabolite (Fig. 5.2; Bjorkhem et al. 1997; Xie et al. 2003). The remaining 36% of cholesterol is secreted from the brain via another, yet unknown pathway that may involve apoE (Lund et al. 2003).

24(*S*)-Hydroxycholesterol is, in contrast with cholesterol, able to traverse the BBB (Bjorkhem et al. 1997, 1998; Lutjohann et al. 1996). At first it looks quite controversially that a compound more polar than cholesterol is able to pass a lipophilic barrier. As with other plasma membranes, the membranes of the endothelial cells are freely permeable to water (Panzenboeck et al. 2002). Introduction of an hydroxyl group in the side-chain of the cholesterol molecule leads to a local reordering of membrane phospholipids such that it is energetically more favourable to expel the oxysterol (Kessel et al. 2001). When a sterol like cholesterol is hydroxylated, there is an increase in the maximal aequous activity that can be achieved, but a reduction in the passive permeability coefficient. However, the increase in solubility is proportionately greater than the reduction in the permeability coefficient, so that the net effect of hydroxylation is to greatly increase the maximal rate of passive diffusion of the molecule across the BBB.

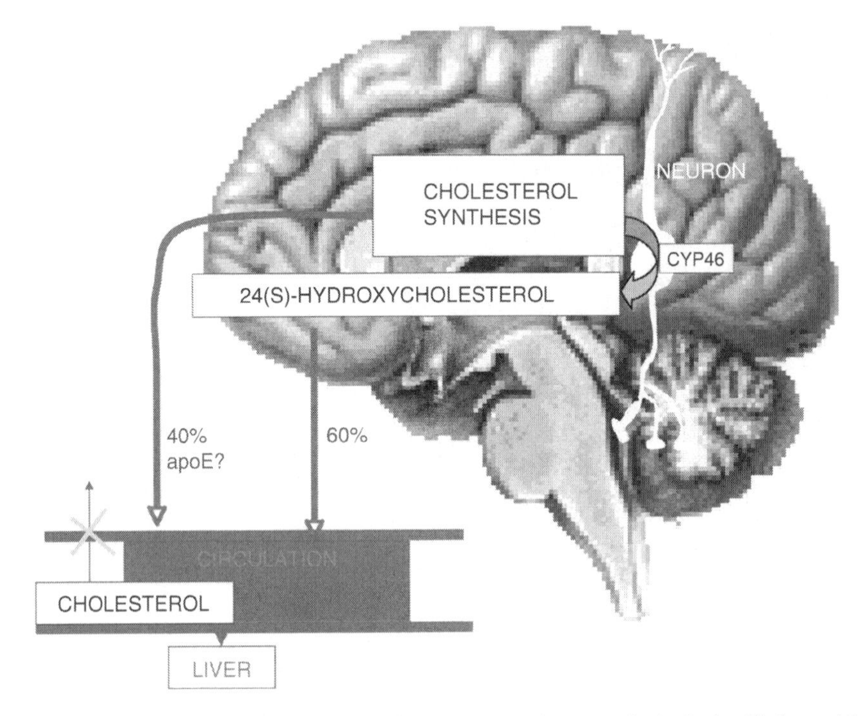

Fig. 5.2 Schematic representation of the overall cholesterol turnover in the brain. Cholesterol is synthesized endogenously predominantly by astrocytes and is converted by the enzyme CYP46A1 that resides in a subset of neurons, whereafter it is released from the brain into the circulation. This pathway is thought to be responsible for about 60% of the secretion of cholesterol from the brain, while the remaining 40% is secreted via a yet unknown pathway that may involve apoE

The enzyme CYP46A1 has been characterized at the molecular level. Its gene contains 15 exons and is located on human chromosome 14q32.1. CYP46A1 is predominantly found in the brain, mainly located in a subset of specific neurons. Deficiency for CYP46A1 in mice results in suppression of cholesterol synthesis in brain by about 25% (Lund et al. 2003; Xie et al. 2003), probably to compensate for the decreased efflux of cholesterol from the brain. Evidence has been obtained indicating a role for CYP46A1 in neurosteroid metabolism (Mast et al. 2003). Besides converting cholesterol into 24(*S*)-hydroxycholesterol, CYP46A1 can also inactivate neurosteroids that may either be synthesized endogenously or derived from the circulation.

Cholesterol in the brain resides in three major compartments with different turnover rates. The largest pool (70–80%, or 260 mg kg^{-1} of the total 330 mg kg^{-1}) has the slowest turnover (>1%) and is present in myelin membranes. Of the remaining 70 mg kg^{-1}, about 10% reside in neurons, which represent about 10% of the brain cells, and therefore contain about 7 mg kg^{-1} of cholesterol. The remaining 63 mg kg^{-1} are present in glial cells (Davison 1965; Muse et al 2001; Xie et al. 2003).

As mentioned, the CYP46A1 enzyme is expressed in a subset of metabolically active neurons such as pyramidal cells of the cortex and Purkinje cells of the cerebellum (Lund et al. 1999). The cholesterol turnover in these neurons must be very high, since they are estimated to contain about 4 mg kg^{-1} of cholesterol while their turnover of 24-hydroxycholesterol is about 0.9 mg kg^{-1} day^{-1}, which is more than 20% day^{-1}. The overall turnover of cholesterol in the body is about 0.8% day^{-1}, i.e. a value similar to the CYP46A1-independent cholesterol turnover in brain, which is about 0.5 mg kg^{-1} day^{-1} of a pool of about 63 mg kg^{-1} day^{-1}.

5.3 Release of 24(S)-Hydroxycholesterol from the Brain into the Circulation

24(*S*)-Hydroxycholesterol release from the brain into the circulation represents the major pathway for the brain to get rid of its cholesterol. From measurements in human tissue, the brain was estimated to contain approximately 80% of the total 24(*S*)-hydroxycholesterol content in the body. Brain samples also contained other oxysterols, but 24(*S*)-hydroxycholesterol was always dominating (Lutjohann et al. 1996). The concentrations of 24(*S*)-hydroxycholesterol in the brain and adrenals were 30- to 1500-fold higher than in any other organ. As judged from the arteriovenous difference over the brain and the levels in cerebrospinal fluid (CSF), 24(*S*)-hydroxycholesterol was the only oxysterol that was transported from the human brain into the circulation. From the concentrations of 24(*S*)-hydroxycholesterol in the CSF and the flux of cerebrospinal fluid into the jugular vein it can be calculated that <1% of the total flux of 24(*S*)-hydroxycholesterol from the brain occurs via the cerebrospinal fluid. Thus, 99% must occur through the BBB. The pool size of 24(*S*)-hydroxycholesterol in rat brain was found to be about 0.3% of that of cholesterol, while the estimated half-life of 24(*S*)-hydroxycholesterol appears to be about 11 h versus about 2–6 months for cholesterol. The high cholesterol turnover in a subset of neurons that produce 24(*S*)-hydroxycholesterol seems to be controversial to the idea that after birth neurons shut down their cholesterol synthesis and rely on astrocytes for their cholesterol supply. Therefore, the source of cholesterol remains to be established: de novo synthesis by neurons or uptake from astrocyte-released cholesterol. Alternatively, a specific subset of metabolically active neurons may display a very high cholesterol turnover rate, while the other metabolically less active neurons largely shut down their cholesterol synthesis.

An alternative pathway for cholesterol release from the brain has been reported. Meaney et al. (2007) have shown that 27-hydroxycholesterol is metabolized into the known C(27) steroidal acid 7alpha-hydroxy-3-oxo-4-cholestenoic acid by neuronal cell models only and can be transferred across the BBB. The contribution of this pathway to the daily release of cholesterol from the brain most likely is restricted and requires further examination (Meaney et al. 2007), particularly since 27-hydroxycholesterol is thought to originate from extracerebral sources.

5.4 Lipoproteins in the Cerebrospinal Fluid

The cerebrospinal fluid (CSF) is a particular compartment of the CNS. Functions of the CSF are to protect the brain from mechanical injury, to provide the brain with nutrients and maintain a constant external milieu for neurons, to secrete waste products via the CSF into the blood and to help the trafficking of hormones. The CSF flows via the ventricles into the subarachnoid space surrounding the brain, where it is reabsorbed into the venous arteries into the circulation. The CSF is produced by the choroid plexus which consists of many capillaries, separated from the ventricles by specific choroid epithelial cells. These are closely connected with each other via tight junctions similar to the endothelial cells of the BBB. Choroid capillary networks are found predominantly in the lateral, the third and the fourth ventricle in the brain. The choroid plexus filters and actively secretes compounds from the plasma into the CSF. Therefore, the BBB at this site predominantly determines the composition of the CSF. Additionally, the CSF compartment is in close anatomical contact with the brain interstitial fluid. Therefore, the composition of the interstitial fluid affects the CSF composition since the two fluids are more or less in equilibrium (Skipor and Thiery 2008).

The presence of spherical lipoprotein-like particles in CSF was first described by Roheim et al. (1979) and later by Pitas et al. (1987). ApoE and apoA-I are the predominant apolipoproteins in human CSF, whereas apoB could not be detected (Borghini et al. 1995; de Vries et al. 1995; Pitas et al. (1987a). Originally, apoE was thought to originate predominantly from astrocytes and possibly also from microglial cells (LaDu et al. 1998; Nakai et al. 1996). ApoA-I present in CSF may be retrieved from the circulation or may be derived from endothelial cells lining the cerebral vessels (de Vries et al. 1995; Elshourbagy et al. 1985; Guyton et al. 1998). Based on an affinity chromatography assay, Koch et al. (2001) described four major classes of CSF-lipoproteins (Lps), all between HDL (10–13 nm) and LDL (22–29 nm) in size, namely: the ones containing apoE, apoA-I, both or none (Koch et al. 2001). All four CSF-Lps groups contained comparable amounts of apoJ. The apoE-containing lipoproteins also contained apoA-IV and ApoD. Lipoproteins containing both apoE and apoA-I resembled those containing apoE exclusively but had an additional presence of apoH. Lipoproteins with apoA-I also contained apoA-II, apo-IV and apoD. Finally, the apoE- and apoA-I-free group was hallmarked by apoD as major apolipoprotein, but additionally contained apo-A-IV, apoH and apoJ. ApoD, H and J are structurally and functionally different from apoE and apoAs. ApoD and apoJ have been suggested to be of neuronal origin (Bassett et al. 2000), while apoAIV is thought to derive from the intestine.

Based on cholesterol content and the protein:lipid ratio, all four lipoprotein classes detected in CSF correspond best with plasma HDL. ApoE is relatively the most abundant apolipoprotein in the CSF. Furthermore, apoE-containing lipoproteins are larger than the other CSF-lipoproteins not containing apoE (Koch et al. 2001; Pitas et al. 1987a). Based on the assumption that the composition of CSF and brain interstitial fluid is very much alike, it can be assumed that similar lipoprotein-like particles are present in interstitial fluid.

5.5 Astrocytes Supply Neurons with Cholesterol

After birth, neurons shut down their own cholesterol synthesis and rely on astrocytes for their cholesterol supply since they require their energy for other processes, such as electrical and chemical signaling and generation of action potentials (Argmann et al. 2005).

Astrocytes are the most abundant cell type in the human brain and are intimately associated with all parts of neurons as well as with cells of the BBB. They provide structural, trophic and metabolic support to neurons. They contribute to brain the BBB, releasing growth factors, buffering of extracellular K^+ and regulating the brain immune response.

Cholesterol in neurons is implicated in a number of processes and a well regulated cholesterol supply therefore of utmost importance. Sufficient amounts of cholesterol may be required for vesicle transport and for maintenance of proper signal transduction pathways and optimal neurotranmitter release. Moreover, it serves as a precursor for neurosteroids (Mauch et al. 2001).

One of the key functions of cholesterol is to regulate membrane fluidity. Cholesterol decreases membrane fluidity, which affects the biophysical properties of the membrane, resulting in a reduced permeability for polar molecules and allowing for maintenance of relevant concentration gradients of ions and other molecules (Ikonen 2008). Neurons, for example, require energy for generation of action potentials, which depends upon the permeability characteristics of the plasma membrane surrounding the axon. These are determined in part by cholesterol and other lipid components of the membrane. Surrounding of axons by myelin, containing large amounts of cholesterol, allows neurons to send electrical signals over long distances with high velocity (Puglielli et al. 2004). Changes in the cholesterol levels dynamically alter the micro-environment of the plasma membrane and thereby the functioning of membrane-bound proteins, such as for example the functions of ion channels. Alterations in cholesterol levels of rat hippocampal neurons has been shown to affect their excitability (Guo et al. 2008).

Notably, neurons require cholesterol for generation of new membranes during regeneration after CNS injury and for the formation of new synapses during a process called synaptic plasticity, a process that is thought to be essential for learning and memory. Synapses of neuronal axons contact the dendrites or the somata of other neurons and upon activation of the neurons send signals electrically or chemically to this neuron. The dynamic and continuous reorganisation of these synaptic contacts is called synaptic plasticity.

Cholesterol delivery may at least in part participate in regulating the number of synapses formed (Slezak and Pfrieger 2003). A continuous turnover of cholesterol in neurons may facilitate the cells' ability for efficient and quick adaptation of cholesterol homeostasis required for dynamic structural changes of neurons, their extensions, and their synapses during synaptic plasticity (Pfrieger 2003).

Indeed, apoE in combination with cholesterol was found to induce the outgrowth of neurites in an isoform-specific manner in neuronal cultures. Cholesterol alone also had such a effect but to a lesser extent (Bellosta et al. 1995; Nathan et al. 1994, 1995,

2002; Pfrieger 2003). There are three common human apoE-isoforms (E2, E3, E4) that differ from each other in only one amino acid (Utermann et al. 1979). However, the apoE-isoforms differentially modulate neurite outgrowth. ApoE3 was found to enhance the outgrowth of neurites from cultured embryonic and adult mouse cortical neurons, while apoE4 decreased neurite outgrowth from these cells (Nathan et al. 2002). In addition, an apoE isoform-specific effect on dendritic spines in the dentate gyrus of brains of transgenic mice and humans was observed (Ji et al. 2003).

As described, astrocytes are considered to be the predominant source of brain cholesterol. Astrocytes secrete cholesterol together with phospholipids and apoE in the form of small, HDL-like particles (Gong et al. 2002). HDL are the smallest of the different lipoprotein classes in the circulation where also larger lipoproteins circulate, i.e., chylomicrons, VLDL, LDL and IDL. ApoE is a major apolipoprotein in the brain and, in lipid trafficking, is thought to exert roles here similar to those in the periphery. ApoE secreted by astrocytes constitutes 1–3% of total secreted protein by these cells (Pitas et al. 1987b). Using in situ atomic force microscopy, it was found that cultured astrocytes secrete lipoproteins with apoE or apoJ that are a little different from HDL in plasma with respect to their size, shape and aggregation properties. They are significantly flatter and smaller than plasma HDL and, in contrast with plasma HDL, do not form ordered arrays on a sillica surface at high concentrations, but rather form amorphous aggregates (Legleiter et al. 2004).

ApoE seems to be required for lipoprotein secretion by astrocytes. Fagan et al. (1999) reported that cultured astrocytes from apoE-deficient mice secrete little phospholipids and free cholesterol in comparison with astrocytes obtained from wild-type mice, despite an unaffected apoJ expression. These authors found that apoE and apoJ normally reside on distinct HDL particles, with apoJ being present on the smallest particles. However, curiously, Karten et al. (2006) found cultured astrocytes obtained from apoE-deficient mice to secrete lipoproteins with a size similar to those secreted by astrocytes obtained from wild-type mice. It is suggested that astrocytes secrete partially lipidated apoE as a major extracellular acceptor of cholesterol released from the cells by an efflux process, mediated by the ABC-transporter (ABC)G1. In contrast with wild-type astrocytes, cholesterol loading of the apoE-deficient astrocytes does not enhance the efflux to apoA-I, but apparently shunts cholesterol into the esterification pathway. Moreover, expression of ABCA1, another cholesterol efflux transporter, was not increased upon cholesterol loading of apoE-deficient astrocytes, unlike that in wild-type astrocytes. Using apoE-deficient mice, evidence was obtained that apoE is not only involved in the delivery of lipids to neurons, but also in the removal of cell debris or membrane fragments, since a decreased clearance of degeneration products was found in the brain of apoE-deficient mice after entorhinal cortex lesioning in comparison with wild-type mice (Fagan et al. 1998). In the presence of apoE these lipids are thought to be re-utilized. In line with this, it was reported that apoE-deficient mice display impaired compensatory sprouting in injured denervated skin (Maysinger et al. 2008).

Since apoE-deficiency does not affect the sterol composition in the murine brain (Jansen et al. 2006), there must be compensatory mechanisms to maintain brain cholesterol homeostasis. In the absence of apoE increased levels of apoD have been found in murine brain. ApoD is a secreted lipocalin that is present in astrocytes, oligodendrocytes and in some scattered neurons; and it has also been suggested to be involved in re-innervation and regeneration (Navarro et al. 2004).

The fact that astrocytes are of major importance for neuronal cholesterol homeostasis is supported by an experiment with npc1$^{-/-}$ mice, a model for Niemann–Pick type C disease, that develop neurodegeneration associated with cholesterol accumulation in neurons due to a defect in cholesterol trafficking. As a consequence of mutations in NPC1, cholesterol does not exit the endocytic pathway, leading to a massive accumulation of unesterified cholesterol in late endosomes and lysosomes after uptake of lipoproteins through the clathrin-coated pit pathway (Xie et al. 2000). Replacement of npc1 in npc1$^{-/-}$ mice, specifically in astrocytes, results in enhanced survival and decreased neuronal storage of cholesterol associated with less accumulation of axonal spheroids and restoration of myelin tracts (Zhang et al. 2008).

5.6 How do Neurons Regulate Their Cholesterol Supply?

If neurons for their cholesterol supply predominantly rely on astrocytes, it can be questioned how neurons communicate with these cells in order to request sufficient amounts of cholesterol for optimal functioning. 24(*S*)-Hydroxycholesterol is specifically formed in neurons and is a natural ligand for the liver X receptors (LXRs), so-called master regulators of cholesterol homeostasis.

In astrocytes cholesterol synthesis is thought to be regulated by two main transcription regulation systems, i.e., sterol regulatory element binding proteins (SREBPs) and LXRs. SREBP activation increases the transcription of genes involved in increasing cellular cholesterol levels (Goldstein et al. 2006), while LXR activation facilitates the disposition of cholesterol from cells (Eckert et al. 2007).

SREBP residing in the ER is activated in sterol-poor conditions and transported to the Golgi complex where it undergoes proteolytic processing. Subsequently, this fragment is imported into the nucleus, where it switches on the transcription of HMG-CoAR and other sterol-regulated genes. Binding of sterols to the two chaperones SCAP and INSIG induces conformational changes resulting in the inhibition of the transport of SREBP from the ER to the Golgi. Binding of cholesterol to SCAP and binding of oxysterols, such as 24-hydroxycholesterol in the brain, to INSIG causes these chaperones to bind one another thereby preventing SCAP from escorting SREBP to the Golgi by making it inaccessible to COPII and preventing the increase in transcription of gene products (Ikonen 2008).

LXRs belong to the nuclear hormone receptor superfamily. Two forms have been identified: LXRα and LXRβ. Both are present in the brain and are thought to be involved in the regulation of brain cholesterol homeostasis (Whitney et al. 2002). This is supported by the observation that aged LXRα/β-deficient mice display several defects in their central nervous system. These include closed ventricles, lipid accumulation in astrocytes and around blood vessels, proliferation of astrocytes and dysorganisation of myelin sheaths (Wang et al. 2002).

Alternatively, cholesterol homeostasis can be regulated at the level of the proteosomal degradation of HMG-CoAR, such as for example by lanosterol (Song et al. 2005).

Evidence now indicates that 24(*S*)-hydroxycholesterol released by neurons signals to astrocytes and induces the secretion of apoE-containing lipoprotein-like particles via the LXR-pathway (Fig. 5.3; Abildayeva et al. 2006). 24(*S*)-Hydroxycholesterol was found to induce apoE transcription, protein synthesis and secretion in a dose- and time-dependent manner in cells of astrocytic but not of neuronal origin. Moreover, 24(*S*)-hydroxycholesterol primes astrocytoma, but not neuroblastoma cells, to mediate cholesterol efflux to apoE. The synthetic LXR agonist GW683965A exerts similar effects, suggesting involvement of an LXR-controlled signaling pathway. The differential effects observed in astrocytoma and

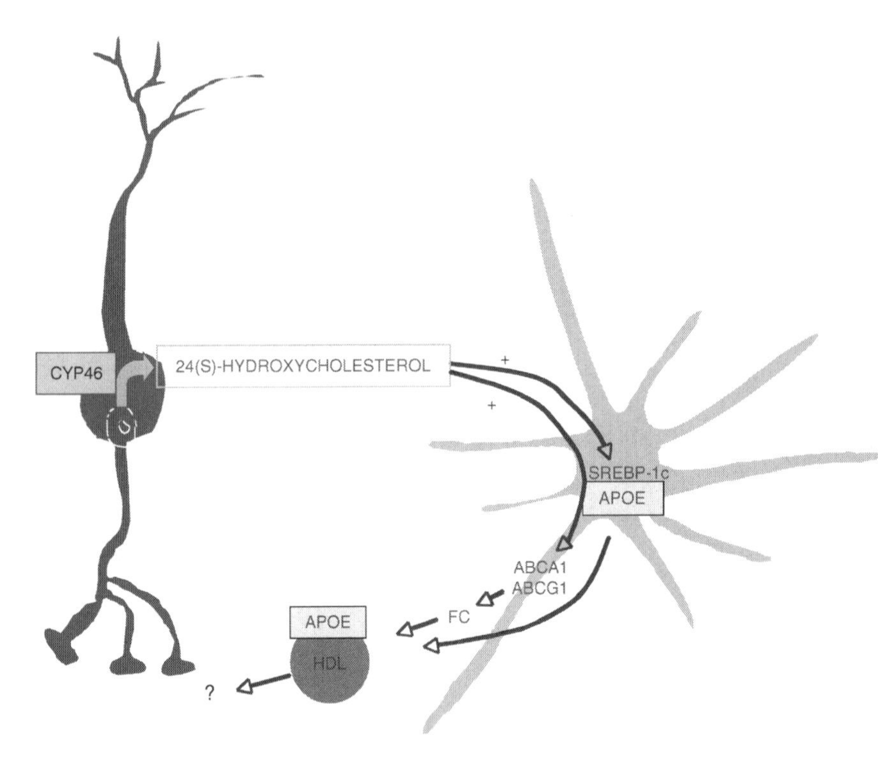

Fig. 5.3 Schematic presentation of the differential effects in astrocytes and neurons of 24(*S*)-hydroxycholesterol on the expression of apoE and ABC transporters expression as well as apoE-mediated cholesterol efflux. 24(*S*)-Hydroxycholesterol may act as a signaling molecule that induces the apoE-mediated cholesterol efflux from astrocytes but not from neurons. ABCA1 and ABCG1 may play a role in mediating cholesterol efflux from astrocytes. Thus, in the intact brain, 24(*S*)-hydroxycholesterol derived from neurons may signal astrocytes to increase production of lipidated apoE particles in order to supply neurons with additional cholesterol during synaptogenesis or neuritic remodeling. Moreover alterations in the transcriptional regulation role of 24(*S*)-hydroxycholesterol on apoE-mediated cholesterol efflux may affect the progression of neurodegenerative diseases, including AD

neuroblastoma cells may be explained by a 10- to 20-fold higher basal LXRα and -β expression level in astrocytoma cells. LXR isoforms differ in their pattern of expression (Steffensen et al. 2003). In the brain LXRβ levels are 2- to 5-fold higher than in the liver, whereas LXRα levels are 3.5- to 14.0-fold lower than in the liver (Apfel et al. 1994; Song et al. 1994) However, 24(*S*)-hydroxycholesterol and GW683965A were found to up-regulate LXRα but not LXRβ expression in astrocytoma cells (data not shown), similar to reports for macrophages and adipocytes, but not liver and muscle (Ulven et al. 2004; Whitney et al. 2001). Therefore, autoregulation of LXRα that has been suggested to occur in adipocytes to coordinate expression of target genes such as APOE (Ulven et al. 2004) may also occur in brain.

The apoE-mediated cholesterol efflux from astrocytoma cells may be controlled by the ATP binding cassette transporters ABCA1 and ABCG1, since their expression was also up-regulated by 24(*S*)-hydroxycholesterol and GW683965A. ABCG4 is most likely not involved, because its expression was induced by LXR-activation only in neuronal cells.

A common feature that astrocytes share with macrophages and adipocytes is their large content of free cholesterol (Krause and Hartman 1984; Vance et al. 2005). In agreement, in macrophages (Argmann et al. 2005), activation of LXR stimulates cellular cholesterol efflux through the co-ordinated induction of ABCA1, ABCG1, and apoE expression. Some aspects of these pathways may be cell-type specific. For example, apoD, which is also synthesized and secreted by astrocytes in lipid-bound form and may compensate functions of apoE in its absence, was unresponsive to LXR agonists in astrocytes, whereas it was induced in adipocytes (Hummasti et al. 2004). In contrast, Patel et al. (1995) found apoD expression in astrocytes to be regulated by 25-hydroxycholesterol and by progesterone.

GW683965A up-regulated the expression of HMG-CoA reductase, LDLR and SREBP2 in astrocytoma cells, supposedly to maintain cellular cholesterol homeostasis during excessive loss by efflux. However, 24(*S*)-hydroxycholesterol down-regulated the expression of these genes. It is known since long that oxysterols reduce the activity of HMG-CoA reductase and are more potent inhibitors of cholesterol synthesis than cholesterol (Kandutsch et al. 1974). However, the mechanisms by which cholesterol and oxysterols reduce cholesterol synthesis differ (Adams et al. 2004). Thus, 24(*S*)-hydroxycholesterol may not exert its effect exclusively via the LXR pathway. Accordingly, 24(*S*)-hydroxycholesterol, but not GW683965A, enhanced the expression of ABCG4 in neurons. Apparently, cholesterol efflux from astrocytes is not driven solely by the rate of cholesterol biosynthesis because an enhanced cholesterol efflux, up-regulation of ABCA1 and ABCG1 and impaired synthesis of cholesterol is observed in astrocytes as a consequence of 24(*S*)-hydroxycholesterol exposure (Abildayeva et al. 2006) and also in macrophages after statin treatment (Argmann et al. 2005).

Numerous studies have demonstrated that ABCA1 is necessary for the efflux of cellular cholesterol to lipid-poor apoA-I (Oram and Lawn 2001). Recently, ABCA1 was found to facilitate the efflux of CNS cholesterol to apoE as the absence of

ABCA1 compromised apoE secretion from both astrocytes and microglia. In addition, apoE that is present in the CSF of ABCA1-deficient animals is poorly lipidated (Hirsch-Reinshagen et al. 2004; Wahrle et al. 2004). A relationship between ABCG1 and the secretion of apoE was suggested by the observation that treatment of macrophages with antisense oligonucleotides to ABCG1 decreased the efflux of cholesterol and phospholipids to HDL and, surprisingly, also the secretion of apoE (Klucken et al. 2000;Wang et al. 2004). Karten et al. (2006) reported that only expression of ABCG1, but not of ABCA1, correlates with cholesterol release by astrocytes. In contrast with ABCA1, ABCG1 and ABCG4 are thought to facilitate the efflux of cholesterol to HDL rather than to lipid-poor apolipoproteins (Klucken et al. 2000; Wang et al. 2004). Although ABCG1 and ABCG4 may function both as homodimers and heterodimers (Oldfield et al. 2002; Rebeck 2004), expression of ABCG1 and ABCG4 overlaps in some but not all tissues assayed (Wang et al. 2004), which may indicate different functions in different tissues. The expression of ABCG4 appears to be largely restricted to nervous tissue (Rebeck 2004). ABCG1 and ABCG4 are thought to regulate cholesterol transport in the brain (Yvan-Charvet et al. 2008). Recently, in vivo evidence from transgenic mouse experiments has been provided for a role for Abcg4 in sterol efflux in the brain and that Abcg1 and Abcg4 have overlapping functions in astrocytes, promoting efflux of cholesterol, desmosterol and possibly other sterol biosynthetic intermediates to HDL (Wang et al. 2008). Moreover, they seem to mediate intracellular vesicular transport of cholesterol/sterols within both neurons and astrocytes (Yvan-Charvet et al. 2008).

Neurons are thought to dispose their cholesterol by conversion into 24(*S*)-hydroxycholesterol (Bjorkhem and Meaney 2004). However, it is not yet known exactly how oxysterols are transported across membranes and through the intracellular water phase. The selective up-regulation of ABCG4 in neuroblastoma cells by 24(*S*)-hydroxycholesterol suggests a possible role for this transporter in oxysterol transport. Alternatively, ABCA1 may be involved. Tam et al. have shown that high-affinity uptake of 25-hydroxycholesterol by membrane vesicles can be mediated by ABCA1, besides mediating its efflux from intact cells (Tam et al. 2006).

5.7 Alternative Pathway for Cholesterol Release from Neurons?

Both 24(*S*)-hydroxycholesterol and also GW683865A enhanced, although to a limited extent, apoA-I-mediated cholesterol efflux from neuronal cells, suggesting this is another neuronal pathway to dispose of cholesterol. However, GW683965A had only a limited effect on ABCA1 expression in these cells. Rebeck et al. (2004) recently reported up-regulation of neuronal ABCA1 expression by the synthetic LXR ligand T0901317. A role for apoA-I in the disposal of cholesterol from neurons is in line with its well known role in so called "reverse cholesterol transport".

ApoA-I is present in brain and in CSF and has been detected in senile plaques in AD brains (Harr et al. 1996; Panin et al. 2000). So far, apoA-I synthesis within the brain has only been ascribed to endothelial cells of the BBB (Mockel et al. 1994). Neurons are always found in close proximity to blood vessels. Considering the total length of the BBB (about 600 km in the human brain), it is assumed that every neuron has its own microvessel. Therefore, the possibility remains that endothelium derived apoA-I participates in the removal of cholesterol and phospholipids from neurons. Evidence supporting the transcytosis of apoA-I across the BBB via the SR-BI was obtained using an in vitro model for the BBB (Kratzer et al. 2007). Upregulation of ABCA1 with oxysterols increased apoA-I binding and internalization. ApoA-I binding, internalization, and transcytosis were reduced by at least 50% after silencing ABCA1 but not after knocking down SR-BI. Thus, it was concluded that ABCA1, but not SR-BI, modulates the transcytosis of apoA-I through endothelial cells (Cavelier et al. 2006). Interestingly, decreased brain levels of total cholesterol have been measured in apoA-I-deficient mice (Fagan et al. 2004).

5.8 Role for cAMP Responsive Element Binding Protein in the Regulation of Neuronal Cholesterol Homeostasis

Besides 24(*S*)-hydroxycholesterol, cAMP responsive element binding protein (CREB) may play a role in the regulation of neuronal cholesterol homeostasis (Lemberger et al. 2008). CREB was found to regulate the induction of specific gene expression patterns in neurons in response to activity. Interestingly, mice deficient for CREB and cAMP responsive element modulator (CREM) displayed altered expression of genes involved in cholesterol metabolism and accumulation of cholesterol in neurons.

How the conversion of cholesterol into 24(*S*)-hydroxycholesterol is regulated remains to be clarified. It was recently reported that the expression of CYP46A1 is regulated by Sp transcription factors (Milagre et al. 2008); and a specific ER-resident ORP8 was also suggested to play a role in oxysterol generation (Yan et al. 2008).

5.9 Internalization of Cholesterol by Neurons

It remains to be established via which pathways neurons internalize cholesterol. Receptors amongst others that have been suggested to be involved are: LDLR, LRP1, LRP8, LR11, apoER2 and VLDLR (Bu et al. 1998; Jaeger and Pietrzik 2008). The LDL-receptor is best known for mediating the internalization of cholesterol in lipoproteins via binding apoE or apoB100. In the brain, the expression of the LDL-receptor is lowest in neurons, where it is expressed in particular in cell bodies and proximal axons, and is highest in astrocytes (Pitas et al. 1987a). LRP is detected predominantly in neurons and in activated astrocytes (Bu et al. 1994; Rebeck et al. 1993, 1995). A number of in vitro and in vivo studies indicate a role for the LDLR-related protein (LRP) in the internalization of cholesterol by neurons (for reviews, see Nathan et al. 2002; Zerbinatti and Bu 2005; Zerbinatti et al. 2004, 2006). Besides the LDLR and LRP neurons express apoER2 mostly in distal axons and LR11 in cell bodies/proximal axons and also SR-BI (Offe et al. 2006; Posse De Chaves et al. 2000). Part of the mitochondrial ATP synthase complex has been identified as a hepatocyte receptor for the uptake of HDL particles (Lyly et al. 2008).

Although the absence of LDLR or apoE does not affect the total levels of cholesterol in the brain or its rate of synthesis, it does affect the transbilayer distribution of cholesterol in synaptic membranes (Igbavboa et al. 1997). Consequently, this may affect neuronal signalling functions. ApoE-deficient mice display severely impaired learning and memory functions (Oitzl et al. 1997). Deficiency for LDLR in mice results in impaired spatial memory, decreased synaptic density in the hippocampus and decreased cell proliferation (Mulder et al. 2004, 2007).

5.10 The Choroid Plexus as an Alternative Source of HDL

Astrocytes are not the only source of HDL-like lipoproteins in the brain. Epithelial cells of the choroid plexus also release cholesterol and thereby play an important role in the regulation of CSF cholesterol homeostasis (Segal 2000). The total surface of the choroid plexus epithelial cells is said to be comparable to the total surface of the BBB. Similar to its effect in astrocytes, 24(*S*)-hydroxycholesterol was found to induce cholesterol release from the apical membrane of choroid plexus epithelial cells in apoE-isoform-dependent manner concommitantly with the induction of ABCA1 and ABCG1 (Fujiyoshi et al. 2007). Using mice expressing apoE coupled to enhanced green fluorescent protein, Xu et al. (2006) have shown that apoE is expressed in cells of the choroid plexus. In line with a role for apoE, cholesterol granulomas have been detected in the choroid plexus in apoE-deficient mice (Owiny and Strandberg 2000). Tachikawa et al. (2005) reported high ABCA1 mRNA levels in choroid plexus, and absence of ABCA1 in mice results in reduced levels of apoE in CSF (Wahrle et al. 2004).

ApoE is also detectable in smooth muscle cells of large blood vessels and in cells surrounding small micro-vessels as well as in oligodendrocytes and ependymal cells, suggesting an involvement in cholesterol efflux from these cells (Boyles et al. 1985; Poirier et al. 1991; Xu et al. 2006). ApoE is normally not detected in neurons except after injury.

Additionally, microglia have also been reported to secrete apoJ-rich and apoE-poor, spherical lipoproteins that are more LDL-like and different from those secreted by astrocytes (Xu et al. 2000).

5.11 Disturbances in Cholesterol Trafficking Between Astrocytes and Neurons in Alzheimer's Disease?

In line with the hypothesis that in AD patients the trafficking of cholesterol from astrocytes to neurons may be compromised, a number of genes that have been associated with AD encode proteins that are involved in this process (for a review, see Carter 2007). APOE4 is the strongest genetic risk factor known for sporadic AD (Corder et al. 1993; Saunders et al. 1993); and it has been hypothesized that apoE may affect the pathogenesis of AD by isoform-specific effects on lipid trafficking between astrocytes and neurons (Michikawa et al. 2000). Moreover, CYP46A1 and ABCA1 have been associated with an enhanced risk of AD. Several studies have suggested a role for 24(*S*)-hydroxycholesterol in the pathogenesis of AD (Reiss 2005; Reiss et al. 2004; Wolozin et al. 2004). Polymorphisms of CYP46A1 have been linked to AD, with the exception of one study which did not find such a relationship (Desai et al. (2002). The expression of this enzyme appeared to have shifted from neurons to glia in AD patients (Bogdanovic et al. 2001). In the early stages of AD, plasma and CSF levels of 24(*S*)-hydroxycholesterol were found to be elevated, but they were reduced in the late stage, possibly due to a loss of CYP46A1-expressing metabolically very active neurons (Heverin et al. 2004; Lutjohann et al. 2000; Papassotiropoulos et al. 2000, 2002). Postmortem cerebrospinal fluid of AD patients contained lower concentrations of cholesterol, phospholipids and fatty acids, while the levels of apoE were only slightly reduced (Bretillon et al. 2000; Heverin et al. 2004; Lutjohann et al. 1996; Mulder et al. 1998; Papassotiropoulos et al. 2002; Schonknecht et al. 2002).

Since the composition of CSF is very similar to that of brain interstitial fluid, it is likely that also in interstitial fluid the levels of HDL-like particles are reduced. This may affect the deposition of Aβ, if these lipoproteins play a role in maintaining it soluble and decreasing its levels by removing it from the brain into the circulation and/or into the CSF.

Although the link between APOE genotypes and AD was established 15 years ago, it remains to be established exactly how apoE4 enhances the risk. ApoE has been suggested to play a role in the still largely unknown, transport and secretion of sterols, other than 24(*S*)-hydroxycholesterol, from the brain across the BBB into the plasma (Mulder and Terwel 1998). Bell et al. (2007) found that lipid-poor apoE is cleared via the BBB, but that lipidated apoE is cleared predominantly via the cerebrospinal fluid and barely via the BBB.

Another indication for a link between cholesterol and AD came from epidemiological studies indicating that cholesterol-lowering drugs, i.e., statins that are widely used for the treatment of hypercholesterolemia, reduced the risk of AD (Wolozin 2004). Although the literature on this topic is rather controversial. Statins were found to reduce the deposition of Aβ in plaques in the brain of AD-mouse models (Eckert et al. 2001). These effects were ascribed to the cholesterol-lowering effects of statins. In agreement, high plasma cholesterol levels and high fat intake were found to be associated with an increased risk of AD (Grant et al. 2002). Ten years ago Sparks et al. (1997) discovered plaques-like structures in brains of patients that died of cardiovascular diseases and not in brain of patients with other causes of death. Via yet unknown mechanisms, high plasma cholesterol concentrations result in an increased depositon of Aβ in the brains of AD-mouse models (Levin-Allerhand et al. 2002; Oksman et al. 2006). Cholesterol itself was found to be present in plaques (Mori et al. 2001). In a mouse model of AD, chronically elevated plasma cholesterol induced increased APOE mRNA in the brain, while these were reduced by administration of statins (Petanceska et al. 2003).

However, no effects of even very high plasma cholesterol on brain Aβ levels were observed in LDLR-deficient mice (Elder et al. 2007).

Since the brain fully relies on de novo synthesis of cholesterol and is thought to be unable to compensate by enhancing the uptake of dietary derived cholesterol from the circulation, it may be very sensitive to cholesterol synthesis reducing agents. High doses of simvastatin were found to reduce cholesterol synthesis in the brain (Thelen et al. 2006a). Therefore, it was initially suggested that the beneficial effects of statins on the development of AD may be the result of its suppresive effects on cholesterol synthesis (Li et al. 2006). However, it is questionable whether this is an advantage. The use of statins has also been associated with memory complaints (Wagstaff et al. 2003). Furthermore, it was found that statins directly inhibit long-term potentiation, which is regarded as a marker for synaptic plasticity (Kotti et al. 2006). In line with this, Kotti et al. (2006) reported that cholesterol synthesis in the brain is essential for learning processes. It may not be cholesterol itself that is required for learning but a non-sterol by-product of the mevalonate pathway, the isoprenoid "geranylgeraniol". The continuous production of small amounts of geranylgeraniol and consequently, a continuous production of cholesterol, in a subgroup of neurons is supposed to be required for spatial, assocative and motor learning. Interesting is the notification that the reported decrease in the cholesterol synthesis rate during aging may be associated with an increase in loss of memory functions (Thelen et al. 2006b).

In contrast with expectations, lovastatin appeared to induce the deposition of amyloid in brain of an AD-mouse model; and George et al. (2004) found that diet-induced hypercholesterolemia reduced brain levels of amyloid in aged mice.

The beneficial effects of statins on the progression of AD may, therefore, not be ascribed to their cholesterol-lowering effect, but to their anti-inflammatory properties or their modulating effects on the cerebral vessel walls.

The observation of an altered processing of cholesterol in fibroblasts from AD patients, suggest that the changes do not remain restricted to the central nervous system (Murphy et al. 2006).

There is evidence to suggest that it is not simply the level of cholesterol in the brain affects the production and deposition of Aβ, but rather its intracellular distribution. In vitro studies show that the cellular amount of cholesterol or the distribution across membranes directly affects the splicing of amyloid from its precursor protein and on its aggregation (Frears et al. 1999; Puglielli et al. 2001; Simons et al. 1998). Cholesterol-depleted neurons produce less amyloid than cholesterol-rich neurons.

Therefore, alterations in brain cholesterol metabolism seem to affect the production and deposition of Aβ. Alternatively, Aβ also seems to directly affect cholesterol synthesis (Hartmann 2006). Liu et al. (1998) reported that Aβ alters intracellular vesicle trafficking and decreases cholesterol esterification and the distribution of cholesterol in neurons. In AD brains, Aβ selectively accumulates in the perikaryon of pyramidal cells as discrete granules that appear to be cathepsin D-positive (Nixon et al. 2001). Cathepsin D has been suggested to play a role in the regulation of composition, trafficking and/or recycling of membrane components. The Aβ-containing granules are thought to be of endosomal/lysosomal origin (Gomez-Ramos et al. 2007).

Interestingly, apoE has been suggested to be involved in the degradation and clearance of deposited Aβ by astrocytes (Koistinaho et al. (2004). It was found that astrocytes can degrade Aβ, but not if: (i) they are deficient for apoE, or (ii) RAP, an antagonist of the LDLR family, is present or is anti-apoE. It can be speculated that the lipoprotein-like particles secreted by astrocytes that are present in the interstitial fluid pick up Aβ from neurons and direct it to astrocytes which internalize the particles via the LDLR or another member of the LDLR family. The latter is supported by the observation that the absence of the LDLR from PDAPP mice, a model for AD, does not affect amyloid levels nor its deposition (Zerbinatti et al. 2006). However, absence of apoE unexpectedly decreased amyloid load in the brain of these PDAPP mice.

Variations in the expression of ABCA1 and ABCG1 have been found to affect amyloid production and/or its deposition. Overexpression of ABCA1 in PDAPP mice results in a decrease in amyloid deposition in the brain (Wahrle et al. 2008), but apparently not by affecting its transport from the brain into the circulation (Akanuma et al. 2008). In vitro evidence was obtained indicating an impact of ABCG1 on amyloid production, but in vivo such an effect was not detectable in PDAPP mice overexpressing ABCG1 (Burgess et al. 2008). Novel evidence showed that ABCA7, a close homolog of ABCA1, belonging to the same ABCA subfamily of full-length ABC transporters, can regulate APP processing (Chan et al. 2008).

LXR agonists may be promising tools in the treatment of AD. Activation of the LXR pathway via synthetic agonists was found to reduce the production of amyloid in cultured neurons (Sun et al. 2003). It was suggested that this resulted from an upregulated expression of ABCA1. Whether or not this needs to be accompagnied by an enhanced neuronal cholesterol efflux remains controversial (Burns et al. 2006; Rebeck 2004; Sun et al. 2003). LXR activation was found to reduce amyloid levels in the brain and its deposition in a AD-mouse model (Koldamova et al. 2005).

5.12 Do Alterations in Systemic Sterol Metabolism Alter Brain Sterol Metabolism?

It remains to be established whether or not variations in extracerebral/systemic cholesterol levels affect brain cholesterol metabolism. Total levels of cholesterol in brain remain unaltered even if a plasma cholesterol are reduced by 84% due to administration of either simvastatin or pravastatin in guinea-pigs (Lutjohann et al. 2004). In line with a strictly and independently regulated brain cholesterol metabolism, dramatic increases in circulating cholesterol levels in apoE-deficient mice due to administration of a high-fat diet did not affect levels of sterols in the brain with the exception of elevated levels of 27-hydroxycholesterol (Paula J. Jansen, Dieter Lütjohann, Karin M. Thelen, Klaus von Bergmann, Fred van Leuven, Frans C.S. Ramaekers and Monique Mulder, unpublished data). Heverin et al. (2005) showed a net flux of 27-hydroxycholesterol from the circulation into the brain using plasma samples collected from the internal jugular vein and an artery of healthy male volunteers. 27-hydroxycholesterol in human brain was distributed consistent with an extracerebral origin, with a concentration gradient from the white to the grey matter – a situation opposite to that of 24S-hydroxycholesterol. 27-Hydroxycholesterol was suggested to be and important link between intra- and extracerebral cholesterol homeostasis (Heverin et al. 2005). Similarly, other circulating cholesterol metabolites were found to enter the brain. For example Panzenboeck et al, (2007) demonstrated an efficient transfer of 7alpha-hydroxy-4-cholesten-3-one, a metabolite of cholestanol, across cultured porcine brain endothelial cells (a model for the blood--brain barrier; Panzenboeck et al. 2007). Accordingly, accumulating evidence indicates an effect of a high-fat/high-cholesterol diet on brain (Mulder et al. 2001). Mateos et al. (2008) reported that a high-fat diet and also 27-hydroxycholesterol down regulates cytoskeleton associated protein (Arc), a protein that is also down regulated in AD brains.

Evidence was obtained that plant sterols, that can be obtained only from the diet and have a structure very similar to that of cholesterol, can end up in the brain of mice (Jansen et al. 2006). The addition of supplements to the diet could therefore provide be an alternative strategy to modulate brain cholesterol metabolism and, thereby, the development and/or the progression of AD. Possible ways to achieve this may be for example via the addition of specific compounds that have been found to affect the LXR pathway (Wagstaff et al. 2003; Yang et al. 2004).

References

Abildayeva K, et al (2006) 24(S)-hydroxycholesterol participates in a liver X receptor-controlled pathway in astrocytes that regulates apolipoprotein E-mediated cholesterol efflux. J Biol Chem 281:12799–12808

Adams CM, et al (2004) Cholesterol and 25-hydroxycholesterol inhibit activation of SREBPs by different mechanisms, both involving SCAP and Insigs. J Biol Chem 279:52772–52780

Akanuma S, et al (2008) ATP-binding cassette transporter A1 (ABCA1) deficiency does not attenuate the brain-to-blood efflux transport of human amyloid-beta peptide (1–40) at the blood-brain barrier. Neurochem Int 52:956–961

Apfel R, et al (1994) A novel orphan receptor specific for a subset of thyroid hormone-responsive elements and its interaction with the retinoid/thyroid hormone receptor subfamily. Mol Cell Biol 14:7025–7035

Argmann CA, et al (2005) Regulation of macrophage cholesterol efflux through hydroxymethylglutaryl-CoA reductase inhibition: a role for RhoA in ABCA1-mediated cholesterol efflux. J Biol Chem 280:22212–22221

Bassett CN, et al (2000) Cerebrospinal fluid lipoproteins in Alzheimer's disease. Microsc Res Tech 50:282–286

Bell RD, et al (2007) Transport pathways for clearance of human Alzheimer's amyloid beta-peptide and apolipoproteins E and J in the mouse central nervous system. J Cereb Blood Flow Metab 27:909–918

Bellosta S, et al (1995) Stable expression and secretion of apolipoproteins E3 and E4 in mouse neuroblastoma cells produces differential effects on neurite outgrowth. J Biol Chem 270:27063–27071

Bjorkhem I, Meaney S (2004) Brain cholesterol: long secret life behind a barrier. Arterioscler Thromb Vasc Biol 24:806–815

Bjorkhem I, et al (1997) Importance of a novel oxidative mechanism for elimination of brain cholesterol. Turnover of cholesterol and 24(S)-hydroxycholesterol in rat brain as measured with 18O2 techniques in vivo and in vitro. J Biol Chem 272:30178–30184

Bjorkhem I, et al (1998) Cholesterol homeostasis in human brain: turnover of 24S-hydroxycholesterol and evidence for a cerebral origin of most of this oxysterol in the circulation. J Lipid Res 39:1594–1600

Bjorkhem I, et al (2001) Oxysterols in the circulation of patients with the Smith–Lemli–Opitz syndrome: abnormal levels of 24S- and 27-hydroxycholesterol. J Lipid Res 42:366–371

Bogdanovic N, et al (2001) On the turnover of brain cholesterol in patients with Alzheimer's disease. Abnormal induction of the cholesterol-catabolic enzyme CYP46 in glial cells. Neurosci Lett 314:45–48

Borghini I, et al (1995) Characterization of subpopulations of lipoprotein particles isolated from human cerebrospinal fluid. Biochim Biophys Acta 1255:192–200

Boyles JK, et al (1985) Apolipoprotein E associated with astrocytic glia of the central nervous system and with nonmyelinating glia of the peripheral nervous system. J Clin Invest 76:1501–1513

Bretillon L, et al (2000) Plasma levels of 24S-hydroxycholesterol in patients with neurological diseases. Neurosci Lett 293:87–90

Brown MS, Goldstein JL (1997) The SREBP pathway: regulation of cholesterol metabolism by proteolysis of a membrane-bound transcription factor. Cell 89:331–340

Bu G, et al (1994) Subcellular localization and endocytic function of low density lipoprotein receptor-related protein in human glioblastoma cells. J Biol Chem 269:29874–29882

Bu G, et al (1998) Nerve growth factor induces rapid increases in functional cell surface low density lipoprotein receptor-related protein. J Biol Chem 273:13359–13365

Burgess BL, et al (2008) ABCG1 influences the brain cholesterol biosynthetic pathway but does not affect amyloid precursor protein or apolipoprotein E metabolism in vivo. J Lipid Res 49:1254–1267

Burns MP, et al (2006) The effects of ABCA1 on cholesterol efflux and Abeta levels in vitro and in vivo. J Neurochem 98:792–800

Carter CJ (2007) Convergence of genes implicated in Alzheimer's disease on the cerebral cholesterol shuttle: APP, cholesterol, lipoproteins, and atherosclerosis. Neurochem Int 50:12–38

Cavelier C, Rohrer L, von Eckardstein A (2006) ATP-binding cassette transporter A1 modulates apolipoprotein A-I transcytosis through aortic endothelial cells. Circ Res 99:1060–1066

Chan SL, et al (2008) ATP-binding cassette transporter A7 regulates processing of amyloid precursor protein in vitro. J Neurochem 106:793–804

Corder EH, et al (1993) Gene dose of apolipoprotein E type 4 allele and the risk of Alzheimer's disease in late onset families. Science 261:921–923

Davison AN (1965) Brain sterol metabolism. Adv Lipid Res 3:171–196

de Vries HE, et al (1995) High-density lipoprotein and cerebral endothelial cells in vitro: interactions and transport. Biochem Pharmacol 50:271–273

Desai P, DeKosky ST, Kamboh MI (2002) Genetic variation in the cholesterol 24-hydroxylase (CYP46) gene and the risk of Alzheimer's disease. Neurosci Lett 328:9–12

Dietschy JM, Turley SD (2001) Cholesterol metabolism in the brain. Curr Opin Lipidol 12:105–112

Eckert GP, Kirsch C, Mueller WE (2001) Differential effects of lovastatin treatment on brain cholesterol levels in normal and apoE-deficient mice. Neuroreport 12:883–887

Eckert GP, et al (2007) Regulation of central nervous system cholesterol homeostasis by the liver X receptor agonist TO-901317. Neurosci Lett 423:47–52

Elder GA, et al (2007) Elevated plasma cholesterol does not affect brain Abeta in mice lacking the low-density lipoprotein receptor. J Neurochem 102:1220–1231

Elshourbagy NA, et al (1985) Expression of rat apolipoprotein A-IV and A-I genes: mRNA induction during development and in response to glucocorticoids and insulin. Proc Natl Acad Sci USA 82:8242–8246

Fagan AM, et al (1998) Evidence for normal aging of the septo-hippocampal cholinergic system in apoE (–/–) mice but impaired clearance of axonal degeneration products following injury. Exp Neurol 151:314–325

Fagan AM, et al (1999) Unique lipoproteins secreted by primary astrocytes from wild type, apoE (–/–), and human apoE transgenic mice. J Biol Chem 274:30001–30007

Fagan AM, et al (2004) ApoAI deficiency results in marked reductions in plasma cholesterol but no alterations in amyloid-beta pathology in a mouse model of Alzheimer's disease-like cerebral amyloidosis. Am J Pathol 165:1413–1422

Frears ER, et al (1999) The role of cholesterol in the biosynthesis of beta-amyloid. Neuroreport 10:1699–1705

Fujiyoshi M, et al (2007) 24S-hydroxycholesterol induces cholesterol release from choroid plexus epithelial cells in an apical- and apoE isoform-dependent manner concomitantly with the induction of ABCA1 and ABCG1 expression. J Neurochem 100:968–978

George AJ, et al (2004) APP intracellular domain is increased and soluble Abeta is reduced with diet-induced hypercholesterolemia in a transgenic mouse model of Alzheimer disease. Neurobiol Dis 16:124–132

Goldstein JL, DeBose-Boyd RA, Brown MS (2006) Protein sensors for membrane sterols. Cell 124:35–46

Gomez-Ramos P, Asuncion Moran M (2007) Ultrastructural localization of intraneuronal Abeta-peptide in Alzheimer disease brains. J Alzheimers Dis 11:53–59

Gong JS, et al (2002) Apolipoprotein E (ApoE) isoform-dependent lipid release from astrocytes prepared from human ApoE3 and ApoE4 knock-in mice. J Biol Chem 277:29919–29926

Grant WB, et al (2002) The significance of environmental factors in the etiology of Alzheimer's disease. J Alzheimers Dis 4:179–189

Guo J, et al (2008) Effects of cholesterol levels on the excitability of rat hippocampal neurons. Mol Membr Biol 25:216–223

Guyton JR, et al (1998) Novel large apolipoprotein E-containing lipoproteins of density 1.006–1.060 g/ml in human cerebrospinal fluid. J Neurochem 70:1235–1240

Harr SD, et al (1996) Brain expression of apolipoproteins E, J, and A-I in Alzheimer's disease. J Neurochem 66:2429–2435

Hartmann T (2006) Role of amyloid precursor protein, amyloid-beta and gamma-secretase in cholesterol maintenance. Neurodegener Dis 3:305–311

Heverin M, et al (2004) Changes in the levels of cerebral and extracerebral sterols in the brain of patients with Alzheimer's disease. J Lipid Res 45:186–193

Heverin M, et al (2005) Crossing the barrier: net flux of 27-hydroxycholesterol into the human brain. J Lipid Res 46:1047–1052

Hirsch-Reinshagen V, et al (2004) Deficiency of ABCA1 impairs apolipoprotein E metabolism in brain. J Biol Chem 279:41197–41207

Hummasti S, et al (2004) Liver X receptors are regulators of adipocyte gene expression but not differentiation: identification of apoD as a direct target. J Lipid Res 45:616–625

Igbavboa U, et al (1997) Transbilayer distribution of cholesterol is modified in brain synaptic plasma membranes of knockout mice deficient in the low-density lipoprotein receptor, apolipoprotein E, or both proteins. J Neurochem 69:1661–1667

Ikonen E (2008) Cellular cholesterol trafficking and compartmentalization. Nat Rev Mol Cell Biol 9:125–138

Jaeger S, Pietrzik CU (2008) Functional role of lipoprotein receptors in Alzheimer's disease. Curr Alzheimer Res 5:15–25

Jansen PJ, et al (2006) Dietary plant sterols accumulate in the brain. Biochim Biophys Acta 1761:445–453
Ji Y, et al (2003) Apolipoprotein E isoform-specific regulation of dendritic spine morphology in apolipoprotein E transgenic mice and Alzheimer's disease patients. Neuroscience 122:305–315
Kandutsch AA, Chen HW (1974) Inhibition of sterol synthesis in cultured mouse cells by cholesterol derivatives oxygenated in the side chain. J Biol Chem 249:6057–6061
Karten B, et al (2006) Expression of ABCG1, but not ABCA1, correlates with cholesterol release by cerebellar astroglia. J Biol Chem 281:4049–4057
Kessel A, Ben-Tal N, May S (2001) Interactions of cholesterol with lipid bilayers: the preferred configuration and fluctuations. Biophys J 81:643–658
Klucken J, et al (2000) ABCG1 (ABC8), the human homolog of the Drosophila white gene, is a regulator of macrophage cholesterol and phospholipid transport. Proc Natl Acad Sci USA 97:817–822
Koch S, et al (2001) Characterization of four lipoprotein classes in human cerebrospinal fluid. J Lipid Res 42:1143–1151
Koistinaho M, et al (2004) Apolipoprotein E promotes astrocyte colocalization and degradation of deposited amyloid-beta peptides. Nat Med 10:719–726
Koldamova RP, et al (2005) The liver X receptor ligand T0901317 decreases amyloid beta production in vitro and in a mouse model of Alzheimer's disease. J Biol Chem 280:4079–4088
Kotti TJ, et al (2006) Brain cholesterol turnover required for geranylgeraniol production and learning in mice. Proc Natl Acad Sci USA 103:3869–3874
Kratzer I, et al (2007) Apolipoprotein A-I coating of protamine-oligonucleotide nanoparticles increases particle uptake and transcytosis in an in vitro model of the blood–brain barrier. J Control Release 117:301–311
Krause BR, Hartman AD (1984) Adipose tissue and cholesterol metabolism. J Lipid Res 25:97–110
LaDu MJ, et al (1998) Nascent astrocyte particles differ from lipoproteins in CSF. J Neurochem 70:2070–2081
Legleiter J, et al (2004) In situ AFM studies of astrocyte-secreted apolipoprotein E- and J-containing lipoproteins. J Colloid Interface Sci 278:96–106
Lemaire-Ewing S, et al (2005) Comparison of the cytotoxic, pro-oxidant and pro-inflammatory characteristics of different oxysterols. Cell Biol Toxicol 21:97–114
Lemberger T, et al (2008) CREB has a context-dependent role in activity-regulated transcription and maintains neuronal cholesterol homeostasis. Faseb J 22:2872–2879
Levin-Allerhand JA, Lominska CE, Smith JD (2002) Increased amyloid levels in APPSWE transgenic mice treated chronically with a physiological high-fat high-cholesterol diet. J Nutr Health Aging 6:315–319
Li L, et al (2006) Simvastatin enhances learning and memory independent of amyloid load in mice. Ann Neurol 60:729–739
Liu Y, Peterson DA, Schubert D (1998) Amyloid beta peptide alters intracellular vesicle trafficking and cholesterol homeostasis. Proc Natl Acad Sci USA 95:13266–13271
Lomnitski L, et al (1999) Distinct alterations in phospholipid metabolism in brains of apolipoprotein E-deficient mice. J Neurosci Res 58:586–592
Lund EG, Guileyardo JM, Russell DW (1999) cDNA cloning of cholesterol 24-hydroxylase, a mediator of cholesterol homeostasis in the brain. Proc Natl Acad Sci USA 96:7238–7243
Lund EG, et al (2003) Knockout of the cholesterol 24-hydroxylase gene in mice reveals a brain-specific mechanism of cholesterol turnover. J Biol Chem 278:22980–22988
Lutjohann D, et al (1996) Cholesterol homeostasis in human brain: evidence for an age-dependent flux of 24S-hydroxycholesterol from the brain into the circulation. Proc Natl Acad Sci USA 93:9799–9804
Lutjohann D, et al (2000) Plasma 24S-hydroxycholesterol (cerebrosterol) is increased in Alzheimer and vascular demented patients. J Lipid Res 41:95–98
Lutjohann D, et al (2004) High doses of simvastatin, pravastatin, and cholesterol reduce brain cholesterol synthesis in guinea pigs. Steroids 69:431–438

Lyly A, et al (2008) Deficiency of the INCL protein Ppt1 results in changes in ectopic F1-ATP synthase and altered cholesterol metabolism. Hum Mol Genet 17:1406–1417

Mateos L, Akterin S, Gil-Bea FJ, Spulber S, Rahman A, Björkhem I, Schultzberg M, Flores-Morales A, Cedazo-Mínguez A. (2008) Activity-regulated cytoskeleton-associated protein in rodent brain is down-regulated by high fat diet in vivo and by 27-hydroxycholesterol in vitro. Brain Pathol (in press)

Marx J (2001) Alzheimer's disease. Bad for the heart, bad for the mind? Science 294:508–509

Mast N, et al (2003) Broad substrate specificity of human cytochrome P450 46A1 which initiates cholesterol degradation in the brain. Biochemistry 42:14284–14292

Mauch DH, et al (2001) CNS synaptogenesis promoted by glia-derived cholesterol. Science 294:1354–1357

Maysinger D, et al (2008) Ceramide is responsible for the failure of compensatory nerve sprouting in apolipoprotein E knock-out mice. J Neurosci 28:7891–7899

Meaney S, et al (2007) Novel route for elimination of brain oxysterols across the blood-brain barrier: conversion into 7alpha-hydroxy-3-oxo-4-cholestenoic acid. J Lipid Res 48:944–951

Michikawa M, et al (2000) Apolipoprotein E exhibits isoform-specific promotion of lipid efflux from astrocytes and neurons in culture. J Neurochem 74:1008–1016

Milagre I, et al (2008) Transcriptional regulation of the human CYP46A1 brain-specific expression by Sp transcription factors. J Neurochem 106:835–849

Mockel B, et al (1994) Expression of apolipoprotein A-I in porcine brain endothelium in vitro. J Neurochem 62:788–798

Moghadasian MH, et al (2002) Cerebrotendinous xanthomatosis: a rare disease with diverse manifestations. Arch Neurol 59:527–529

Mori T, et al (2001) Cholesterol accumulates in senile plaques of Alzheimer disease patients and in transgenic APP(SW) mice. J Neuropathol Exp Neurol 60:778–785

Mulder M, Terwel D (1998) Possible link between lipid metabolism and cerebral amyloid angiopathy in Alzheimer's disease: a role for high-density lipoproteins? Haemostasis 28:174–194

Mulder M, et al (1998) Reduced levels of cholesterol, phospholipids, and fatty acids in cerebrospinal fluid of Alzheimer disease patients are not related to apolipoprotein E4. Alzheimer Dis Assoc Disord 12:198–203

Mulder M, et al (2001) Apolipoprotein E protects against neuropathology induced by a high-fat diet and maintains the integrity of the blood–brain barrier during aging. Lab Invest 81:953–960

Mulder M, et al (2004) Low-density lipoprotein receptor-knockout mice display impaired spatial memory associated with a decreased synaptic density in the hippocampus. Neurobiol Dis 16:212–219

Mulder M, et al (2007) LDL receptor deficiency results in decreased cell proliferation and presynaptic bouton density in the murine hippocampus. Neurosci Res 59:251–256

Murphy EJ, et al (2006) Phospholipid mass is increased in fibroblasts bearing the Swedish amyloid precursor mutation. Brain Res Bull 69:79–85

Muse ED, et al (2001) Parameters related to lipid metabolism as markers of myelination in mouse brain. J Neurochem 76:77–86

Nakai M, et al (1996) Expression of apolipoprotein E mRNA in rat microglia. Neurosci Lett 211:41–44

Nathan BP, et al (1994) Differential effects of apolipoproteins E3 and E4 on neuronal growth in vitro. Science 264:850–852

Nathan BP, et al (1995) The inhibitory effect of apolipoprotein E4 on neurite outgrowth is associated with microtubule depolymerization. J Biol Chem 270:19791–19799

Nathan BP, et al (2002) Apolipoprotein E4 inhibits, and apolipoprotein E3 promotes neurite outgrowth in cultured adult mouse cortical neurons through the low-density lipoprotein receptor-related protein. Brain Res 928:96–105

Navarro A, Del Valle E, Tolivia J (2004) Differential expression of apolipoprotein d in human astroglial and oligodendroglial cells. J Histochem Cytochem 52:1031–1036

Nixon RA, Mathews PM, Cataldo AM (2001) The neuronal endosomal-lysosomal system in Alzheimer's disease. J Alzheimers Dis 3:97–107

Offe K, et al (2006) The lipoprotein receptor LR11 regulates amyloid beta production and amyloid precursor protein traffic in endosomal compartments. J Neurosci 26:1596–1603
Oitzl MS, et al (1997) Severe learning deficits in apolipoprotein E-knockout mice in a water maze task. Brain Res 752:189–196
Oksman M, et al (2006) Impact of different saturated fatty acid, polyunsaturated fatty acid and cholesterol containing diets on beta-amyloid accumulation in APP/PS1 transgenic mice. Neurobiol Dis 23:563–572
Oldfield S, et al (2002) ABCG4: a novel human white family ABC-transporter expressed in the brain and eye. Biochim Biophys Acta 1591:175–179
Oram JF, Lawn RM (2001) ABCA1. The gatekeeper for eliminating excess tissue cholesterol. J Lipid Res 42:1173–1179
Owiny JR, Strandberg JD (2000) Cholesterol granulomas in mice deficient in apolipoprotein E. Contemp Top Lab Anim Sci 39:57–58
Panin LE, Russkikh GS, Polyakov LM (2000) Detection of apolipoprotein A-I, B, and E immunoreactivity in the nuclei of various rat tissue cells. Biochemistry (Mosc) 65:1419–1423
Panzenboeck U, et al (2002) ABCA1 and scavenger receptor class B, type I, are modulators of reverse sterol transport at an in vitro blood-brain barrier constituted of porcine brain capillary endothelial cells. J Biol Chem 277:42781–42789
Panzenboeck U, et al (2007) On the mechanism of cerebral accumulation of cholestanol in patients with cerebrotendinous xanthomatosis. J Lipid Res 48:1167–1174
Papassotiropoulos A, et al (2000) Plasma 24S-hydroxycholesterol: a peripheral indicator of neuronal degeneration and potential state marker for Alzheimer's disease. Neuroreport 11:1959–1962
Papassotiropoulos A, et al (2002) Cerebrospinal fluid levels of beta-amyloid(42) in patients with Alzheimer's disease are related to the exon 2 polymorphism of the cathepsin D gene. Neuroreport 13:1291–1294
Patel SC, et al (1995) Astrocytes synthesize and secrete the lipophilic ligand carrier apolipoprotein D. Neuroreport 6:653–657
Petanceska SS, et al (2003) Changes in apolipoprotein E expression in response to dietary and pharmacological modulation of cholesterol. J Mol Neurosci 20:395–406
Pfrieger FW (2003) Outsourcing in the brain: do neurons depend on cholesterol delivery by astrocytes? Bioessays 25:72–78
Pitas RE, et al (1987a) Lipoproteins and their receptors in the central nervous system. Characterization of the lipoproteins in cerebrospinal fluid and identification of apolipoprotein B,E(LDL) receptors in the brain. J Biol Chem 262:14352–14360
Pitas RE, et al (1987b) Astrocytes synthesize apolipoprotein E and metabolize apolipoprotein E-containing lipoproteins. Biochim Biophys Acta 917:148–161
Poirier J, et al (1991) Astrocytic apolipoprotein E mRNA and GFAP mRNA in hippocampus after entorhinal cortex lesioning. Brain Res Mol Brain Res 11:97–106
Posse De Chaves EI, et al (2000) Uptake of lipoproteins for axonal growth of sympathetic neurons. J Biol Chem 275:19883–19890
Puglielli L, et al (2001) Acyl-coenzyme A: cholesterol acyltransferase modulates the generation of the amyloid beta-peptide. Nat Cell Biol 3:905–912
Puglielli L, et al (2004) Role of acyl-coenzyme a: cholesterol acyltransferase activity in the processing of the amyloid precursor protein. J Mol Neurosci 24:93–96
Rebeck GW (2004) Cholesterol efflux as a critical component of Alzheimer's disease pathogenesis. J Mol Neurosci 23:219–224
Rebeck GW, et al (1993) Apolipoprotein E in sporadic Alzheimer's disease: allelic variation and receptor interactions. Neuron 11:575–580
Rebeck GW, et al (1995) Multiple, diverse senile plaque-associated proteins are ligands of an apolipoprotein E receptor, the alpha 2-macroglobulin receptor/low-density-lipoprotein receptor-related protein. Ann Neurol 37:211–217
Reiss AB (2005) Cholesterol and apolipoprotein E in Alzheimer's disease. Am J Alzheimers Dis Other Demen 20:91–96

Reiss AB, et al (2004) Cholesterol in neurologic disorders of the elderly: stroke and Alzheimer's disease. Neurobiol Aging 25:977–989
Roheim PS, et al (1979) Apolipoproteins in human cerebrospinal fluid. Proc Natl Acad Sci USA 76:4646–4649
Saunders AM, et al (1993) Association of apolipoprotein E allele epsilon 4 with late-onset familial and sporadic Alzheimer's disease. Neurology 43:1467–1472
Schonknecht P, et al (2002) Cerebrospinal fluid 24S-hydroxycholesterol is increased in patients with Alzheimer's disease compared to healthy controls. Neurosci Lett 324:83–85
Segal MB (2000) The choroid plexuses and the barriers between the blood and the cerebrospinal fluid. Cell Mol Neurobiol 20:183–196
Simons M, et al (1998) Cholesterol depletion inhibits the generation of beta-amyloid in hippocampal neurons. Proc Natl Acad Sci USA 95:6460–6464
Skipor J, Thiery JC (2008) The choroid plexus–cerebrospinal fluid system: undervaluated pathway of neuroendocrine signaling into the brain. Acta Neurobiol Exp. (Wars) 68:414–428
Slezak M, Pfrieger FW (2003) New roles for astrocytes: regulation of CNS synaptogenesis. Trends Neurosci 26:531–535
Snipes GJ, Suter U (1997) Cholesterol and myelin. Subcell Biochem 28:173–204
Song BL, Javitt NB, DeBose-Boyd RA (2005) Insig-mediated degradation of HMG CoA reductase stimulated by lanosterol, an intermediate in the synthesis of cholesterol. Cell Metab 1:179–189
Song C, et al (1994) Ubiquitous receptor: a receptor that modulates gene activation by retinoic acid and thyroid hormone receptors. Proc Natl Acad Sci USA 91:10809–10813
Spady DK, Dietschy JM (1983) Sterol synthesis in vivo in 18 tissues of the squirrel monkey, guinea pig, rabbit, hamster, and rat. J Lipid Res 24:303–315
Sparks DL (1997) Coronary artery disease, hypertension, ApoE, and cholesterol: a link to Alzheimer's disease? Ann NY Acad Sci 826:128–146
Steffensen KR, et al (2003) Gene expression profiling in adipose tissue indicates different transcriptional mechanisms of liver X receptors alpha and beta, respectively. Biochem Biophys Res Commun 310:589–593
Sun Y, et al (2003) Expression of liver X receptor target genes decreases cellular amyloid beta peptide secretion. J Biol Chem 278:27688–27694
Tachikawa M, et al (2005) Distinct spatio-temporal expression of ABCA and ABCG transporters in the developing and adult mouse brain. J Neurochem 95:294–304
Tam SP, et al (2006) ABCA1 mediates high-affinity uptake of 25-hydroxycholesterol by membrane vesicles and rapid efflux of oxysterol by intact cells. Am J Physiol Cell Physiol 291:C490–C502
Thelen KM, et al (2006a) Cholesterol synthesis rate in human hippocampus declines with aging. Neurosci Lett 403:15–19
Thelen KM, et al (2006b) Brain cholesterol synthesis in mice is affected by high dose of simvastatin but not of pravastatin. J Pharmacol Exp Ther 316:1146–1152
Travert C, Carreau S, Le Goff D (2006) Induction of apoptosis by 25-hydroxycholesterol in adult rat Leydig cells: protective effect of 17beta-estradiol. Reprod Toxicol 22:564–570
Ulven SM, et al (2004) Tissue-specific autoregulation of the LXRalpha gene facilitates induction of apoE in mouse adipose tissue. J Lipid Res 45:2052–2062
Utermann G, Pruin N, Steinmetz A (1979) Polymorphism of apolipoprotein E. III. Effect of a single polymorphic gene locus on plasma lipid levels in man. Clin Genet 15:63–72
Vance JE, Hayashi H, Karten B (2005) Cholesterol homeostasis in neurons and glial cells. Semin Cell Dev Biol 16:193–212
Wagstaff LR, et al (2003) Statin-associated memory loss: analysis of 60 case reports and review of the literature. Pharmacotherapy 23:871–880
Wahrle SE, et al (2004) ABCA1 is required for normal central nervous system ApoE levels and for lipidation of astrocyte-secreted apoE. J Biol Chem 279:40987–40993
Wahrle SE, et al (2008) Overexpression of ABCA1 reduces amyloid deposition in the PDAPP mouse model of Alzheimer disease. J Clin Invest 118:671–682

Wang L, et al (2002) Liver X receptors in the central nervous system: from lipid homeostasis to neuronal degeneration. Proc Natl Acad Sci USA 99:13878–13883

Wang N, et al (2004) ATP-binding cassette transporters G1 and G4 mediate cellular cholesterol efflux to high-density lipoproteins. Proc Natl Acad Sci USA 101:9774–9779

Wang N, et al (2008) ATP-binding cassette transporters G1 and G4 mediate cholesterol and desmosterol efflux to HDL and regulate sterol accumulation in the brain. FASEB J 22:1073–1082

Whitney KD, et al (2001) Liver X receptor (LXR) regulation of the LXRalpha gene in human macrophages. J Biol Chem 276:43509–43515

Whitney KD, et al (2002) Regulation of cholesterol homeostasis by the liver X receptors in the central nervous system. Mol Endocrinol 16:1378–1385

Wiegand V, et al (2003) Transport of plasma membrane-derived cholesterol and the function of Niemann–Pick C1 Protein. FASEB J 17:782–784

Wolozin B (2004) Cholesterol and the biology of Alzheimer's disease. Neuron 41:7–10

Wolozin B, et al (2004) The cellular biochemistry of cholesterol and statins: insights into the pathophysiology and therapy of Alzheimer's disease. CNS Drug Rev 10:127–146

Xie C, Turley SD, Dietschy JM (2000) Centripetal cholesterol flow from the extrahepatic organs through the liver is normal in mice with mutated Niemann–Pick type C protein. (NPC1). J Lipid Res 41:1278–1289

Xie C, et al (2003) Quantitation of two pathways for cholesterol excretion from the brain in normal mice and mice with neurodegeneration. J Lipid Res 44:1780–1789

Xu Q, et al (2000) Isolation and characterization of apolipoproteins from murine microglia. Identification of a low density lipoprotein-like apolipoprotein J-rich but E-poor spherical particle. J Biol Chem 275:31770–31777

Xu Q, et al (2006) Profile and regulation of apolipoprotein E (ApoE) expression in the CNS in mice with targeting of green fluorescent protein gene to the ApoE locus. J Neurosci 26:4985–4994

Yan D, et al (2008) OSBP-related protein 8 (ORP8) suppresses ABCA1 expression and cholesterol efflux from macrophages. J Biol Chem 283:332–340

Yang C, et al (2004) Disruption of cholesterol homeostasis by plant sterols. J Clin Invest 114:813–822

Yvan-Charvet L, et al (2008) SR-BI inhibits ABCG1-stimulated net cholesterol efflux from cells to plasma HDL. J Lipid Res 49:107–114

Zerbinatti CV, Bu G (2005) LRP and Alzheimer's disease. Rev Neurosci 16:123–135

Zerbinatti CV, et al (2004) Increased soluble amyloid-beta peptide and memory deficits in amyloid model mice overexpressing the low-density lipoprotein receptor-related protein. Proc Natl Acad Sci USA 101:1075–1080

Zerbinatti CV, et al (2006) Apolipoprotein E and low density lipoprotein receptor-related protein facilitate intraneuronal Abeta42 accumulation in amyloid model mice. J Biol Chem 281:36180–36186

Zhang M, Strnatka D, Donohue C, Hallows JL, Vincent I, Erickson RP (2008) Astrocyte-only Npc1 reduces neuronal cholesterol and triples life span of Npc1(–/–) mice. J Neurosci Res 86:2848–2856

Chapter 6
Intracellular Cholesterol Transport

Daniel Wüstner

Abstract Cholesterol is the single most abundant lipid species in mammalian cells. More than 2×10^9 years of evolution designed this molecule to perfectly fit into phospholipid bilayers regulating the fluidity, permeability and bending stiffness of biological membranes. Cholesterol also serves as a precursor of steroid hormones, bile acids and oxysterols, and its cellular synthesis is regulated by a complex machinery. While the molecular mechanisms underlying cholesterol synthesis are known in great detail, knowledge is rather sparse about the inter-compartment transport of cholesterol, including trafficking modes and kinetics, as well as control of endomembrane cholesterol content. This chapter provides an overview of our recent understanding of intracellular transport of cholesterol. It is aimed to create a link between the well characterized biophysical properties of cholesterol in model membranes and its behavior in living cells.

6.1 Biophysical Properties of Cholesterol in Model Membranes

Cholesterol is a very hydrophobic lipid molecule with the C3-connected OH group as the only polar constituent. Consequently, it has a very low solubility in water, forming micelle-like aggregates at concentrations as low as 20–30 nM (Haberland et al. 1973; Loomis et al. 1979). Its very hydrophobic character combined with its planar steroid ring structure and stereochemistry means that cholesterol perfectly fits into phospho- and sphingolipid bilayers. Single-component phospholipid bilayers show a temperature-dependent phase transition characterized by a highly co-operative chain melting process and a transition from the gel (so) to the liquid–crystalline (Lα) phase. To efficiently shield cholesterol from contact with water, it is located underneath the polar lipid head groups and additionally straightens their fatty acyl

D. Wüstner
Department of Biochemistry and Molecular Biology, University of Southern Denmark, Campusvej 55, 5230 Odense M, Denmark
e-mail: wuestner@bmb.sdu.dk

C. Ehnholm (ed.), *Cellular Lipid Metabolism*,
DOI 10.1007/ 978-3-642-00300-4_6, © Springer-Verlag Berlin Heidelberg 2009

chains in the fluid lipid bilayer. This so-called condensing effect of cholesterol decreases the lateral distance of the host phospholipids, resulting in tighter packing in phospholipid/cholesterol bilayers compared to pure phospholipids above the phase transition temperature (Yeagle 1985; Silvius et al. 1996). A consequence of this process is that the lipid bilayer gets thicker, while at the same time its area decreases in such a way that the total bilayer volume remains constant (Lindahl and Edholm 2000; Nagle and Tristram-Nagle 2000; Hofsass et al. 2003). The condensing effect of cholesterol can be measured and is conveniently represented by the ordering of the phospholipid acyl chains determined by NMR spectroscopy. This order parameter can be related to other bilayer properties like bending stiffness and resistance against area dilation – both mechanical properties which increase in the presence of cholesterol (Needham et al. 1988; Henriksen et al. 2004, 2006). Fluorescent cholesterol analogs used for analysis of cholesterol trafficking by microscopy must resemble cholesterol as closely as possible. Importantly, intrinsically fluorescent sterols like dehydroergosterol (DHE) and cholestatrienol (CTL), but not NBD-cholesterol, have a potential to order fatty acyl chains in bilayers comparable to cholesterol at relatively low sterol concentrations (Scheidt et al. 2003), while at the same time raising the bilayer stiffness like cholesterol (Garvik et al. 2009). Due to these similarities to cholesterol, both the sterols DHE and CTL are very suitable for the analysis of intracellular sterol trafficking by sensitive low-light fluorescence microscopy (see Fig. 6.1; Hao et al. 2002; Wüstner et al. 2002, 2005; Hartwig Petersen et al. 2008). At very high cholesterol concentrations (above 30 mol%) a third phase can form in sterol/phospholipid bilayers, called the liquid-ordered (Lo) phase characterized by high acyl chain order combined with rapid lateral diffusion (Vist and Davis 1990; Ipsen et al. 1987; see Fig. 6.2). This Lo phase thus bears properties of both the gel phase (high fatty acyl chain order) and the fluid phase (rapid lateral diffusion), and it attracted great attention in the past decade due to its similarity to lipid fractions isolated from cellular membranes by detergent extraction (Ahmed et al. 1997). However, it has to be emphasized that, by definition, the Lo phase is a property of simple two- or three-component lipid bilayers, and extrapolation to cellular membrane characteristics can lead to wrong, physiologically irrelevant conclusions (see Sect. 6.2). In ternary model systems, the acyl chain-ordering effect of cholesterol in the Lo phase can be directly monitored as a height difference between fluid Lα and ordered Lo domains by atomic force microscopy (see Fig. 6.2). The observed domains appear round, since they are stabilized by line tension energy between the Lo and Lα phases. In other words, lipids in either phase are free to diffuse laterally and organize in such a way as to minimize their contact to the phase "they don't like", so to speak. This gives round phases, since this shape minimizes contact area, similar to the case of fat droplets in a soup, for example. Interestingly, we showed recently that DHE does not only partition preferentially into the Lo phase in ternary mixtures consisting of dioleoylphosphatidylcholine/dipalmitoylphosphatidylcholine and DHE at a 1:1:1 ratio, but can also induce Lo/Lα phase separation with round domains, clearly confirming that DHE behaves very much like cholesterol under the same conditions (Garvik et al. 2009).

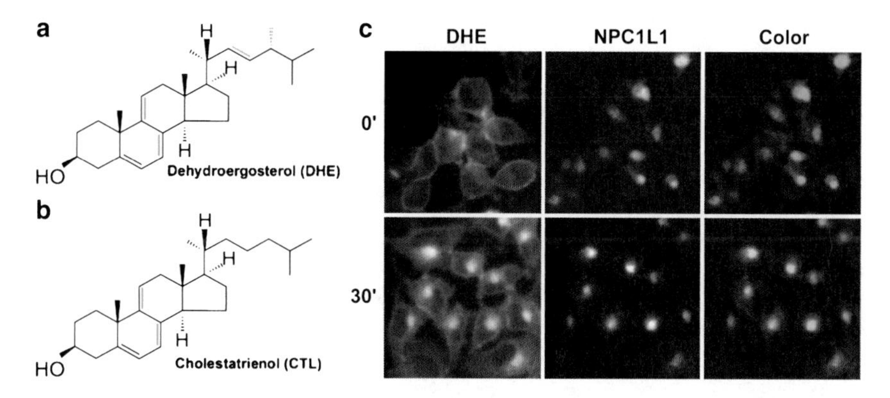

Fig. 6.1 Direct observation of cholesterol transport by microscopy using intrinsically fluorescent sterols. **a**, **b** Chemical structure of dehydroergosterol (*DHE*, **a**) and cholestatrienol (*CTL*, **b**) with differences to cholesterol highlighted in green. While CTL has the same side-chain as cholesterol, the side-chain of DHE is identical to that of ergosterol, i.e., branched methyl and one double bound. The fluorescence properties of both sterols are almost identical due to the fact that the "chromophore" formed by the three conjugated double bonds in the steroid ring system is the same. **c** Distribution of DHE in McArdle7777 rat hepatoma cells expressing enhanced green fluorescent protein-tagged Niemann–Pick C1-like 1 (EGFP-NPC1L1) protein. Cells were pulse-labeled for 1 min with DHE loaded onto cyclodextrin, washed and either imaged immediately (0') or chased for 30 min at 37°C before imaging (30') on a wide-field fluorescence microscope optimized for transmission in the ultraviolet (UV) to detect DHE and additionally equipped with a fluorescein filter set to detect EGFP fluorescence. Right after labeling, DHE is found exclusively in the plasma membrane, while EGFP-NPC1L1 resides in the perinuclear endocytic recycling compartment (ERC) and to a small extent at the cell surface. After 30 min chase, DHE co-localizes very much with EGFP-NPC1L1 in the ERC region. The color overlay shows DHE in red and EGFP-NPC1L1 in green. Co-localization appears yellow to orange. For further details, see Hartwig Petersen et al. (2008) and text

It has been suggested that the cell regulates plasma membrane protein function by modulating the transverse pressure profile, e.g., via altering the lipid composition and distribution or the cholesterol content (Cantor 1999). Indeed, modulating the cholesterol content of cellular membranes has been shown to directly affect protein function, like inhibition of calcium ATPases and protein translocation machinery in the ER (Nilsson et al. 2001; Li et al. 2004). The proper function of rhodopsin and several G protein-coupled receptors depends on cholesterol. For example, modification of membrane cholesterol and sphingolipids alters the conformation and function but not intracellular trafficking of the G protein-coupled cholecystokinin receptor, while cholesterol is required for maintenance of the specific agonist binding activity.of the serotonin(1A) receptor (Chattopadhyay et al. 2005; Harikumar et al. 2005; see Fig. 6.3). Much effort has been spent in the past and there is still considerable focus on the many facets of cholesterol's properties in membranes, but not all of them can be discussed here. Beside the ability to order lipid acyl chains and to induce the Lo phase, two other properties of cholesterol and

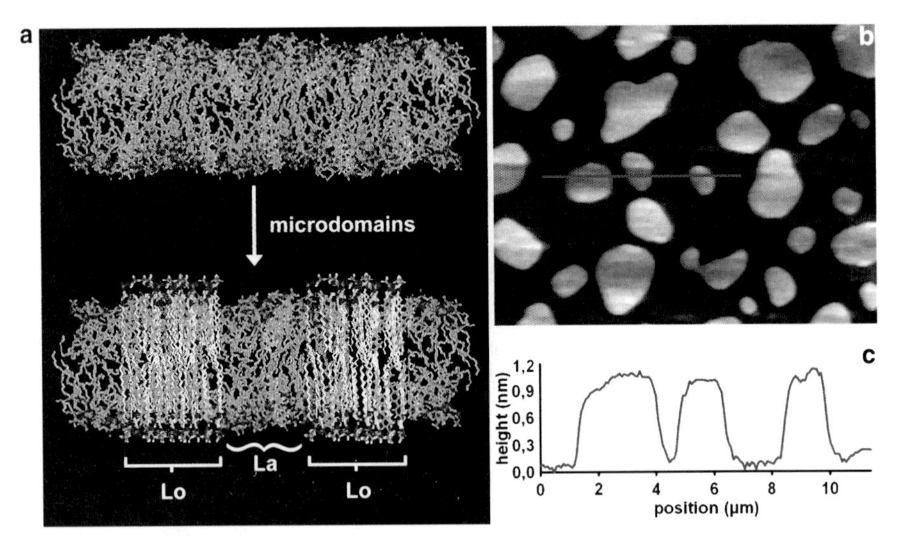

Fig. 6.2 Cholesterol induces fluid–fluid immiscibility in phospholipid membranes. **a** Cartoon of a lipid bilayer made from molecular simulation of phosphatidylcholine (*PC*) with added cholesterol molecules in green. *Upper panel* shows the phospholipid bilayer in the fluid Lα phase with characteristic disordered fatty acid chains. Carbon atoms are shown in light gray, phosphorus in yellow, oxygen in red and nitrogen in blue, hydrogen atoms are omitted for clarity. *Lower panel* shows fluid–fluid immiscibility as in **a**, with two small domains in the ordered Lo phase surrounded by the fluid Lα phase. Due to straightening of the fatty acyl chains and tighter packing with cholesterol, the Lo phase appears to be thicker than the surrounding Lα phase. **b** Atomic force microscopy of a lipid bilayer consisting of three components, dipalmitoylphosphatidylcholine (DPPC)/dioleoylphosphatidylcholine (DOPC)/cholesterol at a molar ratio of 40:40:20 at 20°C on mica support. Round thick Lo domains (light color) surrounded by a continuous Lα phase (brown color) can be clearly distinguished. The Lo phase contains most of the saturated DPPC and probably most cholesterol. **c** Line scan with the AFM tip along the blue line in **b**, indicating the height difference between the Lα phase (set to zero height) and the Lo phase (approx. 1 nm higher). Note that, while **b** and **c** are based on experimental data, **a** is only a cartoon to illustrate cholesterol-induced phase separation in lipid membranes. This illustration was generated from an atomistic Monte Carlo simulation performed with the AMBER force field on a Linux machine by the author in collaboration with Dr. Heinz Sklenar (Max Delbrück Centrum, Berlin, Germany). **b** and **c** are courtesy of Dr. Adam Cohen Simonsen from the Department of Physics and Chemistry, Center for membrane physics (MEMPHYS) at the University of Southern Denmark

related sterols are of importance when looking at intracellular cholesterol transport. First, the transbilayer orientation: cholesterol (and DHE) show a rapid flip-flop from one leaflet to the other in model membranes, with a half-time of a few seconds (Leventis and Silvius 2001; John et al. 2002). The same has been reported for cholesterol in red blood cell membranes (Steck et al. 2002) and can be attributed to the low free energy barrier associated with moving a single polar OH group through the hydrocarbon chains of the bilayer (Hamilton 2003; Róg et al. 2008). Second, cholesterol shows a reasonable mobility out of the membrane, perpendicular to the bilayer plane (Gliss et al. 1999; Wüstner 2007; Lange and Steck 2008). This might explain how the relatively slow release of cholesterol from membranes could be

Fig. 6.3 Possible impact of cholesterol on receptor dynamics in the plasma membrane. Cholesterol (shown on the *left* in green) could affect lateral diffusion of a symbolic receptor molecule (green zig-zag *arrow*) consisting of several transmembrane helices and connecting loops (shown in red; *I*). Cholesterol could bind directly to the receptor and thereby alter its conformation (curved blue arrow; *II*) and/or induce receptor clustering (green double arrow; *III*) at the cell surface. In addition, cholesterol could change binding properties of the receptor to a ligand (blue) either due to change of receptor structure or mobility (*IV*). Note that this is an idealized receptor and ligand embedded in a schematically drawn plasma membrane. See text for more details

overcome by cytoplasmic carriers: these proteins might pick up individual sterol molecules during their transient protrusion from the bilayer (see Sect. 6.5).

Finally, I want to mention that there is an upper limit of cholesterol solubility in lipid membranes. Above 50–60 mol% cholesterol in the bilayer the host phospholipids are no longer able to shield cholesterol from contacting water, causing the precipitation of cholesterol in form of monohydrate crystals from the membrane (Huang and Feigenson 1999; Huang et al. 1999). Cholesterol crystals likely formed by this mechanism have been detected in atherosclerotic lesions and in macrophage cell culture under free cholesterol loading conditions (Small et al. 1974; Tangirala et al. 1994; Kellner-Weibel et al. 1999).

6.2 Molecular Organization and Function of Cholesterol in the Plasma Membrane

The plasma membrane harbours most cellular cholesterol, since tight lipid packing as mediated by cholesterol's condensing effect is important to create a permeability barrier against ions, small metabolites and even oxygen in the plasma membrane (Subczynski et al. 1989; Krylov et al. 2001; Maxfield and Wüstner 2002; Khan et al. 2003). It has been shown that the two membrane leaflets independently create a diffusion barrier and that a major role is played by cholesterol to maintain this barrier function (Subczynski et al. 1989; Hill and Zeidel 2000; Krylov et al. 2001). Phospho- and sphingolipids are unevenly distributed between the two plasma membrane leaflets (op den Kamp 1979; Maxfield and Wüstner 2002). While most sphingolipids and phosphatidylcholine (PC) reside in the exoplasmic leaflet, aminophospholipids and phosphatidylinositols (PI) are enriched on the inner, cytoplasmic membrane leaflet. This phenomenon was first described for red blood cells and is now a commonly accepted fact for mammalian cells in general. The lipid transbilayer asymmetry is actively maintained, in the case of aminophospholipids, for example, by a translocase probably belonging to the type P-ATPases (Pomorski et al. 2004). While much has been learned about transbilayer phospho- and sphingolipid transport in the past decade, the distribution of cholesterol across the plasma membrane leaflets remains obscure. Some studies state that cholesterol is enriched in the outer lamelle, others provide evidence for preferred sterol localization to the inner membrane leaflet (Maxfield and Wüstner 2002). Finally, it has been proposed that cholesterol flip-flop across the plasma membrane of red blood cells occurs within seconds, suggesting that its transbilayer dynamics are very high (Steck et al. 2002). Such rapid cholesterol flip-flop, together with reasonable mobility perpendicular to the membrane plane, could play an important role in coupling the two membrane leaflets (Gliss et al. 1999; Wüstner 2007a). Based on cholesterol's well known partition preference into membranes enriched in PC and sphingomyelin (SM) compared to aminophospholipids or PI, it is more likely that cholesterol accumulates in the outer but not in the inner leaflet of the plasma membrane (Ohvo-Rekila et al. 2002; Niu and Litman 2002). A very recent study by Maxfield and co-workers, however, found the fluorescent sterol DHE residing preferentially (i.e., by 60–70%) in the cytoplasmic leaflet of the plasma membrane in CHO cells (Mondal et al. 2008). The uncertainty regarding cholesterol distribution between the two membrane leaflets is due to methodological limitations (Maxfield and Wüstner 2002).

It was soon realized that membrane protein functions like signal transduction upon ligand binding to receptors require functional and probably also spatial compartmentation of proteins in the plasma membrane (reviewed by Karnovsky et al. 1982; Edidin 2003). In some cell types, like endothelial cells, specialized plasma membrane invaginations named caveolae were identified as important cholesterol-dependent "sub-compartments" of the cell surface being involved in signal transduction. Originally, it was proposed that the plasma membrane is a two-dimensional fluid with embedded proteins in a "sea" of fluid lipids (see Edidin 2003). However, a large body of evidence has accumulated during the past 15 years showing that this fluid–mosaic model of the plasma membrane is oversimplified (Subczynski and Kusumi 2003). One very popular model assumes that proteins show lateral aggregation in small microdomains in the plasma membrane together with certain lipids. According to this "raft" model, cholesterol clusters specifically with sphingolipids in the exoplasmic leaflet of the plasma membrane in so-called rafts which might play an important role in signal transduction and protein sorting (Ahmed et al. 1997; Simons and Ikonen 1997; Maxfield 2002). Raft microdomains were originally characterized by their resistance to solubilization at low temperature, using non-ionic detergents like Triton X-100, as well as by single-particle tracking techniques (Simons and van Meer 1988; Pralle et al. 2000). Detergent treatment itself, however, can induce membrane domain formation and clustering of certain proteins in the plasma membrane (Heerklotz 2002; Mayor and Maxfield 1995). Moreover, more than 70% of the plasma membrane surface area is detergent-resistant and putative raft proteins show no difference in their (bulk) surface diffusion properties compared to non-raft proteins (Maxfield 2002; Kenworthy et al. 2004). Another model, called the "membrane skeleton fence model", is based on extensive single particle tracking and fluorescence recovery after photobleaching studies and assumes that the membrane-attached cytoskeleton plays a crucial role in compartmentalization of the plasma membrane (Koppel et al. 1981; Sheetz 2001; Subczynski and Kusumi 2003). Fences for lateral diffusion have been characterized for the transferrin receptor, band 3 and other membrane proteins in living mammalian cells (Sako and Kusumi 1995; Tomishige et al. 1998). Even for fluorescent phospholipids, "hop-diffusion" between plasma membrane compartments has been demonstrated (Fujiwara et al. 2002). Thus, while membrane compartmentation is mainly "lipid-driven" in the "raft" model, the "membrane skeleton fence model" assumes "preformed" compartments due to the dynamic interaction of the cytoskeleton with the plasma membrane (Sheetz 2001; Maxfield 2002; Subczynski and Kusumi 2003; Mukherjee and Maxfield 2004). Without reviewing all the arguments for or against different membrane models (additionally, see Maxfield and Wüstner 2002; Munro 2003; Kenworthy 2008), I want to bring the reader's attention to the most recent results using fluorescence nanoscopy of living cells, like stimulated emission depletion (STED) and photoactivatable localization microscopy (PALM; Hell 2007). These techniques provide images of the cell surface with high resolution below the diffraction limit of 200 nm, suggesting that these techniques are most suitable to distinguish between competing membrane models (Sieber et al. 2006, 2007; Hess 2007). For example, in a recent study using STED, it was shown that self-association of EGFP-tagged syntaxin 1 expressed in PC12 cells depends on weak homophilic protein–protein interactions without the need for rafts (Sieber et al. 2006, 2007). Importantly, syntaxin clusters require normal cholesterol levels in the plasma membrane, but this is not related to rafts defined by detergent resistance, since these clusters are Triton X-100 soluble and do not co-patch with raft markers (Lang et al. 2001). It is likely that many of the observed effects of cholesterol on lateral diffusion or interaction of proteins in the plasma membrane depend on cholesterol-mediated membrane interaction with the underlying actin cytoskeleton (Sheetz 2001; Kwik et al. 2003; Sun et al. 2007; Ganguly et al. 2008). The molecular mechanisms remain to be deciphered but likely involve cholesterol-regulated levels and distribution of phosphatidylinositols, specifically phosphatidyl-4,5-bisphosphate (PIP2) in the inner leaflet of the plasma membrane (Kwik et al. 2003). In fact, atomic force and fluorescence microscopy recently demonstrated that cholesterol depletion by incubating cells with cyclodextrin redistributed the PIP2 and F-actin, accompanied by a significant increase in the adhesion energy between the membrane and the cytoskeleton and a decrease in the membrane diffusion constants of both raft and non-raft proteins, as well as fluorescent lipid probes (Kwik et al. 2003; Kenworthy et al. 2004; Goodwin et al. 2005; Shvartsman et al. 2006; Sun et al. 2007).

While it is evident that the plasma membrane has a heterogeneous lateral organization at the nanometer to micron scale, no agreement exists on the lateral distribution of cholesterol in the plasma membrane. Early attempts to characterize sterol membrane organisation were based on filipin staining in freeze–fracture electron micrographs. Evidence for lateral heterogeneity of cholesterol based on this method with a preferred enrichment in areas of cell attachment, caveolae and filopodia has been provided (Montesano 1979; Bridgman and Nakajima 1983). Filipin, a sterol binding polyene antifungal antibiotic with a pentaene chromophore, binds non-esterified cholesterol and related sterols, and this binding to membrane cholesterol can be detected by fluorescence microscopy or by electron microscopy, where characteristic bumps show up on membrane sheets (Maxfield and Wüstner 2002). Due to the fact that filipin extracts sterols from membranes and thereby changes membrane lipid organization, using this probe gives no absolute measure of membrane cholesterol content or distribution in cells (Behnke et al. 1984a, b). Moreover, when properly used, cell fixation is required, which also is a source of artefacts. Without reviewing all results based on various biophysical methods in fixed cells (for a detailed discussion, see Wüstner 2007a, b), I will shortly review recent data obtained by wide-field UV imaging and three-photon microscopy of DHE in living cells. Both methods showed that the plasma membrane staining of DHE is heterogeneous and non-random, while the interpretation of this observation was very different (Wüstner 2005, 2007b; Zhang 2005). Schroeder and colleagues analyzed three-photon images and suggested that DHE clusters in certain lateral domains in the plasma membrane (Zhang 2005). In contrast, we used highly sensitive multicolor deconvolution wide-field microscopy to show that the same heterogeneous membrane staining is found for DHE as well as for fluid lipid probes (Wüstner 2007b). We demonstrated in an independent study that DHE completely segregates from these probes in model giant unilamellar vesicles (Garvik et al. 2009). Thus, the absence of segregation of DHE and fluid phase lipid markers in living cell membranes suggests that the heterogeneous membrane staining and co-patching of both probes is caused by the rough surface topography of living cells but not by lateral sterol-rich and -poor domains in the bilayer plane (Wüstner 2007b). We could confirm this conclusion in subsequent studies by dynamic wide-field imaging as well as in cholesterol-loaded macrophage foam cells and membrane blebs (Wüstner 2008; Wüstner and Færgeman 2008). Importantly, we could not detect higher amounts of DHE in membrane cavoelae visualized by GFP-tagged caveolin in adipocytes and fibroblasts (Wüstner and Færgeman 2008). This result, obtained on living cells, suggests that caveolae have the same cholesterol content as the remaining plasma membrane and are not cholesterol-enriched membrane microdomains. In a recent analysis investigating dynamic plasma membrane staining patterns of DHE by wide-field UV and three-photon microscopy, we come to the same conclusion, namely rough surface topography as the cause for heterogeneous sterol staining (Wüstner et al. 2009). Thus, we can rule out that the two different methods of visualizing DHE in living cells provide different results.

6.3 Overview of Membrane Traffic Along the Endocytic and Secretory Pathways and its Dependence on Cholesterol

Our understanding of membrane traffic along the endocytic and secretory pathways is largely based on studies of protein transport. This chapter does not serve as a general reference to the endocytic or secretory pathways, since we touch only those aspects relevant to cholesterol trafficking. I begin this discussion therefore with a short introduction into clathrin-dependent endocytosis (see Fig. 6.4). For a general review of membrane traffic, the interested reader is referred to recent reviews (Mukherjee et al. 1997; Bonifacino and Lippincott-Schwartz 2003; Maxfield and McGraw 2004; Pfeffer 2007). Ligands, like LDL or iron-transporting transferrin, bind to their respective receptors at the cell surface; those receptor–ligand complexes become collected in clathrin-coated pits (CCP) which trigger – together with linking of the receptor to adaptor proteins – internalization of the complex. The internalized vesicle looses rapidly its clathrin coat by ATP-dependent uncoating, followed by fusion of the endocytic vesicle with a population of early endosomes, called sorting endosomes (SE). Due to the slightly acidic pH, ligands like LDL dissociate from their receptor within the SE and follow the degradative pathway, i.e., shuttling to late endosomes and lysosomes for degradation. Similarly, iron is released from transferrin in the SE, while apotransferrin bound to its receptor recycles to the cell surface, either directly from SE in a rapid PI3-kinase dependent circuit (half-time about 2 min) or after transit though the long-living perinuclear ERC with a half-time of 9–12 min (Mukherjee et al. 1997; van Dam et al. 2002; Maxfield and McGraw 2004). This classic, well accepted scheme was actually developed over the years, based on the pioneering studies of Brown and Goldstein, who first described LDL trafficking and metabolism in fibroblasts (Goldstein and Brown 1974). There are several sorting mechanisms at play in the SE, including segregation of ubiquitinated receptors destined for down-regulation into clathrin-coated regions of the SE and recruitment of Hsr and ESCRT proteins to Rab7-positive late endosomes, followed by lysosomal targeting and digestion of cargo (Maxfield and McGraw 2004; van der Goot and Gruenberg 2006). Ligands dissociated from their receptor due to the pH drop in the SE, like LDL, can be sorted towards the degradative pathway by a simple geometric sorting mechanism: tubules containing a high surface-to-volume ratio bud repeatedly from the SE towards the ERC thereby segregating membrane-bound (LDL) receptors from luminal cargo (Maxfield and McGraw 2004). The SE lose fusion competence for incoming vesicles within 6–8 min and mature into late endosomes, thereby acquiring the machinery for digesting internalized cargo. One characteristic of this maturation is a switch in associated rab GTPases: SE contain rab5, being important for fusion, vesicle motility and recruitment of PI3-kinases, as well as rab4, being involved in rapid recycling to the cell surface. In contrast, newly forming late endosomes acquire rab7, which plays a role in motor-mediated transport of these organelles along microtubule tracks (Miaczynska and Zerial 2002; Murray and Wolkoff 2003;

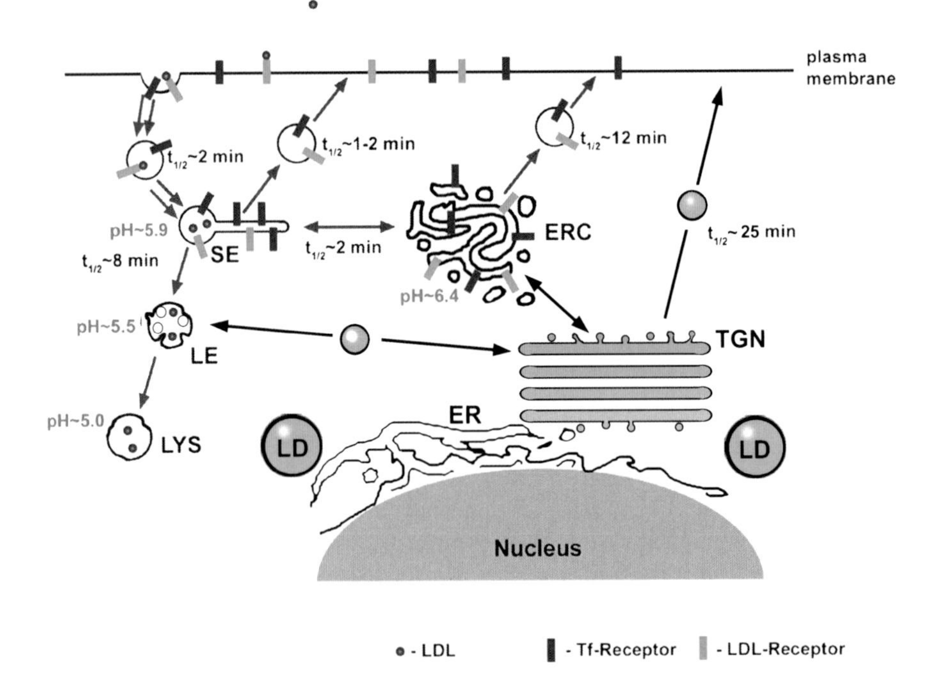

SE - sorting endosome, LE - late endosome, LYS - lysosome
ERC - endocytic recycling compartment
TGN - trans-Golgi network, ER - endoplasmic reticulum
LD - lipid droplet

Fig. 6.4 Overview of membrane transport between endocytic and secretory compartments. Ligands like LDL (dark blue sphere) bind on the cell surface to their receptor, which after collection in clathrin-coated pits is internalized as vesicle and targeted to early sorting endosomes (*SE*). There, LDL dissociates from the receptor and shuttles to late endosomes (*LE*), while the LDL receptor recycles to the cell surface, either directly from SE, or after passage through the endocytic recycling compartment (*ERC*). The fate of the LDL receptor is shared by apotransferrin bound to the transferrin receptor, a prominent marker of the SE and ERC. Maturation of SE into LE and further conversion of LE into lysosomes is accompanied by formation of internal vesicles in LE, acquisition of lysobisphosphatidic acid, a drop in pH and import of hydrolases, proteases and other lysosomal enzymes from the Golgi and specifically from the trans-Golgi network (*TGN*). The endoplasmic reticulum (*ER*) is the major site of protein and lipid synthesis including synthesis of cholesterol. Fatty acids and cholesterol are esterified in the ER and stored in cytoplasmic lipid droplets (*LD*) being derived from the ER membrane. Vesicles containing new proteins and lipids form continuously in the ER and shuttle though Golgi stacks (shown in light green). During this passage, proteins are modified (e.g., glycosylated) and exported to the plasma membrane in course of secretory membrane transport. Some Golgi-resident proteins take complex trafficking routes also through the endocytic system and reach the TGN either from LE (like Furin) or from the ERC (like mannose-6-phosphate receptor and TGN38). All of these pathways are probably involved in cellular cholesterol transport. Specifically the cholesterol derived from hydrolysis of cholesteryl esters in the LE or lysosomes is delivered either directly to the cell surface or probably via the TGN and ERC to the ER, where the sterol-sensing machinery resides. The released cholesterol can also be re-esterified by acetyl-CoA-acyltransferase being probably located in a sub-compartment of the ER in close proximity to TGN and ERC. Approximate half-times ($t_{1/2}$) of inter-compartment vesicle transport and pH values of endosomal compartments were taken from the literature. See text for further details

Rink 2005). Mature late endosomes are characterized by a lot of internal vesicles, the unusual lipid lysobisphosphatidic acid (LBPA), enrichment of mannose-6-phosphate receptor (M6PR) and the acquisition of rab9 (van der Goot and Gruenberg 2006; Pfeffer 2007). This small rab GTPase is central for shuttling M6PR back to the trans-Golgi network (TGN), thereby linking the endocytic to the secretory pathway. Rab9 is also essential for exporting cholesterol hydrolyzed from LDL-cholesteryl esters by acidic cholesteryl esterase from late endosomes and lysosomes (Narita et al. 2005; Ganley and Pfeffer 2006). There exist several clathrin-independent endocytic pathways in mammalian cells, which are regulated by cholesterol and sphingolipids (Mayor and Pagano 2007; Sandvig et al. 2008). This is a rapidly expanding field, but it is already clear that these pathways require different machineries, import different cargo and might start from different regions at the cell surface including caveolae or membrane ruffles. For example, one clathrin-independent way involves RhoA and Cdc42 and serves a role in internalizing GPI-anchored proteins, while another requires caveolae and mediates uptake of glycosphingolipid analogs like BODIPY-lactosylceramide (BODIPY-LacCer; Mayor and Pagano 2007).

Importantly, a well defined amount of cholesterol in organelle membranes is crucial for endocytic trafficking of proteins and lipids. Acute cholesterol depletion using cyclodextrin inhibits clathrin-dependent endocytosis, while mild cholesterol extraction for shorter periods interferes with other, clathrin-independent pathways (Mayor and Pagano 2007). Stimulation of macropinocytosis and membrane ruffling via activation of Rac was found upon cholesterol-loading of macrophages, while both processes become blocked by cholesterol-depletion of cells (Grimmer et al. 2002; Pierini et al. 2003; Qin et al. 2006; Nagao et al. 2007). Lowering the cellular cholesterol content or inhibiting sphingolipid synthesis slows recycling of GPI-anchored proteins from the ERC, while return of transferrin to the cell surface occurs with unchanged kinetics (Mayor et al. 1998; Chatterjee et al. 2001). Sorting of fluorescent lipid probes with varying acyl chain length between degradative and recycling pathways depends on cholesterol (Hao et al. 2004), while caveolae-like endocytosis of sphingolipid probes is stimulated by cholesterol-loading of cells (Sharma et al. 2004). Cholesterol-loading of endosomes disturbs rab4-dependent recycling to the cell surface (Choudhury et al. 2004), rab9-dependent export of cargo from late endosomes to the TGN (Ganley and Pfeffer 2006) and rab7-dependent late endosome motility (Lebrand et al. 2002). Similarly, cholesterol-loading blocks secretory transport from the TGN to the cell surface and triggers cPLA2- and dynamin-dependent Golgi vesiculation (Ying et al. 2003; Grimmer et al. 2005). Statins, which block HMG-CoA reductase and thereby cholesterol synthesis, in combination with growing cells in cholesterol-depleted medium, retard the endoplasmic reticulum (ER)-to-Golgi transport of vesicular stomatitis virus glycoprotein and scavenger receptor A (Ridsdale et al. 2006; Runz et al. 2006). Cholesterol depletion causes lowering of the lateral mobility of these two secretory membrane proteins in the ER and thereby decreases packaging efficiency into transport vesicles (Ridsdale et al. 2006; Runz et al. 2006). The lowered ER-to-Golgi trafficking of the secretory protein ts-O45-G in response to cholesterol depletion is accompanied or maybe even caused by delayed accumulation of this protein at

ER-exit sites (ERES; Runz et al. 2006). Raising the amount of cholesterol in the ER reversibly inhibits an early step in protein translocation across the ER membrane (Nilsson et al. 2001). Similarly, cholesterol-loading of macrophages inhibits sarcoplasmic-endoplasmic reticulum calcium ATPase-2b and increases lipid order in the ER membrane (Li et al. 2004). Together, these results clearly demonstrate that secretory membrane traffic depends on well defined amounts of cholesterol in the ER and Golgi compartments.

What might be the molecular mechanisms underlying cholesterol requirements of endocytic and secretoric membrane transport? This question remains open for most of the experimental observations. Based on the biophysical properties of cholesterol several hypotheses can be put forward:

1. Cholesterol in endosomes might be required for maintaining the ion permeability barrier and thereby the transmembrane electrical potential in these organelles.
2. At the same time, endosomal and Golgi membranes need to be flexible in order to bend and form tubules for inter-compartment trafficking, setting an upper limit for membrane cholesterol content, since too much cholesterol would profoundly lower bending flexibility (Henriksen et al. 2004).
3. Protein diffusion must enable formation of protein scaffolds for budding and scission decisions.

These competing demands require well defined amounts of cholesterol, and this can explain why cholesterol-depletion as well as -loading perturbs various membrane trafficking pathways. In a few cases clear experimental evidence for a molecular mechanism underlying cholesterol's function has been provided. For example, cholesterol loading of late endosomes, as occurs in Niemann–Pick type C (NPC) disease (see below, Sect. 6.6), has been reported to stabilize the prenylated rab9 at the membrane, thereby inhibiting the GDI-mediated exchange of rab9 to its GDP-bound form in the cytosol. This process, though only shown on artificial model liposomes, might underlie the observed retardation of Rab9-mediated late-endosome-to-Golgi trafficking of M6PR in response to cholesterol loading (Ganley and Pfeffer 2006). Inhibition of clathrin-dependent endocytosis by acute cholesterol depletion might be caused by a reorganization of F-actin underling the plasma membrane due to a changed distribution or content of PIP2, both effects have been observed upon cholesterol depletion (Kwik et al. 2003). In fact, reduced plasma membrane PIP2 has a similar inhibiting effect on clathrin-dependent endocytosis as acute cholesterol depletion (Jost et al. 1998; Boucrot et al. 2006; Zoncu et al. 2007).

6.4 Function of Various Organelles in Cellular Cholesterol Metabolism and Transport

Cholesterol synthesis starts by formation of hydroxymethylglutaryl (HMG)-CoA from condensation of acetyl-CoA with acetoacetyl-CoA in the cytoplasm (in contrast to the reverse reaction taking place in mitochondria to recover acetyl-CoA units from ketone bodies). Next, the HMG-CoA is reduced to mevalonate by ER-resident

HMG-CoA reductase. The activity of this enzyme is highly regulated by: (a) feed-back inhibition by cholesterol and lanosterol, via ubiquitination and degradation of the enzyme, and (b) by glucagon and insulin, acting via a cycle of phosphorylations/ dephosphorylations of hormone-sensitive kinases and phosphatases. At the transcriptional level, the amount of HMG-CoA reductase is strictly controlled by sterol regulatory response element (SREBP) containing a transcription factor which – upon proteolytic cleavage in the Golgi – induces transcription of the genes for HMG-CoA reductase and other sterol-regulated genes in the nucleus (for a detailed review, see Goldstein et al. 2006). A complex of SREBP, SREBP cleavage activating protein (SCAP) and INSIG resides in the ER at normal cholesterol levels, while SCAP containing a sterol sensing domain (SSD) senses a decrease in ER cholesterol under starvation conditions. This causes INSIG to dissociate from SREBP/SCAP, and the latter protein complex becomes incorporated into COPII coated vesicles shuttling SREBP/SCAP to the Golgi, where two proteases cleave the N-terminal transcription factor from SREBP. The released transcription factor moves to the nucleus, where it initiates synthesis of HMG-CoA reductase (as well as the LDL receptor and other proteins) thereby triggering synthesis of new cholesterol as well as uptake of circulating plasma cholesterol (via LDL receptor). Importantly, INSIG controls expression of HMG-CoA reductase via SCAP/SREBP as well as ubiquitination and degradation of the enzyme. In both cases, INSIG binds to a SSD of either SCAP or HMG-CoA reductase, suggesting a competitive mechanism regulated by cholesterol and lanosterol as well as by oxysterols (Goldstein et al. 2006). In order to sense even slight changes in ER cholesterol content, the ER contains only 0.5–1.0% of total cellular cholesterol, even though the surface area of the ER exceeds that of the plasma membrane in many cells. Thus, de novo synthesized cholesterol is rapidly exported from the site of synthesis; and one pathway characterized so far comprises shuttling of new cholesterol to the plasma membrane within 10 min, or so. The ER-to-plasma membrane route of newly synthesized cholesterol is studied using radioactive ^{3}H- or ^{14}C-acetate and by measuring the appearance of ^{3}H- or ^{14}C-cholesterol at the cell surface (DeGrella and Simoni 1982; Lange and Matthies 1984; Kaplan et al. 1985; Cruz and Chang 2000; Heino et al. 2000). This pathway largely bypasses the secretory organelles, including the Golgi, but requires ATP and ceases below 15°C, suggesting involvement of vesicular transport (DeGrella and Simoni 1982; Kaplan et al. 1985; Heino et al. 2000). In contrast, maintenance of the high concentration gradient of cholesterol between plasma membrane and ER did not require metabolic energy (Kaplan et al. 1985). Mitochondria have a low cholesterol content as well, despite the fact that they are the target organelles for cholesterol being metabolized to steroid hormones in steroidogenic cells (Maxfield and Wüstner 2002). The plasma membrane contains most, i.e., about 60% of total cellular cholesterol corresponding to about 40% of the lipids in this membrane (Maxfield and Wüstner 2002; Warnock et al. 1993). While early endosomes including the ERC harbor about 30–35% of total, the remaining cellular cholesterol of about 5–10% is distributed between late endosomes, lysosomes, Golgi apparatus and other organelles, including mitochondria, ER and peroxisomes. This does not preclude that certain compartments, like internal vesicles of multivesicular bodies/late endosomes contain relatively large amounts of cholesterol when normalized to the total membrane area of these organelles (Möbius et al. 2003). Similarly, due to their close apposition and continuous inter-compartment membrane traffic, the estimate for the ERC might include to some extent the TGN, even if we and others failed to detect high amounts of DHE or cholesterol in the latter compartment (Hao et al. 2002; Wüstner et al. 2002, 2005; Möbius et al. 2003). Moreover, it has to be emphasized that the values of compartmental cholesterol given above are associated with a reasonable degree of uncertainty. In fact, the standard deviation of these values as provided in the literature can be as large as 30% of the reported mean value (Warnock et al. 1993; Hao et al. 2002; Möbius et al. 2003). Beside technical limitations reviewed by Maxfield and Wüstner (2002), this uncertainty is due to variations between investigated cell types as well as due to a reported dependence of the cholesterol-to-phospholipid ratio in the plasma membrane on cell density (Lange et al. 1989; Corvera et al. 2000; Takahashi et al. 2007). It is remarkable, however, that most late endosomes and lysosomes have a low cholesterol content despite the fact that cholesterol-rich LDL particles are digested in these organelles. It is a central open question, as to how ingested LDL cholesterol is released from LDL after hydrolysis of cholesteryl esters and to which sites it is subsequently targeted (see Sects. 6.4–6.6). The fact that internal vesicles of some late endosomes are cholesterol-rich while lysosomes are cholesterol-poor is one hint that release of LDL-derived cholesterol occurs in late endosomes. Parallel measurements have been performed for the appearance of radioactive cholesterol at the cell surface, using cyclodextrin as acceptor, and for the re-esterification as a measure of transport to the ER after incubating cells with ^{3}H-cholesteryl ester containing LDL (Neufeld et al. 1996). In such experiments, cholesterol was first enriched in the lysosomes using progesterone, and it was found that, upon wash-out of progesterone, 70% of cholesterol released from LDL traffic first to the plasma membrane and subsequently to the ER, where most esterification takes place (Neufeld et al. 1996). The remaining 30% of LDL-derived cholesterol shuttle directly from the late endosome/lysosome to the ER bypassing the plasma membrane (Neufeld et al. 1996; Underwood et al. 1998). Molecular details of the cholesterol trafficking pathway from plasma membrane to the ER are lacking. The arrival of cholesterol in the ER is often measured by cholesterol esterification, based on the assumption that acyl-CoA-acyltransferase (ACAT) – the enzyme catalyzing linkage of activated fatty acids to the OH group of cholesterol to form cholesteryl esters – is located in the ER (Lange et al. 1999). Recent data, however, questioned the reliability of this assay for monitoring cholesterol transport to the ER. First, it was shown that ACAT resides in just a small sub-compartment of the ER located adjacent to the ERC and TGN (Khelef et al. 1998, 2000). Second, based on the well known effects of the oxysterol 25hydroxycholesterol on cholesterol esterification and sensing in the ER, it has been demonstrated that cholesterol esterification by ACAT is dissociable from cholesterol transport to the SREBP-based sensing machinery in the ER (Du et al. 2004). Third, activity of a second cholesterol esterase with an activity optimum at acidic pH, in contrast to ACAT, which works best at neutral pH, has been detected in endosomes or lysosomes of macrophages and other cells (Hornick et al. 1997; Wang et al. 2005). Thus, current knowledge of cholesterol trafficking to the ER (based solely on cholesterol esterification) may need to be critically re-assessed (Du et al. 2004). It is likely that measurement of cholesterol esterification monitors to a large extent cholesterol trafficking to a sub-compartment of the ER, where ACAT is located.

There is evidence for involvement of the Golgi in targeting LDL derived cholesterol to this ER sub-compartment. First, treatment of cells with brefeldin A, a drug which induces rapid mixing of Golgi and ER, strongly increased re-esterification of cholesterol released from LDL (Neufeld et al. 1996). Second, electron microscopy of filipin indicated enrichment of Golgi membranes with cholesterol after LDL ingestion (Coxey et al. 1993). Third, TGN specific SNARE proteins, like VAMP4, syntaxin 6 and 16, are involved in re-esterification of LDL-derived cholesterol (Urano et al. 2008). A problem with brefeldin is that it interferes with many trafficking steps in cells, including recycling from the ERC to the cell surface (Lippincott-Schwartz et al. 1991; van Dam et al. 2002). The above-mentioned SNARE machinery regulates also early/recycling endosome to TGN transport of Shiga toxin B (Mallard et al. 2002). This transport step, i.e., early endosome/ERC to TGN, was inhibited by overexpression of wild-type or dominant-negative rab11, and in these cells free cholesterol accumulation in the ERC and cholesterol esterification were reduced (Wilcke et al. 2000; Hölttä-Vuori et al. 2002). Given that the ERC is a cholesterol-rich organelle (Hornick et al. 1997; Hao et al. 2002; Hartwig Petersen et al. 2008), it is likely that transport of LDL-derived cholesterol to ACAT involves sorting and recycling endosomes in addition to the TGN. This conclusion is fully in line with the data described above. Again, the close apposition of ERC, TGN and the ACAT-containing sub-compartment of the ER makes this also very likely (Khelef et al. 2000). Importantly, the reduced cholesterol esterification in rab11-overexpressing cells could be rescued by incubating the cells with cholesterol loaded onto cyclodextrin, further supporting the view that ACAT is fed by different pathways, one probably directly from the plasma membrane (Hölttä-Vuori et al. 2002). Supporting this notion, enhanced non-vesicular transport of DHE to ER associated lipid droplets was found in cholesterol-loaded macrophages, and ACAT stimulation is regulated by a cholesterol oxidase accessible pool (Tabas et al. 1988; Wüstner et al. 2005). Another way to assess ER targeting of LDL derived cholesterol beside measurement of re-esterification by ACAT (with all its limitations) is by quantifying the processing of SREBP (Du et al. 2004; Kristiana et al. 2008). Strikingly, both methods provided very different kinetic results, suggesting that there are either different cholesterol pools and/or transport pathways feeding ACAT and SREBP within the ER (Kristiana et al. 2008). In line with this conclusion is recent morphological data demonstrating high sub-compartmentalization of the ER beyond the simple discrimination of smooth and rough ER (Snapp et al. 2003).

6.5 Vesicular and Non-Vesicular Transport of Cholesterol: Targets, Kinetics and Regulation

Although it is well accepted that the heterogeneous distribution of cholesterol among various organelles must be tightly regulated, the molecular mechanisms underlying intracellular cholesterol transport are largely unknown. There is accumulating evidence that sterols, like cholesterol in mammalian cells and ergosterol in yeast cells, move by vesicular modes following general membrane traffic but also by non-vesicular mechanisms (Hao et al. 2002; Wüstner et al. 2002, 2005; Li et al. 2004; Baumann et al. 2005). ATP-depletion or low temperature, both treatments which inhibit vesicular transport, only partially inhibit the targeting of DHE to the ERC (Hao et al. 2002), while the trafficking of newly synthesized cholesterol to the plasma membrane of mammalian cells largely bypasses the secretory pathway (Kaplan et al. 1985). Similarly, ergosterol transport between plasma membrane and ER occurs by a non-vesicular pathway (Li et al. 2004; Baumann et al. 2005).

Spontaneous cholesterol transfer between lipid membranes occurs with a characteristic half-time of 2–3 h (Bar et al. 1986, 1987). Partition coefficients measured between liposomes consisting of various host lipids clearly established that cholesterol prefers membranes being enriched in SM and/or phospholipids bearing saturated fatty acyl chains (Bar et al. 1987; Leventis and Silvius 2001; Niu and Litman 2002). Some 25 years ago, Wattenberg and Silbert (1983) observed that cholesterol partitions preferentially into plasma membrane fractions, compared to membrane fractions made of mitochondria or ER. The authors did another smart experiment and measured cholesterol partition between lipids extracted from each of these membrane fractions and got the same result. Thus, the relative enrichment of cholesterol in plasma membrane compared to ER or mitochondria is a result of the different lipid composition of the isolated membrane fractions (Wattenberg and Silbert 1983). The authors could also show that an important lipid property causing this preferred cholesterol enrichment is the high content of SM and saturated lipids in the plasma membrane, compared to the intracellular organelle membranes. Finally, it was established in this pioneering study that the differences in cholesterol partition preference between isolated membrane fractions are not as high as the observed in vivo differences in organelle and plasma membrane cholesterol content (Wattenberg and Silbert 1983). The latter conclusion is very important. It emphasizes that intracellular cholesterol distribution is maintained by active transport processes in living cells, such as vesicular transport of cholesterol itself. Supporting this assumption, Ikonen and co-workers have shown that the ERC-associated rab-GTPase, rab11, regulates recycling and esterification of plasma membrane cholesterol, while rab8 is involved in cholesterol egress from late endosomes after ingestion of LDL (Hölttä-Vuori et al. 2002; Linder et al. 2006). Using fluorescence imaging of DHE, we demonstrated that cholesterol (analogs) can be internalized by several endocytic patwhays and recycle from the ERC in a manner dependent on EHD/rme1, known to regulate recycling of transferrin and other receptors (Hao et al. 2002; Wüstner and Færgeman 2008). Further, active processes can also involve maintenance of specific organelle lipid and protein composition by vesicular, ATP-dependent membrane traffic. Such active transport might generate favorable conditions for relative cholesterol accumulation, like in the ERC, or an environment characteristic for sterol-poor organelle membranes, like in the ER. Although nobody knows exactly what those membrane environmental conditions are, a hypothesis has been put forward based on the chemical potential of free cholesterol in lipid monolayer and bilayer studies (Huang and Feigenson 1999; Radhakrishnan and McConnell 2000; Maxfield and Menon 2006; Lange and Steck 2008). This model proposes the existence of two pools of cholesterol in various intracellular membranes – a phospholipid-complexed and a free pool, the latter being in non-vesicular exchange with other membranes. The proportion of free to complexed cholesterol is characteristic for every organelle and specific lipid composition and reflects the specific cholesterol affinity of the host phospho- and sphingolipids. Again, this specific lipid environment and thereby the ratio of free and complexed cholesterol could be a result of active membrane transport processes. Now, the key point of the model is that despite large differences in total membrane cholesterol content, the chemical potential of free, non-complexed cholesterol in for example the ER could be identical to that found in the plasma membrane (Radhakrishnan and McConnell 2000; Maxfield and Menon 2006; Lange and Steck 2008). Slight changes in the chemical potential of the free sterol pool, for example by adding extra cholesterol to the plasma membrane or by changing the phospholipid composition, could trigger non-vesicular cholesterol efflux to extracellular acceptors or uptake and targeting to intracellular acceptor membranes like the ER. The nature of the cholesterol in complexes is a matter of debate. While some physico-chemists argue that cholesterol indeed forms stoichiometric complexes with phospholipids (McConnell and Radhakrishnan 2003), others state that cholesterol keeps shielded from the water interface under phospholipid head groups, requiring multibody interactions and even formation of superlattice-like arrangements (Huang et al. 1999). Both ideas have in common that the capacity of phospholipids to either bind cholesterol or to shield it from water in superlattice-like structures (Huang and Feigenson 1999; Somerharju et al. 1999), is limited and specific for certain phospholipid compositions. Accordingly, raising cholesterol in a membrane beyond this capacity will shift cholesterol towards the free exchangeable pool. Independent of the exact underlying physico-chemical mechanisms for the limited bilayer capacity to accommodate cholesterol, this model is attractive because it explains, or is at least in line with many experimental observations. For example, the sigmoid dependence between plasma membrane and ER cholesterol can be explained by rapid influx of cell surface cholesterol above a critical threshold (Lange et al. 1999; Lange and Steck 2008). Similarly, activity of HMG-CoA reductase drops rapidly (half-time 10–20 min) and almost completely when plasma membrane cholesterol is raised slightly (by about 10%) above normal levels (Lange and Steck 2008). We observed a strongly increased non-vesicular uptake of DHE from the plasma membrane upon cholesterol loading of macrophages, demonstrating that the plasma membrane has a limited capacity to accommodate cholesterol (Wüstner et al. 2005). The latter observation could even be the underlying cause of the well known threshold phenomenon of ACAT stimulation in these cells above certain levels of cellular cholesterol (see Sect. 6.6). Using kinetic fluorescence imaging of DHE, we established that sterol moves in a non-vesicular, ATP-independent manner from the basolateral to the apical canalicular membrane of polarized epithelial HepG2 cells and that the forward rate constant is higher than that of the transport back from the canalicular to the basolateral domain (Wüstner et al. 2002). This is likely due to the favorable environment in the apical membrane being enriched in sphingolipids, for example. Continuous vesicle traffic between the subapical recycling compartment and the canalicular membrane might be crucial for maintenance of the specific lipid composition of the apical membrane with higher sterol affinity, compared to the basolateral domain. The chemical potential theory is basically an extension of early considerations on the different sterol affinity of cellular membrane lipids by Wattenberg and Silbert (1983) described above. Its new point is the proposal that membrane cholesterol is always close to a physiological set-point given by the distinct lipid and protein composition of the organelles and the plasma membrane (Maxfield and Menon 2006; Lange and Steck 2008). Slight changes in the free cholesterol pool near the set-point can trigger non-vesicular cholesterol exchange between organelles and thereby large cellular responses. Of course, much remains to be done to really prove this idea; in fact, the major task is to identify these different cholesterol pools and to characterize proteins mediating the responses to cellular cholesterol perturbation. The observed relations, however, are indeed intriguing and might pave the way for future research in this direction.

What are the mechanisms underlying non-vesicular cholesterol transport in cells? In the case of polarized epithelial cells, non-vesicular sterol transport between the plasma membrane domains, as observed for hepatoma HepG2 cells, might occur by lateral diffusion (Wüstner et al. 2002). While tight junctions would prevent diffusion of lipids in the exoplasmic leaflet of the plasma membrane (Dragsten et al. 1981), sterols can rapidly flip-flop across the bilayer and thereby circumvent this diffusion barrier (see Fig. 6.5). Rapid non-vesicular

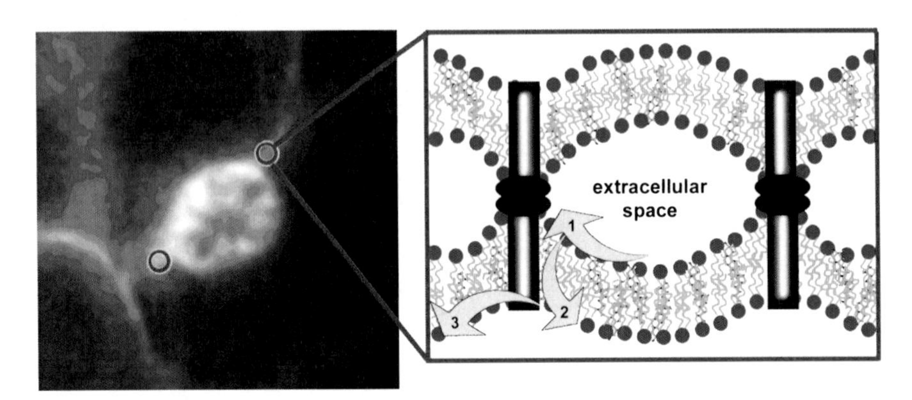

Fig. 6.5 Transport of cholesterol to the apical canalicular membrane of hepatic cells. Polarized human hepatoma HepG2 cells were pulse-labeled with DHE loaded onto cyclodextrin for 1 min, washed and imaged on a wide-field fluorescence microscope optimized for transmission in the ultraviolet (UV) to detect DHE. The canalicular membrane forming the biliary canaliculus is strongly labelled with DHE including several microvilli (*left* panel). Tight junctions seal the canalicular from the basolateral membrane by several rows of proteins including occludins (symbolized in black, *right* panel). DHE can be rapidly shuttled from the basolateral to the canalicular membrane by ATP-independent, non-vesicular transport. This trafficking mode probably involves lateral diffusion in the exoplasmic membrane leaflet (*1*), transbilayer migration of sterol to the inner membrane half (*2*), when DHE encounters the diffusion barrier generated by tight junctions proteins in the outer layer and (*3*) continued lateral diffusion in the cytoplasmic leaflet of the plasma membrane (see text for details)

shuttling of cholesterol from the basolateral to the apical, bile canaliculi-forming membrane of hepatocytes could be an efficient way to deliver cholesterol from the plasma to the bile compartment during the course of reverse cholesterol transport (Robins and Fasulo 1999; Wüstner et al. 2002, 2004). In the case of intracellular non-vesicular trafficking, cholesterol has to somehow cross the aqueous phase between two compartments. The rate-limiting step in passive cholesterol transport between two bilayers is cholesterol's desorption from the donor membrane (Bar et al. 1986; Steck et al. 1988). Due to its low aqueous solubility, partition of cholesterol into the water phase is thermodynamically unfavorable. However, once released or just partially protruded from a membrane, cholesterol can be caught by an acceptor, like a transfer protein or another membrane, if located in sufficient proximity. The nebulous free cholesterol pool in the chemical potential model explained above could in fact resemble cholesterol with a high out-of-plane mobility, partly protruding from the bilayer (cf. Sect. 6.1). This could raise the probability of a protruded cholesterol molecule being transferred to a lipid-binding protein. A few candidates for cytoplasmic cholesterol carriers have been described and their potential to transfer cholesterol between membranes has been shown in vitro. The definitive proof of their intracellular function as a cholesterol transporter, however, is still missing. The most prominent example is the steroidogenic acute regulatory protein (StAR), which mediates cholesterol import into mitochondria for steroid hormone synthesis in steroidogenic tissues. StAR is the prototype of the StAR-related lipid transfer family of proteins; other members are the phosphatidylcholine transfer protein (PC-TP/StarD2), the ceramide transfer protein (CERT/StarD11) and the late endosomal transmembrane protein MLN64 (StarD3; Alpy and Tomasetto 2005). While PC-TP might be involved in many process involving lipid transfer steps, CERT mediates specifically ER-to-Golgi trafficking of ceramide (Alpy and Tomasetto 2005). MLN64 is probably part of the machinery moving cholesterol released from ingested LDL cholesteryl esters out of the late endosome and lysosome (Alpy and Tomasetto 2005; Hölttä-Vuori et al. 2005). Knock-down of MLN64 results in the dispersion of late endosomes and lysosomes, due to a missing recruitment of actin to these organelles (Hölttä-Vuori et al. 2005). Members of the StART family share structural similarities, including a hydrophobic pocket limited by alpha helices and a flexible lid which can open and close for the binding and releasing of a cholesterol molecule (Alpy and Tomasetto 2005; Murcia et al. 2006). Other soluble proteins with the ability to bind and even transfer cholesterol in model membranes are oxysterol-binding proteins and Niemann–Pick C protein 2 (NPC2; see Sect. 6.6). Once expelled from the donor membrane and being bound to a soluble protein with a hydrophobic pocket, the next problem is to create specificity by transferring the bound cholesterol or other lipid to the correct target membrane. Some lipid-binding proteins bear targeting sequences for two compartments, like CERT, which associates with the ER as well as with the Golgi (Alpy and Tomasetto 2005; Levine and Loewen 2006). It has been speculated that sites of close membrane apposition observable by electron microscopy play a central role in inter-compartment, non-vesicular lipid transfer via proteins like CERT and others (Levine and Loewen 2006). A central role in this hypothesis involves the ER-making membrane contact zones with the plasma membrane (Becker et al. 2005), lipid droplets (Prattes et al. 2000), mitochondria (Rusiñol et al. 1994), Golgi (Ladinsky et al. 1999) and eventually ERC and other endosomes (Khelef et al. 2000). The close proximity of donor and acceptor membrane in these contact zones plus the existence of targeting sequences in the (putative) transfer protein for donor and target compartment could provide a highly efficient and specific way of non-vesicular lipid/cholesterol transport between organelles and plasma membrane. These ideas are largely based on observations in yeast cells and certainly need experimental back up for mammalian cells.

6.6 Alterations in Intracellular Cholesterol Trafficking in Atherosclerosis and Lipid Storage Diseases

Much has been learned about normal intracellular cholesterol trafficking from disorders causing alterations in lipid transport or metabolism. The classic example for a genetic disease of cholesterol transport is familial hypercholesterolemia, studied by Joseph Goldstein and Michael Brown, leading to the discovery of the central role played by the LDL receptor system in the regulation of cellular cholesterol uptake (Goldstein and Brown 1974). Another example is Tangier disease, where strongly reduced plasma HDL and increased cholesteryl ester accumulation in macrophages could be ascribed to a defect in ATP-binding cassette (ABC) transporter A1 (ABCA1; Tall et al. 2002). This defect resulted in dramatically lowered capacity of macrophages to efflux cholesterol and thereby to reduced HDL and progression of atherosclerosis. Plant sterols accumulate in the body due to a defect in the plant sterol efflux ABC transporters ABCG5/8 (Graf et al. 2002; Yu et al. 2002). This leads to reduced extrusion of these sterols from the enterocytes in the intestine or hepatocytes in the liver of affected patients in sitosterolemia (Maxfield and Tabas 2005). Like in the case of Tangier disease and hypercholesterolemia, a cholesterol transport system and its regulation have been identified by studying the underlying causes of the disease sitosterolemia; the two half-transporters ABCG5/8 together mediate the biliary secretion of cholesterol (Yu et al. 2002).

Another group of diseases, the lysosomal storage disorders like Tay–Sachs, Fabry, Niemann–Pick and Sandhoff disease, are characterized by the accumulation of sphingolipids and cholesterol in late endosomes and lysosomes (Maxfield and Tabas 2005). In Tay–Sachs, Fabry and Sandhoff disease, the primary cause of lysosomal lipid storage is a defect in sphingolipid degradation (Futerman and van Meer 2004). The same is true for Niemann–Pick disease types A and B, also called sphingomyelin lipidosis, while the exact etiology of Niemann–Pick disease type C is not known (Futerman and van Meer 2004). It is believed that defective export of LDL-derived cholesterol from the lysosomal compartment causes sterol and sphingolipid accumulation in late endosomes and lysosomes also containing NPC1 (Puri et al. 1999; Mukherjee and Maxfield 2004). Since the transport of bulk-phase markers like sucrose or membrane proteins like M6PR or transferrin is also affected in NPC disease, it is possible that mutation of the NPC1 protein causes a general traffic jam in the endocytic system (Kobayashi et al. 1999; Neufeld et al. 1999; Liscum and Sturley 2004; Mukherjee and Maxfield 2004; Pipalia et al. 2006). Importantly, knock-out of late endosomal/lysosomal Lamp1/Lamp2 or expression of dominant-negative dynamin as well as treatment with anti-lysobisphosphatidic (anti-LBPA) antibodies causes a similar phenotype with lysosomal or late endosomal cholesterol accumulation (Kobayashi et al. 1999; Eskelinen et al. 2004; Robinet et al. 2006). In contrast, overexpression of the small, late endosome-associated GTPases rab8 and rab9 can restore the trafficking defect and lower cholesterol accumulation (Narita et al. 2005; Linder et al. 2006). Very recent results support the idea that mutations of NPC1 cause a traffic jam in the endosomal/lysosomal system. Lloyd-Evans et al. (2008) reported that an initiating factor in NPC1 disease pathogenesis is the accumulation of sphingosine, a precursor of sphingolipids, which by an unknown mechanism causes the depletion of calcium in acidic endosomal compartments. This, in turn, disturbs normal trafficking along the degradative pathway, since the formation of lysosomes via a late endosome-lysosome hybrid organelle is calcium-regulated (Piper and Luzio 2004; Luzio et al. 2007). The authors found that drugs which elevate the cytoplasmic calcium concentration could not correct the low calcium content of late endosomes/lysosomes in NPC1 mutant cells, but they could compensate for a lack of calcium release from the acidic compartments in these cells, thereby rescuing normal transport to lysosomes (Lloyd-Evans et al. 2008). Strikingly, such drugs, like curcumin and tharpsigargin, increased life span and slowed disease progression in the NPC1 knock-out mouse, paving the way for a new therapeutic approach. Although Lloyd-Evans et al. (2008) could demonstrate reduced cholesterol levels in NPC1 mutant fibroblasts upon thapsigargin treatment, the molecular mechanisms linking lysosomal calcium to cholesterol transport remain to be established. Another very interesting observation in this direction was made recently by Kaufmann and Krise (2008): they observed that NPC1 functions in regulating late endosomal/lysosomal content of amines. Based on their data, the authors suggest that NPC1-mediated clearance of amine containing molecules from lysosomes involves a vesiclular transport pathway and eventually NPC1-mediated fusion of late endosomes with lysosomes (Kaufmann and Krise 2008). Most research groups report that initial export of cholesterol derived from hydrolysis of LDL-associated esters to the plasma membrane or ER is inhibited by the NPC1 mutation (Liscum and Munn 1999; Wojtanik and Liscum 2003; Liscum and Sturley 2004; Millard et al. 2005; Infante et al. 2008). Fibroblasts bearing this mutation but also cells from NPC1null individuals do not down-regulate LDL-receptor expression via the SCAP/SREBP2 system upon cholesterol loading and have a strongly reduced ability to re-esterify LDL-derived cholesterol by ACAT in the ER (Wojtanik and Liscum 2003; Liscum and Sturley 2004; Millard et al. 2005; Kristiana et al. 2008; Urano et al. 2008). The same, however, has been found in HeLa cells overexpressing dominant-negative dynamin, suggesting that several proteins which regulate membrane traffic from late endosomes interfere with cholesterol egress after LDL ingestion (Robinet et al. 2006). It has been suggested that NPC1's function as a cholesterol exporter out of late endosomes/lysosomes depends on the mobility of the

NPC1 protein in organelles (Ko et al. 2001; Zhang et al. 2001). While functional NPC1 protein traffics bidirectionally in late endosomal vesicles along microtubules, mutated NPC1 (e.g., with a point mutation P692S in NPC1's sterol-sensing domain) shows no directed movement and strongly increases the accumulation of LDL and free cholesterol in the lysosomal compartments (Ko et al. 2001; Zhang et al. 2001; Millard et al. 2005). Purified NPC1 binds cholesterol with low affinity on a luminal loop, but the role played by this process in living cells remains to be determined (Infante et al. 2008). It has been suggested that NPC1 functions as a fatty acid transporter due to its similarity to bacterial RMD permeases; however, fatty acid flux through late endosomes was not affected by NPC1 mutation (Davies et al. 2000; Passeggio and Liscum 2005). Some evidence suggests that the NPC1 protein is involved in delivering cholesterol released from LDL in late endosomes to the TGN (Coxey et al. 1993; Urano et al. 2008). This would be supported by defect trafficking of M6PR from late endosomes to TGN in NPC disease (Ganley and Pfeffer 2006), inhibited transport of fluorescent sphingolipids to the TGN in NPC mutant fibroblasts (Puri et al. 1999, 2001) as well as by the observed transient association of the NPC1 protein with Golgi elements (Higgins et al. 1999; Garver et al. 2002). However, the close association of early and recycling endosomes with the TGN, as stated above, as well as some conflicting results regarding NPC1's intracellular location, leave this issue open for further studies. While defects in NPC1 protein are responsible for 95% of all observed cases of NPC disease, the remaining 5% are caused by defects in NPC2 protein with identical phenotype (Naureckiene et al. 2000). The gene defective in NPC1 disease encodes a 1278-amino-acid protein containing 13 putative transmembrane domains, whereas NPC2 is a soluble, heterogeneously N-glycosylated 131-amino-acid protein (Liscum and Sturley 2004). Human NPC2 is modified by a terminal mannose-6-phosphate group targeting the protein to lysosomes (Willenborg et al. 2005). NPC2 is also present in high amounts in mammalian epididymal fluid and bovine milk, it binds cholesterol and DHE with high affinity in an equimolar complex and it was shown to transfer both sterols between model membranes (Naureckiene et al. 2000; Friedland et al. 2003; Cheruku et al. 2006; Babalola et al. 2007; Infante et al. 2008; Xu et al. 2008). Importantly, sterol binding and inter-membrane transfer are significantly stimulated in the presence of LBPA and PI and work optimally at acidic pH, as found in late endosomes (Cheruku et al. 2006; Babalola et al. 2007; Xu et al. 2008). Antibodies against LBPA block NPC2-mediated transfer of the fluorescent sterol CTL between membranes (Xu et al. 2008). Based on these and other observations, a working model has been developed which proposes that NPC2 shuttles cholesterol released via hydrolysis of LDL-derived cholesteryl esters and maybe other sterols from internal LBPA-rich vesicles to the limiting membrane of late endosomes. From here, NPC1 transports the cholesterol to other organelles like ER or TGN and maybe plasma membrane and early/recycling endosomes (Sleat et al. 2004; Liou et al. 2006; Infante et al. 2008). The recent observation that sphingosine accumulation precedes cholesterol build-up in late endosomes/lysosomes and that calcium and amine homeostasis are intimately linked to NPC1 function; and lysosomal biogenesis might open the research field to new directions (Kaufmann and Krise 2008; Lloyd-Evans et al. 2008).

Atherosclerosis and coronary heart disease (CHD) are the major cause of death in Western society. Although cholesterol accumulation in the body during the life-span of a normal individual is an important factor in development of atherosclerosis, the molecular mechanisms underlying this disease are manifold and very complex. In fact, there exist genetic and patho-biochemical connections between atherosclerosis, CHD, insulin resistance, obesity and metabolic syndrome, demonstrating the overwhelming complexity of these diseases. One model of etiology of atherosclerosis based on lipoprotein modification, retention and aggregation in the subendothelium (intima) of the vessel wall has been put forward (Tabas 2000). Based on this model, development of the disease can be divided into an early and a late phase, as excellently described in a recent review (Maxfield and Tabas 2005). Briefly, the first visible sign in atherogenesis is the retention of otherwise circulating lipoproteins in the intima of the vessel wall, followed by recruitment of monozytes into the area. These cells differentiate into macrophages and try to engulf the aggregated and retained lipoproteins in a process with similarities to phagocytosis and clathrin-dependent endocytosis. Importantly, during this initial prolonged contact of the modified lipoproteins with the macrophages, cholesterol uptake into cells dramatically exceeds the internalization of the apoprotein of oxidized and/or aggregated LDL (Buton et al. 1999). Uptake of soluble, oxidized LDL by macrophages is mediated by various scavenger receptors, while the receptor(s) mediating internalization of aggregated and retained LDL are not known (Tabas 2000). It is important to emphasize that, in contrast to the LDL receptor, the known receptors for atherogenic, modified LDL, like scavenger receptor A/CD36, are not under control of the SREBP/SCAP system, and their expression is therefore not down-regulated in response to cellular cholesterol loading. Accordingly, macrophages continue to internalize cholesterol beyond physiological levels. Once cellular cholesterol content is expanded above a threshold level of about 25% of normal, ACAT is activated by an unknown mechanism and catalyzes a massive synthesis of cholesteryl esters out of the excess free cholesterol (Xu and Tabas 1991). These cholesteryl esters are stored by the formation of lipid droplets; and the large number of these organelles gives the cholesterol-loaded macrophages a foamy appearance, hence the name foam cells. Early atherosclerotic lesions containing foam cells are not large enough to compromise blood flow and remain therefore mostly undetected. During the progression of atherosclerosis, however, the ability of macrophages to esterify the excess cholesterol declines, and the cells begin to accumulate free non-esterified cholesterol. This is a hallmark of late lesional macrophages, which try to accommodate to this situation by enhancing the synthesis of phospholipids, in particular PC via activation of CTP:phosphocholine cytidyl transferase (CT) (Shiratori et al. 1994; Tabas 2000). This cellular response to free cholesterol loading is likely an adaptation to overcome cholesterol's limited solubility in lipid bilayers of about 50–55 mol%, as explained in Sect. 6.1. In the tissue culture of macrophages, this stage is easily recognized by many internal membrane whorls containing the excess phospholipids and cholesterol (Shiratori et al. 1994; Tabas 2000). Cell-culture macrophages loaded with non-esterified cholesterol additionally show increased cell ruffling and spreading, Rac activation and actin re-organization, as well as strongly inhibited migration capability (Qin et al. 2006;

Nagao et al. 2007). If this also happens in vivo, actin-mediated membrane protrusions could further enhance cholesterol uptake from aggregated lipoproteins, while the lowered ability to migrate prevents egress of these cholesterol-loaded macrophages from the lesional area (Maxfield and Tabas 2005).

A turning point in late-stage atherogenesis is macrophage death as a consequence of continuing free cholesterol loading of the cells. The unfolded protein response (UPR) in the ER is activated upon acute free cholesterol loading, which stimulates expression of the cell death effector CHOP, leading to apoptosis (Feng et al. 2003a). Nanomolar concentrations of hydrophobic amines – which specifically block the egress of LDL-derived cholesterol from late endosomes and its targeting to the ER – inhibited UPR and protected macrophages from apoptosis (Feng et al. 2003a). Similarly, macrophages with a heterozygous mutation in the NPC1 protein showed no free-cholesterol-induced UPR or apoptosis (Feng et al. 2003b). These results demonstrate that cholesterol derived from ingested atherogenic lipoproteins in macrophages travels through late endosomes like normal LDL cholesterol, indicating that the overall cholesterol trafficking route is functional in lesional macrophage foam cells. Macrophage death is accompanied by an increase in necrotic areas in lesions, plaque rupture and acute thrombosis (Maxfield and Tabas 2005).

6.7 Future Prospects

This chapter summarizes our current understanding of intracellular cholesterol trafficking in health and disease. It emphasizes the two different transport modes of cholesterol in cells, vesicular and non-vesicular, and highlights known regulators of both trafficking mechanisms. Future work might concentrate on the following aspects:

1. Understanding the relationship between transport and metabolism of cholesterol in cells: how do cholesterol precursors move in cells? Newly synthesized zymosterol, a precursor of cholesterol, traffics from the ER to the plasma membrane with a half-time of 9 min and turns over faster than cholesterol in the latter compartment (Lange et al. 1991). Thus, trafficking of cholesterol's precursors might be important in establishing the steady-state distribution of cholesterol among various organelles. Similarly, one might ask, what role is played by continuous sterol modification, like oxidation, in establishing the heterogeneous intracellular cholesterol distribution?
2. What exactly are the biophysical properties of cholesterol determining its cellular targeting and its impact on membrane trafficking in general? Is the ability to induce a Lo phase in ternary model membranes indeed related to cholesterol trafficking in cells?
3. What is the transbilayer distribution of cholesterol in the plasma membrane and in organelle membranes? New methods based on imaging of fluorescent sterols and side-specific quenchers might be used to address this question (Mondal et al. 2008).

4. How is non-vesicular cholesterol transport regulated in cells, and what are the exact proportions of vesicular and non-vesicular trafficking modes under normal and patho-biological conditions? It remains to be determined whether there are many cytoplasmic cholesterol transport proteins with low specificity for donor and target membrane, or whether inter-membrane shuttling of sterols is indeed mediated by a few specific transfer proteins. Elucidating the molecular mechanisms underlying specific interaction/binding of cholesterol to carrier and transmembrane proteins is crucial to answer this question.
5. How is cholesterol transported from the late endosomal/lysosomal pool after digestion of LDL and from the plasma membrane to the sterol-sensing machinery and to ACAT in the ER? Deciphering the molecular mechanisms underlying concerted action of NPC1 and NPC2 might clarify this issue.

Answers to the questions raised above certainly require further improvements in technology to monitor inter-organelle sterol trafficking by either microscopy in living cells or by new in vitro reconstitution approaches.

Acknowledgements I acknowledge funding by grants from the Danish Heart Association Hjerteforeningen, the Diabetes Foundation Diabetesforeningen and the Danish Research Agency Forskningsstyrelsen, Forskningsrådet for Natur og Univers. Dr. Adam Cohen Simonsen from the Department of Physics and Chemistry, Center for Membrane Physics (MEMPHYS) at the University of Southern Denmark is acknowledged for providing the AFM images shown in Fig. 6.2.

References

Ahmed SN, Brown DA, London E (1997) On the origin of sphingolipid/cholesterol-rich detergent-insoluble cell membranes: physiological concentrations of cholesterol and sphingolipid induce formation of a detergent-insoluble, liquid-ordered lipid phase in model membranes. Biochemistry 36:10944–10953

Alpy F, Tomasetto C (2005) Give lipids a START: the StAR-related lipid transfer (START) domain in mammals. J Cell Sci 118:2791–2801

Babalola JO, Wendeler M, Breiden B, Arenz C, Schwarzmann G, Locatelli-Hoops S, Sandhoff K (2007) Development of an assay for the intermembrane transfer of cholesterol by Niemann-Pick C2 protein. Biol Chem 388:617–626

Bar LK, Barenholz Y, Thompson TE (1986) Fraction of cholesterol undergoing spontaneous exchange between small unilamellar phosphatidylcholine vesicles. Biochemistry 25:6701–6705

Bar LK, Barenholz Y, Thompson TE (1987) Dependence on phospholipid composition of the fraction of cholesterol undergoing spontaneous exchange between small unilamellar vesicles. Biochemistry 26:5460–5465

Baumann NA, Sullivan DP, Ohvo-Rekila H, Simonot C, Pottekat A, Klaassen Z, Beh CT, Menon AK (2005) Transport of newly synthesized sterol to the sterol-enriched plasma membrane occurs via nonvesicular equilibration. Biochemistry 44:5816–5826

Becker T, Volchuk A, Rothman JE (2005) Differential use of endoplasmic reticulum membrane for phagocytosis in J774 macrophages. Proc Natl Acad Sci USA 102:4022–4026

Behnke O, Tranum-Jensen J, van Deurs B (1984a) Filipin as a cholesterol probe. I. Morphology of filipin-cholesterol interaction in lipid model systems. Eur J Cell Biol 35:189–199

Behnke O, Tranum-Jensen J, van Deurs B (1984b) Filipin as a cholesterol probe. II. Filipin-cholesterol interaction in red blood cell membranes. Eur J Cell Biol 35:200–215

Bonifacino JS, Lippincott-Schwartz J (2003) Coat proteins: shaping membrane transport. Nat Rev Mol Cell Biol 4:409–414

Boucrot E, Saffarian S, Massol R, Kirchhausen T, Ehrlich M (2006) Role of lipids and actin in the formation of clathrin-coated pits. Exp Cell Res 312:4036–4048

Bridgman PC, Nakajima Y (1983) Distribution of filipin-sterol complexes on cultured muscle cells: cell-substratum contact areas associated with acetylcholine clusters. J Cell Biol 96:363–372

Buton X, Mamdouh Z, Ghosh RN, Du H, Kuriakose G, Beatini N, Grabowski GA, Maxfield FR, Tabas I (1999) Unique cellular events occurring during the initial interaction of macrophages with matrix-retained or methylated aggregated low density lipoprotein (LDL). Prolonged cell-surface contact during which LDL-cholesteryl ester hydrolysis exceeds LDL protein degradation. J Biol Chem 274:32112–32121

Cantor RS (1999) Lipid composition and the lateral pressure profile in bilayers. Biophys J 76:2625–2639

Chatterjee S, Smith ER, Hanada K, Stevens VL, Mayor S (2001) GPI anchoring leads to sphingolipid-dependent retention of endocytosed proteins in the recycling endosomal compartment. EMBO J 20:1583–1592

Chattopadhyay A, Jafurulla M, Kalipatnapu S, Pucadyil TJ, Harikumar KG (2005) Role of cholesterol in ligand binding and G-protein coupling of serotonin(1A) receptors solubilized from bovine hippocampus. Biochem Biophys Res Commun 327:1036–1041

Cheruku SR, Xu Z, Dutia R, Lobel P, Storch J (2006) Mechanism of cholesterol transfer from the Niemann-Pick type C2 protein to model membranes supports a role in lysosomal cholesterol transport. J Biol Chem 281:31594–31604

Choudhury A, Sharma DK, Marks DL, Pagano RE (2004) Elevated endosomal cholesterol levels in Niemann-Pick cells inhibit rab4 and perturb membrane recycling. Mol Biol Cell 15:4500–4511

Corvera S, DiBonaventura C, Shpetner HS (2000) Cell confluence-dependent remodeling of endothelial membranes mediated by cholesterol. J Biol Chem 275:31414–31421

Coxey RA, Pentchev PG, Campbell G, Blanchette-Mackie EJ (1993) Differential accumulation of cholesterol in Golgi compartments of normal and Niemann-Pick type C fibroblasts incubated with LDL: a cytochemical freeze-fracture study. J Lipid Res 34:1165–1176

Cruz JC, Chang TY (2000) Fate of endogenously synthesized cholesterol in Niemann-Pick type C1 cells. J Biol Chem 275:41309–41316

Davies JP, Chen FW, Ioannou YA (2000) Transmembrane molecular pump activity of Niemann-Pick C1 protein. Science 290:2295–2298

DeGrella RF, Simoni RD (1982) Intracellular transport of cholesterol to the plasma membrane. J Biol Chem 257:14256–14262

Dragsten PR, Blumenthal R, Handler JS (1981) Membrane asymmetry in epithelia: is the tight junction a barrier to diffusion in the plasma membrane. Nature 294:718–722

Du X, Pham YH, Brown AJ (2004) Effects of 25-hydroxycholesterol on cholesterol esterification and sterol regulatory element-binding protein processing are dissociable: implications for cholesterol movement to the regulatory pool in the endoplasmic reticulum. J Biol Chem 279:47010–47016

Edidin M (2003) Lipids on the frontier: a century of cell-membrane bilayers. Nat Cell Biol 4:414–418

Eskelinen EL, Schmidt CK, Neu S, Willenborg M, Fuertes G, Salvador N, Tanaka Y, Lüllmann-Rauch R, Hartmann D, Heeren J, von Figura K, Knecht E, Saftig P (2004) Disturbed cholesterol traffic but normal proteolytic function in LAMP-1/LAMP-2 double-deficient fibroblasts. Mol Biol Cell 15:3132–3145

Feng B, Yao PM, Li Y, Devlin CM, Zhang D, Harding HP, Sweeney M, Rong JX, Kuriakose G, Fisher EA, Marks AR, Ron D, Tabas I (2003a) The endoplasmic reticulum is the site of cholesterol-induced cytotoxicity in macrophages. Nat Cell Biol 5:781–792

Feng B, Zhang D, Kuriakose G, Devlin CM, Kockx M, Tabas I (2003b) Niemann-Pick C heterozygosity confers resistance to lesional necrosis and macrophage apoptosis in murine atherosclerosis. Proc Natl Acad Sci USA 100:10423–10428

Friedland N, Liou H-L, Lobel P, Stock AM (2003) Structure of a cholesterol-binding protein deficient in Niemann-Pick type C2 disease. Proc Natl Acad Sci USA 100:2512–2517

Fujiwara T, Ritchie K, Murakoshi H, Jacobson K, Kusumi A (2002) Phospholipids undergo hop diffusion in compartmentalized cell membrane. J Cell Biol 157:1071–1081

Futerman AH, van Meer G (2004) The cell biology of lysosomal storage disorders. Nat Rev Mol Cell Biol 5:554–565
Ganguly S, Pucadyil TJ, Chattopadhyay A (2008) Actin cytoskeleton-dependent dynamics of the human serotonin(1A) receptor correlates with receptor signaling. Biophys J 95:451–463
Ganley IG, Pfeffer SR (2006) Cholesterol accumulation sequesters Rab9 and disrupts late endosome function in NPC1-deficient cells. J Biol Chem 281:17890–17899
Garver WS, Krishnan K, Gallagos JR, Michikawa M, Francis GA, Heidenreich RA (2002) Niemann-Pick C1 protein regulates cholesterol transport to the trans-Golgi network and plasma membrane caveolae. J Lipid Res 43:579–589
Garvik O, Benediktson P, Ipsen JH, Simonsen AC, Wüstner D (2009) The fluorescent cholesterol analog dehydroergosterol induces liquid-ordered domains in model membranes. Chem Phys Lipids (in press)
Gliss C, Randel O, Casalta H, Sackmann E, Zorn R, Bayerl T (1999) Anisotropic motion of cholesterol in oriented DPPC bilayers studied by quasielastic neutron scattering: the liquid-ordered phase. Biophys J 77:331–340
Goldstein JL, Brown MS (1974) Binding and degradation of low density lipoproteins by cultured human fibroblasts. Comparison of cells from a normal subject and from a patient with homozygous familial hypercholesterolemia. J Biol Chem 249
Goldstein JL, DeBose-Boyd RA, Brown MS (2006) Protein sensors for membrane sterols. Cell 124:35–46
Goodwin JS, Drake KR, Remmert CL, Kenworthy AK (2005) Ras diffusion is sensitive to plasma membrane viscosity. Biophys J 89:1398–1410
Graf GA, Li WP, Gerard RD, Gelissen I, White A, Cohen JC, Hobbs HH (2002) Coexpression of ATP-binding cassette proteins ABCG5 and ABCG8 permits their transport to the apical surface. J Clin Invest 110:659–669
Grimmer S, van Deurs B, Sandvig K (2002) Membrane ruffling and macropinocytosis in A431 cells require cholesterol. J Cell Sci 115:2953–2962
Grimmer S, Ying M, Wälchli S, van Deurs B, Sandvig K (2005) Golgi vesiculation induced by cholesterol occurs by a dynamin- and cPLA2-dependent mechanism. Traffic 6:144–156
Haberland ME, Reynolds JA (1973) Self-association of cholesterol in aqueous solution. Proc Natl Acad Sci USA 70:2313–2318
Hamilton JA (2003) Fast flip-flop of cholesterol and fatty acids in membranes: implications for membrane transport proteins. Curr Opin Lipidol 14:263–271
Hao M, Mukherjee S, Sun Y, Maxfield FR (2004) Effects of cholesterol depletion and increased lipid unsaturation on the properties of endocytic membranes. J Biol Chem 279:14171–14178
Hao M, Lin SX, Karylowski OJ, Wüstner D, McGraw TE, Maxfield FR (2002) Vesicular and non-vesicular sterol transport in living cells. The endocytic recycling compartment is a major sterol storage organelle. J Biol Chem 277:609–617
Harikumar KG, Puri V, Singh RD, Hanada K, Pagano RE, Miller LJ (2005) Differential effects of modification of membrane cholesterol and sphingolipids on the conformation, function, trafficking of the G protein-coupled cholecystokinin receptor. J Biol Chem 280:2176–2185
Hartwig Petersen N, Færgeman NJ, Yu L, Wüstner D (2008) Kinetic imaging of NPC1L1 and sterol trafficking between plasma membrane and recycling endosomes in hepatoma cells. J Lipid Res 49:2023–2037
Heerklotz H (2002) Triton promotes domain formation in lipid raft mixtures. Biophys J 83:2693–2701
Heino S, Lusa S, Somerharju P, Ehnholm C, Olkkonen VM, Ikonen E (2000) Dissecting the role of the golgi complex and lipid rafts in biosynthetic transport of cholesterol to the cell surface. Proc Natl Acad Sci USA 97:8375–8380
Hell SW (2007) Far-field optical nanoscopy. Science 316:1153–1158
Henriksen J, Rowat AC, Ipsen JH (2004) Vesicle fluctuation analysis of the effects of sterols on membrane bending rigidity. Eur Biophys J 33:732–741
Henriksen J, Rowat AC, Brief E, Hsueh YW, Thewalt JL, Zuckermann MJ, Ipsen JH (2006) Universal behavior of membranes with sterols. Biophys J 90:1639–1649

Hess ST, Gould TJ, Gudheti MV, Maas SA, Mills KD, Zimmerberg J (2007) Dynamic clustered distribution of hemagglutinin resolved at 40 nm in living cell membranes discriminates between raft theories. Proc Natl Acad Sci USA 104:17370–17375

Higgins ME, Davies JP, Chen FW, Ioannou YA (1999) Niemann-Pick C1 is a late endosome-resident protein that transiently associates with lysosomes and the trans-Golgi network. Mol Genet Metab 68:1–13

Hill WG, Zeidel ML (2000) Reconstituting the barrier properties of a water-tight epithelial membrane by design of leaflet-specific liposomes. J Biol Chem 275:30176–30185

Hofsass, C, E. Lindahl, O. Edholm (2003) Molecular dynamics simulations of phospholipid bilayers with cholesterol. Biophys. J. 84:2192–2206.- 6.

Hölttä-Vuori M, Tanhuanpää K, Möbius W, Somerharju P, Ikonen E (2002) Modulation of cellular cholesterol transport and homeostasis by Rab11. Mol Biol Cell 13:3107–3122

Hölttä-Vuori M, Alpy F, Tanhuanpaa K, Jokitalo E, Mutka AL, Ikonen E (2005) MLN64 is involved in actin-mediated dynamics of late endocytic organelles. Mol Biol Cell 16:3873–3886

Hornick CA, Hui DY, DeLamatre JG (1997) A role for retrosomes in intracellular cholesterol transport from endosomes to the plasma membrane. Am J Physiol 273:C1075–C1081

Huang J, Feigenson GW (1999) A microscopic interaction model of maximum solubility of cholesterol in lipid bilayers. Biophys J 76:2142–2157

Huang J, Buboltz JT, Feigenson GW (1999) Maximum solubility of cholesterol in phosphatidylcholine and phosphatidylethanolamine bilayers. Biochim Biophys Acta 1417:89–100

Infante RE, Radhakrishnan A, Abi-Mosleh L, Kinch LN, Wang ML, Grishin NV, Goldstein JL, Brown MS (2008) Purified NPC1 protein: II. Localization of sterol binding to a 240-amino-acid soluble luminal loop. J Biol Chem 283:1064–1075

Infante RE, Abi-Mosleh L, Radhakrishnan A, Dale JD, Brown MS, Goldstein JL (2008) Purified NPC1 protein. I. Binding of cholesterol and oxysterols to a 1278-amino-acid membrane protein. J Biol Chem 283:1052–1063

Infante RE, Wang ML, Radhakrishnan A, Kwon HJ, Brown MS, Goldstein JL (2008) NPC2 facilitates bidirectional transfer of cholesterol between NPC1 and lipid bilayers, a step in cholesterol egress from lysosomes. Proc Natl Acad Sci USA 105:15287–15292

Ipsen JH, Karlstrom G, Mouritsen OG, Wennerstrom H, Zuckermann MJ (1987) Phase equilibria in the phosphatidylcholine-cholesterol system. Biochim Biophys Acta 905:162–172

John K, Kubelt J, Müller P, Wüstner D, Herrmann A (2002) Rapid transbilayer movement of the fluorescent sterol dehydroergosterol in lipid membranes. Biophys J 83:1525–1534

Jost M, Simpson F, Kavran JM, Lemmon MA, Schmid SL (1998) Phosphatidylinositol-4,5-bisphosphate is required for endocytic coated vesicle formation. Curr Biol 8:1399–1402

Kaplan MR, et al (1985) Transport of cholesterol from the endoplasmic reticulum to the plasma membrane. J Cell Biol 101:446–453

Karnovsky MJ, Kleinfeld AM, Hoover RL, Klausner RD (1982) The concept of lipid domains in membranes. J Cell Biol 94:1–6

Kaufmann AM, Krise JP (2008) Niemann-Pick C1 functions in regulating lysosomal amine content. J Biol Chem 283:24584–24593

Kellner-Weibel G, Yancey PG, Jerome WG, Walser T, Mason RP, Phillips MC, Rothblat GH (1999) Crystallization of free cholesterol in model macrophage foam cells. Arterioscler Thromb Vasc Biol 19:1891–1898

Kenworthy AK (2008) Have we become overly reliant on lipid rafts? Talking Point on the involvement of lipid rafts in T-cell activation. EMBO Rep 9:531–535

Kenworthy AK, Nichols BJ, Remmert CL, Hendrix GM, Kumar M, Zimmerberg J, Lippincott-Schwartz J (2004) Dynamics of putative raft–associated proteins at the cell surface. J Cell Biol 165:735–746

Khan N, Shen J, Chang TY, Chang CC, Fung PC, Grinberg O, Demidenko E, Swartz H (2003) Plasma membrane cholesterol: a possible barrier to intracellular oxygen in normal and mutant CHO cells defective in cholesterol metabolism. Biochemistry 42:23–29

Khelef N, Soe TT, Quehenberger O, Beatini N, Tabas I, Maxfield FR (2000) Enrichment of acyl coenzyme A:cholesterol O-acyltransferase near trans-Golgi network and endocytic recycling compartment. Arterioscler Thromb Vasc Biol 20:1769–1776

Khelef N, Buton X, Beatini N, Wang H, Meiner V, Chang TY, Farese RVJ, Maxfield FR, Tabas I (1998) Immunolocalization of acyl-coenzyme A:cholesterol O-acyltransferase in macrophages. J Biol Chem 273:11218–11224

Ko DC, Gordon MD, Jin JY, Scott MP (2001) Dynamic movements of organelles containing Niemann-Pick C1 protein: NPC1 involvement in late endocytic events. Mol Biol Cell 12:601–614

Kobayashi T, Beuchat MH, Lindsay M, Frias S, Palmiter RD, Sakuraba H, Parton RG, Gruenberg J (1999) Late endosomal membranes rich in lysobisphosphatidic acid regulate cholesterol transport. Nat Cell Biol 1:113–118

Koppel DE, Sheetz MP, Schindler M (1981) Matrix control of protein diffusion in biological membranes. Proc Natl Acad Sci USA 78:6981–6985

Kristiana I, Yang H, Brown AJ (2008) Different kinetics of cholesterol delivery to components of the cholesterol homeostatic machinery: implications for cholesterol trafficking to the endoplasmic reticulum. Biochim Biophys Acta. 1781:724–730

Krylov AV, Pohl P, Zeidel ML, Hill WG (2001) Water permeability of asymmetric planar lipid bilayers: leaflets of different composition offer independent and additive resistance to permeation. J Gen Physiol 118:333–339

Kwik J, Boyle S, Fooksman D, Margolis L, Sheetz MP, Edidin M (2003) Membrane cholesterol, lateral mobility, the phosphatidylinositol 4,5-bisphosphate-dependent organization of cell actin. Proc Natl Acad Sci USA 100:13964–13969

Ladinsky MS, Mastronarde DN, McIntosh JR, Howell KE, Staehelin LA (1999) Golgi structure in three dimensions: functional insights from the normal rat kidney cell. J Cell Biol 144:1135–1149

Lang T, Bruns D, Wenzel D, Riedel D, Holroyd P, Thiele C, Jahn R (2001) SNAREs are concentrated in cholesterol-dependent clusters that define docking and fusion sites for exocytosis. EMBO J 20:2202–2213

Lange Y, Matthies HJ (1984) Transfer of cholesterol from its site of synthesis to the plasma membrane. J Biol Chcm 259:14624–14630

Lange Y, Steck TL (2008) Cholesterol homeostasis and the escape tendency (activity) of plasma membrane cholesterol. Prog Lipid Res 47:319–332

Lange Y, Echevarria F, Steck TL (1991) Movement of Zymosterol, a precursor of cholesterol, among three membranes in human fibroblasts. J Biol Chem 266:21439–21443

Lange Y, Ye J, Rigney M, Steck TL (1999) Regulation of endoplasmic reticulum cholesterol by plasma membrane cholesterol. J Lipid Res 40:2264–2270

Lange Y, Swaisgood MH, Ramos BV, Steck TL (1989) Plasma membranes contain half the phospholipid and 90% of the cholesterol and sphingomyelin in cultured human fibroblasts. J Biol Chem 264:3786–3793

Lebrand C, Corti M, Goodson H, Cosson P, Cavalli V, Mayran N, Fauré J, Gruenberg J (2002) Late endosome motility depends on lipids via the small GTPase Rab7. EMBO J 21:1289–1300

Leventis R, Silvius JR (2001) Use of cyclodextrins to monitor transbilayer movement and differential lipid affinities of cholesterol. Biophys J 81:2257–2267

Levine T, Loewen C (2006) Inter-organelle membrane contact sites: through a glass, darkly. Curr Opin Cell Biol 18:371–378

Li Y, Prinz WA (2004) ATP-binding cassette (ABC) transporters mediate nonvesicular, raft-modulated sterol movement from the plasma membrane to the endoplasmic reticulum. J Biol Chem 279:45226–45234

Li Y, Ge M, Ciani L, Kuriakose G, Westover EJ, Dura M, Covey DF, Freed JH, Maxfield FR, Lytton J, Tabas I (2004) Enrichment of endoplasmic reticulum with cholesterol inhibits sarcoplasmic-endoplasmic reticulum calcium ATPase-2b activity in parallel with increased order of membrane lipids: implications for depletion of endoplasmic reticulum calcium stores and apoptosis in cholesterol-loaded macrophages. J Biol Chem 279:37030–37039

Lindahl E, Edholm O (2000) Spatial and energetic-entropic decomposition of surface tension in lipid bilayers from molecular dynamics simulations. J Chem Phys 113:3882–3893

Linder MD, Uronen RL, Holtta-Vuori M, van der Sluijs P, Peranen J, Ikonen E (2006) Rab8-dependent recycling promotes endosomal cholesterol removal in normal and sphingolipidosis cells. Mol Biol Cell 18:47–56

Liou HL, Dixit SS, Xu S, Tint GS, Stock AM, Lobel P (2006) NPC2, the protein deficient in Niemann-Pick C2 disease, consists of multiple glycoforms that bind a variety of sterols. J Biol Chem 281:36710–36723
Lippincott-Schwartz J, Yuan L, Tipper C, Amherdt M, Orci L, Klausner RD (1991) Brefeldin A's effects on endosomes, lysosomes, the TGN suggest a general mechanism for regulating organelle structure and membrane traffic. Cell 67:601–616
Liscum L, Munn NJ (1999) Intracellular cholesterol transport. Biochim Biophys Acta 1438:19–37
Liscum L, Sturley SL (2004) Intracellular trafficking of Niemann-Pick C proteins 1 and 2: obligate components of subcellular lipid transport. Biochim Biophys Acta 1685:22–27
Lloyd-Evans E, Morgan AJ, He X, Smith DA, Elliot-Smith E, Sillence DJ, Churchill GC, Schuchman EH, Galione A, Platt FM (2008) Niemann-Pick disease type C1 is a sphingosine storage disease that causes deregulation of lysosomal calcium. Nat Med. 14:1247–1255
Loomis CR, Shipley GG, Small DM (1979) The phase behavior of hydrated cholesterol. J Lipid Res 20:525–535
Luzio JP, Bright NA, Pryor PR (2007) The role of calcium and other ions in sorting and delivery in the late endocytic pathway. Biochem Soc Trans 35:1088–1091
Mallard F, Tang BL, Galli T, Tenza D, Saint-Pol A, Yue X, Antony C, Hong W, Goud B, Johannes L (2002) Early/recycling endosomes-to-TGN transport involves two SNARE complexes and a Rab6 isoform. J Cell Biol 156:653–664
Maxfield FR (2002) Plasma membrane microdomains. Curr Opin Cell Biol 14:483–487
Maxfield FR, Menon AK (2006) Intracellular sterol transport and distribution. Curr Opin Cell Biol 18:379–385
Maxfield FR, Wüstner D (2002) Intracellular cholesterol transport. J Clin Invest 110:891–898
Maxfield FR, Tabas I (2005) Role of cholesterol and lipid organization in disease. Nature 438:612–621
Maxfield FR, McGraw TE (2004) Endocytic recycling. Nat Rev Mol Cell Biol 5:121–132
Mayor S, Maxfield FR (1995) Insolubility and redistribution of GPI-anchored proteins at the cell surface after detergent treatment. Mol Biol Cell 6:929–944
Mayor S, Pagano RE (2007) Pathways of clathrin-independent endocytosis. Nat Rev Mol Cell Biol 8:603–612
Mayor S, Sabharanjak S, Maxfield FR (1998) Cholesterol-dependent retention of GPI-anchored proteins in endosomes. EMBO J 17:4626–4638
McConnell HM, Radhakrishnan A (2003) Condensed complexes of cholesterol and phospholipids. Biochim et Biophys Acta 1610:159–173
Miaczynska M, Zerial M (2002) Mosaic organization of the endocytic pathway. Exp Cell Res 272:8–14
Millard EE, Gale SE, Dudley N, Zhang J, Schaffer JE, Ory DS (2005) The sterol-sensing domain of the Niemann-Pick C1 (NPC1) protein regulates trafficking of low density lipoprotein cholesterol. J Biol Chem 280:28581–28590
Möbius W, van Donselaar E, Ohno-Iwashita Y, Shimada Y, Heijnen HF, Slot JW, Geuze HJ (2003) Recycling compartments and the internal vesicles of multivesicular bodies harbor most of the cholesterol found in the endocytic pathway. Traffic 64:222–231
Montesano R (1979) Inhomogeneous distribution of filipin-sterol complexes in smooth muscle cell plasma membrane. Nature 280:328–329
Mukherjee S, Maxfield FR (2004) Lipid and cholesterol trafficking in NPC. Biochim Biophys Acta 1685:28–37
Mukherjee S, Maxfield FR (2004) Membrane domains. Annu Rev Cell Dev Biol 20:839–866
Mukherjee S, Ghosh RN, Maxfield FR (1997) Endocytosis. Physiol Rev 77:759–803
Munro S (2003) Lipid rafts: elusive or illusive. Cell 115:377–388
Murcia M, Faraldo-Gomez JD, Maxfield FR, Roux B (2006) Modeling the structure of the START domains of MLN64 and StAR proteins in complex with cholesterol. J Lipid Res 47:2614–2630
Murray JW, Wolkoff AW (2003) Roles of the cytoskeleton and motor proteins in endocytic sorting. Adv Drug Deliv Rev 55:1385–1403

Nagao T, Qin C, Grosheva I, Maxfield FR, Pierini LM (2007) Elevated cholesterol levels in the plasma membranes of macrophages inhibit migration by disrupting RhoA regulation. Arterioscler Thromb Vasc Biol 27:1596–1602

Nagle JF, Tristram-Nagle S (2000) Structure of lipid bilayers. Biochim Biophys Acta 1469:159–195

Narita K, Choudhury A, Dobrenis K, Sharma DK, Holicky EL, Marks DL, Walkley SU, Pagano RE (2005) Protein transduction of Rab9 in Niemann-Pick C cells reduces cholesterol storage. FASEB J 19:1558–1560

Naureckiene S, Sleat DE, Lackland H, Fensom A, Vanier MT, Wattiaux R, Jadot M, Lobel P (2000) Identification of HE1 as the second gene of Niemann-Pick C disease. Science 290:2298–2301

Needham D, McIntosh TJ, Evans E (1988) Thermomechanical and transition properties of dimyristoylphosphatidylcholine/cholesterol bilayers. Biochemistry 27:4668–4673

Neufeld EB, Cooney AM, Pitha J, Dawidowicz EA, Dwyer NK, Pentchev PG, Blanchette-Mackie EJ (1996) Intracellular trafficking of cholesterol monitored with a cyclodextrin. J Biol Chem 271:21604–21613

Neufeld EB, Wastney M, Patel S, Suresh S, Cooney AM, Dwyer NK, Roff CF, Ohno K, Morris JA, Carstea ED, Incardona JP, Strauss JFR, Vanier MT, Patterson MC, Brady RO, Pentchev PG, Blanchette-Mackie EJ (1999) The Niemann-Pick C1 protein resides in a vesicular compartment linked to retrograde transport of multiple lysosomal cargo. J Biol Chem 274:9627–9635

Nilsson I, Ohvo-Rekila H, Slotte JP, Johnson AE, von Heijne G (2001) Inhibition of protein translocation across the endoplasmic reticulum membrane by sterols. J Biol Chem 276:41748–41754

Niu S-L, Litman BJ (2002) Determination of membrane cholesterol partition coefficient using a lipid vesicle-cyclodextrin binary system: effect of phospholipid acyl chain unsaturation and headgroup composition. Biophys J 83:3408–3415

Ohvo-Rekila H, Ramstedt B, Leppimaki P, Slotte JP (2002) Cholesterol interactions with phospholipids in membranes. Prog Lipid Res 41:66–97

op den Kamp JAF (1979) Lipid asymmetry in membranes. Biochemistry 48:47–71

Passeggio J, Liscum L (2005) Flux of fatty acids through NPC1 lysosomes. J Biol Chem 280:10333–10339

Pfeffer SR (2007) Unsolved mysteries in membrane traffic. Annu Rev Biochem 76:629–645

Pierini LM, Eddy RJ, Fuortes M, Seveau S, Casulo C, Maxfield FR (2003) Membrane lipid organization is critical for human neutrophil polarization. J Biol Chem 278:10831–10841

Pipalia NH, Hao M, Mukherjee S, Maxfield FR (2006) Sterol, protein and lipid trafficking in Chinese hamster ovary cells with Niemann-Pick type C1 defect. Traffic 8:130–141

Piper RC, Luzio JP (2004) CUPpling calcium to lysosomal biogenesis. Trends Cell Biol 14:471–473

Pomorski T, Holthuis JC, Herrmann A, van Meer G (2004) Tracking down lipid flippases and their biological functions. J Cell Sci 117:805–813

Pralle A, Keller P, Florin EL, Simons K, Horber JK (2000) Sphingolipid-cholesterol rafts diffuse as small entities in the plasma membrane of mammalian cells. J Cell Biol 148:997–1008

Prattes S, Horl G, Hammer A, Blaschitz A, Graier WF, Sattler W, Zechner R, Steyrer E (2000) Intracellular distribution and mobilization of unesterified cholesterol in adipocytes: triglyceride droplets are surrounded by cholesterol-rich ER-like surface layer structures. J Cell Sci 113:2977–2989

Puri V, Watanabe R, Dominguez M, Sun X, Wheatley CL, Marks DL, Pagano RE (1999) Cholesterol modulates membrane traffic along the endocytic pathway in sphingolipid-storage diseases. Nat Cell Biol 1:386–388

Puri V, Watanabe R, Singh RD, Dominguez M, Brown JC, Wheatley CL, Marks DL, Pagano RE (2001) Clathrin-dependent and -independent internalization of plasma membrane sphingolipids initiates two Golgi targeting pathways. J Cell Biol 154:535–547

Qin C, Nagao T, Grosheva I, Maxfield FR, Pierini LM (2006) Elevated plasma membrane cholesterol content alters macrophage signaling and function. Arterioscler Thromb Vasc Biol 26:372–378

Radhakrishnan A, McConnell HM (2000) Chemical activity of cholesterol in membranes. Biochemistry 39:8119–8124

Ridsdale A, Denis M, Gougeon PY, Ngsee JK, Presley JF, Zha X (2006) Cholesterol is required for efficient endoplasmic reticulum-to-Golgi transport of secretory membrane proteins. Mol Biol Cell 17:1593–1605

Rink J, Ghigo E, Kalaidzidis Y, Zerial M (2005) Rab conversion as a mechanism of progression from early to late endosomes. Cell 122:735–749
Robinet P, Fradagrada A, Monier MN, Marchetti M, Cogny A, Moatti N, Paul JL, Vedie B, Lamaze C (2006) Dynamin is involved in endolysosomal cholesterol delivery to the endoplasmic reticulum: role in cholesterol homeostasis. Traffic 7:811–823
Robins SJ, Fasulo JM (1999) Delineation of a novel hepatic route for the selective transfer of unesterified sterols from high-density lipoproteins to bile: studies using the perfused rat liver. Hepatology 29:1541–1548
Róg T, Stimson LM, Pasenkiewicz-Gierula M, Vattulainen I, Karttunen M (2008) Replacing the cholesterol hydroxyl group with the ketone group facilitates sterol flip-flop and promotes membrane fluidity. J Phys Chem B 112:1946–1952
Runz H, Miura K, Weiss M, Pepperkok R (2006) Sterols regulate ER-export dynamics of secretory cargo protein ts-O45-G. EMBO J 25:2953–2965
Rusiñol AE, Cui Z, Chen MH, Vance JE (1994) A unique mitochondria-associated membrane fraction from rat liver has a high capacity for lipid synthesis and contains pre-Golgi secretory proteins including nascent lipoproteins. J Biol Chem 269:27494–27502
Sako Y, Kusumi A (1995) Barriers for lateral diffusion of transferrin receptor in the plasma membrane as characterized by receptor dragging by laser tweezers: fence versus tether. J Cell Biol 129:1559–1574
Sandvig K, Torgersen ML, Raa HA, van Deurs B (2008) Clathrin-independent endocytosis: from nonexisting to an extreme degree of complexity. Histochem. Cell Biol 129:267–276
Scheidt HA, Müller P, Herrmann A, Huster D (2003) The potential of fluorescent and spin-labeled steroid analogs to mimic natural cholesterol. J Biol Chem 278:45563–45569
Sharma DK, Brown JC, Choudhury A, Peterson TE, Holicky E, Marks DL, Simari R, Parton RG, Pagano RE (2004) Selective stimulation of caveolar endocytosis by glycosphingolipids and cholesterol. Mol Biol Cell 15:3114–3122
Sheetz MP (2001) Cell control by membrane cytoskeleton adhesion. Nat Rev Mol Cell Biol 2:392–396
Shiratori Y, Okwu AK, Tabas I (1994) Free cholesterol loading of macrophages stimulates phosphatidylcholine biosynthesis and up-regulation of CTP: phosphocholine cytidylyltransferase. J Biol Chem 269:11337–11348
Shvartsman DE, Gutman O, Tietz A, Henis YI (2006) Cyclodextrins but not compactin inhibit the lateral diffusion of membrane proteins independent of cholesterol. Traffic 7:917–926
Sieber JJ, Willig KI, Kutzner C, Gerding-Reimers C, Harke B, Donnert G, Rammner B, Eggeling C, Hell SW, Grubmüller H, Lang T (2007) Anatomy and dynamics of a supramolecular membrane protein cluster. Science 317:1072–1076
Sieber JJ, Willig KI, Heintzmann R, Hell SW, Lang T (2006) The SNARE motif is essential for the formation of syntaxin clusters in the plasma membrane. Biophys J 90:2843–2851
Silvius JR, del Giudice D, Lafleur M (1996) Cholesterol at different bilayer concentrations can promote or antagonize lateral segregation of phospholipids of differing acyl chain length. Biochemistry 35:15198–15208
Simons K, Ikonen E (1997) Functional rafts in cell membranes. Nature 387:569–572
Simons K, van Meer G (1988) Lipid sorting in epithelial cells. Biochemistry 27:6197–6202
Sleat DE, Wiseman JA, El-Banna M, Price SM, Verot L, Shen MM, Tint GS, Vanier MT, Walkley SU, Lobel P (2004) Genetic evidence for nonredundant functional cooperativity between NPC1 and NPC2 in lipid transport. Proc Natl Acad Sci USA 101:5886–5891
Small et al (1974) Physical-chemical basis of lipid deposition in atherosclerosis. Science 185:222–229
Snapp EL, Hegde RS, Francolini M, Lombardo F, Colombo S, Pedrazzini E, Borgese N, Lippincott-Schwartz J (2003) Formation of stacked ER cisternae by low affinity protein interactions. J Cell Biol 163:257–269
Somerharju P, Virtanen JA, Cheng KH (1999) Lateral organisation of membrane lipids. The superlattice view. Biochim Biophys Acta 1440:32–48
Steck TL, Kezdy FJ, Lange Y (1988) An activation-collision mechanism for cholesterol transfer between membranes. J Biol Chem 263:13023–13031

Steck TL, Ye J, Lange Y (2002) Probing red cell membrane cholesterol movement with cyclodextrin. Biophys J 83:2118–2125

Subczynski WK, Kusumi A (2003) Dynamics of raft molecules in the cell and artificial membranes: approaches by pulse EPR spin labeling and single molecule optical microscopy. Biochim Biophys Acta 1610:231–243

Subczynski WK, Hyde JS, Kusumi A (1989) Oxygen permeability of phosphatidylcholine-cholesterol membranes. Proc Natl Acad Sci USA 86:4474–4478

Sun M, Northup N, Marga F, Huber T, Byfield FJ, Levitan I, Forgacs G (2007) The effect of cellular cholesterol on membrane-cytoskeleton adhesion. J Cell Sci 120:2223–2231

Tabas I (2000) Cholesterol and phospholipid metabolism in macrophages. Biochim Biophys Acta 1529:164–174

Tabas I, Rosoff WJ, Boykow GC (1988) Acyl coenzyme A:cholesterol acyl transferase in macrophages utilizes a cellular pool of cholesterol oxidase-accessible cholesterol as substrate. J Biol Chem 263:1266–1272

Takahashi M, Murate M, Fukuda M, Sato SB, Ohta A, Kobayashi T (2007) Cholesterol controls lipid endocytosis through Rab11. Mol Biol Cell 18:2667–2677

Tall AR, Costet P, Wang N (2002) Regulation and mechanisms of macrophage cholesterol efflux. J Clin Invest 110:899–904

Tangirala RK, Jerome WG, Jones NL, Small DM, Johnson WJ, Glick JM, Mahlberg FH, Rothblat GH (1994) Formation of cholesterol monohydrate crystals in macrophage-derived foam cells. J Lipid Res 35:93–104

Tomishige MY, Sako Y, Kusumi A (1998) Regulation mechanism of the lateral diffusion of band 3 in erythrocyte membranes by membrane cytoskeleton. J Cell Biol 142:989–1000

Underwood KW, Jacobs NL, Howley A, Liscum L (1998) Evidence for a cholesterol transport pathway from lysosomes to endoplasmic reticulum that is independent of the plasma membrane. J Biol Chem 273:4266–4274

Urano Y, Watanabe H, Murphy SR, Shibuya Y, Geng Y, Peden AA, Chang CC, Chang TY (2008) Transport of LDL-derived cholesterol from the NPC1 compartment to the ER involves the trans-Golgi network and the SNARE protein complex. Proc Natl Acad Sci USA 105:16513–16518

van Dam EM, Ten Broeke T, Jansen K, Spijkers P, Stoorvogel W (2002) Endocytosed transferrin receptors recycle via distinct dynamin and phosphatidylinositol 3-kinase-dependent pathways. J Biol Chem 277:48876–48883

van der Goot FG, Gruenberg J (2006) Intra-endosomal membrane traffic. Trends Cell Biol 16:514–521

Vist MR, Davis JH (1990) Phase equilibria of cholesterol/dipalmitoylphosphatidylcholine mixtures: 2H nuclear magnetic resonance and differential scanning calorimetry. Biochemistry 29:451–464

Wang Y, Castoreno AB, Stockinger W, Nohturfft A (2005) Modulation of endosomal cholesteryl ester metabolism by membrane cholesterol. J Biol Chem 280:11876–11886

Warnock DE, Roberts C, Lutz MS, Blackburn WA, Young WWJ, Baenziger JU (1993) Determination of plasma membrane lipid mass and composition in cultured Chinese hamster ovary cells using high gradient magnetic affinity chromatography. J Biol Chem 268:10145–10153

Wattenberg BW, Silbert DF (1983) Sterol partitioning among intracellular membranes. Testing a model for cellular sterol distribution. J Biol Chem 258:2284–2289

Wilcke M, Johannes L, Galli T, Mayau V, Goud B, Salamero J (2000) Rab11 regulates the compartmentalization of early endosomes required for efficient transport from early endosomes to the trans-Golgi network. J Cell Biol 151:1207–1220

Willenborg M, Schmidt CK, Braun P, Landgrebe J, von Figura K, Saftig P, Eskelinen EL (2005) Mannose 6-phosphate receptors, Niemann-Pick C2 protein, lysosomal cholesterol accumulation. J Lipid Res 46:2559–2569

Wojtanik KM, Liscum L (2003) The transport of low density lipoprotein-derived cholesterol to the plasma membrane is defective in NPC1 cells. J Biol Chem 278:14850–14856

Wüstner D (2005) Improved visualization and quantitative analysis of fluorescent membrane sterol in polarized hepatic cells. J Microsc 220:47–64

Wüstner D (2007) Fluorescent sterols as tools in membrane biophysics and cell biology. Chem Phys Lipids 146:1–25
Wüstner D (2007) Plasma membrane sterol distribution resembles the surface topography of living cells. Mol Biol Cell 18:211–228
Wüstner D (2008) Free-cholesterol loading does not trigger phase separation of the fluorescent sterol dehydroergosterol in the plasma membrane of macrophages. Chem Phys Lipids 154:129–136
Wüstner D, Herrmann A, Hao M, Maxfield FR (2002) Rapid nonvesicular transport of sterol between the plasma membrane domains of polarized hepatic cells. J Biol Chem 277:30325–30336
Wüstner D, Færgeman NJ (2008) Spatiotemporal analysis of endocytosis and membrane distribution of fluorescent sterols in living cells. Histochem Cell Biol 130:891–908
Wüstner D, Mondal M, Huang A, Maxfield FR (2004) Different transport routes for high density lipoprotein and its associated free sterol in polarized hepatic cells. J Lipid Res 45:427–437
Wüstner D, Mondal M, Tabas I, Maxfield FR (2005) Direct observation of rapid internalization and intracellular transport of sterol by macrophage foam cells. Traffic 6:396–412
Wüstner D, Landt Larsen A, Sage D, Bagatolli LA, Færgeman NJ, Brewer JR (2009) One- and three-photon fluorescence microscopy for imaging of dehydroergosterol in living cells and organisms. (Submitted for publication)
Xu XX, Tabas I (1991) Lipoproteins activate acyl-coenzyme A:cholesterol acyltransferase in macrophages only after cellular cholesterol pools are expanded to a critical threshold level. J Biol Chem 266:17040–17048
Xu Z, Farver W, Kodukula S, Storch J (2008) Regulation of sterol transport between membranes and NPC2. Biochemistry 47:11134–11143
Yeagle PL (1985) Cholesterol and the cell membrane. Biochim Biophys Acta 822:267–287
Ying M, Grimmer S, Iversen TG, van Deurs B, Sandvig K (2003) Cholesterol loading induces a block in the exit of VSVG from the TGN. Traffic 4:772–784
Yu L, Li-Hawkins J, Hammer RE, Berge KE, Horton JD, Cohen JC, Hobbs HH (2002) Overexpression of ABCG5 and ABCG8 promotes biliary cholesterol secretion and reduces fractional absorption of dietary cholesterol. J Clin Invest 110:671–680
Zhang M, Dwyer NK, Love DC, Cooney A, Comly M, Neufeld E, Pentchev PG, Blanchette-Mackie EJ, Hanover JA (2001) Cessation of rapid late endosomal tubulovesicular trafficking in Niemann–Pick type C1 disease. Proc Natl Acad Sci USA 98:4466–4471
Zhang W, McIntosh AL, Xu H, Wu D, Gruninger T, Atshaves B, Liu JC, Schroeder F (2005) Structural analysis of sterol distributions in the plasma membrane of living cells. Biochemistry 44:2864–2884
Zoncu R, Perera RM, Sebastian R, Nakatsu F, Chen H, Balla T, Ayala G, Toomre D, De Camilli PV (2007) Loss of endocytic clathrin-coated pits upon acute depletion of phosphatidylinositol 4,5-bisphosphate. Proc Natl Acad Sci USA 104:3793–3798

Chapter 7
Role of the Endothelium in Lipoprotein Metabolism

Arnold von Eckardstein and Lucia Rohrer

Abstract For a long time the endothelium was considered as a passive exchange barrier of lipoproteins between plasma and extravascular tissues. During the past two decades many data from clinical studies, cell culture, and animal experiments have shown that endothelial cells are a target of physiological and pathological actions of lipoproteins: Whereas lysosphingolipids and apolipoprotein (apo)A-I in native high-density lipoproteins (HDL) exert protective effects on the integrity and function of endothelial cells, modified low-density lipoproteins (LDL) and remnants of lipoproteins tend to disturb endothelial function. One central function of the endothelium is the control of protein trafficking between intravascular and extravascular compartments. Both LDL and HDL can pass the intact endothelium through transcytosis by processes which involve caveolin-1 and the LDL-receptor (for LDL) or ATP-binding cassette (ABC) transporters and scavenger receptor (SR)-BI for apoA-I and HDL, respectively. Finally the endothelium has evolved as a regulator of lipoprotein metabolism: By expressing or presenting lipases [lipoprotein lipase (LPL), hepatic lipase (HL), endothelial lipase (EL)] as well as LPL-receptors (glycerophosphatidyl inositol anchored HDL binding protein 1; GPIHBP1) the endothelium contributes to the remodelling of all lipoprotein classes. Selective conditional knock-outs of widely expressed genes like peroxisome proliferator agent receptor gamma (PPARγ) in mice are starting to reveal additional specific effects of the endothelium on lipid and lipoprotein metabolism.

7.1 Introduction

A monolayer of polarized endothelial cells lines the intima of blood vessels. Locally the intact endothelium forms a barrier for the exchange of molecules and cells between intravascular and extravascular compartments. Thereby and in

---

A. von Eckardstein and L. Rohrer
Institut für Klinische Chemie, Universitätsspital Zürich, Rämistrasse 100, 8091 Zürich, Switzerland
e-mail: arnold.voneckardstein@usz.ch

C. Ehnholm (ed.), *Cellular Lipid Metabolism*.
DOI 10.1007/978-3-642-00300-4_7,

addition the endothelium regulates many systemic processes, including tissue fluid homeostasis, vascular tone, angiogenesis, hemostasis, and host defense. As the clinical consequence endothelial dysfunction contributes to many pathological conditions, such as atherosclerosis, diabetic microangiopathy, thrombosis, and inflammation, as well as cancer growth and metastasis.

Lipoproteins have three principal relationships with the endothelium:

1. Endothelial cells express or expose several proteins which regulate lipoprotein metabolism, notably lipases.
2. The endothelium forms a selectively permissive barrier for the passage of lipoproteins between intra- and extravascular compartments.
3. The endothelium is a target of physiological and pathological actions of lipoproteins.

7.2 Expression of Proteins Involved in Lipoprotein Metabolism

Endothelial cells express many proteins that play an important role in intravascular lipoprotein metabolism including lipoprotein receptors [e.g. low-density lipoprotein (LDL) receptor, scavenger receptor B1 (SR-B1)], apolipopoteins (e.g. apoA-I in porcine brain capillary endothelial cells), lipid transfer proteins (e.g. phospholipid transfer protein) and lipases. Except for the lipases (Hasham and Pillarisetti 2006; Wong and Schotz 2002) and the recently discovered glycosylphosphatidylinositol-anchored HDL binding protein 1 (GPIHBP1; Beigneux et al. 2008), endothelial cells are believed to contribute only minor amounts of each respective protein to the whole-body pools. However, in view of the huge surface of the endothelium and the resulting large exposure area to lipoproteins, the endothelium may in fact play a much bigger role in lipoprotein metabolism than usually acknowledged. This hypothesis can only be addressed by the analysis of conditional knock-out mice with selective and endothelium specific inactivation of the various lipoprotein genes. In fact, the targeted knock-out of the peroxisome proliferator activator receptor gamma (PPARγ) in the endothelium revealed an as yet unappreciated role of the endothelium in the regulation of metabolism, as these mice presented with elevated plasma concentrations of triglycerides and free fatty acids but increased insulin sensitivity (Kanda et al. 2009).

In the current absence of such conditional knock-out models for lipoprotein genes, one can only discuss the contribution of the endothelium to lipoprotein metabolism by referring to proteins that are predominantly produced or exposed by endothelial cells, namely lipoprotein lipase (LPL), hepatic lipase (HL), and endothelial lipase (EL). These three lipases, which are attached to the glycocalix of the endothelium, mediate the hydrolysis of triglycerides or phospholipids of lipoproteins and thereby contribute to the remodelling of lipoproteins (Wong and Schotz 2002). The primary products of their lipolytic activities are free fatty acids, diacyl-, and monoacylglycerols or lysophospholipids. Notably lysophosphatidylcholine and unesterified polyunsaturated fatty acids are biologically active because they themselves or their metabolic products bind to cognate cell surface or nuclear receptors, for example lysophosphatidylcholine receptors and peroxisome proliferator activator receptors, respectively. Because of their close proximity to lipoprotein hydrolysis, endothelial cells are the immediate targets of any pro- or anti-inflammatory/atherogenic effects exerted by these lipids (Hasham and Pillarisetti 2006; see also Sect. 7.3).

7.2.1 Lipoprotein Lipase and GPIHBP1

LPL is predominantly synthesized by the parenchymal cells of adipose tissue and striated skeletal or cardiac muscle and then translocated by basolateral-to-apical transcytosis through a process which requires both heparan sulfate proteoglycans and the very low density lipoprotein (VLDL) receptor (Obunike 2001). Attached onto the luminal surface of capillaries, LPL-dimers hydrolyse triglycerides into chylomicrons and VLDL and thereby provide adipocytes and myocytes with fatty acids for energy storage and production, respectively (Wong and Schotz 2002). LPL activity depends on the presence of apoC-II and is modulated by apoA-V, apoC-III, and apoE (Forte et al. 2008; Pollin et al. 2008; Wong and Schotz 2002). Previously the enzyme was postulated to be attached to negatively charged heparan sulfate proteoglycans of the endothelial glycocalix (Wong and Schotz 2002). Most recent data however indicate that GPIHBP1 (expressed by endothelial cells) binds both the enzyme and its lipoprotein substrates (Beigneux et al. 2008). It has been proposed that GPIHBP1 binds the fast heparin-releasable LPL pool on the luminal surface of endothelial cells, whereas heparansulfate proteoglycans bind a slow heparin-releasable pool of LPL in the subendothelial space (Weinstein et al. 2008). The limiting role of GPIHBP1 for lipolysis is highlighted by the hypertriglyceridemia of humans and mice with GPIHBP1 deficiency, which is as severe as that of patients or mice deficient in LPL or apoC-II (Beigneux et al. 2007, 2008; Kanda et al. 2009). The lipolysis of chylomicrons and VLDL results in the formation of surface remnants, which contain water-soluble apolipoproteins and phospholipids and contribute to the maturation of HDL, as well as core remnants, which contain apoB and apoE and are enriched in cholesteryl esters. In addition to its enzymatic activity LPL also plays a role in the catabolism of the core remnants, since LPL monomers that are detached from the endothelium associate with these remnants and act together with apoE or apoB as ligands for their binding to heparansulfate proteoglycans, the LDL-receptor or the LDL-receptor related protein (LRP) in the liver (Bishop et al. 2008; Wong and Schotz 2002).

7.2.2 Hepatic Lipase

HL is produced by the liver and attached to heparansulfate proteoglycans on the surface of sinusoidal endothelial cells and hepatic parenchymal cells in the space of Disse (Perret et al. 2002; Wong and Schotz 2002). HL contributes to the remodelling of almost all lipoprotein classes by hydrolysing triglycerides and phospholipids of

VLDL, IDL, LDL, and HDL (Perret et al. 2002). The hydrolysis of phospholipids in HDL liberates lipid-poor or lipid-free apoA-I, which induces cellular lipid efflux by interaction with the ATP-binding cassette transporter (ABC) A1 and can thereby initiate the de novo formation of HDL (Barrans et al. 1994; Perret et al. 2002; Rye and Barter 2004). In addition and like LPL, it acts as a co-receptor for hepatic uptake of IDL and LDL by members of the LDL receptor gene family and of HDL by an as yet unknown holoparticle receptor (Guendouzi et al. 1998; Perret et al. 2002). The crucial role of HL for lipoprotein metabolism is highlighted by the severe mixed hyperlipidemia affecting all lipoprotein classes, including HDL in both patients and mice with HL deficiency (Zambon et al. 2003).

7.2.3 Endothelial Lipase

As indicated by its name, EL is synthesized mainly by the endothelium but also by liver, lung, macrophages, testis, ovary, and placenta. Like LPL and HL, EL contains heparin-binding domains which are needed for the attachment of EL to the glycocalix of endothelial cells. In contrast to LPL and HL, EL exerts almost exclusively phospholipase activity (Badellino and Rader 2004; Jaye and Krawiec 2004; Lamarche and Paradis 2007; Wong and Schotz 2002). Both humans and mice with EL deficiency have elevated levels of HDL cholesterol (Ishida et al. 2004; Ko et al. 2005). In contrast, transgenic overexpression of EL was found to decrease HDL cholesterol levels (Ishida et al. 2003). The impact of EL on HDL metabolism is also highlighted by the inverse correlation of EL levels in postheparin plasma with HDL cholesterol concentrations (Badellino et al. 2006). A closer analysis by using NMR spectroscopy revealed an inverse correlation between postheparin plasma concentration of EL and large HDL particles and a positive correlation with small HDL particles (Badellino et al. 2006). Interestingly EL concentrations do not correlate with apoA-I levels (Paradis et al. 2006). Taken together the data indicate that EL-mediated phospholipids hydrolysis remodels large HDL into small ones. Interestingly these smaller HDL particles formed by EL have a diminished ability to interact with scavenger receptor B type I (SR-BI; Gauster et al. 2004). Whether HDL remodelling by EL increases or decreases cholesterol efflux capacity is controversial (Gauster et al. 2004; Qiu and Hill 2009). Both in vitro and in vivo data from mice indicate that EL also hydrolyses phospholipids in apoB containing lipoproteins and thereby enhances catabolism of LDL (Lamarche and Paradis 2007). However, in humans positive correlations between postheparin EL concentration and levels of triglyceride and apoB were found, possibly as the result of increased EL expression in obesity (Badellino et al. 2006; Paradis et al. 2006). Therefore it is questionable whether EL plays any role in the remodelling of apoB-containing lipoproteins in humans (Lamarche and Paradis 2007). Data on the role of EL on atherosclerosis are controversial. EL deficiency was found to modulate atherosclerosis in atherosclerosis-prone mice in one study (Ishida et al. 2004) but not in another (Ko et al. 2005). By contrast, concentrations of EL in postheparin plasma correlated with coronary calcification in humans (Badellino et al. 2006).

7.3 Lipoprotein Transport Through the Endothelium

7.3.1 General Aspects of Transendothelial Lipoprotein Transport

During their metabolism, lipoproteins pass endothelial barriers on several occasions, namely after secretion from the liver and intestine into the blood and lymph, respectively, from the blood into extravascular compartments, where they exert their (patho)physiological functions (e.g. in the arterial wall), from the extravascular compartments back into the circulation (e.g. through vasa vasorum and lymphatic vessels), and finally from the circulation into the catabolic organs (notably liver). These transendothelial transport processes have been studied to a little degree, probably because they have been largely envisioned to occur through fenestrae of sinusoidal vessels or through discontinuities of a damaged endothelium.

Excess transendothelial lipoprotein transport plays a pivotal role in the pathogenesis of atherosclerosis. In fact, according to the response-to-retention hypothesis, the entry of remnants of lipid-rich lipoproteins (i.e. chylomicron remnants, IDL, LDL) from the blood stream into the intima of large arteries initiates and then perpetuates atherosclerosis (Tabas et al. 2007). HDL must also pass the endothelial barrier of arteries to exert their anti-atherosclerotic properties, for example to induce cholesterol efflux from lipid laden macrophages. Even more so, to continue the reverse transport of cholesterol from these foam cells in the arterial intima to the liver, HDL must again leave the arterial wall, probably via vasa vasorum which have grown into the thickened intima. It hence crosses an endothelial barrier a second time, this time from basolateral-to-apical. The accumulation of both pro-atherogenic and anti-atherogenic lipoproteins observed in atherosclerotic arteries is hence the result of increased influx and decreased efflux as well as enhanced binding to the extracellular matrix of the subendothelial intima (Simionescu 2007; Tabas et al. 2007). Although it is generally stated that subendothelial retention rather than unbalanced influx and efflux of lipoproteins is the major determinant of lipoprotein accumulation in the intima (Schwenke and Carew 1989; Tabas et al. 2007), it is important to note that the fluxes of lipoproteins into and out of the arterial wall have not been much investigated.

Initial studies in animals and men demonstrated that influx of lipoproteins into the vascular wall increases with the plasma concentration and decreases with the lipoprotein size (Nordestgaard et al. 1995). It has therefore been believed for a long time that lipoproteins enter the vascular wall by leakage through physiological fenestrae or damaged parts of the endothelium. However, since apoptosis of endothelial cells and disintegration of the endothelial layer occur in late and advanced atherosclerotic lesions and since lipoprotein entry into the subendothelial space is an early if not the first event in the pathogenesis of atherosclerosis (Simionescu 2007; Tabas et al. 2007), there is little reason to believe that endothelial injury is a prerequisite for lipoprotein entry into the vascular wall. In agreement with this, recent clinical studies did not find any effect of atherosclerosis or microalbuminuria on the fractional escape rate of radiolabeled LDL or albumin, which is considered as an index of transendothelial exchange (Jensen et al. 2004; Kornerup et al. 2004). However, transendothelial exchange rates of LDL and albumin correlated with one another and were increased in patients with diabetes mellitus or non-diabetic patients with hyperinsulinemia suggesting that disturbed insulin action precedes and contributes to the increased arterial entry of LDL leading to atherosclerosis (Jensen et al. 2004; Kornerup et al. 2003). Moreover, interventions such as blood pressure lowering have been found to reduce the entry of LDL into the subendothelial space of carotid arteries (Born et al. 2003). All these findings argue against the concept of unspecific and unregulated lipoprotein transudation into the subendothelium and rather suggest that transendothelial transport of lipoproteins is a regulated process. This statement does not exclude that endothelial dysfunction (e.g. as the result of shear stress) modulates the permeability of the endothelium for lipoproteins. For example confocal laser scanning microscopy studies in mice revealed that the endothelial glycocalix is thinner at arterial sections which are exposed to increased shear stress. The thinner glycocalix at the carotid bifurcation as compared to the common carotid region was associated with a two- to three-fold increase in intimal LDL accumulation (van den Berg et al. 2009).

In general the transport of molecules through a barrier is determined by their water solubility, size, and charge. However, systematic permeability studies in endothelial monolayers revealed that this correlation is true only for water-soluble molecules with a diameter below 6 nm. The transport of larger molecules is much slower and does not show any correlation with size (Mehta and Malik 2006). Thus the transendothelial transport of lipoproteins ranging in diameter between 8 nm (HDL) and hundreds of nanometers (chylomicrons) requires special prerequisites. In fact morphological, biochemical, and physiological studies provide strong evidence that proteins pass the intact endothelium by both paracellular and transcellular routes which involve the regulated opening and closure of interendothelial junctions and vesicular pathways, respectively.

7.3.2 Paracellular (Lipo)protein Transport

The barrier function of the endothelium is primarily formed by the intercellular connection of adjacent cells through adherens junctions, gap junctions, and tight junctions. It is secondarily enforced by the glycocalix formed by negatively charged surface proteins (proteoglycans, glycosaminoglycans) and the extracellular matrix on the abluminal (i.e. basolateral) side which is predominantly produced by endothelial cells themselves and constituted of collagen type IV, fibronectin, laminin, entactin, and proteoglycans. Adherens junctions are composed of vascular endothelial cadherin molcules (VE-cadherin) which are connected with the actin cytoskeleton through associated α-, β- and p120-catenins (Komarova et al. 2007; Mehta and Malik 2006). Also the tight junction proteins occludin and claudin are connected with the actin cytoskeleton, however via zona occludens proteins (Mehta and Malik 2006). The number and composition of the different junctions vary among endothelia of different vascular beds. In general arteries and capillaries of the blood–brain barrier contain many tight junctions, whereas endothelial cells of postcapillary venules do not express any (Mehta and Malik 2006).

The passage of molecules with a diameter larger than 6 nm requires the opening of interendothelial junctions. The opening and closing of adherens junctions are regulated by two antagonistic signalling cascades elicited by the binding of thrombin and sphingosine-1-phosphate (S1P) to their cognate G protein-coupled receptors, namely the protease activated receptor 1 (PAR1) and the sphingosine-1-phosphate receptor 1 (S1P1) (Komarova et al. 2007; Mehta and Malik 2006).

Thrombin cleaves and thereby activates the protease activated receptor PAR1 which thereby couples to the α-subunits of the heterotrimeric G proteins $G\alpha_q$ and Gα12/13. $G\alpha_q$ activates phospholipase C β- which in turn generates inositol-1,4,5-triphosphate (IP3) and diacylglycerol (DAG). These second messengers activate protein kinase C (PKC)α and serine/threonine phosphatase by stimulating both Ca^{2+} release from the endoplasmic reticulum and extracellular Ca^{2+} uptake. The PKCα-mediated phosphorylation of p120-catenin and the dephosphorylation of VE-cadherin and β-catenin by the two phosphatases finally disrupt the VE-cadherin–catenin complex. The increase in intracellular Ca^{2+} leads to the activation of the myosin light-chain kinase (MLCK) either directly or indirectly via the tyrosine kinase src (Hu and Minshall 2009). In parallel, PAR1-mediated recruitment of Gα12/13 activates the GTPase RhoA which activates downstream kinases ROCK and LIM and thereby inhibits both actin polymerization (via cofilin activation) and crosslinking of actin with the plasma membrane [via ezrin/radixin/moesin (ERM) proteins]. Both myosin light-chain phosphorylation by MLCK and inhibition of actin polymerization finally destabilize the microtubule actin cytoskeleton and thereby increase paracellular permeability. Thus PAR1-mediated signalling opens adherens junctions by disassembling the junction proteins and destabilizing the cytoskeleton (Komarova et al. 2007; Mehta and Malik 2006).

S1P is generated by the enzymatic breakdown of sphingomyelin and the subsequent phosphorylation of sphingosin by sphingosin kinase in erythrocytes and platelets and is transported in plasma by HDL (50%), albumin (40%), and LDL (10%). It binds to at least five G protein-coupled receptors, four of which are expressed in endothelial cells (Argraves and Argraves 2007; Yatomi 2008). Activation of S1P1 recruits $G\alpha_i$ which in turn activates Rac1 and Cdc42. This induces the reorganization of actin into cortical bundles which favor the reassembly of adherens junctions (Argraves et al. 2008; Komarova et al. 2007).

As yet this regulated paracellular transport has not been investigated for lipoproteins. However, since HDL transports about 50% of S1P in plasma, it is very likely that this lipoprotein contributes to the regulation of endothelial integrity and hence transendothelial macromolecule transport. In fact, HDL was previously described to increase the endothelial barrier integrity as measured by electric cell substrate impedance sensing (Argraves et al. 2008). Pertussis toxin, an inhibitor of Gi-coupled S1P receptors, as well as antagonists of the S1P receptor, S1P1, inhibited barrier enhancement by HDL, so it is very likely that S1P mediates this protective effect, at least in part (Argraves et al. 2008). HDL-associated S1P is known to stimulate the Erk1/2 and Akt signaling pathways in endothelial cells (for details, see Sect. 7.3). However, both HDL-induced barrier enhancement and HDL-induced motility showed a greater dependence on Akt activation as compared with Erk1/2 activation. In addition to these direct effects of HDL on endothelial permeability, HDL was previously described to favor endothelial integrity by inducing several endothelial repair mechanisms, such as proliferation and migration of endothelial cells as well as the recruitment of endothelial progenitor cells (see Sect. 7.3.3). This involves the interaction of HDL with its receptor SR-BI which is localized in caveolae of endothelial cells and signals via a carboxy-terminal PDZ domain to the multi-PDZ domain-containing adaptor protein PDZK1 which in turn phosphorylates src and activates Akt and Erk1/2 (Mineo and Shaul 2007; Zhu et al. 2008).

7.3.3 Transendothelial (Lipo)protein Transport

The transendothelial vesicular transport has been termed transcytosis and consists of endocytosis, vesicular transport, and exocytosis of the cargo at the opposite site (Mehta and Malik 2006; Simionescu 2007). This process can be receptor-dependent and then frequently involves caveolae or clathrin coated pits, or receptor-independent (i.e. fluid phase-mediated; Mehta and Malik 2006; Simionescu 2007). Until recently the physiological relevance of transcytosis through the endothelium was controversial, as no limiting factor was known (Rippe et al. 2002). This general scepticism is diminishing after detailed description of transendothelial albumin transport and the delineation of caveolin-1 as a rate-limiting factor (Frank et al. 2009; Mehta and Malik 2006). Since albumin binds lipids, notably fatty acids and lysolipids, transendothelial albumin transport not only serves as a model but is probably very relevant for transendothelial lipid transport. In addition biochemical and morphological experiments provide evidence for transcytosis of LDL, HDL or lipid-free apoA-I through microvascular endothelial cells of lymphatic vessels and the blood–brain barrier, as well as through the macrovascular endothelial cells of arteries.

Morphological studies have indicated that the transendothelial transport of LDL involves caveolae rather than clathrin-coated pits (Frank et al. 2009; Rippe et al. 2002). In agreement with a limiting role of caveolin-1 in LDL transcytosis, aortic rings of caveolin-1 knock-out mice showed a 50% reduction in the uptake of radiolabeled LDL at 37°C compared to aortic rings of wild-type mice (Frank et al. 2008). In vivo caveolin-1 knock-out mice showed a reduced fast-phase clearance (45 min) of radiolabeled LDL (which reflects transendothelial loss of the label into the extravascular space) but no alteration in LDL clearance after 2 h (which predominantly reflects hepatic uptake). At 24 h after injection caveolin-1 knock-out mice showed a 50% reduced uptake of radiolabeled LDL into arteries but a 30% increased uptake into liver (Frank et al. 2008). Taken together the results suggest that caveolin-1 plays an important role for the transendothelial uptake of LDL into the arterial wall. Interestingly, upon a Western-type diet apoE/caveolin-1 double knock-out mice had less atherosclerosis than apoE-only knock-out animals although they had more severely elevated LDL and VLDL plasma levels. However, it is not yet known whether this atheroprotective effect of caveolin-1 ablation is due to reduced arterial LDL influx or due to reduced VLDL production and increased HDL levels, which are also observed in these animals (Frank et al. 2004).

Transport studies in porcine brain capillary endothelial cells identified the LDL receptor as a rate-limiting factor for the transcytosis of LDL through an in vitro blood–brain barrier model (Goti et al. 2002; Dehouck et al. 1997). However, it is important to note that the brain does not contain apoB, so it is questionable whether the LDL receptor mediates transcytosis of LDL holoparticles through the blood–brain barrier in vivo. It is more likely that, after internalization, LDL is dissociated so that only some LDL-associated components like cholesterol, α-tocopherol, and water-soluble apolipoproteins but not apoB are transported into the brain (Goti et al. 2002). Also the LDL receptor-related proteins LRP1 and LRP2 were found to mediate transport of macromolecules, for example of the receptor-associated protein RAP, through the blood–brain barrier (Pan et al. 2004). Since the blood–brain barrier is very tight for the transendothelial transport of macromolecules and since the expression of many multi-drug resistance proteins by brain capillary endothelial cells also impedes the delivery of small molecules into the brain, the LDL receptor and both LRP1 and LRP2 have become interesting targets for therapeutic drug and protein delivery into the brain (Chao et al. 2003; Spencer and Verma 2007).

Less experimental data have been obtained on the transendothelial transport of HDL. Immunohistochemical studies of bovine aortic endothelial cells recovered DiI-labeled HDL in caveolae and co-localized gold-labeled HDL with caveolin-1 (Chao et al. 2003). Our own laboratory has investigated the transport of HDL and its main protein constituent apoA-I through monolayers of bovine aortic endothelial cells (Cavelier et al. 2006; Rohrer et al. 2006, 2009). Endothelial cells bound and associated ^{125}I-apoA-I and ^{125}I-HDL with high affinity and in a saturable and specific manner. Biotinylation experiments, fluorescence microscopy, and immunoelectronmicroscopy found that endothelial cells internalize labeled apoA-I and HDL. Only minor amounts of the internalized ^{125}I-apoA-I and ^{125}I-HDL were degraded. Cultivated in a Transwell system, the cells transported ^{125}I-apoA-I or ^{125}I-HDL from the apical to the basolateral compartment in a competitive and temperature-sensitive manner (Rohrer et al. 2006, 2009). Furthermore, after specific transport the originally lipid-free and preβ-mobile apoA-I molecules were recovered as lipidated particles which have electrophoretic α-mobility (Rohrer et al. 2006). Using pharmacological inhibitors and RNA interference, we also showed that ABCA1 but not SR-BI modulate cell surface binding, internalization and transport of ^{125}I-apoA-I (Cavelier et al. 2006). In contrast, inhibition of SR-BI and ABCG1 but not inhibition of ABCA1 decreased the binding and transport of ^{125}I-HDL (Rohrer et al. 2009). Taken together the data suggest that arterial endothelial cells transcytose apoA-I and HDL by distinct pathways which involve transcytosis.

Also transport studies in a blood–brain barrier cell culture model provided evidence for the apical-to-basolateral transcytosis of HDL-holoparticles as well as associated α-tocopherol and apoA-I through porcine brain capillary endothelial cells by a process which involves SR-BI and caveolin-1 (Balazs et al. 2004). ApoA-I also facilitates the transport of protamine-oligonucleotide nanoparticles through brain capillary endothelial cells by an SR-BI-regulated pathway (Kratzer et al. 2007). Thus, similar to ligands of the LDL receptor and LRPs, apoA-I may help to target therapeutic molecules into the brain (Kratzer et al. 2007; Spencer and Verma 2007).

7.4 Target for Physiological and Pathological Effects of Lipoproteins

As exchange barrier between plasma and tissue, endothelial cells are exposed to lipoproteins to a higher degree than any other cell type in the organism and, consequently, are the target of many physiological and pathological effects which influence endothelial function and survival. The lipoprotein components, receptors and signaling pathways which mediate the various protective or harmful effects of lipoproteins on endothelial cells are very heterogenous. From a general and hence oversimplified view one can summarize most of the present findings that (modified) apoB-containing lipoproteins [i.e. remnants of chylomicrons, LDL, Lp(a)] favor endothelial dysfunction whereas HDL mediate endothelial protection and repair. However, in several clinical conditions (e.g. inflammation) HDL can become dysfunctional and lose its protective effects on the endothelium or even gain paradoxical and harmful effects (Ansell et al. 2007).

7.4.1 Regulation of the Vascular Tone

The vascular tone of arteries is regulated by endothelium-dependent and endothelium-independent mechanisms. Endothelium-dependent vasoreactivity depends on several hormones and mediators which are secreted by the endothelium, for example nitric oxide (NO), prostacyclin (PGI_2), C-type natriuretic peptide (CNP), and endothelin-1 (Félétou and Vanhoutte 2007). Disturbed endothelium-dependent vasodilation is considered as a hallmark of endothelial dysfunction and an early sign of (presymptomatic) atherosclerosis (Deanfield et al. 2007; Félétou and Vanhoutte 2007). Endothelium-dependent vasorelaxation has been assessed in clinical studies by various invasive and non-invasive methods (Deanfield et al. 2007). Classically, endothelial vasoreactivity is assessed by angiographic measurement of lumen changes of coronary or brachial arteries in response to acetylcholine and other drugs which activate endothelial NO synthase (eNOS) and thereby elicit NO release. Less invasive and therefore also applicable to healthy volunteers is the measurement of flow-mediated brachial artery vasoreactivity which is monitored by high-resolution Doppler ultrasonography. In this assay hyperemia, which is induced by rapid deflation of a blood pressure cuff occluding the upper arm, leads to shear stress and thereby NO release in the forearm arteries, which results in vasodilation (Deanfield et al. 2007). Independently of the method used, many clinical studies observed that elevated plasma concentrations of

LDL cholesterol, triglycerides or chylomicron remnants, as well as low concentrations of HDL cholesterol, are associated with decreased endothelium-dependent vasodilation of coronary or brachial arteries (Brunner et al. 2005; Davignon and Ganz 2004; Norata and Catapano 2005; Wheeler-Jones 2007; Zhang et al. 2008; Zheng and Liu 2008). Even acute changes in the lipoprotein composition had strong effects on endothelium-dependent vasoreactivity: consumption of a single high-fat meal but not consumption of a low-fat meal was found to transiently impair flow-mediated vasodilation for 4 h. The change in this endothelium-dependent vasodilation correlated with the 2-h change in postprandial triglyceride or remnant lipoprotein levels (Plotnick et al. 1997). Conversely rapid lowering of LDL cholesterol by LDL apheresis (Tamai et al. 1997) as well as the infusion of artificially reconstituted HDL consisting of apoA-I and phosphatidylcholine were found to improve endothelium-dependent vasodilation in patients with hypercholesterolemia or low HDL-cholesterol (Bisoendial et al. 2003; Spieker et al. 2002).

Both the inhibitory effects of chylomicron remnants and the stimulatory effects of HDL on endothelium-dependent vasorelaxation are at least in part mediated by changes in NO production (Norata and Catapano 2005; Zheng and Liu 2007): chylomicron remnants were found to suppress eNOS phosphorylation by inducing the focal adhesion kinase FAK and downstream activation of the phosphatidlyinositol-3-kinase (PI3K)/Akt (protein kinase B) pathway (Kawakami et al. 2002). The structural component of chylomicron remnants mediating its adverse effect on eNOS activation is not known. In addition modified lipoproteins and chylomicron remnants stimulate NADPH oxidase to synthesize superoxides which may inactivate NO by peroxynitrite formation (Dunn et al. 2008; Shin et al. 2004).

Conversely to chylomicron remnants, HDL stimulates eNOS phosphorylation by activating the PI3K/Akt signaling cascade (Norata and Catapano 2005). eNOS activation by HDL requires SR-BI (Yuhanna et al. 2001). Mineo et al. proposed that eNOS activation via binding of HDL to SR-BI involves MAP kinase signaling and subsequent Akt phosphorylation (Mineo and Shaul 2007; Mineo et al. 2003). Nofer et al. (2004) however showed that the stimulatory effect of HDL is not exerted by lipid-free apoA-I or reconstituted HDL consisting of apoA-I and phosphatidylcholine but by lysosphingolipids, including S1P, sphingosylphosphorylcholin (SPC), and lysosulfatide (LSF). It has therefore been proposed that the binding of HDL to SR-BI increases the abundance of HDL on the endothelial surface and hence the presentation of the agonist S1P to its cognate receptors (Nofer and Assmann 2005). In fact the physiological relevance of S1P and the need of its receptors for the vasodilatory effects of HDL have been proven in experiments on isolated arteries and mice which do not express the receptor S1P3 (Nofer et al. 2004). HDL-induced phosphorylation of Akt and eNOS as well as vasorelaxation of pre-contracted aortic rings from S1P3-deficient mice were reduced by 60% (Nofer et al. 2004). The residual ability of HDL to induce vasorelaxation may be mediated by other S1P receptors or even by completely different pathways and/or mediators. In agreement with the former explanation S1P was also found to mediate vasoconstriction by as yet unidentified S1P receptors expressed in vascular smooth muscle cells (Argraves and Argraves 2007).

In agreement with the latter explanation, HDL was found to stimulate the endothelial secretion of PGI2 and CNP which both induce smooth muscle cell relaxation and hence vasodilation and to inhibit abluminal secretion of endothelin-1 which stimulates vasoconstriction (Norata and Catapano 2005). The stimulatory effect of HDL on PGI2 release is dose-dependent and involves induction of cyclooxy-genase 2 (COX-2) via the MAP kinase pathway (Norata et al. 2004). It is also exerted by delipidated HDL apolipoproteins although at a weaker degree than by intact HDL (Norata et al. 2004). Interestingly chylomicron remnants also induce COX-2 expression by activation of nuclear factor kappa B (NFκB) and MAP kinase signaling pathways, namely ERK1/2 and P38MAPK (Evans et al. 2004; Wheeler-Jones 2007). The common effects of HDL and chylomicron remnants on MAP kinase signaling and COX-2 induction suggest the action of a joint agonist. In fact, ω-6-fatty acids carried by both chylomicron remnants and HDL have been suggested to account for at least some of the effects from either lipoprotein class on COX-2 induction (Norata and Catapano 2005; Wheeler-Jones 2007). The most likely sources of arachidonic or linoleic acids are phospholipids, from which they are liberated by endothelial phospholipases such as EL. Whether COX-2 activation in endothelial cells is protective or harmful is a matter of controversy, since COX-2 products exert both pro- and anti-inflammatory properties. The preponderance of physiological or pathological effects may hence depend on the local environment and the expression of downstream enzymes such as prostacyclin synthase.

7.4.2 Leukocyte Adhesion and Extravasation

Endothelial activation by inflammatory stimuli induces the expression of E-selectins, vascular cellular adhesion molecule 1(VCAM-1), and intercellular adhesion molecule 1 (ICAM-1), which mediate reversible (rolling) and then irreversible binding (adhesion) of leukocytes to the endothelium (van Gils et al. 2009). Monocyte chemotactic protein 1 (MCP1; which is also produced by endothelial cells) ultimately stimulates the extravasation (diapedesis) of leukocytes into the extravascular tissue. The induction of cell adhesion molecules as well as MCP-1 mainly depends on the expression of the transcription factor NFκB, which is activated by cytokines and reactive oxygen species (ROS; Pantano et al. 2006; van Gils et al. 2009). Native LDL (and even more so oxidatively modified LDL) as well as chylomicron remnants were found to stimulate the production of ROS (e.g. by NADPH-oxidase) to activate NFκB, and to thereby induce the expression of cell adhesion molecules in endothelial cells (Gleissner et al. 2007; Wheeler-Jones 2007; Zheng and Liu 2007). The binding of modified LDL or chylomicron remnants to the lectin-like oxidized LDL receptor 1 (LOX-1) appears to play a pivotal role in this scenario, since this receptor is known to signal to NADPH oxidase and NFκB (Dunn et al. 2008; Gleissner et al. 2007; Wheeler-Jones 2007; Zheng and Liu 2007) and since inhibition of LOX-1 by monoclonal antibodies or antisense oligoncleotides abrogates the stimulatory effect of oxidized LDL and chylomicron remnants on NADPH-oxidase activation and cell adhesion molecule expression (Dunn et al. 2008). Interestingly, the binding of oxidized LDL itself increases the expression of LOX-1 in a positive feedback loop (Dunn et al. 2008). At least a part of the proinflammatory effects of oxidized LDL and chylomicron remnants are exerted by lysophosphatidylcholine (i.e. a lipolytic product of phosphatidylcholine generated by endothelial phospholipases; Gleissner et al. 2007; Wheeler-Jones 2007; Zheng and Liu 2007).

Several research groups demonstrated that HDL inhibits the expression of E-selectin, VCAM-1, ICAM-1, and MCP-1 by endothelial cells as well as by the endothelial diapedesis of monocytes (Mineo et al. 2006; Norata and Catapano 2005; Zhang et al. 2008). In addition HDLs were found to inhibit ROS formation by NADPH oxidase as well as NFκB activation (Robbesyn et al. 2003). The inhibitory effects of HDL on the transmigration of leukocytes through the endothelium appear to be exerted by different components and pathways. Barter and colleagues initially showed that reconstituted HDL but not lipid-free apolipoproteins inhibit VCAM-1 and E-selectin expression (Baker et al. 2000). The inhibitory effect of reconstituted HDL is strongly influenced by its phosphatidylcholine moiety. Phosphatidylcholines containing poly-unsaturated fatty acids have the strongest inhibitory effect (Baker et al. 2000). In addition Barter and colleagues showed that the inhibitory effect of native and reconstituted HDL on cytokine-induced VCAM-1 expression involves the inhibition of sphingosin-1-kinase and NFκB, so that they concluded that HDL prevents endothelial monocyte adhesion by inhibiting endogenous S1P synthesis (Xia et al. 1999). In contrast to this harmful effect of intracellular S1P, Kimura et al. (2003) and Nofer et al. (2003) demonstrated that extracellular S1P carried by HDL inhibits NADPH oxidase, subsequent ROS formation, and the expression of E-selectin and cell adhesion molecule.Again this protective effect appears to be exerted by the dual interactions of HDL and S1P with two receptors, namely the lipoprotein receptor SR-BI and the sphingolipid receptor S1P1 (Kimura et al. 2006).

7.4.3 Platelet Aggregation, Coagulation, and Fibrinolysis

The endothelium plays an important role in hemostasis by providing a surface for platelet aggregation and coagulation and by secreting several pro- and anti-aggregatory small molecules, such as NO, PGI_2 and platelet-activating factor (PAF), as well as pro-coagulant and anti-fibrinolytic proteins such as tissue factor (TF), von Willebrand factor (vWF), and plasminogen activator inhibitor type 1 (PAI-1).

As described above, both HDL and chylomicron remnants stimulate PGI2 production by endothelial cells, whereas NO production is inhibited by LDL and chylomicron remnants but stimulated by HDL (Evans et al. 2004; Mineo et al. 2006; Norata and Catapano 2005; Norata et al. 2004; Wheeler-Jones 2007). The pro- and anti-aggregatory effects of apoB-containing lipoproteins and HDL, respectively, is also mediated by their stimulating and inhibiting effects, respectively, on the expression of selectins (Baker et al. 2000; Gleissner et al. 2007; Kimura et al. 2006; Mineo et al. 2006). In addition HDL has been shown to inhibit thromboxan A2 formation (Oravec et al. 1998). The concerted effects on the endothelium and platelet activities make HDL act anti-platelet aggregatory (Mineo et al. 2006).

Native and even more so oxidized LDL and VLDL were reported to stimulate the secretion of TF and PAI-1 by endothelial cells and to thereby favor the prothrombotic state (Bai et al. 2006; Rosenson and Lowe 1998; Zhao et al. 2008) known especially for patients with metabolic syndrome and hypertriglyceridemia. HDL rather inhibits the VLDL-stimulated TF and oxLDL-stimulated PAI-1 secretion (Bai et al. 2006; Ren and Shen 2000; Viswambharan et al. 2004). These data as well as other data on the effects of lipoproteins on coagulation and fibrinolysis factors suggest that HDL inhibits coagulation and supports fibrinolysis, whereas triglyceride rich lipoproteins and their remnants exert opposite prothrombotic effects.

7.4.4 Endothelial Survival and Repair

Due to their exposure to much chemical and physical harm, endothelial cells have a high risk of death. To maintain the integrity of the vasculature the endothelium needs factors that support cellular survival and the replacement of damaged endothelial cells. The latter may occur via proliferation and migration of neighboring cells or via repopulation with blood-borne endothelial progenitor cells (EPCs) which are ultimately derived from bone marrow (Zampetaki et al. 2008). In more exaggerated situations such as organ growth and tissue repair the additional need for blood supply leads to angiogenesis and neovascularization. Lipoproteins, notably HDL, affect many aspects of endothelial cell apoptosis, proliferation, migration, and differentiation (Argraves and Argraves 2007; Mineo and Shaul 2007; Mineo et al. 2006; Nofer and Assmann 2005).

Many local mediators, including cytokines, activated killer cells, and oxidized LDL, can cause the apoptosis of endothelial cells. At least some of the pro-apoptotic effects of modified LDL are exerted via binding to LOX-1, which increases the Bax/Bcl2 ratio via NFκB activation (Li and Mehta 2000). Also triglyceride-rich lipoproteins promote endothelial cell death by NFκB activation (Norata et al. 2003; Shin et al. 2004). In contrast, HDL was found to inhibit endothelial cell death being induced by oxidized LDL, TNFα, the terminal complement complex or growth factor deprivation (Argraves and Argraves 2007; Mineo and Shaul 2007; Mineo et al. 2006; Nofer and Assmann 2005; Norata and Catapano 2005). Interestingly and depending on the pro-apoptotic stimuli, the anti-apoptotic activity of HDL has been assigned to different components of HDL, namely to apoA-I for cell death induced by oxidized LDL, VLDL or cytokines (Suc et al. 1997; Speidel et al. 1990; Sugano et al. 2000), to lysosphingolipids for protection from cell death by growth factor deprivation (Kimura et al. 2003; Nofer et al. 2001; Kimura et al. 2001), and to apoJ for the inhibition of the terminal complement complex (Rosenfeld et al. 1983). The lysosphingolipids S1P, SPC, and LSF interfere with apoptosis by activating the PI3K/Akt pathway, which in turn prevents the activation of caspases 3 and 9 (Nofer et al. 2001).

S1P has also been identified as the agonist of HDL which stimulates endothelial cell proliferation by activating the small G potein Ras and the P42/44 MAP kinase pathway

(Miura et al. 2003a; Nofer et al. 2000; von Otte et al. 2006). Interestingly the same pertussis-toxin-sensitive signaling cascade has been made responsible for the S1P-stimulated migration of endothelial cells and the formation of endothelial tubes (i.e. capillary-like structures; Kimura et al. 2003; Miura et al. 2003b). Antisense oligonucleotides against the S1P receptors S1P1 and S1P3 blocked both the HDL-induced endothelial migration and tube formation, suggesting that S1P is the only agonist for these endothelial repair activities (Miura et al. 2003b).

However in another laboratory pertussis toxin did not inhibit endothelial cell migration stimulated by HDL, so that Mineo, Shaul and colleagues postulated the presence of another pathway and agonist which induces the migration of endothelial cells by HDL (Mineo and Shaul 2007; Mineo et al. 2006). In fact, reconstituted HDL consisting of apoA-I, palmitoyloleylphosphatidylcholine, and cholesterol were also able to induce endothelial cell migration (Seetharam et al. 2006). In this case interaction of HDL with SR-BI is thought to activate the small G protein Rac via src-kinase, PI3K, Akt, and MAP kinase, which in turn stimulates the formation of lamellipodia, a prerequisite for cell migration (Seetharam et al. 2006). Interestingly and in contrast to VEGF-induced endothelial cell migration, this process does not involve eNOS and NO formation (Seetharam et al. 2006). Studies in genetic mouse models also indicate that HDL and SR-BI are involved in re-endothelialization. ApoA-I-deficient and hence HDL-deficient mice show decreased re-endothelialization as compared to wild-type mice. Normalization of apoA-I and HDL levels by liver-directed gene apoA-I transfer restored the re-endothelialization capacity. SR-BI knock-out mice also showed decreased re-endothelialization despite elevated HDL cholesterol levels supporting the critical role of this HDL receptor in endothelial repair (Seetharam et al. 2006).

As pointed out in the beginning of this section, re-endothelialization can occur as the result of proliferation and migration of neighboring cells as well as of repopulation with blood-borne EPCs. Several data from animal studies point to the modulatory activity of HDL in the latter process. In atherosclerosis-prone animal models, the infusion of reconstituted HDL as well as apoA-I gene transfer increased the number of EPCs in the blood (Tso et al. 2006). In an allograft heart transplant model, apoA-I gene transfer also promoted incorporation of EPCs in allografts and attenuated transplant arteriosclerosis (Feng et al. 2008a; Sumi et al. 2007). Interestingly, the beneficial effect of apoA-I gene transfer was offset in mice with SR-BI knock-out (Feng et al. 2008b, 2009). The accompanying in vitro studies also showed that the beneficial effect of HDL on EPC migration involved signaling via SR-BI and extracellular signal-regulated kinases (ERK) as well as increased NO production in EPCs (Noor et al. 2007).

Also, data from humans point to the improvement of EPC function by HDL: In patients with coronary artery disease a positive correlation between circulating EPC and HDL concentrations was demonstrated (Noor et al. 2007; Petoumenos et al. 2008). Ex vivo HDL was observed to enhance the differentiation of human monocytic cells into EPCs, inhibit apoptosis of EPCs, increase eNOS protein expression in EPCs, but decrease pro-MMP-9, stimulate proliferation of early outgrowth colonies after extended cell cultivation, and improve adhesion EPCs on human coronary artery endothelial cells by up-regulation of β2 and α4 integrins (Petoumenos et al. 2008). Pilot studies showed an increase in EPCs in the blood of diabetic patients treated with reconstituted HDL (van Oostrom et al. 2007).

References

Ansell BJ, Fonarow GC, Fogelman AM (2007) The paradox of dysfunctional high-density lipoprotein. Curr Opin Lipidol 18:427– 434

Argraves KM, Argraves WS (2007) HDL serves as a S1P signaling platform mediating a multitude of cardiovascular effects. J Lipid Res 48:2325–2333

Argraves KM, Gazzolo PJ, Groh EM, Wilkerson BA, Matsuura BS, Twal WO, Hammad SM, Argraves WS (2008) High density lipoprotein-associated sphingosine 1-phosphate promotes endothelial barrier function. J. Biol Chem 283:25074–25081

Badellino KO, Rader DJ (2004) The role of endothelial lipase in high-density lipoprotein metabolism. Curr Opin Cardiol 19:392–395

Badellino KO, Wolfe ML, Reilly MP, Rader DJ (2006) Endothelial lipase concentrations are increased in metabolic syndrome and associated with coronary atherosclerosis. PLoS Med 3:e22

Bai H, Liu BW, Deng ZY, Shen T, Fang DZ, Zhao YH, Liu Y (2006) Plasma very-low-density lipoprotein, low-density lipoprotein, and high-density lipoprotein oxidative modification induces procoagulant profiles in endogenous hypertriglyceridemia. Free Radic Biol Med 40:1796–1803

Baker PW, Rye KA, Gamble JR, Vadas MA, Barter PJ (2000) Phospholipid composition of reconstituted high density lipoproteins influences their ability to inhibit endothelial cell adhesion molecule expression. J Lipid Res 41:1261–1267

Balazs Z, Panzenboeck U, Hammer A, Sovic A, Quehenberger O, Malle E, Sattler W (2004) Uptake and transport of high-density lipoprotein (HDL) and HDL-associated alpha-tocopherol by an in vitro blood-brain barrier model. J Neurochem 89:939–950

Barrans A, Collet X, Barbaras R, Jaspard B, Manent J, Vieu C, Chap H, Perret B (1994) Hepatic lipase induces the formation of pre-beta 1 high density lipoprotein (HDL) from triacylglycerol-rich HDL2. A study comparing liver perfusion to in vitro incubation with lipases. J Biol Chem 269:11572–11577

Beigneux AP, Davies BS, Gin P, Weinstein MM, Farber E, Qiao X, Peale F, Bunting S, Walzem RL, Wong JS, Blaner WS, Ding ZM, Melford K, Wongsiriroj N, Shu X, de Sauvage F, Ryan RO, Fong LG, Bensadoun A, Young SG (2007) Glycosylphosphatidylinositol-anchored high-density lipoprotein-binding protein 1 plays a critical role in the lipolytic processing of chylomicrons. Cell Metab 5:279–291

Beigneux AP, Davies BS, Bensadoun A, Fong LG, Young SG (2008) GPIHBP1 – a GPI-anchored protein required for the lipolytic processing of triglyceride-rich lipoproteins. J Lipid Res (in press)

Bishop JR, Stanford KI, Esko JD (2008) Heparan sulfate proteoglycans and triglyceride-rich lipoprotein metabolism. Curr Opin Lipidol 19:307–313

Bisoendial RJ, Hovingh GK, Levels JH, Lerch PG, Andresen I, Hayden MR, Kastelein JJ, Stroes ES (2003) Restoration of endothelial function by increasing high-density lipoprotein in subjects with isolated low high-density lipoprotein. Circulation 107:2944–2948

Born GV, Medina R, Shafi S, Cardona-Sanclemente LE (2003) Factors influencing the uptake of atherogenic plasma proteins by artery walls. Biorheology 40:13–22

Brunner H, Cockcroft JR, Deanfield J, Donald A, Ferrannini E, Halcox J, Kiowski W, Lüscher TF, Mancia G, Natali A, Oliver JJ, Pessina AC, Rizzoni D, Rossi GP, Salvetti A, Spieker LE, Taddei S, Webb DJ (2005) Working group on endothelins and endothelial factors of the european society of hypertension. Endothelial function and dysfunction. Part II: association with cardiovascular risk factors and diseases. A statement by the working group on endothelins and endothelial factors of the european society of hypertension. J Hypertens 23:233–246

Cavelier C, Rohrer L, von Eckardstein A (2006) ATP-binding cassette transporter A1 modulates apolipoprotein A-I transcytosis through aortic endothelial cells. Circ Res 99:1060–1066

Chao WT, Fan SS, Chen JK, Yang VC (2003) Visualizing caveolin-1 and HDL in cholesterol-loaded aortic endothelial cells. J Lipid Res 44:1094–1099

Davignon J, Ganz P (2004) Role of endothelial dysfunction in atherosclerosis. Circulation 109[Suppl 1]:III27–32

Deanfield JE, Halcox JP, Rabelink TJ (2007) Endothelial function and dysfunction: testing and clinical relevance. Circulation 115:1285–1295

Dehouck B, Fenart L, Dehouck MP, Pierce A, Torpier G, Cecchelli R (1997) A new function for the LDL receptor: transcytosis of LDL across the blood–brain barrier. J Cell Biol 138:877–889

Dunn S, Vohra RS, Murphy JE, Homer-Vanniasinkam S, Walker JH, Ponnambalam S (2008) The lectin-like oxidized low-density-lipoprotein receptor: a pro-inflammatory factor in vascular disease. Biochem J 409:349–355

Evans M, Berhane Y, Botham KM, Elliott J, Wheeler-Jones CP (2004) Chylomicron-remnant-like particles modify production of vasoactive mediators by endothelial cells. Biochem Soc Trans 32:110–112

Félétou M, Vanhoutte PM (2007) Endothelium-dependent hyperpolarizations: past beliefs and present facts. Ann Med 39:495–516

Feng Y, Jacobs F, Van Craeyveld E, Brunaud C, Snoeys J, Tjwa M, Van Linthout S, De Geest B (2008a) Human ApoA-I transfer attenuates transplant arteriosclerosis via enhanced incorporation of bone marrow-derived endothelial progenitor cells. Arterioscler Thromb Vasc Biol 28:278–283

Feng Y, van Eck M, Van Craeyveld E, Jacobs F, Carlier V, Van Linthout S, Erdel M, Tjwa M, De Geest B (2009) Critical role of scavenger receptor-BI expressing bone marrow-derived endothelial progenitor cells in the attenuation of allograft vasculopathy after human apo A-I transfer. Blood 113:755–764

Forte TM, Shu X, Ryan RO (2008) The ins (cell) and outs (plasma) of apolipoprotein A-V. J Lipid Res (in press)

Frank PG, Lee H, Park DS, Tandon NN, Scherer PE, Lisanti MP (2004) Genetic ablation of caveolin-1 confers protection against atherosclerosis. Arterioscler Thromb Vasc Biol 24:98–105

Frank PG, Pavlides S, Cheung MW, Daumer K, Lisanti MP (2008) Role of caveolin-1 in the regulation of lipoprotein metabolism. Am J Physiol Cell Physiol 295:C242–C248

Frank PG, Pavlides S, Lisanti MP (2009) Caveolae and transcytosis in endothelial cells: role in atherosclerosis. Cell Tissue Res 335:41–47

Gauster M, Oskolkova OV, Innerlohinger J, Glatter O, Knipping G, Frank S (2004) Endothelial lipase-modified high-density lipoprotein exhibits diminished ability to mediate SR-BI (scavenger receptor B type I)-dependent free-cholesterol efflux. Biochem J 382:75–82

Gleissner CA, Leitinger N, Ley K (2007) Effects of native and modified low-density lipoproteins on monocyte recruitment in atherosclerosis. Hypertension 50:276–283

Goti D, Balazs Z, Panzenboeck U, Hrzenjak A, Reicher H, Wagner E, Zechner R, Malle E, Sattler W (2002) Effects of lipoprotein lipase on uptake and transcytosis of low density lipoprotein (LDL) and LDL-associated alpha-tocopherol in a porcine in vitro blood–brain barrier model. J Biol Chem 277:28537–28544

Guendouzi K, Collet X, Perret B, Chap H, Barbaras R (1998) Remnant high density lipoprotein2 particles produced by hepatic lipase display high-affinity binding and increased endocytosis into a human hepatoma cell line (HEPG2). Biochemistry 37:14974–14980

Hasham SN, Pillarisetti S (2006) Vascular lipases, inflammation and atherosclerosis. Clin Chim Acta 372:179–183

Hu G, Minshall RD (2009) Regulation of transendothelial permeability by Src kinase. Microvasc Res 77:21–25

Ishida T, Choi S, Kundu RK, et al (2003) Endothelial lipase is a major determinant of HDL level. J Clin Invest 111:347–355

Ishida T, Choi SY, Kundu RK, Spin J, Yamashita T, Hirata K, Kojima Y, Yokoyama M, Cooper AD, Quertermous T (2004) Endothelial lipase modulates susceptibility to atherosclerosis in apolipoprotein-E-deficient mice. J Biol Chem 279:45085–45092

Jaye M, Krawiec J (2004) Endothelial lipase and HDL metabolism. Curr Opin Lipidol 15:183–189

Jensen JS, Feldt-Rasmussen B, Jensen KS, Clausen P, Scharling H, Nordestgaard BG (2004) Transendothelial lipoprotein exchange and microalbuminuria. Cardiovasc Res 63:149–154

Kanda T, Brown JD, Orasanu G, Vogel S, Gonzalez FJ, Sartoretto J, Michel T, Plutzky J (2009) PPARgamma in the endothelium regulates metabolic responses to high-fat diet in mice. J Clin Invest. 119:110–124

Kawakami A, Tanaka A, Nakajima K, Shimokado K, Yoshida M (2002) Atorvastatin attenuates remnant lipoprotein-induced monocyte adhesion to vascular endothelium under flow conditions. Circ Res 91:263–271

Kimura T, Sato K, Kuwabara A, Tomura H, Ishiwara M, Kobayashi I, Ui M, Okajima F (2001) Sphingosine 1-phosphate may be a major component of plasma lipoproteins responsible for the cytoprotective actions in human umbilical vein endothelial cells. J Biol Chem 276:31780–31785

Kimura T, Sato K, Malchinkhuu E, Tomura H, Tamama K, Kuwabara A, Murakami M, Okajima F (2003) High-density lipoprotein stimulates endothelial cell migration and survival through sphingosine 1-phosphate and its receptors. Arterioscler Thromb Vasc Biol 23:1283–1288

Kimura T, Tomura H, Mogi C, Kuwabara A, Damirin A, Ishizuka T, Sekiguchi A, Ishiwara M, Im DS, Sato K, Murakami M, Okajima F (2006) Role of scavenger receptor class B type I and sphingosine 1-phosphate receptors in high density lipoprotein-induced inhibition of adhesion molecule expression in endothelial cells. J Biol Chem 281:37457–37467

Ko KW, Paul A, Ma K, Li L, Chan L (2005) Endothelial lipase modulates HDL but has no effect on atherosclerosis development in apoE–/– and LDLR–/– mice. J Lipid Res 46:2586–2594

Komarova YA, Mehta D, Malik AB (2007) Dual regulation of endothelial junctional permeability. Sci STKE 2007:re8

Kornerup K, Nordestgaard BG, Feldt-Rasmussen B, Borch-Johnsen K, Jensen KS, Jensen JS (2003) Increased transvascular low density lipoprotein transport in insulin dependent diabetes: a mechanistic model for development of atherosclerosis. Atherosclerosis 170:163–168

Kornerup K, Nordestgaard BG, Jensen TK, Feldt-Rasmussen B, Eiberg JP, Jensen KS, Jensen JS (2004) Transendothelial exchange of low-density lipoprotein is unaffected by the presence of severe atherosclerosis. Cardiovasc Res 64:337–345

Kratzer I, Wernig K, Panzenboeck U, Bernhart E, Reicher H, Wronski R, Windisch M, Hammer A, Malle E, Zimmer A, Sattler W (2007) Apolipoprotein A-I coating of protamine-oligonucleotide nanoparticles increases particle uptake and transcytosis in an in vitro model of the blood–brain barrier. J Control Release 117:301–311

Lamarche B, Paradis ME (2007) Endothelial lipase and the metabolic syndrome. Curr Opin Lipidol 18:298–303

Li D, Mehta JL (2000) Upregulation of endothelial receptor for oxidized LDL (LOX-1) by oxidized LDL and implications in apoptosis of human coronary artery endothelial cells: evidence from use of antisense LOX-1 mRNA and chemical inhibitors. Arterioscler Thromb Vasc Biol 20:1116–1122

Mehta D, Malik AB (2006) Signaling mechanisms regulating endothelial permeability. Physiol Rev 86:279–367

Mineo C, Shaul PW (2007) Role of high-density lipoprotein and scavenger receptor B type I in the promotion of endothelial repair. Trends Cardiovasc Med 17:156–161

Mineo C, Yuhanna IS, Quon MJ, Shaul PW (2003) High density lipoprotein-induced endothelial nitric-oxide synthase activation is mediated by Akt and MAP kinases. J Biol Chem 278:9142–9149

Mineo C, Deguchi H, Griffin JH, Shaul PW (2006) Endothelial and antithrombotic actions of HDL. Circ Res 98:1352–1364

Miura S, Fujino M, Matsuo Y, Kawamura A, Tanigawa H, Nishikawa H, Saku K (2003a) High density lipoprotein-induced angiogenesis requires the activation of Ras/MAP kinase in human coronary artery endothelial cells. Arterioscler Thromb Vasc Biol 23:802–808

Miura S, Tanigawa H, Matsuo Y, Fujino M, Kawamura A, Saku K (2003b) Ras/Raf1-dependent signal in sphingosine-1-phosphate-induced tube formation in human coronary artery endothelial cells. Biochem Biophys Res Commun 306:924–929

Nofer JR, Assmann G (2005) Atheroprotective effects of high-density lipoprotein-associated lysosphingolipids. Trends Cardiovasc Med 15:265–271

Nofer JR, Fobker M, Höbbel G, Voss R, Wolinska I, Tepel M, Zidek W, Junker R, Seedorf U, von Eckardstein A, Assmann G, Walter M (2000) Activation of phosphatidylinositol-specific phospholipase C by HDL-associated lysosphingolipid. Involvement in mitogenesis but not in cholesterol efflux. Biochemistry 39:15199–15207

Nofer JR, Levkau B, Wolinska I, Junker R, Fobker M, von Eckardstein A, Seedorf U, Assmann G (2001) Suppression of endothelial cell apoptosis by high density lipoproteins (HDL) and HDL-associated lysosphingolipids. J Biol Chem 276:34480–34485

Nofer JR, Geigenmüller S, Göpfert C, Assmann G, Buddecke E, Schmidt A (2003) High density lipoprotein-associated lysosphingolipids reduce E-selectin expression in human endothelial cells. Biochem Biophys Res Commun 310:98–103

Nofer JR, van der Giet M, Tölle M, Wolinska I, von Wnuck Lipinski K, Baba HA,Tietge UJ, Gödecke A, Ishii I, Kleuser B, Schäfers M, Fobker M, Zidek W, Assmann G, Chun J, Levkau B (2004) HDL induces NO-dependent vasorelaxation via the lysophospholipid receptor S1P3. J Clin Invest 113:569–581

Noor R, Shuaib U, Wang CX, Todd K, Ghani U, Schwindt B, Shuaib A (2007) High-density lipoprotein cholesterol regulates endothelial progenitor cells by increasing eNOS and preventing apoptosis. Atherosclerosis 192:92–99

Norata GD, Catapano AL (2005) Molecular mechanisms responsible for the antiinflammatory and protective effect of HDL on the endothelium. Vasc Health Risk Manag 1:119–129

Norata GD, Pirillo A, Callegari E, Hamsten A, Catapano AL, Eriksson P (2003) Gene expression and intracellular pathways involved in endothelial dysfunction induced by VLDL and oxidised VLDL. Cardiovasc Res 59:169–180

Norata GD, Callegari E, Inoue H, Catapano AL (2004) HDL3 induces cyclooxygenase-2 expression and prostacyclin release in human endothelial cells via a p38 MAPK/CRE-dependent pathway: effects on COX-2/PGI-synthase coupling. Arterioscler Thromb Vasc Biol 24:871–877

Nordestgaard BG, Wootton R, Lewis B (1995) Selective retention of VLDL, IDL, and LDL in the arterial intima of genetically hyperlipidemic rabbits in vivo. Molecular size as a determinant of fractional loss from the intima-inner media. Arterioscler Thromb Vasc Biol 15:534–542

Obunike JC, Lutz EP, Li Z, Paka L, Katopodis T, Strickland DK, Kozarsky KF, Pillarisetti S, Goldberg IJ (2001) Transcytosis of lipoprotein lipase across cultured endothelial cells requires both heparan sulfate proteoglycans and the very low density lipoprotein receptor. J Biol Chem 276:8934–8941

Oravec S, Demuth K, Myara I, Hornych A (1998) The effect of high density lipoprotein subfractions on endothelial eicosanoid secretion. Thromb Res 92:65–71

Pan W, Kastin AJ, Zankel TC, van Kerkhof P, Terasaki T, Bu G (2004) Efficient transfer of receptor-associated protein (RAP) across the blood–brain barrier. J Cell Sci 117: 5071–5078

Pantano C, Reynaert NL, van der Vliet A, Janssen-Heininger YM (2006) Redox-sensitive kinases of the nuclear factor-kappaB signaling pathway. Antioxid Redox Signal 8:1791–1806

Paradis ME, Badellino KO, Rader DJ, et al (2006) Visceral adiposity and endothelial lipase. J Clin Endocrinol Metab 91:3538–3543

Perret B, Mabile L, Martinez L, Tercé F, Barbaras R, Collet X (2002) Hepatic lipase: structure/function relationship, synthesis, and regulation. J Lipid Res 43:1163–1169

Petoumenos V, Nickenig G, Werner N (2008) High density lipoprotein exerts vasculoprotection via endothelial progenitor cells. J Cell Mol Med (in press)

Plotnick GD, Corretti MC, Vogel RA (1997) Effect of antioxidant vitamins on the transient impairment of endothelium-dependent brachial artery vasoactivity following a single high-fat meal. JAMA 278:1682–1686

Pollin TI, Damcott CM, Shen H, Ott SH, Shelton J, Horenstein RB, Post W, McLenithan JC, Bielak LF, Peyser PA, Mitchell BD, Miller M, O'Connell JR, Shuldiner AR (2008) A null mutation in human APOC3 confers a favorable plasma lipid profile and apparent cardioprotection. Science 322:1702–1705

Qiu G, Hill JS (2009) Endothelial lipase promotes apolipoprotein AI-mediated cholesterol efflux in THP-1 macrophages. Arterioscler Thromb Vasc Biol 29:84–91

Ren S, Shen GX (2000) Impact of antioxidants and HDL on glycated LDL-induced generation of fibrinolytic regulators from vascular endothelial cells. Arterioscler Thromb Vasc Biol 20:1688–1693

Rippe B, Rosengren BI, Carlsson O, Venturoli D (2002) Transendothelial transport: the vesicle controversy. J Vasc Res 39:375–390

Robbesyn F, Garcia V, Auge N, Vieira O, Frisach MF, Salvayre R, Negre-Salvayre A (2003) HDL counterbalance the proinflammatory effect of oxidized LDL by inhibiting intracellular reactive oxygen species rise, proteasome activation, and subsequent NF-kappaB activation in smooth muscle cells. FASEB J 17:743–745

Rohrer L, Cavelier C, Fuchs S, Schlüter MA, Völker W, von Eckardstein A (2006) Binding, internalization and transport of apolipoprotein A-I by vascular endothelial cells. Biochim Biophys Acta 1761:186–194

Rohrer L, Ohnsorg PM, Lehner M, Häusler F, Rinninger F and von Eckardstein A (2009) HDL transport through aortic endothelial cells is mediated by Scavenger Receptor BI and ATP Binding Cassette Transporter G1. Circ Res. In press

Rosenfeld SI, Packman CH, Leddy JP (1983) Inhibition of the lytic action of cell-bound terminal complement components by human high density lipoproteins and apoproteins. J Clin Invest 71:795–808

Rosenson RS, Lowe GD (1998) Effects of lipids and lipoproteins on thrombosis and rheology. Atherosclerosis 140:271–280

Rye KA, Barter PJ (2004) Formation and metabolism of prebeta-migrating, lipid-poor apolipoprotein A-I. Arterioscler Thromb Vasc Biol 24:421–428

Schwenke DC, Carew TE (1989) Initiation of atherosclerotic lesions in cholesterol-fed rabbits. II. Selective retention of LDL vs selective increases in LDL permeability in susceptible sites of arteries. Arteriosclerosis 9:908–18

Seetharam D, Mineo C, Gormley AK, Gibson LL, Vongpatanasin W, Chambliss KL, Hahner LD, Cummings ML, Kitchens RL, Marcel YL, Rader DJ, Shaul PW (2006) High-density lipoprotein promotes endothelial cell migration and reendothelialization via scavenger receptor-B type I. Circ Res 98:63–72

Shin HK, Kim YK, Kim KY, Lee JH, Hong KW (2004) Remnant lipoprotein particles induce apoptosis in endothelial cells by NAD(P)H oxidase-mediated production of superoxide and cytokines via lectin-like oxidized low-density lipoprotein receptor-1 activation: prevention by cilostazol. Circulation 109:1022–1028

Simionescu M (2007) Implications of early structural–functional changes in the endothelium for vascular disease. Arterioscler Thromb Vasc Biol 27:266–274

Speidel MT, Booyse FM, Abrams A, Moore MA, Chung BH (1990) Lipolyzed hypertriglyceridemic serum and triglyceride-rich lipoprotein cause lipid accumulation in and are cytotoxic to cultured human endothelial cells High density lipoproteins inhibit this cytotoxicity. Thromb Res 58:251–264

Spencer BJ, Verma IM (2007) Targeted delivery of proteins across the blood–brain barrier. Proc Natl Acad Sci USA 104:7594–7599

Spieker LE, Sudano I, Hürlimann D, Lerch PG, Lang MG, Binggeli C, Corti R, Ruschitzka F, Lüscher TF, Noll G (2002) High-density lipoprotein restores endothelial function in hypercholesterolemic men. Circulation 105:1399–1402

Suc I, Escargueil-Blanc I, Troly M, Salvayre R, Nègre-Salvayre A (1997) HDL and ApoA prevent cell death of endothelial cells induced by oxidized LDL. Arterioscler Thromb Vasc Biol 17:2158–2166

Sugano M, Tsuchida K, Makino N (2000) High-density lipoproteins protect endothelial cells from tumor necrosis factor-alpha-induced apoptosis. Biochem Biophys Res Commun 272:872–876

Sumi M, Sata M, Miura S, Rye KA, Toya N, Kanaoka Y, Yanaga K, Ohki T, Saku K, Nagai R (2007) Reconstituted high-density lipoprotein stimulates differentiation of endothelial progenitor cells and enhances ischemia-induced angiogenesis. Arterioscler Thromb Vasc Biol 27:813–818

Tabas I, Williams KJ, Borén J (2007) Subendothelial lipoprotein retention as the initiating process in atherosclerosis: update and therapeutic implications. Circulation 116:1832–1844

Tamai O, Matsuoka H, Itabe H, Wada Y, Kohno K, Imaizumi T (1997) Single LDL apheresis improves endothelium-dependent vasodilatation in hypercholesterolemic humans. Circulation 95:76–82

Tso C, Martinic G, Fan WH, Rogers C, Rye KA, Barter PJ (2006) High-density lipoproteins enhance progenitor-mediated endothelium repair in mice. Arterioscler Thromb Vasc Biol 26:1144–1149

van den Berg BM, Spaan JA, Vink H (2009). Impaired glycocalyx barrier properties contribute to enhanced intimal low-density lipoprotein accumulation at the carotid artery bifurcation in mice. Pflugers Arch 457:1199–1206

van Gils JM, Zwaginga JJ, Hordijk PL (2009). Molecular and functional interactions among monocytes, platelets, and endothelial cells and their relevance for cardiovascular diseases. J Leukoc Biol 85:195–204

van Oostrom O, Nieuwdorp M, Westerweel PE, Hoefer IE, Basser R, Stroes ES, Verhaar MC (2007) Reconstituted HDL increases circulating endothelial progenitor cells in patients with type 2 diabetes. Arterioscler Thromb Vasc Biol 27:1864–1865

Viswambharan H, Ming XF, Zhu S, Hubsch A, Lerch P, Vergères G, Rusconi S, Yang Z (2004) Reconstituted high-density lipoprotein inhibits thrombin-induced endothelial tissue factor expression through inhibition of RhoA and stimulation of phosphatidylinositol 3-kinase but not Akt/endothelial nitric oxide synthase. Circ Res 94:918–925

von Otte S, Paletta JR, Becker S, König S, Fobker M, Greb RR, Kiesel L, Assmann G, Diedrich K, Nofer JR (2006) Follicular fluid high density lipoprotein-associated sphingosine 1-phosphate is a novel mediator of ovarian angiogenesis. J Biol Chem 281:5398–5405

Weinstein MM, Yin L, Beigneux AP, Davies BS, Gin P, Estrada K, Melford K, Bishop JR, Esko JD, Dallinga-Thie GM, Fong LG, Bensadoun A, Young SG (2008) Abnormal patterns of lipoprotein lipase release into the plasma in GPIHBP1-deficient mice. J Biol Chem 283:34511–34518

Wheeler-Jones CP (2007) Chylomicron remnants: mediators of endothelial dysfunction? Biochem Soc Trans 35:442–445

Wong H, Schotz MC (2002) The lipase gene family. J Lipid Res 43:993–999

Xia P, Vadas MA, Rye KA, Barter PJ, Gamble JR (1999) High density lipoproteins (HDL) interrupt the sphingosine kinase signalling pathway. A possible mechanism for protection against atherosclerosis by HDL. J Biol Chem 274:33143–33147

Yatomi Y (2008) Plasma sphingosine 1-phosphate metabolism and analysis. Biochim Biophys Acta 1780:606–611

Yuhanna IS, Zhu Y, Cox BE, Hahner LD, Osborne-Lawrence S, Lu P, Marcel YL, Anderson RG, Mendelsohn ME, Hobbs HH, Shaul PW (2001) High-density lipoprotein binding to scavenger receptor-BI activates endothelial nitric oxide synthase. Nat Med 7:853–857

Zambon A, Bertocco S, Vitturi N, Polentarutti V, Vianello D, Crepaldi G (2003) Relevance of hepatic lipase to the metabolism of triacylglycerol-rich lipoproteins. Biochem Soc Trans 31:1070–1074

Zampetaki A, Kirton JP, Xu Q (2008) Vascular repair by endothelial progenitor cells. Cardiovasc Res 78:413–421

Zhang Q, Liu L, Zheng XY (2008) Protective roles of HDL, apoA-I and mimetic peptide on endothelial function: through endothelial cells and endothelial progenitor cells. Int J Cardiol (in press)

Zhao R, Ma X, Shen GX (2008) Transcriptional regulation of plasminogen activator inhibitor-1 in vascular endothelial cells induced by oxidized very low density lipoproteins. Mol Cell Biochem 317:197–204

Zheng XY, Liu L (2007) Remnant-like lipoprotein particles impair endothelial function: direct and indirect effects on nitric oxide synthase. J Lipid Res 48:1673–1680

Zhu W, Saddar S, Seetharam D, Chambliss KL, Longoria C, Silver DL, Yuhanna IS, Shaul PW, Mineo C (2008) The scavenger receptor class B type I adaptor protein PDZK1 maintains endothelial monolayer integrity. Circ Res 102:480–487

Chapter 8
Receptor-Mediated Endocytosis and Intracellular Trafficking of Lipoproteins

Joerg Heeren and Ulrike Beisiegel

Abstract Members of the low-density lipoprotein receptor (LDLR) gene family are structurally related receptors involved in receptor-mediated endocytosis and signal transduction that regulate a wide range of physiological processes. Receptor-mediated endocytosis of cholesterol-rich LDL and triglyceride-rich lipoproteins (TRL) into the liver via the LDLR and the LDLR-related protein 1 (LRP1) determine the plasma concentrations of proatherogenic lipoproteins. Recent studies indicate that LDLR-mediated internalisation of LDL and very (V)LDL engages a differentially regulated intracellular sorting machinery, suggesting that the LDLR is more than a simple constitutive endocytotic receptor. The binding and internalisation of TRL via hepatic LRP1 is even more complex. After internalisation, LDLR and LRP1 facilitate a different intracellular fate of their ligands. Whereas LDL follows the classical pathway for degradation, TRL disintegrate in late and peripheral endosomes, allowing a differential sorting of TRL components. This chapter summarises current understanding of the molecular mechanisms which are important for the internalisation and subsequent intracellular transport of LDL and TRL mediated by the LDLR and LRP1.

8.1 Lipoproteins and Their Receptors

Unravelling the mechanisms regulating the clearance of proatherogenic lipoproteins such as triglyceride-rich lipoproteins (TRL) and low-density lipoprotein (LDL) are of great importance, as the understanding of these cellular processes may lead to

J. Heeren
Institute for Biochemistry and Molecular Biology II: Molecular Cell Biology, University Medical Center Hamburg – Eppendorf, Martinistrasse 52, 20246, Hamburg, Germany
e-mail: heeren@uke.uni-hamburg.de

U. Beisiegel
Institute for Biochemistry and Molecular Biology II: Molecular Cell Biology, University Medical Center Hamburg – Eppendorf, Martinistrasse 52, 20246 Hamburg, Germany
e-mail: beisiegel@uke.uni-hamburg.de

C. Ehnholm (ed.), *Cellular Lipid Metabolism*,
DOI 10.1007/978-3-642-00300-4_8, © Springer-Verlag Berlin Heidelberg 2009

the development of new therapies for dyslipidemia. Therefore we first introduce the metabolism of LDL and TRL and the physiological importance of lipoprotein internalisation for extrahepatic organs such as the adrenals and bone.

8.1.1 Metabolism of LDL

LDL is generated from liver-derived very (V)LDL by intravascular lipolytic processing mediated by lipoprotein lipase (LPL), hepatic lipase (HL) and lipid exchange proteins, such as phospholipid transfer protein (PLTP; Huskonen et al. 2004) and cholesteryl ester transfer protein (CETP; de Grooth et al. 2004), which are described in more detail in Chap. 9. Importantly, the conversion of VLDL into intermediate-density lipoproteins (IDL) and further to LDL reduces the binding affinity of apolipoprotein E (apoE) and other apolipoproteins (such as apoC) to the LDL particle. IDL to LDL conversion finally results in LDL particles that contain a single copy of the non-exchangeable apoB100 molecule only, thereby prolonging the short half-life of IDL from a few minutes to hours for LDL. The low-density lipoprotein receptor (LDLR) plays a central role in co-ordinating the internalisation of circulating LDL. Demonstrating the importance of the LDLR pathway, mutations in the LDLR cause severe hypercholesterolemia and premature atherosclerosis (Brown and Goldstein 1986). LDLR-mediated removal of LDL from the circulation occurs mainly in peripheral cells and hepatocytes. The liver removes approximately 50% of the circulating LDL (Vance 2002). Interestingly, the LDLR is not only able to interact with LDL-associated apoB100 at the basolateral surface of parenchymal liver cells. Several studies (Twisk et al. 2000; Gillian-Daniel et al. 2002; Larsson et al. 2004) have demonstrated that the LDLR is also directly involved in the secretory pathway of VLDL, at least in vitro. In general, apoB100 lipidation and VLDL secretion is regulated by the amount of hepatocellular lipids. Under conditions of low lipid availability, substantial amounts of newly synthesised apoB100 are degraded. Thus, in general VLDL secretion is reduced when the amount is of hepatocellular lipids is low. However, the absence of the LDLR results in higher apoB100 secretion even under conditions of low hepatocellular lipids (Twisk et al. 2000; Gillian-Daniel et al. 2002; Larsson et al. 2004). This indicates that binding of the LDLR to immature VLDL can occur intracellularly and thereby influence the secretion of apoB100-containing lipoproteins. In contrast, independent in vivo studies in the mouse system show normal apoB100 production and secretion in the absence of the LDLR after the injection of Triton WR 1339 (Jones et al. 2007). There could be various reasons for the discrepancy of results obtained in vitro and in vivo. In particular the utilisation of Triton derivatives to measure lipoprotein synthesis in vivo might influence not only the interaction of lipoproteins with the cell surface. These detergents also have an impact on the structure and functionality of internal membranes, possibly altering intracellular membrane and ligand-sorting events. Future experiments using novel approaches without the co-injection of detergents will probably help to clarify whether binding of apoB100 to the LDLR in the endoplasmatic reticulum is of physiological importance for the secretion of lipoproteins.

The extrahepatic sites of native LDL uptake in vivo have been characterised by the "trapped ligand" approach, which determines the accumulation of radioactivity after injection of tyramine-cellobiose-labelled LDL (Pittman and Steinberg 1984). Although the liver is the dominant site of LDL uptake, steroid-producing tissues such as the adrenals and the ovaries show a higher specific LDLR activity compared to the liver. Interestingly, LDLR deficiency in mice did not impair corticosterone synthesis in adrenals, indicating that de novo cholesterol production or selective cholesterol uptake from HDL via scavenger receptor class B1 (SR-B1) may be equally important and can compensate for lack of LDLR activity in adrenal steroidogenesis (Kraemer 2007). In the human system it has also been shown that patients with homozygous familial hypercholesterolemia have normal corticosteroid production (Illingworth et al. 1984). However, these studies were carried out in the absence of the LDLR, which might underestimate the importance of LDL uptake into adrenals under physiological conditions. It has been shown that LDL can be a significant source for cholesterol delivery into human adrenals in the presence of an active LDLR (Carr and Simpson 1981; Liu et al. 2000).

8.1.2 Metabolism of Triglyceride-Rich Lipoproteins

Intestinal chylomicrons (CM) and liver-derived VLDL represent the two classes of TRL that are responsible for the transport of lipids to the various body cells. CM mediate the transport of dietary lipids, including lipophilic vitamins, whereas VLDL carry lipids from the liver to peripheral tissues. In humans, TRL can be distinguished by their apoB composition. apoB48, synthesised as a result of apoB mRNA editing (Anant and Davidson 2001), is the major apolipoprotein of intestinal-derived CM, while apoB100 is required for the assembly of VLDL in the liver (Kane 2001). In contrast to humans, mice exhibit hepatic apoB mRNA-editing activity, which results in the formation of apoB100- and apoB48-containing TRL in the liver.

In the bloodstream TRL are first hydrolysed by lipoprotein lipaseLPL and subsequently by HL, leading to the formation of TRL remnants (Mahley and Ji 1999; Merkel et al. 2002). During lipolysis the particles simultaneously become enriched with HDL-derived apoE. After hydrolysis LPL and HL remain associated predominantly with postprandial apoB48 containing CM remnants (CR; Dichek et al. 1999, 2004; Xiang et al. 1999; Heeren et al. 2002). The majority of the resulting CR and VLDL remnants (also called IDL), which are depleted in triglycerides, are then rapidly internalised into the liver through binding of apoE, LPL and HL to the LDLR-related protein 1 (LRP1; Beisiegel et al. 1989; Rohlmamn et al. 1998; Mahley and Ji 1999; Verges et al. 2004) or apoB100 and apoE interacting with the LDLR (Ishibashi at al. 1994; Mahley and Ji 1999). The current knowledge of remnant internalisation and subsequent intracellular processing is presented in Sect. 8.3.

Next to the liver, extrahepatic tissues have the capacity for uptake of TRL (Hussain et al. 1989). To permit such an uptake into bone cells, the endothelium in the bone is fenestrated and thereby allows direct contact with the TRL particles and the cells. Recently it could be demonstrated that osteoblasts participate in CR metabolism and that CR components are relevant for osteoblast function (Niemeier et al. 2008). ApoE and LRP1 are expressed by osteoblasts (Niemeier et al. 2005; Schilling et al. 2005); and interestingly apoE deficient mice display a high bone formation rate (Schilling et al. 2005). LRP1 mediates the uptake of CR-associated vitamin K1 into human osteoblasts (Niemeier et al. 2005) and thereby this process has a direct impact on the degree of osteocalcin carboxylation and secretion by osteoblasts in mice (Niemeier et al. 2008). The adipose tissue is also involved in lipoprotein receptor-mediated uptake of TRL particles. In rodent adipocytes, LRP1 activation mediated by insulin is associated with an increased uptake of TRL remnants in vitro (Corvera et al. 1983; Descamps et al. 1993). In adipose-specific LRP1-deficient mice, adipose tissue mass was substantially reduced and TRL internalisation was abolished in cultivated LRP1-deficient adipocytes. It was suggested that reduced adipose tissue mass can be explained by reduced particle uptake in vivo (Hofmann et al. 2007). Since most functional studies on adipocyte function have been performed in the mouse system, there is a strong demand to study lipoprotein metabolism in human adipocytes. In a recent report we described that immortalised mesenchymal stem cells can be stably differentiated into human adipocytes (Prawitt et al. 2008). After differentiation, these cells maintained numerous morphologic and functional features of mature adipocytes. Interestingly, adipocyte differentiation is associated with a concomitant up-regulation of LRP1 and apoE indicating that these proteins are also important for TRL particle uptake into human adipose tissue. However, at least in this human adipocyte cell model an inverse regulation of receptor expression and remnant uptake was observed. Therefore, LRP1 and apoE might have an additional role next to TRL uptake in human adipose tissue that is independent of lipoprotein internalisation. Such a putative alternative function could include the regulation of signal transduction cascades involved in adipocyte differentiation and proliferation.

The VLDL receptor (VLDLR), expressed in adipose tissue, heart and muscle, also functions as a receptor mediating CR internalisation into adipocytes in vitro (Niemeier et al. 1996; Takahashi et al. 2004). However, VLDLR is expressed on endothelial cells rather than in the underlying parenchymal cells (Wyne et al. 1996). This observation probably indicates that the VLDLR primarily supports LPL-mediated lipolysis, as suggested by Goudriaan et al. (2004). Furthermore, in contrast to the fenestrated endothelium of liver and bone, the capillary endothelium of adipose tissue is continuous and therefore impermeable for lipoprotein particles. Thus, the role of LRP1 and VLDLR for the internalisation of lipoprotein particles in adipose tissue in vivo is unclear and need to be addressed in future work.

8.2 Receptor-Mediated Endocytosis of LDL

The pathway of receptor-mediated LDL uptake and the negative feed-back regulation of LDLR gene expression mediated by cellular cholesterol levels is one of the best studied cellular processes and was rewarded by the Nobel Prize in 1985 (Goldstein and Brown 1986). However, recent studies have revealed novel insights in the cell physiology of the LDLR which have great impact on the intracellular targeting of internalised LDL. The next section gives an overview of the **classical pathway** and presents new findings in the LDLR pathway.

8.2.1 Structure and Function of the LDL Receptor

The mature LDLR is a type-I integral transmembrane glycoprotein (Fig. 8.1) containing five different domains: (1) the ligand-binding domain, (2) the EGF-precursor homology domain, (3) the O-linked sugar domain, (4) the transmembrane domain and (5) the cytosolic domain. The ligand-binding domain is localised at the N-terminal region of the LDLR and consists of seven "LDLR class A repeats" (Sudhof et al. 1986). Each repeat is encoded by a single exon and contains approximately 40 amino acids. Six cysteine residues within the LDLR class A repeats form characteristic disulfide bonds, which facilitate the presentation of a characteristic tripeptide acidic signature at the surface of the domain, thought to mediate the binding of positively charged residues to apoB100 and apoE.

Three epidermal growth factor (EGF) repeats, together with a so-called β-propeller domain, form the EGF-precursor homology region. The EGF domains contain characteristic disulfide bonds similar to the ligand-binding repeats. Two of these repeats are located next to the ligand-binding repeats whereas one repeat lays C-terminal to the β-propeller. Thus, the EGF repeats are adjacent to an approximately 260 amino acid domain containing six YWTD motifs forming the β-propeller domain (Jeon et al. 2001). This structure is responsible for the pH-dependent release of LDL from the LDLR in endosomal compartments. The elegant model illustrated by Jeon and Blacklow (2005) clarified that, at neutral pH, LDL-derived apoB100 bind to LDLR class A repeats 3 and 7 of the ligand-binding domain. At the acidic pH found in endosomes, LDLR class A repeats 4 and 5 interact with the β-propeller domain, which results in the release of LDL from its receptor.

The O-linked sugar domain is highly enriched in serine and threonine residues that serve as anchor for O-linked sugars. Although detailed structural knowledge is available, the precise function of this region is not known. The deletion of this region does not impair the function of LDLR with respect to ligand binding, internalisation and degradation. It has been postulated (Gent and Braakman 2004) that this domain might: (1) modulate the presentation of the ectodomain, (2) stabilise the LDLR at endosomal pH and/or (3) prevent proteolytic cleavage of the ectodomain by metalloproteases at the cell surface. This so-called "shedding" of the ectodomain has been described for other members of the LDLR family which lack an O-linked sugar domain such as LRP1 and SorLA. However, the physiological function of this domain remains unclear. Next to the O-linked sugar domain, a hydrophobic transmembrane domain (22–24 amino acids in length) ensures the LDLR is anchored in the plasma membrane.

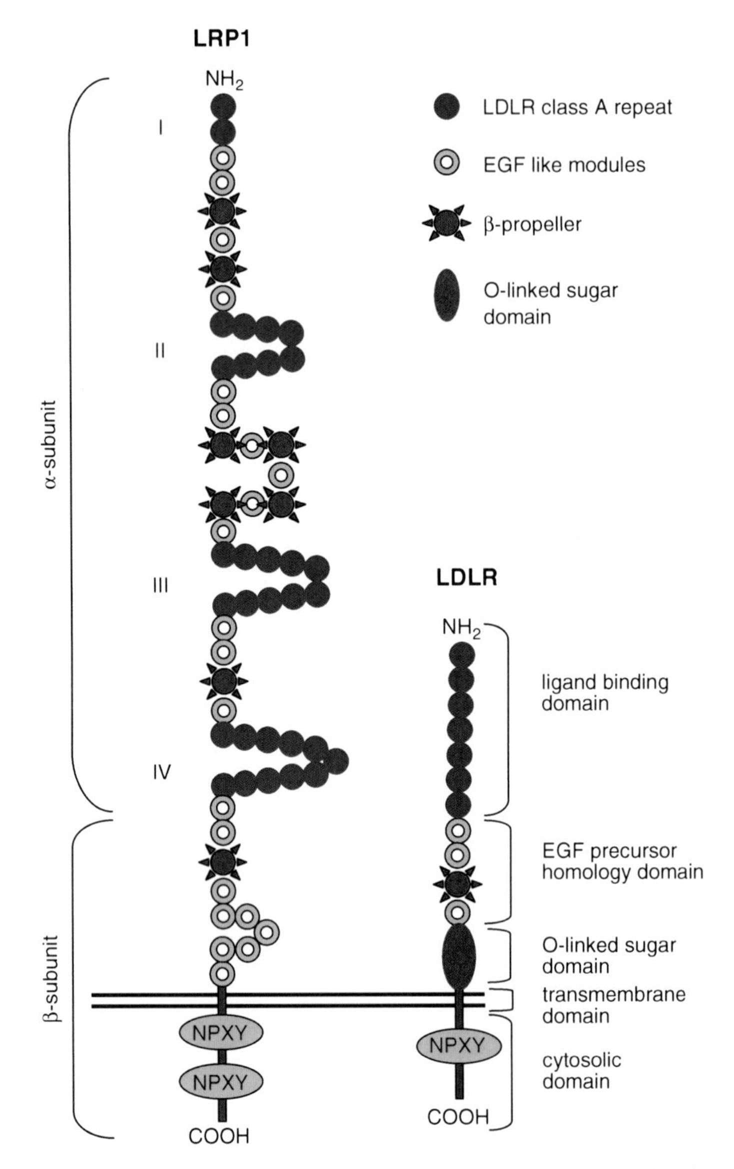

Fig. 8.1 Schematic domain organisation of LRP1 and LDLR. The *LDLR* contains five different domains: (1) the ligand-binding domain, (2) the EGF-precursor homology domain, (3) the O-linked sugar domain, (4) the transmembrane domain and (5) the cytosolic domain. The ligand-binding domain consists of seven "LDLR class A repeats" and mediate the binding to apoB100 and apoE. Three epidermal growth factor (EGF) repeats together with a so-called β-propeller domain form the EGF-precursor homology region. The O-linked sugar domain is highly enriched in serine and threonine residues that serve as anchor for O-linked sugars. The hydrophobic transmembrane domain ensures the anchoring the LDLR in the plasma membrane. The cytoplasmic tail of the LDLR containing the characteristic NPXY sequence interact with phospho-tyrosine binding domains of cellular adaptors proteins which are important for LDL endocytosis and subsequent intracellular transport. The *LRP1* is proteolytically cleaved within the Golgi complex togenerate two subunits: (1) the N-terminal 515-kDa α-subunit containing the ligand-binding domains and (2) the C-terminal 85-kDa β-subunit containing an extracellular part, the transmembrane spanning domain and the cytoplasmic intracellular domain. The N-terminal domain lacks a membrane-spanning region but

The cytoplasmic tail of the LDLR consists of approximately 50 amino acids and is responsible for endocytosis and intracellular transport. Site-directed mutagenesis revealed several amino acids which are important for correct LDLR function. The most prominent motif within the LDLR tail is the characteristic NPXY sequence (NPVY in the case of the LDLR), which is known to bind phospho-tyrosine-binding (PTB) domains of adaptors proteins. The NPVY sequence is, at least in hepatocytes, important for the clustering of LDLR within coated pits (Keyel et al. 2006). A more detailed view of the importance of this sequence is provided in Sect. 8.2.3. A comparison of the domain organisation between LDLR and LRP1 is illustrated in Fig. 8.1.

8.2.2 *Ligands of the LDL Receptor*

The LDLR binds and mediates the internalisation of lipoproteins that contain apoB100 and/or apoE. The most important lipoproteins that are internalised by the LDLR are LDL and their respective precursors VLDL and IDL (Goldstein et al. 1985). Liver-derived VLDL contain a single copy of apoB100 and multiple copies of apoE (Shelness and Sellers 2001), which together facilitate high-affinity binding to the LDLR. The interaction of apoE to the LDLR is highly dependent on the association of apoE with lipids. As determined by different spectroscopic techniques, lipidated apoE has an approximately 500-fold higher affinity for LDLR compared to the delipidated form (Wilson et al. 1991). The lipidation of apoE probably causes a conformational rearrangement and increased exposure of basic residues at the surface of the apoE molecule (Sehayek et al. 1991). Furthermore, LPL-mediated conversion of VLDL into IDL leads to a higher ratio of apoE to total particle protein (Goldberg 1996). Both effects explain the higher binding affinity of IDL to the LDLR compared to VLDL. The basic residues arginine and lysine of apoE located between amino acids 120–160 are responsible for the binding to the LDLR (Saito et al. 2004). The significance of these basic residues for LDLR binding is supported by the differential internalisation rate of apoE isoforms. ApoE exists in three isoforms which affect their binding properties to the LDLR. ApoE3 and E4, but not apoE2, exhibit a strong binding affinity to the LDLR (Schneider et al. 1981). In contrast to apoE3 and apoE4, the arginine residue at position 158 is replaced by a cysteine residue in the isoform apoE2. This amino acid change modulates the receptor binding domain and inhibits the LDLR-mediated internalisation of apoE2-TRL in the liver.

Fig. 8.1 (continued) remain non-covalently associated with the smaller C-terminal β-subunit. The α-subunit contains four clusters of ligand-binding repeats which are encircled by eight β-propeller domains, each adjacent to EGF-like repeats. Both LRP1 subunits lack the O-linked sugar domain. The cytoplasmatic LRP1 α-subunit contains diverse potential endocytosis and signaling motifs: two NPXY motifs whereas the distal NPXY sequence overlaps with the endocytosis signal XYYL and two dileucine motifs. The structures are illustrated according to Herz and Strickland (2001), Jeon and Blacklow (2005) and Gent and Braakman (2004)

This leads to the accumulation of remnant particles in plasma, thereby contributing to dyslipidemia in homozygous apoE2 carriers, and is an essential factor for the development of type III hyperlipoproteinemia (Mahley et al. 1999).

Enzymatic conversion of IDL to LDL is associated with the loss of all exchangeable apolipoproteins including apoE. This results in an LDL particle with one single molecule of apoB100, which compared to apoE-containing IDL has a lower binding affinity for the LDLR. The binding domain of apoB100 is located at the C-terminal end explaining why TRL-derived apoB48 cannot bind to the LDLR. Boren et al. (1998) identified residues 3359–3367 in apoB100 as those interacting with the LDLR. However, since the mutation R3500Q immediately upstream of the LDLR binding region in apoB100 disrupts binding to the LDL receptor and causes hypercholesterolemia, additional sequences must contribute to apoB100 binding capability. To investigate the entire sequences required for effective binding of apoB100 to the LDLR, the same group reconstituted human wild-type and mutated apoB100 into LDL by the use of several transgenic mouse lines (Boren et al. 2000). They concluded that normal receptor binding in apoB100 depends on an interaction between arginine 3500 and tryptophan 4369, because both mutations disrupt the binding of LDL to the LDLR. Thus, the ability of apoB100 to interact with the LDLR depends on the conformation in the C-terminal domain, which is probably essential for the correct exposure of the arginine- and lysine-rich binding motif between residues 3359 and 3367 on the surface of LDL particles.

8.2.3 *Intracellular Processing of LDL*

After endocytosis within clathrin-coated vesicles, LDL enter the lysosome where the cholesteryl esters are hydrolysed to cholesterol and free fatty acids and the apoB100 moiety is degraded to free amino acids (see Fig. 8.2; Goldstein et al. 1985). New aspects in the early internalisation events which influence sorting and targeting of receptors and their respective ligands along the lysosomal pathway have recently been described. Initially, sorting endosomes within the early endosomal compartment were considered as the primary location that disintegrates the receptor/ligand complex to enable receptor recycling back to the cell surface (LDLR) and lysosomal targeting of the ligand (LDL). This model has now been challenged, as ligands for clathrin-mediated endocytosis are differentially sorted into distinct populations of early endosomes (Lakadamyali et al. 2006). Live cell microscopy to follow individual endosomes and ligand particles in real time showed that early endosomes comprise two distinct populations: (1) a dynamic population that matures rapidly toward late endosomes and (2) a static population that matures much more slowly. LDL was targeted preferentially into the dynamic pool of early endosomes, a process associated with the recruitment of specific adaptor proteins. The importance of the adaptor protein ARH1 for uptake and subsequent intracellular sorting LDL is described in more detail in Sect. 8.2.4. Interestingly, a fraction of endocytosed LDL escapes lysosomal degradation and is secreted in an unaltered

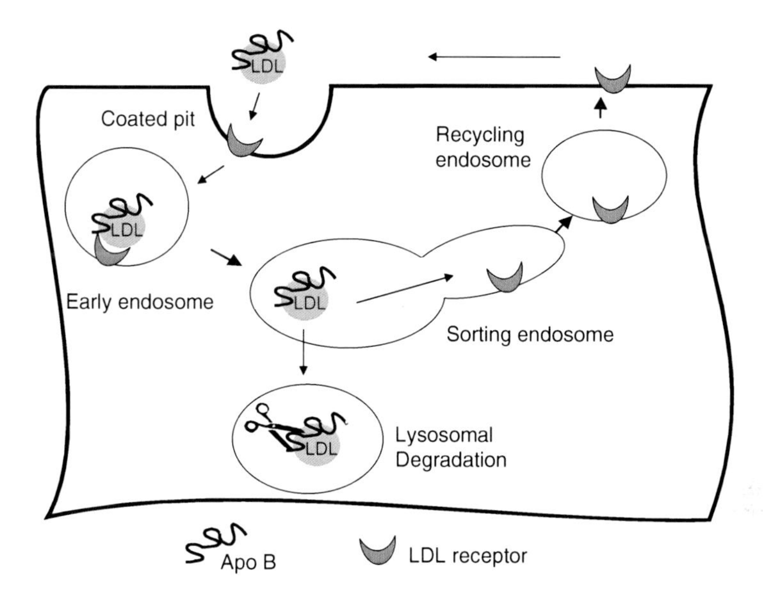

Fig. 8.2 Uptake and intracellular processing of LDL. After LDL receptor-mediated internalisation of LDL, the change in pH within the lumen of early endosomes causes a conformational change in the receptor. This results in the release of bound LDL in early endosomes. Whereas LDL receptors are recycled back to the plasma membrane via tubular extensions of the sorting endosomes, the content of the endosomal lumen including LDL follows the lysosomal pathway (adapted from Goldstein and Brown 1985)

or only slightly altered form by a process termed retroendocytosis (Greenspan and St Clair 1984; Aulinskas et al. 1985). Edge et al. (1986) observed that, in hepatocytes, much less endocytosed apoB100 is degraded compared to skin fibroblasts. Although these observations indicate that the degradation of apoB100 may not be the sole metabolic fate of the protein moiety of endocytosed LDL in the liver, its potential implications for the homeostasis of the apoB containing lipoproteins in the circulation have not yet been addressed in further detail.

8.2.4 Regulation of LDL Receptor Function

LDLR gene expression is transcriptionally regulated via sterol regulatory element-binding proteins (SREBP) in dependence of cellular cholesterol levels. These ER-derived transcription factors are the major regulatory element that determines LDLR activity and are the subject of excellent reviews (Brown et al. 2000; Rawson 2003). Since defects in the LDLR gene were not detected in all patients with familial hypercholesterolemia, great efforts were made to identify further candidate genes that can cause inherited forms of severe hypercholesterolemia (Rader et al. 2003). Indeed, gene defects in proprotein convertase subtilisin-like kexin type 9 (PCSK9) and the LDLR adaptor protein ARH1 correlate with elevated LDL plasma levels (Cohen et al. 2003; Horton et al. 2007). LDL-elevating PCSK9 mutations cause a autosomal dominant hypercholesterolemia (ADH), leading to reduced levels of LDLR protein and consequently to high plasma levels of LDL cholesterol. Since overexpression of wild-type PCSK9 specifically accelerates the degradation of LDLR protein, these findings are indicative of gain-of-function mutations within the PCSK9 gene promoting lysosomal targeting of LDLR, thereby reducing the amounts of LDLR at the cell surface being available for LDL internalisation and clearance, ultimately causing ADH. The posttranslational regulation of LDLR protein levels by PCSK9 as well as its clinical importance has been summarised in detail elsewhere (Attie and Seidah 2005; Horton et al. 2007).

Autosomal recessive hypercholesterolemia (ARH) results from a mutated gene located on the short arm of chromosome 1. The respective gene product, a protein with a phospho-tyrosine binding (PTB) domain, is called ARH1. This adaptor protein binds to the internalisation sequence NPVY within the cytoplasmatic tail of the LDLR and connects the LDLR to the endocytotic machinery of the cell. Retroviral expression of normal ARH1 in transformed lymphocytes from an affected individual with ARH restored LDLR internalisation, confirming that ARH is caused by defective adaptor protein function (Eden et al. 2002). The C-terminal domain of ARH1 directly interacts with clathrin and adaptor protein 2 (AP2), which are both essential components for receptor-mediated endocytosis via clathrin coated pits (He et al. 2002; Mishra et al. 2002). Interestingly, ARH1 is only required for hepatic LDLR-dependent LDL uptake (Keyel et al. 2006; Jones et al. 2007). In fibroblasts and probably also in other tissues, the adaptor protein disabled 2 (dab2) can also mediate LDLR-dependent endocytosis. Since hepatic dab2 levels are very low, defective ARH1 protein resulted in a reduced LDL clearance only into the liver. Somewhat unexpectedly, VLDL clearance via the LDLR was not affected in ARH1-deficient mice (Jones et al. 2007). The authors concluded that the clearance of VLDL and IDL mediated by the LDLR is independent of ARH1, which could explain the lower LDL-cholesterol levels in patients with ARH1 deficiency compared to patients lacking the LDLR. These findings also indicate that the NPVY internalisation sequence within the cytoplasmatic tail of the LDLR interacting with ARH1 is differentially involved in VLDL and LDL internalisation; and a report by the same investigators support this assumption (Michaely et al. 2007). Although the exact determinants for lipoprotein internalisation have yet to be identified, the presence of specific uptake mechanisms for different lipoproteins (LDL, VLDL) via the same receptor (LDLR) suggest that: (1) the differential expression and activity of adaptor proteins in the various tissues can modulate LDLR function and (2) the ligands themselves determine the internalisation process mediated by the LDLR. The relevance of these observations in regard to intravascular VLDL and LDL metabolism and intracellular processing of VLDL/LDL constituents needs to be addressed in the future and will probably provide new therapeutic options to lower the plasma levels of these proatherogenic lipoproteins.

8.3 Receptor-Mediated Endocytosis of Chylomicron Remnants

Hepatic lipoprotein uptake is a complex but well regulated process involving multiple different players. Although not all steps are completely understood, recent advances from in vitro studies and mouse models have allowed the development of a general concept for the internalisation and processing of postprandial lipoproteins in the liver. From the liver sinusoids, CR enter the space of Disse through the fenestrated endothelium. The vast majority of these particles first attaches to cell surface structures such as heparin sulfate proteoglycans (HSPG; for reviews, see Havel and Hamilton 2004; Mahley and Huang 2007; Bishop et al. 2008). Recent studies in genetically engineered mice showed that a liver-specific knock-out of N-deacetylase/N-sulfotransferase (resulting in significant changes in liver heparin sulfate composition) led to an increase in plasma triglyceride levels under both fasted and fed conditions. This study proves that HSPG have a prominent role for the binding of TRL to the surface of parenchymal liver cells (McArthur 2007). The authors suggested that HSPG can also be considered as an endocytotic receptor which is directly involved in the internalisation process. However, it has been shown that postprandial apoB_{48} containing CR are internalised into the liver via binding of apoE and LPL to LRP1 (Beisiegel et al. 1989; Rohlmann et al. 1998; Heeren et al. 2002; Verges et al. 2004) and via apoE interacting with the LDLR (Mahley and Ji 1999; Ishibashi et al. 1994). LDLR knock-out mice accumulated LDL-sized particles (Ishibashi et al. 1993), whereas in fasted animals the liver-specific loss of LRP1 expression did not influence lipoprotein profiles (Rohlmann et al. 1998). This is due to the up-regulation of hepatic LDLR which compensates for the loss of LRP1 to ensure clearance of fasting TRL. Only double-knockout mice, deficient for LDLR and hepatic LRP1, showed an accumulation of remnant lipoproteins. Thus, both receptors are essential for the clearance of TRL in vivo (Rohlmann et al. 1998).

8.3.1 Structure and Function of LRP1

LRP1 is one of the largest members of the LDLR family and is a multifunctional type I cell surface glycoprotein that participates in endocytosis, phagocytosis and signalling pathways (Herz and Strickland 2001; Lillis et al. 2005). It is synthesised in the ER as a 600-kDa precursor protein which is proteolytically cleaved within the Golgi complex to generate two subunits: (1) the N-terminal 515-kDa α-subunit containing the ligand-binding domains and (2) the C-terminal 85-kDa β-subunit containing an extracellular part, the transmembrane-spanning domain and the cytoplasmic intracellular domain (Herz et al. 1990). The N-terminal domain lacks a membrane-spanning region but remain non-covalently associated with the smaller C-terminal β-subunit. Compared to the extracellular part of the LDLR, the extracellular domain organisation of LRP1 is much more complex. The α-subunit contains four clusters of ligand-binding repeats, the sizes of which differ due to variations in the number of LDL class A repeats (cluster 1 has two, cluster 2 has eight, cluster 3 has ten, cluster 4 has 11). Most of the ligands interact only with ligand-binding clusters 2 and 4. Interestingly, only apoE and the chaperone RAP bind to cluster 3 (Herz and Strickland 2001). The ligand-binding domains are encircled by eight β-propeller domains, each adjacent to EGF-like repeats, suggesting that pH-dependent release might be regulated in dependence of the internalised ligands. In contrast to the LDLR, the extracellular parts of both LRP1 subunits lack the O-linked sugar domain. The extracellular element of the α-subunit contains only EGF-like repeats which are located next to transmembrane domain. The cytoplasmatic LRP1 α-subunit contains diverse potential endocytosis and signalling motifs: two NPXY motifs whereas the distal NP**XY** sequence overlaps with the endocytosis signal **XY**YL and two dileucine motifs (Li at al. 2000). The tyrosine residue within the distal NPXY motif is phosphorylated upon treatment with platelet-derived growth factor (PDGF), influencing the binding of adaptor proteins as described below. In addition, serine and threonine phosphorylation, mediated by protein kinase A and C, generates a cytoplasmatic receptor tail that has increased affinity for adaptor proteins such as **d**is**ab**led **1** (Dab1) and **ce**ll **d**eath abnormal/en**gul**fment ada**p**ter protein (CED-6/GULP) which is associated with a reduced internalisation rate (Li et al. 2001; Ranganathan et al. 2004). Thus, phosphorylation of the α-subunit represents a molecular mechanism to modulate the different functions mediated by LRP1 (see Sect. 8.3.4).

8.3.2 Ligands of LRP1

LRP1 is widely expressed in several tissues and interacts with more than 30 unrelated ligands (Lillis et al. 2005) including apoE-containing lipoproteins, proteinase/proteinase inhibitor complexes, bacterial toxins, viruses, growth factors, matrix proteins and intracellular proteins such as the ER-resident chaperon receptor-associated protein (RAP). Efficient binding to LRP1 makes RAP a suitable tool for investigating LRP1 function. Studies with "apoB48-only" and "apoB100-only" mice on a LDLR-deficient background identified that RAP-mediated inhibition of LRP1 resulted in the accumulation of apoB48 containing remnants in the plasma (Veniant et al. 1998). These findings imply that LRP1 facilitates the clearance of postprandial CR but plays no significant role for the catabolism of apoB100-containing lipoproteins. As mentioned above, not only CR-derived apoE, but also LPL binding to LRP1 efficiently initiates the internalisation of CR (Beisiegel et al. 1991; Heeren et al. 2002). Furthermore, LRP1 mediates the uptake of LPL into the liver (Verges et al. 2004) and hepatic LRP1 deficiency leads to the accumulation of LPL in plasma (Espirito Santo et al. 2004). This non-enzymatic function of LPL as a LRP1 ligand probably compensates for other defects that impair TRL clearance, such as apoE variants (Mann et al. 1999), but may also contribute to the delayed clearance of TRL in patients with dyslipidemias caused by LPL mutations (Merkel et al. 2002). Some evidence also points to a potential role of hepatic lipase in LRP1-mediated CR internalisation (Krapp et al. 1999; Gonzalez-Navarro et al. 2004;

Verges et al. 2004). Thus, both LRP1 and LDLR are involved in the endocytosis of postprandial lipoproteins, suggesting that HSPG are predominantly involved in tethering lipoproteins at the cell surface. Even though LRP1 has been shown to be responsible for the hepatic clearance of CR, only a very small proportion of the endocytotic receptor is present at the surface of liver cells under normal conditions in vitro and in vivo (Harasaki et al. 2005; Tamaki et al. 2007). However, the postprandial hormone is known to stimulate a rapid LRP1 translocation to the plasma membrane of rodent adipocytes (Corvera et al. 1989; Descamps et al. 1993) and in hepatocytes (Tamaki et al. 2007). Recent data indicate that insulin-stimulated translocation of LRP1 correlates with an increased LRP1-mediated uptake (Laatsch et al. 2008). In wild-type mice, insulin triggers the hepatic uptake of LRP1 ligands, whereas in the absence of hepatic LRP1 expression or in mice with impaired hepatic insulin-signalling, insulin-triggered uptake of LRP1-specific ligands was abolished (Laatsch et al. 2008). Thus, this suggests that impaired hepatic LRP1 translocation can contribute to the postprandial hyperlipidemia observed in subjects with insulin resistance.

8.3.3 *Intracellular Processing of Chylomicron Remnants*

After internalisation of CR, lipoproteins are immediately disintegrated in late and peripheral endosomes, which allows a differential sorting of CR components. While core lipids and apolipoprotein B (apoB48) are targeted via late endosomes to lysosomes, the majority of CR-derived apoE together with surface lipids remain in peripheral recycling endosomes. The current model proposes (see Fig. 8.3) that the pool of CR-derived apoE is then mobilised by high-density lipoproteins (HDL) or HDL-derived apoA-I to be recycled back to the plasma membrane, followed by apoE re-secretion and the concomitant formation of apoE containing HDL. The HDL-induced recycling of apoE is accompanied by cholesterol efflux and involves the internalisation and targeting of HDL-derived apoA-I to endosomes containing both apoE and cholesterol (Heeren et al. 2006). The cellular mechanisms of apoE recycling and its connection to cholesterol efflux is described in more detail in the following sections.

8.3.3.1 Recycling of Apolipoprotein E

In initial studies it was observed that internalised VLDL particles, but not different from LDL, were poorly degraded in HepG2 hepatoma cells (Lombardi et al. 1993, Rensen et al. 2000). When analysing the internalisation of rabbit β-VLDL in mouse macrophages, apoE was resistant to lysosomal degradation and accumulated in widely distributed vesicles (Tabas et al. 1990). Similarly, VLDL-sized apoE containing triglyceride-rich emulsion particles were more resistant to degradation in HepG2 than LDL-derived apoB (Rensen et al. 2000) suggesting that, following internalisation by liver cells, apoE can escape degradation and can be re-secreted.

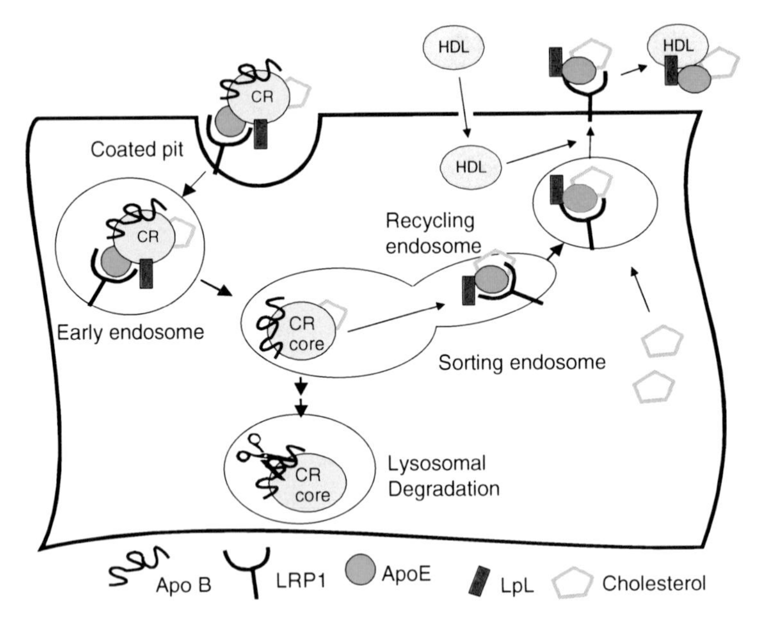

Fig. 8.3 ApoE recycling initiates HDL particle formation. After LRP1-mediated internalisation the disintegration of CR within sorting endosomes resulted in two different intracellular pathways. The lipid core is directed to lysosomal compartments where it is hydrolysed and its constituents can be used for metabolic requirements. The CR-derived "surface remnants" enriched in apoE and LPL and cellular cholesterol are transferred back to the plasma membrane via a peripheral recycling compartment. The intracellular association of apoE with HDL generates HDL_E, which are re-secreted and can probably serve as apoE supply for remnant enrichment with apoE in the postprandial period

These findings indicated that TRL are disintegrated after internalisation and that some TRL constituents are targeted along the degradative pathway, whereas others, such as apoE, are re-secreted. Similar to these TRL model particles, CR components were disintegrated and differentially sorted in a peripheral cellular compartment after internalisation into human hepatoma cells (Heeren et al. 1999). Whereas CR-derived lipids and apoB were directed to lysosomes, CR-derived apoE, apoC and LPL were recycled back to the cell surface, where re-secretion could occur (Heeren et al. 1999, 2001). Since evidence has accumulated that the complex processing of internalised TRL also exists in vivo, the intracellular stability of apoE after hepatic uptake was studied in C57BL/6 mice (Rensen et al. 2000). Only 15–20% of apoE was degraded in hepatocytes, whereas 75% of the cholesteryl oleate was hydrolysed. Fazio and co-workers injected radiolabelled mouse VLDL particles into C57BL/6 mice and consistently identified iodinated apoE in Golgi enriched fractions (Fazio et al. 1999). Similar results were obtained in LDLR-deficient mice, indicating that apoE recycling involves LRP1-mediated TRL internalisation. Up to 60% of internalised apoE appeared to be re-utilised in the plasma, providing further evidence of a unique intracellular pathway to process apoE-containing remnant lipoproteins in vivo (Heeren et al. 2001; Swift et al. 2001). Subsequent analysis of re-secreted proteins revealed that recycled intact apoE was found in plasma associated with HDL (Heeren et al. 2001). These findings provided the first evidence that recycling of TRL-derived apoE plays an important role in the modulation of HDL particles in vivo. Fazio and co-workers identified internalised VLDL-derived apoE to be associated with nascent lipoprotein particles in liver Golgi fractions (Fazio et al. 1999) and they concluded that recycling of apoE could be mediated in part by Golgi-derived secretory vesicles. However, the majority of TRL-derived apoE is probably re-secreted from peripheral endosomal compartments, where apoE recycling is connected with HDL metabolism. The intracellular link between TRL-derived apoE and HDL metabolism in early endosomes was described by several groups. In these studies, HDL or purified apoA-I stimulated the release of internalised apoE in hepatocytes and fibroblasts (Hasty et al. 2003; Heeren et al. 2003). Similarly, HDL or apoA-I-induced apoE recycling also occurs in mouse macrophages and human neuronal cells (Farkas et al. 2005; Rellin et al. 2008) and therefore it has been investigated whether apoE recycling could also be linked to intracellular cholesterol transport in other cells (see Sect. 8.3.2.2).

Since epidemiological data show that apoE3 and apoE4 have differential effects on HDL metabolism, we investigated whether the intracellular processing of TRL-derived apoE4 differs from apoE3-TRL (Heeren et al. 2004). ApoE4 correlates with high LDL-cholesterol, elevated triglyceride and low HDL levels (Davignon et al. 1988; Dallongevielle et al. 1992) and is associated with atherosclerosis and Alzheimer's disease (Dallongeville et al. 1992; Roses 1996). But despite the importance of apoE4 as a risk factor, the cellular mechanisms which are responsible for the differences between apoE3 and apoE4 are not well understood. Cell surface binding, internalisation, endosomal transport and disintegration of apoE3-TRL and apoE4-TRL in early endosomes have been shown to be identically comparable in hepatoma cells (Heeren et al. 2004). Thus, a dissimilar recycling of apoE isoforms could contribute to the development of apoE4-associated diseases. Along these lines, we could demonstrate that HDL-induced recycling of TRL-derived apoE4 is impaired and associated with a decreased cholesterol efflux (Heeren et al. 2004). Although the molecular mechanisms and the potential involvement of members of the LDLR family are yet unclear, the biophysical characteristics of apoE3 and apoE4 might provide an explanation for the reduced efficiency of intracellular apoE4 processing. ApoE3 binds preferentially to HDL, while apoE4 has a higher affinity to VLDL (Dong et al. 1994), suggesting that apoE isoforms interact differently with lipids. TRL-derived apoE3 might therefore associate more rapidly with internalised apoA-I and cholesterol to form HDL-containing apoE (HDL_E) particles.

8.3.3.2 Apolipoprotein E Recycling and Cholesterol Efflux

In human hepatoma cells and fibroblasts, HDL-induced apoE recycling is accompanied by cholesterol efflux (Heeren et al. 2003). Large amounts of internalised TRL-derived apoE co-localises with early endosome antigen 1 (EEA1), a marker for early endosomes. ApoE and EEA1-positive endosomes also contain cholesterol and, upon HDL incubation, simultaneous apoE recycling and cholesterol efflux occurs simultaneously from peripheral early endosomes. These results point to a direct link between HDL-induced apoE recycling and cholesterol efflux from early endosomal compartments. Although internalised TRL-derived apoE was not detected in AP1-positive Golgi vesicles (Heeren et al. 2001) and apoE recycling was unaffected in the absence of an intact Golgi apparatus (Farkas et al. 2004), current data does not completely rule out that apoE recycling concomitantly affects cholesterol efflux via the Golgi network where cholesterol-rich membranes are assembled and moved forward to the plasma membrane. At first, we and others favoured HDL to act at the plasma membrane to promote apoE recycling and cholesterol efflux. New findings addressing this question clearly challenge this model (Heeren et al. 2003). In hepatoma cells, HDL-derived apoA-I was shown to co-localise with apoE and cholesterol in EEA1-positive endosomes, strongly suggesting that HDL-derived apoA-I is internalised and targeted to pre-existing apoE/cholesterol-containing endosomes to promote apoE recycling and cholesterol efflux. Time-lapse confocal microscopy confirmed that HDL-derived apoA-I is targeted to TRL-derived apoE/cholesterol-containing endosomes in the periphery (Heeren et al. 2003). Similarly, FPLC analysis of secreted proteins from macrophages identified recycled apoE in fractions that contained small HDL, indicating that TRL-derived apoE exits cells in a remodelled, moderately lipidated form (Hasty et al. 2005).

ApoE/cholesterol-containing endosomes are positive for EEA1, a marker for early endosomes. EEA1 is absent from the plasma membrane and restricted to a subcompartment of early endosomes that contain Rab5 and Rab5 effectors (Rubino et al. 2000). These Rab5/EEA1-positive endosomes are connected to the recycling compartment by tubular structures but clearly distinct from recycling endosomes that are targeted to the cell surface. Rather other Rab proteins, such as Rab4 and Rab11, are characteristic for recycling endosomes in the cellular periphery. This suggests that, after TRL disassembly in early sorting endosomes, apoE/cholesterol complexes are retained to allow complex formation with internalised HDL/apoA-I before entering the recycling compartment. HDL-induced apoE3 recycling also occurs in neuronal cells, but here apoE recycling is not accompanied by enhanced cholesterol efflux (Rellin et al. 2008). In summary, the association of internalised apoA-I with intracellular apoE/cholesterol complexes leads to the formation of apoE-enriched HDL particles. In macrophages, HDL-induced apoE recycling may prevent foam cell formation by enhancing the removal of cholesterol from the periphery. In liver, this mechanism secures the maintenance of high levels of HDL_E in plasma in the postprandial phase and accelerates the enrichment of CR with apoE. After a high fat meal, the re-utilisation of internalised apoE would overcome any shortage of HDL_E due to insufficient de novo synthesis of apoE and thereby assure the efficient hepatic clearance of CR.

8.3.4 Regulation of LRP1 Function

LRP1 regulates the endocytosis of a variety of different ligands, modulates PDGF, TGFβ and Wnt-dependent signalling pathways (Herz et al. 2001; Lillis et al. 2005) and, with regard to lipoprotein metabolism, LRP1 is probably responsible for the differential intracellular targeting and recycling of internalised apoE. The various functions likely involve the association of LRP1 with distinct intracellular adaptor proteins containing phospho-tyrosine-binding domains (Gotthardt et al. 2000; Uhlik et al. 2005); and therefore organ-specific functions of LRP1 might be regulated by the expression pattern of the respective adaptor proteins and/or depend on the phosphorylation status of the cytoplasmatic tail of LRP1. In general, the two NPXY motifs within the cytoplasmatic tail are able to interact with adaptor proteins in a phospho-tyrosine-independent or dependent manner. Although the importance of growth factor (e.g. PDGF-mediated LRP1 tyrosine phosphorylation) has been shown by several investigators in vitro and in vivo (Loukinova et al. 2002; Boucher et al. 2003; for reviews, see Lillis et al. 2005; Stolt and Bock 2006), the role of adaptor proteins for LRP1-mediated internalisation of triglyceride-rich lipoproteins remains elusive. Since after a meal there is a special need for effective hepatic clearance of postprandial lipoproteins, it is interesting to note that the endocytotic activity of LRP1 is at least partly regulated by the metabolic condition within the cell. In the fasted state, LRP1 is phosphorylated in dependence of the hunger signal cAMP at threonine and serine residues by activated protein kinase A (PKA). LRP1 phosphorylation results in a decrease in the endocytosis rate of LRP1 and the efficiency of ligand delivery for degradation (Li et al. 2000). Serine/threonine phosphorylation has been shown to modulate adaptor protein binding to the distal NPXY motif and reduce the association of LRP1 with adaptor molecules of the endocytotic machinery such as AP-2 (Ranganathan et al. 2004). This might explain the slower endocytosis rate of LRP1 in the presence of high cellular cAMP levels. In the postprandial phase, the drop in hepatocellular cAMP concentration mediated by increased insulin levels phase would consequently increase LRP1 endocytotic activity, which probably facilitates the LRP1-mediated clearance of postprandial lipoproteins, as shown by Laatsch et al. (2008).

The proximal NPXY within the cytoplasmatic tail of LRP1 has been identified as the binding motif for sorting nexin 17 (Snx17) and this interaction is important for the efficient recycling and basolateral targeting of LRP1 but not for the internalisation process (van Kerkhof et al. 2005). Interestingly, analysis of mutant LRP1 knock-in mice revealed that the reinsertion of an inactivating mutation in the proximal NPXY motif caused prenatal death, whereas a mutation in the distal NPXY motif did not cause any obvious pathological effects (Roebroek et al. 2006).

In summary, endocytosis and diverse signal transduction processes mediated by LRP1 are critically dependent on adaptor proteins interacting with NPXY motifs. However, the physiological importance of adaptor protein association with the cytoplasmatic tail of hepatic LRP1 for the internalisation and subsequent recycling of apoE-containing lipoproteins are yet unclear.

References

Anant S, Davidson NO. (2001) Molecular mechanisms of apolipoprotein B mRNA editing. Curr Opin Lipidol 12:159–165

Attie AD, Seidah NG. (2005) Dual regulation of the LDL receptor–some clarity and new questions. Cell Metab 1:290–292

Aulinskas T, Oram J, Bierman E, Coetzee G, Gevers W, van der Westhuyzen D. (1985) Retroendocytosis of low density lipoprotein by cultured human skin fibroblasts. Arteriosclerosis 5:45–54

Beisiegel U, Weber W, Ihrke G, Herz J, Stanley KK. (1989) The LDL-receptor-related protein, LRP, is an apolipoprotein E-binding protein. Nature 341:162–164

Bishop JR, Stanford KI, Esko JD. (2008) Heparan sulfate proteoglycans and triglyceride-rich lipoprotein metabolism. Curr Opin Lipidol 19:307–313

Boren J, Lee I, Zhu W, Arnold K, Taylor S, Innerarity TL. (1998) Identification of the low density lipoprotein receptor-binding site in apolipoprotein B100 and the modulation of its binding activity by the carboxyl terminus in familial defective apo-B100. J Clin Invest 101:1084–1093

Borén J, Ekström U, Agren B, Nilsson-Ehle P, Innerarity TL. (2001) The molecular mechanism for the genetic disorder familial defective apolipoprotein B100. J Biol Chem 276:9214–9218

Boucher P, Gotthardt M, Li WP, Anderson RG, Herz J. (2003) LRP: role in vascular wall integrity and protection from atherosclerosis. Science 300:329–332

Brown MS, Goldstein JL. (1986) A receptor-mediated pathway for cholesterol homeostasis. Science 232:34–47

Carr BR, Simpson ER. (1981) Lipoprotein utilization and cholesterol synthesis by the human fetal adrenal gland. Endocr Rev 2:306–326

Chambenoit O, Hamon Y, Marguet D, Rigneault H, Rosseneu M, Chimini G. (2001) Specific docking of apolipoprotein A-I at the cell surface requires a functional ABCA1 transporter. J Biol Chem 276:9955–9960

Cohen JC, Kimmel M, Polanski A, Hobbs HH. (2003) Molecular mechanisms of autosomal recessive hypercholesterolemia. Curr Opin Lipidol 14:121–127

Corvera S, Graver DF, Smith RM. (1989) Insulin increases the cell surface concentration of alpha 2-macroglobulin receptors in 3T3-L1 adipocytes. Altered transit of the receptor among intracellular endocytic compartments. J Biol Chem 264:10133–10138

Dallongeville J, Lussier-Cacan S, Davignon J. (1992) Modulation of plasma triglyceride levels by apoE phenotype: a meta-analysis. J Lipid Res 33:447–454

Davignon J, Gregg RE, Sing CF. (1988) Apolipoprotein E polymorphism and atherosclerosis. Arteriosclerosis 8:1–21

de Grooth GJ, Klerkx AH, Stroes ES, Stalenhoef AF, Kastelein JJ, Kuivenhoven JA. (2004) A review of CETP and its relation to atherosclerosis. J Lipid Res 45:1967–1974

Descamps O, Bilheimer D, Herz J. (1993) Insulin stimulates receptor-mediated uptake of apoE-enriched lipoproteins and activated alpha 2-macroglobulin in adipocytes. J Biol Chem 268:974–981

Dichek HL, Brecht W, Fan J, Ji Z-S, McCormick SPA, Akeefe H, Conzo L, Sanan DA, Weisgraber KH, Young SG, Taylor JM, Mahley RW. (1998) Overexpression of hepatic lipase in transgenic mice decreases apolipoprotein B-containing and high density lipoproteins. J Biol Chem 273:1896–1903

Dichek HL, Qian K, Agrawal N. (2004) The bridging function of hepatic lipase clears plasma cholesterol in LDL receptor-deficient "apoB-48-only" and "apoB-100-only" mice. J Lipid Res 45:551–560

Dong LM, Wilson C, Wardell MR, Simmons T, Mahley RW, Weisgraber KH, Agard DA. (1994) Human apolipoprotein E. Role of arginine 61 in mediating the lipoprotein preferences of the E3 and E4 isoforms. J Biol Chem 269:22358–22365

Eden ER, Patel DD, Sun XM, Burden JJ, Themis M, Edwards M, Lee P, Neuwirth C, Naoumova RP, Soutar AK. (2002) Restoration of LDL receptor function in cells from patients with autosomal recessive hypercholesterolemia by retroviral expression of ARH1. J Clin Invest 110:1695–1702

Edge S, Hoeg J, Triche T, Schneider P, Brewer H Jr. (1986) Cultured human hepatocytes. Evidence for metabolism of low density lipoproteins by a pathway independent of the classical low density lipoprotein receptor. J Biol Chem 261:3800–3806

Espirito Santo SM, Pires NM, Boesten LS, Gerritsen G, Bovenschen N, van Dijk KW, Jukema JW, Princen HM, Bensadoun A, Li WP, Herz J, Havekes LM, van Vlijmen BJ. (2004) Hepatic low-density lipoprotein receptor-related protein deficiency in mice increases atherosclerosis independent of plasma cholesterol. Blood 103:3777–3782

Farkas MH, Swift LL, Hasty AH, Linton MF, Fazio S. (2003) The recycling of apolipoprotein E in primary cultures of mouse hepatocytes. Evidence for a physiologic connection to high density lipoprotein metabolism. J Biol Chem 278:9412–9417

Farkas MH, Weisgraber KH, Shepherd VL, Linton MF, Fazio S, Swift LL. (2004) The recycling of apolipoprotein E and its amino-terminal 22 kDa fragment: evidence for multiple redundant pathways. J Lipid Res 45:1546–1554

Fazio S, Linton MF, Hasty AH, Swift LL. (1999) Recycling of apolipoprotein E in mouse liver. J Biol Chem 274:8247–8253

Gent J, Braakman I. (2004) Low-density lipoprotein receptor structure and folding. Cell Mol Life Sci 61:2461–2470

Gillian-Daniel DL, Bates PW, Tebon A, Attie AD. (2002) Endoplasmic reticulum localization of the low density lipoprotein receptor mediates presecretory degradation of apolipoprotein B. Proc Natl Acad Sci USA 99:4337–4342

Goldberg IJ. (1996) Lipoprotein lipase and lipolysis: central roles in lipoprotein metabolism and atherogenesis. J Lipid Res 37:693–707

Goldstein J, Brown M, Anderson R, Russell D, Schneider W. (1985) Receptor-mediated endocytosis: concepts emerging from the LDL receptor system. Annu Rev Cell Biol 1:1–39

Gonzalez-Navarro H, Nong Z, Amar MJ, Shamburek RD, Najib-Fruchart J, Paigen BJ, Brewer HB Jr, Santamarina-Fojo S. (2004) The ligand-binding function of hepatic lipase modulates the development of atherosclerosis in transgenic mice. J Biol Chem 279:45312–45321

Gotthardt M, Trommsdorff M, Nevitt MF, Shelton J, Richardson JA, Stockinger W, Nimpf J, Herz J. (2000) Interactions of the low density lipoprotein receptor gene family with cytosolic adaptor and scaffold proteins suggest diverse biological functions in cellular communication and signal transduction. J Biol Chem 275:25616–25624

Goudriaan JR, Espirito Santo SM, Voshol PJ, Teusink B, van Dijk KW, van Vlijmen BJ, Romijn JA, Havekes LM, Rensen PC. (2004) The VLDL receptor plays a major role in chylomicron metabolism by enhancing LPL-mediated triglyceride hydrolysis. J Lipid Res 45:1475–1481

Greenspan P, St Clair R. (1984) Retroendocytosis of low density lipoprotein. Effect of lysosomal inhibitors on the release of undegraded 125I-low density lipoprotein of altered composition from skin fibroblasts in culture. J Biol Chem 259:1703–1713

Harasaki K, Lubben NB, Harbour M, Taylor MJ, Robinson MS. (2005) Traffic 6:1014–1026

Hasty AH, Plummer MR, Weisgraber KH, Linton MF, Fazio S, Swift LL. (2005) The recycling of apolipoprotein E in macrophages: Influence of HDL and apolipoprotein AI. J Lipid Res 46:1433–1439

Havel RJ, Hamilton RL. (2004) Hepatic catabolism of remnant lipoproteins: where the action is. Arterioscler Thromb Vasc Biol 24:213–215

He G, Gupta S, Yi M, Michaely P, Hobbs HH, Cohen JC. (2002) ARH is a modular adaptor protein that interacts with the LDL receptor, clathrin, and AP-2. J Biol Chem 277:44044–44049

Heeren J, Grewal T, Jackle S, Beisiegel U. (2001) Recycling of apolipoprotein E and lipoprotein lipase through endosomal compartments in vivo. J Biol Chem 276:42333–42338

Heeren J, Weber W, Beisiegel U. (1999) Intracellular processing of endocytosed triglyceride-rich lipoproteins comprises both recycling and degradation. J Cell Sci 112:349–359

Heeren J, Niemeier A, Merkel M, Beisiegel U. (2002) Endothelial-derived lipoprotein lipase is bound to postprandial triglyceride-rich lipoproteins and mediates their hepatic clearance in vivo. J Mol Med 80:576–584

Heeren J, Grewal T, Laatsch A, Rottke D, Rinninger F, Enrich C, Beisiegel U. (2003) Recycling of apoprotein E is associated with cholesterol efflux and high density lipoprotein internalization. J Biol Chem 278:14370–14378

Heeren J, Grewal T, Laatsch A, Becker N, Rinninger F, Rye KA, Beisiegel U. (2004) Impaired recycling of apolipoprotein E4 is associated with intracellular cholesterol accumulation. J Biol Chem 279:55483–55492

Heeren J, Beisiegel U, Grewal T. (2006) Apolipoprotein E recycling: implications for dyslipidemia and atherosclerosis. Arterioscler Thromb Vasc Biol 26:442–448

Herz J, Strickland DK. (2001) LRP: a multifunctional scavenger and signaling receptor. J Clin Invest 108:779–784

Herz J, Kowal RC, Goldstein JL, Brown MS. (1990) Proteolytic processing of the 600 kd low density lipoprotein receptor-related protein (LRP) occurs in a trans-Golgi compartment. EMBO J 9:1769–1776

Hofmann SM, Zhou L, Perez-Tilve D, Greer T, Grant E, Wancata L, Thomas A, Pfluger PT, Basford JE, Gilham D, Herz J, Tschöp MH, Hui DY. (2007) Adipocyte LDL receptor-related protein-1 expression modulates postprandial lipid transport and glucose homeostasis in mice. J Clin Invest 117:3271–3282

Horton JD, Cohen JC, Hobbs HH. (2007) Molecular biology of PCSK9: its role in LDL metabolism. Trends Biochem Sci 32:71–77

Illingworth DR, Alam NA, Lindsey S. (1984) Adrenocortical response to adrenocorticotropin in heterozygous familial hypercholesterolemia. J Clin Endocrinol Metab 58:206–211

Ishibashi S, Herz J, Maeda N, Goldstein JL, Brown MS. (1994) The two-receptor model of lipoprotein clearance: tests of the hypothesis in "knockout" mice lacking the low density lipoprotein receptor, apolipoprotein E, or both proteins. Proc Natl Acad Sci USA 91:4431–4435

Hussain MM, Mahley RW, Boyles JK, Lindquist PA, Brecht WJ, Innerarity TL. (1989) Chylomicron metabolism. Chylomicron uptake by bone marrow in different animal species. J Biol Chem 264:17931–17938

Huuskonen J, Olkkonen VM, Jauhiainen M, Ehnholm C. (2001) The impact of phospholipid transfer protein (PLTP) on HDL metabolism. Atherosclerosis 155:269–281

Jeon H, Meng W, Takagi J, Eck MJ, Springer TA, Blacklow SC. (2001) Implications for familial hypercholesterolemia from the structure of the LDL receptor YWTD-EGF domain pair. Nat Struct Biol 8:499–504

Jones C, Garuti R, Michaely P, Li WP, Maeda N, Cohen JC, Herz J, Hobbs HH. (2007) Disruption of LDL but not VLDL clearance in autosomal recessive hypercholesterolemia. J Clin Invest 117:165–174

Kane JP. (1996) Structure and function of the plasma lipoproteins and their receptors. In: Fuster V, Ross R, Topol EJ (eds) Atherosclerosis and coronary disease. Lippincott–Raven, Philadelphia, pp 89–103

Keyel PA, Mishra SK, Roth R, Heuser JE, Watkins SC, Traub LM. (2006) A single common portal for clathrin-mediated endocytosis of distinct cargo governed by cargo-selective adaptors. Mol Biol Cell 17:4300–4317

Kraemer FB. (2007) Adrenal cholesterol utilization. Mol Cell Endocrinol 265/266:42–45

Krapp A, Ahle S, Kersting S, Hua Y, Kneser K, Nielsen M, Gliemann J, Beisiegel U. (1996) Hepatic lipase mediates the uptake of chylomicrons and beta-VLDL into cells via the LDL receptor-related protein (LRP). J Lipid Res 37:926–936

Laatsch A, Merkel M, Beisiegel U, Heeren J. (2008) Insulin stimulates the hepatic low density lipoprotein receptor-related protein 1 (LRP1) in vivo. Atherosclerosis (in press)

Lakadamyali M, Rust MJ, Zhuang X. (2006) Ligands for clathrin-mediated endocytosis are differentially sorted into distinct populations of early endosomes. Cell 124:997–1009

Larsson SL, Skogsberg J, Björkegren J. (2004) The low density lipoprotein receptor prevents secretion of dense apoB100-containing lipoproteins from the liver. J Biol Chem 279:831–836

Li Y, Marzolo MP, van Kerkhof P, Strous GJ, Bu G. (2000) The YXXL motif, but not the two NPXY motifs, serves as the dominant endocytosis signal for low density lipoprotein receptor-related protein. J Biol Chem 275:17187–17194

Li Y, van Kerkhof P, Marzolo MP, Strous GJ, Bu G. (2001) Identification of a major cyclic AMP-dependent protein kinase A phosphorylation site within the cytoplasmic tail of the low-density lipoprotein receptor-related protein: implication for receptor-mediated endocytosis. Mol Cell Biol 21:1185–1195

Lillis AP, Mikhailenko I, Strickland DK. (2005) Beyond endocytosis: LRP function in cell migration, proliferation and vascular permeability. J Thromb Haemost 3:1884–1893

Liu J, Heikkilä P, Meng QH, Kahri AI, Tikkanen MJ, Voutilainen R. (2000) Expression of low and high density lipoprotein receptor genes in human adrenals. Eur J Endocrinol 142:677–682

Lombardi P, Mulder M, van der BH, Frants RR, Havekes LM. (1993) Inefficient degradation of triglyceride-rich lipoprotein by HepG2 cells is due to a retarded transport to the lysosomal compartment. J Biol Chem 268:26113–26119

Loukinova E, Ranganathan S, Kuznetsov S, Gorlatova N, Migliorini MM, Loukinov D, Ulery PG, Mikhailenko I, Lawrence DA, Strickland DK. (2002) Platelet-derived growth factor (PDGF)-induced tyrosine phosphorylation of the low density lipoprotein receptor-related protein (LRP). Evidence for integrated co-receptor function betwenn LRP and the PDGF. J Biol Chem 277:15499–15506

MacArthur JM, Bishop JR, Stanford KI, Wang L, Bensadoun A, Witztum JL, Esko JD. (2007) Liver heparan sulfate proteoglycans mediate clearance of triglyceride-rich lipoproteins independently of LDL receptor family members. J Clin Invest 117:153–164

Mahley RW, Ji ZS. (1999) Remnant lipoprotein metabolism: key pathways involving cell-surface heparan sulfate proteoglycans and apolipoprotein E. J Lipid Res 40:1–16

Mahley RW, Huang Y, Rall SC Jr. (1999) Pathogenesis of type III hyperlipoproteinemia (dysbetalipoproteinemia). Questions, quandaries, and paradoxes. J Lipid Res 40:1933–1949

Mahley RW, Huang Y. (2007) Atherogenic remnant lipoproteins: role for proteoglycans in trapping, transferring, and internalizing. J Clin Invest 117:94–98

Mann WA, Meyer N, Berg D, Greten H, Beisiegel U. (1999) Lipoprotein lipase compensates for the defective function of apo E variants in vitro by interacting with proteoglycans and lipoprotein receptors. Atherosclerosis 145:61–69

Merkel M, Eckel RH, Goldberg IJ. (2002) Lipoprotein lipase: genetics, lipid uptake, and regulation. J Lipid Res 43:1997–2006

Michaely P, Zhao Z, Li WP, Garuti R, Huang LJ, Hobbs HH, Cohen JC. (2007) Identification of a VLDL-induced, FDNPVY-independent internalization mechanism for the LDLR. EMBO J 26:3273–3282

Mishra SK, Watkins SC, Traub LM. (2002) The autosomal recessive hypercholesterolemia. (ARH) protein interfaces directly with the clathrin-coat machinery. Proc Natl Acad Sci USA 99:16099–16104

Myers JN, Tabas I, Jones NL, Maxfield FR. (1993) Beta-very low density lipoprotein is sequestered in surface-connected tubules in mouse peritoneal macrophages. J Cell Biol 123:1389–1402

Niemeier A, Gàfvels M, Heeren J, Meyer N, Angelin B, Beisiegel U. (1996) VLDL receptor mediates the uptake of human chylomicron remnants in vitro. J Lipid Res 37:1733–1742

Niemeier A, Kassem M, Toedter K, Wendt D, Ruether W, Beisiegel U, et al. (2005) Expression of LRP1 by human osteoblasts: a mechanism for the delivery of lipoproteins and vitamin K1 to bone. J Bone Miner Res 20:283–293

Niemeier A, Niedzielska D, Secer R, Schilling A, Merkel M, Enrich C, Rensen PCN, Heeren J. (2008) Uptake of postprandial lipoproteins into bone in vivo: Impact on Osteoblast Function. Bone (in press)

Prawitt J, Niemeier A, Kassem M, Beisiegel U, Heeren J. (2008) Characterization of lipid metabolism in insulin-sensitive adipocytes differentiated from immortalized human mesenchymal stem cells. Exp Cell Res 314:814–824

Rader DJ, Cohen J, Hobbs HH. (2003) Monogenic hypercholesterolemia: new insights in pathogenesis and treatment. J Clin Invest 111:1795–1803

Ranganathan S, Liu CX, Migliorini MM, Von Arnim CA, Peltan ID, Mikhailenko I, Hyman BT, Strickland DK. (2004) Serine and threonine phosphorylation of the low density lipoprotein receptor-related protein by protein kinase Calpha regulates endocytosis and association with adaptor molecules. J Biol Chem 279:40536–40544

Rellin L, Heeren J, Beisiegel U. (2008) Recycling of apolipoprotein E is not associated with cholesterol efflux in neuronal cells. Biochim Biophys Acta 1781:232–238

Rensen PC, Jong MC, van Vark LC, van der BH, Hendriks WL, Van Berkel TJ, Biessen EA, Havekes LM. (2000) Apolipoprotein E is resistant to intracellular degradation in vitro and in vivo. Evidence for retroendocytosis. J Biol Chem 275:8564–8571

Roebroek AJ, Reekmans S, Lauwers A, Feyaerts N, Smeijers L, Hartmann D. (2006) Mutant Lrp1 knock-in mice generated by recombinase-mediated cassette exchange reveal differential importance of the NPXY motifs in the intracellular domain of LRP1 for normal fetal development. Mol Cell Biol 26:605–616

Rohlmann A, Gotthardt M, Hammer RE, Herz J. (1998) Inducible inactivation of hepatic LRP gene by cre-mediated recombination confirms role of LRP in clearance of chylomicron remnants. J Clin Invest 101:689–695

Roses AD. (1996) Apolipoprotein E alleles as risk factors in Alzheimer's disease. Annu Rev Med 47:387–400

Rubino M, Miaczynska M, Lippe R, Zerial M. (2000) Selective membrane recruitment of EEA1 suggests a role in directional transport of clathrin-coated vesicles to early endosomes. J Biol Chem 275:3745–3748

Saito H, Lund-Katz S, Phillips MC. (2004) Contributions of domain structure and lipid interaction to the functionality of exchangeable human apolipoproteins. Prog Lipid Res 43:350–380

Schilling AF, Schinke T, Munch C, Gebauer M, Niemeier A, Priemel M et al. (2005) Increased bone formation in mice lacking apolipoprotein E. J Bone Miner Res 20:274–282

Schneider WJ, Kovanen PT, Brown MS, Goldstein JL, Utermann G, Weber W, Havel RJ, Kotite L, Kane JP, Innerarity TL, Mahley RW. (1981) Familial dysbetalipoproteinemia. Abnormal binding of mutant apoprotein E to low density lipoprotein receptors of human fibroblasts and membranes from liver and adrenal of rats, rabbits, and cows. J Clin Invest 68:1075–1085

Sehayek, E, Lewin-Velvert, U, Chajek-Shaul, T, Eisenberg, S. (1991) Lipolysis exposes unreactive endogenous apolipoprotein E-3 in human and rat plasma very low density lipoprotein. J Clin Invest 88: 553–560

Shelness GS, Sellers JA. (2001) Very-low-density lipoprotein assembly and secretion. Curr Opin Lipidol 12:151–157

Smith JD, Waelde C, Horwitz A, Zheng P. (2002) Evaluation of the role of phosphatidylserine translocase activity in ABCA1-mediated lipid efflux. J Biol Chem 277:17797–17803

Stolt PC, Bock HH. (2006) Modulation of lipoprotein receptor functions by intracellular adaptor proteins. Cell Signal 18:1560–1571

Südhof TC, Goldstein JL, Brown MS, Russell DW. (1985) The LDL receptor gene: a mosaic of exons shared with different proteins. Science 228:815–822

Swift LL, Farkas MH, Major AS, Valyi-Nagy K, Linton MF, Fazio S. (2001) A recycling pathway for resecretion of internalized apolipoprotein E in liver cells. J Biol Chem 276:22965–22970

Tabas I, Lim S, Xu XX, Maxfield FR. (1990) Endocytosed beta-VLDL and LDL are delivered to different intracellular vesicles in mouse peritoneal macrophages. J Cell Biol 111:929–940

Tamaki C, Ohtsuki S, Terasaki T. (2007) Mol Pharmacol 72:850–855

Takahashi S, Sakai J, Fujino T, Hattori H, Zenimaru Y, Suzuki J, Miyamori I, Yamamoto TT. (2004) The very low-density lipoprotein. (VLDL) receptor: characterization and functions as a peripheral lipoprotein receptor. J Atheroscler Thromb 11:200–208

Twisk J, Gillian-Daniel DL, Tebon A, Wang L, Barrett PH, Attie AD. (2000) The role of the LDL receptor in apolipoprotein B secretion. J Clin Invest 105:521–532

Uhlik MT, Temple B, Bencharit S, Kimple AJ, Siderovski DP, Johnson GL. (2005) Structural and evolutionary division of phosphotyrosine binding. (PTB) domains. J Mol Biol 345:1–20

van Kerkhof P, Lee J, McCormick L, Tetrault E, Lu W, Schoenfish M, Oorschot V, Strous GJ, Klumperman J, Bu G. (2005) Sorting nexin 17 facilitates LRP recycling in the early endosome. EMBO J 24:2851–2861

Vance J. (2002) Assembly and secretion of lipoproteins. In: Vance D, Vance J. (eds) Biochemistry of lipids, lipoproteins and membranes. Elsevier, Amsterdam, pp 505–526

Verges M, Bensadoun A, Herz J, Belcher JD, Havel RJ. (2004) Endocytosis of hepatic lipase and lipoprotein lipase into rat liver hepatocytes in vivo is mediated by the low density lipoprotein receptor-related protein. J Biol Chem 279:9030–9036

von Eckardstein A, Nofer JR, Assmann G. (2001) High density lipoproteins and arteriosclerosis. Role of cholesterol efflux and reverse cholesterol transport. Arterioscler Thromb Vasc Biol 21:13–27

Wilson C, Wardell MR, Weisgraber KH, Mahley RW, Agard DA. (1991) Three-dimensional structure of the LDL receptor-binding domain of human apolipoprotein E. Science 252:1817–1822

Wyne KL, Pathak K, Seabra MC, Hobbs HH. (1996) Expression of the VLDL receptor in endothelial cells. Arterioscler Thromb Vasc Biol 16:407–415

Xiang SQ, Cianflone K, Kalant D, Sniderman AD. (1999) Differential binding of triglyceride-rich lipoproteins to lipoprotein lipase. J Lipid Res 40:1655–1663

Zaiou M, Arnold KS, Newhouse YM, Innerarity TL, Weisgraber KH, Segall ML, Phillips MC, Lund-Katz S. (2000) Apolipoprotein E–low density lipoprotein receptor interaction. Influences of basic residue and amphipathic alpha-helix organization in the ligand. J Lipid Res 41:1087–1095

Chapter 9
Angiopoietin-Like Proteins and Lipid Metabolism

Sander Kersten

Abstract Lipoprotein lipase, hepatic lipase, and endothelial lipase play a pivotal role in the clearance and remodelling of plasma lipoproteins. Their activity is governed at the level of gene transcription as well as via changes in enzyme activity by specific activator or inhibitor proteins. The latter group includes two members of the Angiopoietin family: Angiopoietin-like proteins 3 and 4 (Angptl3, Angptl4). Angptl4 is expressed ubiquitously and is a direct target gene of the peroxisome proliferator activated receptors, whereas Angptl3 is expressed exclusively in liver and is regulated by the liver X receptor. For both Angptl4 and Angptl3, injection of recombinant protein, adenoviral over-expression, and transgene-mediated overexpression have been shown to cause marked hypertriglyceridemia. Conversely, inactivation of the *Angptl4* or *Angptl3* gene is associated with lower plasma triglyceride levels. Recent human genetic studies suggest that rare or common sequence variants in the *ANGPTL4* and *ANGPTL3* gene, respectively, impact plasma TG levels, supporting a role for ANGPTL4 and ANGPTL3 in lipoprotein metabolism in humans. The present chapter provides an overview of the role of Angptl4 and Angptl3 in the regulation of lipoprotein metabolism. It is postulated that alterations in Angptl4 and Angptl3 signalling might be involved in dyslipidemia.

9.1 Introduction

Changes in the plasma level of lipoproteins are known to affect atherosclerosis and associated coronary heart disease (CHD). Indeed, it is well established that elevated plasma low-density lipoprotein (LDL) levels increase the risk for CHD, whereas elevated high-density lipoprotein (HDL) levels are considered atheroprotective. Besides LDL, elevated plasma triglycerides (TG) are increasingly recognized as an

S. Kersten
Nutrition, Metabolism and Genomics group, Division of Human Nutrition, Wageningen University, Bomenweg 2, 6703 HD Wageningen, the Netherlands; and Nutrigenomics Consortium, TI Food and Nutrition, Nieuwe Kanaal 9A, 6709 PA Wageningen, the Netherlands
e-mail: sander.kersten@wur.nl

C. Ehnholm (ed.), *Cellular Lipid Metabolism*,
DOI 10.1007/978-3-642-00300-4_9,

important, independent risk factor for CHD. High TG levels frequently occur together with low HDL levels, often with normal levels of LDL, and are referred to as abnormalities of the TG–HDL axis. This atherogenic dyslipidemia is connected with central obesity characterized by excess fat tissue in and around the abdomen.

Depending on nutritional status, TG are present in blood plasma mainly packaged in chylomicrons or in very low-density lipoproteins (VLDL). Accordingly, plasma TG levels are determined by the balance between the rate of production of chylomicrons and VLDL in intestine and liver, respectively, and their rate of clearance in peripheral organs such as skeletal muscle, heart, and adipose tissue. Clearance of TG-rich lipoproteins is mediated by the enzyme lipoprotein lipase (LPL), which is anchored into the capillary endothelium via heparin sulfate proteoglycans and catalyses TG hydrolysis. Other lipases that target plasma lipoproteins include hepatic lipase, which promotes clearance of remnant particles in liver, and endothelial lipase, which preferentially hydrolyses (phospho)lipids within HDL.

While LPL gene expression is known to respond to changes in nutritional status, LPL is also extensively regulated at the level of enzyme activity (Merkel et al. 2002). Several modulators of LPL activity are known, including apolipoproteins Apoc3, Apoc2, Apoc1, and Apoa5. In addition to apolipoproteins, in the past few years it has become evident that two proteins belonging to the family of angiopoietin-like proteins (Angptls), Angptl3 and Angptl4, can modulate LPL activity. Here, a detailed overview is provided of the role of Angptl3 and Angptl4 in the regulation of lipid metabolism.

9.2 Angptl4 and Lipid Metabolism

9.2.1 Discovery and Structure of Angptl4

Angptl4 was identified by Kim et al. (2000) as hepatic fibrinogen/angiopoietin-related protein (HFARP) while searching for additional members of the angiopoietin family (Kim et al. 2000). In parallel, Yoon discovered *Angptl4* by a subtractive cloning strategy to identify target genes of the nuclear hormone receptor PPARγ in adipose tissue (Yoon et al. 2000). Finally, Kersten et al. (2000) identified *Angptl4* as a PPARα target gene by comparing mRNA from livers of wild-type and PPARα-deficient mice (Kersten et al. 2000). They named the protein fasting-induced adipose factor (FIAF), which in the literature is used interchangeably with Angptl4.

In mouse, the gene sequence encoding *Angptl4* covers 6.6 kb and consists of 7 exons (Kersten et al. 2000; Yoon et al. 2000). The human gene shows a similar organization and is located on chromosome 19, in a region close to a locus associated with atherosclerosis susceptibility (Yoon et al. 2000). The open reading frame derived from the cDNA sequence spans about 1.2 kb and gives rise to a 406- and 410-amino-acid protein in human and mouse, respectively (Kersten et al. 2000; Kim et al. 2000), that undergoes glycosylation and is secreted (Ge et al. 2004a; Kersten et al. 2000; Kim et al. 2000). Within its angiopoietin/fibrinogen-like domain,

Angptl4 contains several conserved cysteine residues that contribute to intermolecular disulfide bonding, resulting in the formation of variable-sized multimeric structures (Ge et al. 2004a; Mandard et al. 2004; Yoon et al. 2000). In addition, Angptl4 is proteolytically processed, releasing N- and C-terminal fragments that can be detected in human serum. The exact site of cleavage has yet to be exactly defined but may vary between R162 and R229. While the N-terminal fragment of Angptl4 impacts lipid metabolism, the C-terminal fragment is a structural and likely regulatory component of the extra-cellular matrix that so far has been shown to affect a number of different processes including angiogenesis, endothelial cell function, vascular leakage, and cell adhesion (Cazes et al. 2006; Galaup et al. 2006; Hermann et al. 2005; Ito et al. 2003; Le Jan et al. 2003; Padua et al. 2008). However, currently there is a lack of consensus on the direction of the impact of Angptl4 on these processes.

9.2.2 Regulation of Angptl4 Expression

In mouse, *Angptl4* is expressed in a variety of tissues, but highest levels are found in white and brown adipose tissue, followed by ovary, liver, lung, heart, and intestine (Kersten et al. 2000; Yoon et al. 2000). In human, expression of *ANGPTL4* is particularly high in liver, followed by adipose tissue and pancreatic islets (Zandbergen et al. 2006). Befitting its alternative name "fasting-induced adipose factor", expression of *Angptl4* is up-regulated by fasting in a variety of tissues (Dutton and Trayhurn 2008; Ge et al. 2005; Kersten et al. 2000; Mandard et al. 2006; Sukonina et al. 2006). The effect of fasting is likely mediated by elevated plasma free fatty acids, which stimulate *Angptl4* gene transcription via peroxisome proliferator activated receptors (PPARs). Indeed, treatment with synthetic agonists for PPARα potently induces *Angptl4* expression in liver, skeletal muscle, heart, and intestine (Bunger et al. 2007; Ge et al. 2005; Kersten et al. 2000; Mandard et al. 2004). Furthermore, expression of *Angptl4* is induced by PPARα agonists in white adipose tissue, liver, heart, and muscle (Kersten et al. 2000, unpublished data; Yoon et al. 2000), and by PPARβ/δ agonists in heart, muscle, keratinocytes, and liver (Akiyama et al. 2004; Kersten et al., unpublished data; Schmuth et al. 2004; Schug et al. 2007). Using transactivation assays, electrophoretic mobility shifts, and in vivo chromatin immunoprecipitations, *Angptl4* was shown to be a direct PPAR target gene, containing a conserved functional PPAR response element in its third intron (Mandard et al. 2004). In addition to PPARs, synthetic and endogenous ligands for RXR and possibly FXR induce Angptl4 expression (Fig. 9.1).

Regulation of Angptl4 mRNA or protein has been studied in several mouse models and under a variety of physiological stimuli. Expression of *Angptl4* is highly induced by hypoxia as observed in cardiomyocytes, brain, endothelial cells, and adipocytes (Belanger et al. 2002; Gustavsson et al. 2007; Le Jan et al. 2003; Wang et al. 2007; Wiesner et al. 2006), suggesting a role for Angptl4 in angiogenesis. Alternatively, upregulation of Angptl4 under hypoxic conditions may be linked to the inhibitory effect of Angptl4 on cellular fat uptake (see below) and is aimed at favouring anaerobic metabolism (e.g. glucose) over aerobic metabolism (e.g. fatty acids).

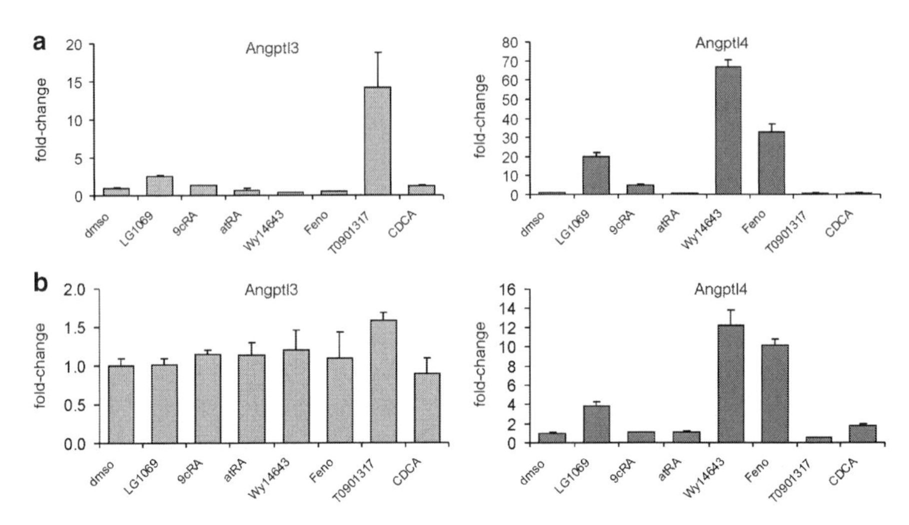

Fig. 9.1 Angptl3 and Angptl4 expression are upregulated by different nuclear receptor agonists. Rat FAO hepatoma cells (**a**) or primary rat hepatocytes (**b**) were treated for 24 h with different nuclear receptor agonists [*LG1069* 1 μM, *9cRA* 0.5 μM, *atRA* 0.5 μM, *Wy14643* 5 μM, fenofibrate (*feno*) 50 μM, *T0901317* 1 μM, chenodeoxycholic acid (*CDCA*) 50 μM]. Gene expression of Angptl3 and Angptl4 was analysed by qPCR

Other conditions in which adipose *Angptl4* expression is elevated include pregnancy and lactation (Josephs et al. 2007) and obesity and diabetes (Yoon et al. 2000). In the pituitary gland but not in the hypothalamus, *Angptl4* expression is increased by food restriction, suggesting a role for Angptl4 in the neuroendocrine response to food deprivation (Wiesner et al. 2004). In the intestinal epithelium of germ-free mice lacking gut microbes, *Angptl4* expression increases during the suckling–weaning transition. This increase is absent in germ-free mice given a normal microbiota derived from conventionally raised mice, suggesting that *Angptl4* expression is suppressed by the intestinal microbiota (Backhed et al. 2004).

9.2.3 Role of Angptl4 in Lipid Metabolism

Regulation of Angptl4 by PPARs, fasting, high-fat feeding, and obesity first suggested that Angptl4 might be involved in lipid metabolism (Kersten et al. 2000; Yoon et al. 2000). In 2002, Yoshida and colleagues (2002) were the first to demonstrate a direct and immediate stimulatory effect of recombinant Angptl4 on plasma levels of TG, FFA, and non-HDL cholesterol in mice (Yoshida et al. 2002). Induction of plasma TG by Angptl4 has been confirmed in several transgenic and adenoviral mouse models of *Angptl4* over-expression, as well as in *Angptl4* knock-out mice (Backhed et al. 2004; Ge et al. 2004b; Koster et al. 2005; Mandard et al. 2006; Xu et al. 2005; Yu et al. 2005). Also, repeated injections of mice with a monoclonal antibody against Angptl4 lowers plasma TG levels (Desai et al. 2007). Hypertriglyceridemia induced by *Angptl4* is dependent on its oligomeric state, as rendering Angptl4 defective in oligomerization resulted in a reduced hypertriglyceridemic effect (Ge et al. 2004b). Angptl4 raises plasma TG by suppressing LPL-mediated clearance of plasma TG-rich lipoproteins, (Ge et al. 2004b; Lichtenstein et al. 2007), thereby inhibiting cellular fatty acid uptake from that specific plasma fatty acid pool, and perhaps to a minor extent by raising hepatic VLDL production (Desai et al. 2007; Mandard et al. 2006). Angptl4 directly inhibits the activity of lipoprotein lipase (LPL), as shown by numerous in vitro and in vivo studies (Koster et al. 2005; Lichtenstein et al. 2007; Yoshida et al. 2002; Yu et al. 2005). In addition, Angptl4 inhibits activity of hepatic lipase, contributing to a reduced hepatic uptake of cholesteryl-esters and triglycerides from remnant particles, which in turn leads to induction of hepatic cholesterol synthesis (Lichtenstein et al. 2007). Recent in vitro studies indicate that the N-terminal fragment of Angptl4 inhibits LPL by promoting the conversion of catalytically active LPL dimers into catalytically inactive LPL monomers, thereby permanently inactivating LPL (Sukonina et al. 2006). This mechanism is supported in vivo, although it may not account for the full extent of inhibition of LPL in *Angptl4*-transgenic mice (Lichtenstein et al. 2007; Fig. 9.2). We showed that, in mouse blood, Angptl4 is physically associated with LPL monomers, which in turn are bound to LDL. Possibly, by binding LPL monomers Angptl4 pulls the equilibrium between LPL dimers and monomers toward monomers, which effectively results in an inhibition of LPL activity (Lichtenstein et al. 2007).

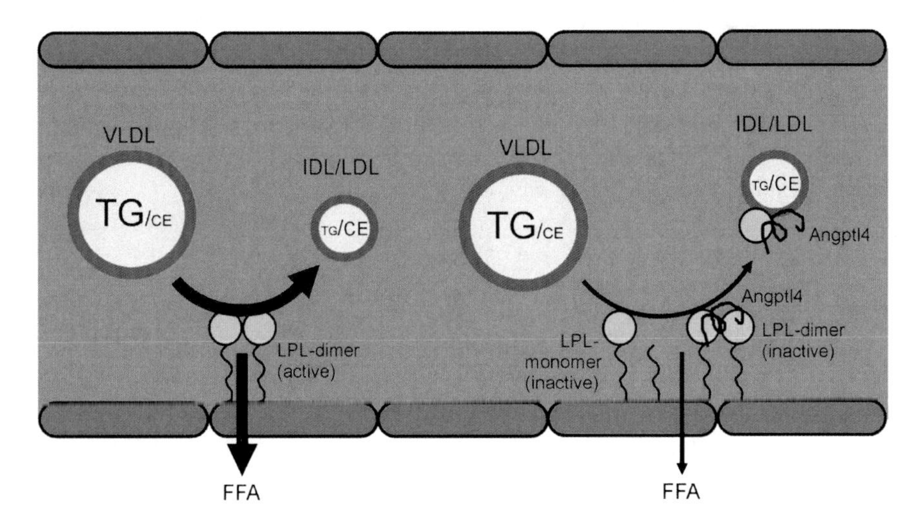

Fig. 9.2 Inhibition of LPL-mediated plasma TG lipolysis by Angptl4. Catalytically active LPL dimers are bound to endothelial cells lining the capillaries via heparin-sulfate proteoglycans. Angptl4 inhibits the activity of LPL by promoting dissociation of LPL dimers into catalytically inactive LPL monomers. LPL monomers dissociate from the vessel wall and associate with circulating IDL/LDL particles. In addition, Angptl4 may inhibit LPL independently of dimer dissociation. Decreased LPL activity leads to impaired clearance of TG from the plasma and decreased uptake of fatty acids into the cell

It has been suggested that the induction of Angptl4 mRNA levels in adipose tissue by fasting may explain the known rapid decrease of LPL activity in adipose tissue upon fasting (Sukonina et al. 2006). While this may be true, it should be emphasized that fasting also markedly induces expression of *Angptl4* in skeletal muscle, concurrent with an increase in LPL activity (Lichtenstein et al. 2007; Ruge et al. 2005).

One of the key questions regarding Angptl4 function is to what extent Angptl4 acts locally by selectively inhibiting LPL in the tissue in which it is produced. The specific inhibition of cardiac LPL activity in mice transgenic mice over-expressing *Angptl4* in heart is supportive of a paracrine action of Angptl4 (Yu et al. 2005). In contrast, the marked elevation of plasma TG in mice over-expressing *Angptl4* specifically in liver, which lacks LPL, suggests that Angptl4 acts as a circulating LPL inhibitor that mediates inter-organ communication (Xu et al. 2005).

By inhibiting activity of LPL and as a consequence cellular uptake of plasma TG-derived fatty acids, Angptl4 would be expected to decrease fat storage in adipose tissue. Surprisingly, bodyweight and composition were reported to be unaltered in *Angptl4* knock-out mice, either on chow diet or after a high-fat/high-carbohydrate feeding for 15 weeks (Koster et al. 2005). In contrast, using the same Angptl4 knock-out mice on a pure C57Bl/6 background, we found a significant increase in epididymal fat mass compared to wild-type mice. Furthermore, in mice over-expressing *Angptl4* in liver and peripheral tissues, epididymal and perirenal fat mass was decreased by 50% compared to their wild-type littermates (Mandard et al. 2006). While the reduction in fat mass by Angptl4 may be explained by decreased uptake of plasma TG, other mechanisms likely contribute as well. Plasma levels of FFA and glycerol were elevated in mice over-expressing *Angptl4*, suggesting that Angptl4 stimulates adipose tissue lipolysis (Mandard et al. 2006; Lichtenstein et al. 2007). In agreement with this notion, injection of recombinant Angptl4 abruptly raises plasma FFA (Yoshida et al. 2002). Angptl4 may induce lipolysis via upregulation of Pnpla2 (adipose triglyceride lipase). Other mechanisms that may contribute to reduced fat mass in Angptl4 transgenic mice include adipose upregulation of Ucp1 as well as increased fatty acid oxidation, suggesting an increased abundance of brown adipocytes (Mandard et al. 2006).

In addition to lipoprotein metabolism, Angptl4 has also been implicated in regulation of glucose homeostasis. Adenoviral-mediated over-expression of *Angptl4* markedly lowered blood glucose and improved glucose tolerance, while inducing hepatic steatosis and (transient) hypertriglyceridemia (Xu et al. 2005). In addition, in primary rat hepatocytes, adenoviral-mediated *Angptl4* over-expression decreased glucose production and enhanced insulin-mediated inhibition of gluconeogenesis. In contrast, plasma glucose levels were unchanged in *Angptl4* knock-out mice and in transgenic mice over-expressing *Angptl4* in liver (Koster et al. 2005). Furthermore, blood glucose levels were unaltered in transgenic mice with general *Angptl4* over-expression, while glucose tolerance was worsened, especially after high-fat feeding (Mandard et al. 2006). Using hyperinsulinemic–euglycemic clamp techniques it was shown that *Angptl4* over-expression causes insulin-resistance in the periphery, but improves insulin-sensitivity at the hepatic level (Lichtenstein et al. 2007). At the present time, it is difficult to reconcile these various findings regarding the effects of Angptl4 on glucose homeostasis. Clearly, the dramatic effect of Angptl4 on plasma glucose levels as reported by Xu et al. (2005) has not been reproduced in other studies using Angptl4 transgenic or knock-out mice.

9.2.4 Role of Angptl4 in Human

In human, highest expression levels for *ANGPTL4* are found in liver, followed by adipose tissue (Zandbergen et al. 2006). Similar to Angptl4, ANGPTL4 is cleaved into N- and C-terminal fragments, which together with full-length ANGPTL4 are detectable in human blood plasma (Mandard et al. 2004).

Very little is known about possible determinants of plasma ANGPTL4, likely because assays to quantitatively assess ANGPTL4 levels in human plasma are presently not commercially available. Using a homemade ELISA or by semi-quantitative Western blot, it has been shown that the plasma ANGPTL4 concentration increases after treatment with rosiglitazone or fenofibrate, which serve as synthetic agonists for PPARγ and PPARα, respectively (Mandard et al. 2004; Xu et al. 2005). In obese patients with type 2 diabetes, serum levels of Angptl4 were found to be lower than in healthy subjects with or without obesity (Xu et al. 2005). No correlation seems to exist between plasma ANGPTL4 levels and body mass index (Mandard et al. 2004; Xu et al. 2005). In a very recent study, plasma ANGPTL4 levels were negatively correlated with HDL-C and positively correlated with plasma glucose and TG. No differences in plasma ANGPTL4 were found between healthy subjects and diabetic patients (Stejskal et al. 2008).

In support of a role of ANGPTL4 in regulation of lipoprotein metabolism in human, carriers of a rare sequence variant of *ANGPTL4* (E40K) were shown to have decreased plasma TG and elevated plasma HDL-C levels (Romeo et al. 2007). In contrast, no association was found between four common genetic variants within the *ANGPTL4* gene and anthropometric data, family history of diabetes, plasma triglyceride and FFA levels, insulin sensitivity, or insulin secretion (Staiger et al. 2008). Also, in genome-wide association studies, no sequence variants in loci near the *ANGPTL4* gene were associated with plasma lipid parameters, including TG. These data suggest that variations within the ANGPTL4 gene may only minimally account for inter-individual variations in plasma lipid levels.

9.3 Angptl3 and Lipid Metabolism

9.3.1 Discovery and Structure of Angptl3

Angptl3 was discovered while searching EST databases for signal sequences and amphipathic helices (Conklin et al. 1999). Mouse *Angptl3* contains 455 amino acids and is encoded by 7 exons on mouse chromosome 4, spanning 7–8 kb. The human *ANGPTL3* gene is located on chromosome 1 and encodes a protein that is slightly larger (460 amino acids). Angptl3 is structurally highly homologous to Angptl4 and shares the modular structure consisting of a signal sequence, a small unique region, a coiled-coil domain and a large fibrinogen/angiopoietin-like domain. Also, similar to Angptl4, it contains several conserved cysteine residues involved in intermolecular disulfide bonding to form multimeric structures and is proteolytically processed at R221 and/or R224 (Ono et al. 2003).

9.3.2 Regulation of Angptl3 Expression

Angptl3 is expressed almost exclusively in liver. While Angptl4 is under transcriptional control of PPARs, *Angptl3* gene expression is governed by liver X receptor (LXR), a nuclear hormone receptor involved in the regulation of hepatic TG and cholesterol metabolism. Activation of LXR using the synthetic agonist T0901317 or the endogenous agonist 22(*R*)-hydroxycholesterol markedly increases Angptl3 mRNA in mouse liver (Inaba et al. 2003; Ge et al. 2005), human HepG2 cells (Kaplan et al. 2003), rat FAO cells (Ge et al. 2005), and rat primary hepatocytes (Fig. 9.1). In mouse liver, *Angptl3* expression is also upregulated by cholesterol feeding (Kaplan et al. 2003). A functional LXR response element is present in the promoter of the human *ANGPTL3* gene, establishing *ANGPTL3* as a direct target gene of LXR (Inaba et al. 2003; Kaplan et al. 2003). Expression of *Angptl3* is suppressed by thyroid hormone, which is mediated by the thyroid hormone receptor beta. Down-regulation of *Angptl3* is achieved by TRβ-dependent interference with the HNF1α signaling pathway (Fugier et al. 2006). *Angptl3* expression is also suppressed by PPARβ/δ, which likely competes with LXR for the common heterodimeric partner RXRα (Matsusue et al. 2006). Although the nuclear receptor FXR is highly expressed in liver and plays a major role in hepatic lipid metabolism, no effect of FXR activation on *Angptl3* expression is observed in either primary rat hepatocytes or rat FAO hepatoma cells (Fig. 9.1). Finally, *Angptl3* gene expression and plasma protein levels are increased in insulin-deficient streptozotocin-treated mice, in leptin-resistant C57BL/6J(db/db) mice, and in leptin-deficient C57BL/6J(ob/ob) mice, indicating that insulin and leptin inhibit *Angptl3* expression in hepatocytes (Inukai et al. 2004; Shimamura et al. 2004).

9.3.3 Role of Angptl3 in Lipid Metabolism

The connection between Angptl3 and lipoprotein metabolism was first made by the finding that KK/San mice have a 4-bp insertion in exon 6 of the *Angptl3* gene, thereby explaining their markedly reduced plasma FFA, cholesterol and TG levels

(Koishi et al. 2002). Conversely, adenoviral-mediated over-expression and intravenous injection of recombinant Angplt3 leads to increased plasma TGs, FFAs, and cholesterol levels (Koishi et al. 2002; Shimizugawa et al. 2002). Similar to Angptl4, the dramatic effect of Angptl3 on fasting plasma TG levels can be ascribed to suppression of VLDL-TG clearance via inhibition of the activity of LPL, rather than stimulation of hepatic VLDL production, which is only minimally affected (Ando et al. 2003; Shimizugawa et al. 2002). Targeted deletion of the *Angptl3* gene gives rise to a similar phenotype as observed in the KK/San mice: markedly decreased plasma TG levels associated with elevated heparin-releasable LPL activity (Fujimoto et al. 2006; Koster et al. 2005). A N-terminal portion of the protein corresponding to amino acids (AA) 17–165 is sufficient to elicit the hypertriglyceridemic effect, while deletion mutants of Angptl3 containing AA 207–460 or AA 79–207 are ineffective. The former region is present in the N-terminal fragment that is produced upon proteolysis of Angptl3. Mutating the putative cleavage sites reduces the potency of Angptl3 towards elevating plasma TG, suggesting that regulated proteolysis of Angptl3 may be an important mechanism to govern its activity and thus the activity of LPL (Ono et al. 2003). The plasma TG-raising effect of Angptl3 is supposedly mediated by a putative heparin-binding domain present in the N-terminal region (Ono et al. 2003).

Besides plasma TG, plasma HDL-cholesterol and HDL-phospholipids are also decreased in *Angptl3* knock-out mice. These alterations are likely explained by an inhibitory effect of Angptl3 on endothelial lipase, as demonstrated by in vitro and in vivo studies (Shimamura et al. 2007). Compared to LPL and HL, EL has a higher specificity towards hydrolysing phospholipids in HDL. Recently, Angptl3 was suggested to mediate the effect of proprotein convertases on EL activity and consequently plasma HDL levels (Jin et al. 2007). According to the mechanism proposed, proprotein convertases cleave Angptl3 to release a N-terminal fragment. This fragment inhibits EL, resulting in increased plasma HDL levels. As Angptl3 is particularly active toward LPL, this mechanism would be expected to impact plasma TG as well, although no data are currently available.

There is some ambiguity regarding the effects of Angptl3 on HL activity. In vitro studies suggest that Angptl3 is only marginally active towards HL (Shimizugawa et al. 2002). In contrast, in mice Angptl3 deletion was associated with a significant increase in post-heparin HL activity (Fujimoto et al. 2006). Similar observations were made in *Angptl3* knock-out mice on an *Apoe* knock-out background (Ando et al. 2003).

Similar to Angptl4, recombinant Angptl3 or adenovirus-mediated over-expression of Angptl3 causes elevated plasma FFA levels (Koishi et al. 2002), while Angptl3 inactivation leads to decreased plasma FFA and glycerol levels (Fujimoto et al. 2006; Shimamura et al. 2003). In line with these data, recombinant Angptl3 stimulates release of FFA and glycerol from differentiated 3T3-L1 adipocytes, probably via direct interaction with adipocytes. The data provide compelling evidence that Angptl3 stimulates adipose tissue lipolysis. At the present time, the mechanism behind this effect remains unclear.

9.3.4 Role of Angptl3 in Human

Recent genome-wide association studies have shown that a common sequence variant at a locus near the *ANGPTL3* gene is associated with plasma concentration of TG, but not other lipoprotein parameters (Kathiresan et al. 2008; Willer et al. 2008). These data suggest that genetic variation within the *ANGPTL3* gene contributes to genetic variation in plasma TG levels within specific populations; and they validate the importance of ANGPTL3 in the regulation of lipoprotein metabolism in humans.

Information on plasma ANGPTL3 levels in humans and possible correlation with biochemical, metabolic, or clinical parameters is limited. In a recent report, plasma ANGPTL3 levels in healthy individuals averaged 225 ng/ml (Stejskal et al. 2007). A significant correlation was found between plasma ANGPTL3 and systolic blood pressure, plasma LDL, and plasma A-FABP. In contrast, no correlation was found between ANGPTL3 and plasma TG. Remarkably, the plasma ANGPTL3 concentration was about 3-fold higher in another study. In that study, a significant positive correlation was found between plasma ANGPTL3 and carotid artery intima–media thickness in healthy human subjects (Hatsuda et al. 2007).

9.5 Conclusion

Overall, the effects of Angptl3 and Angptl4 on plasma lipoprotein metabolism are highly similar, which indicates a common mechanism of action. Evidence abounds that Angptl4 and Angptl3 raise plasma TG levels in mice by suppressing clearance of plasma VLDL and chylomicrons via inhibition of LPL and HL activity. In addition, Angptl3 inhibits endothelial lipase. While these effects are very well supported, the overall role of Angptl4 and Angptl3 in the physiological regulation of fuel selection as well as its impact on cellular metabolism are poorly understood. Furthermore, the physiological determinants of plasma ANGPTL4 and ANGPTL3 levels in humans remain to be identified. Finally, the impact of genetic variants of the human *ANGPTL4* and *ANGPTL3* gene on plasma lipoprotein levels requires further investigation.

References

Akiyama TE, Lambert G, Nicol CJ, Matsusue K, Peters JM, Brewer HB Jr, Gonzalez FJ (2004) Peroxisome proliferator-activated receptor beta/delta regulates very low density lipoprotein production and catabolism in mice on a Western diet. J Biol Chem 279:20874–20881

Ando Y, Shimizugawa T, Takeshita S, Ono M, Shimamura M, Koishi R, Furukawa H (2003) A decreased expression of angiopoietin-like 3 is protective against atherosclerosis in apoE-deficient mice. J Lipid Res 44:1216–1223

Backhed F, Ding H, Wang T, Hooper LV, Koh GY, Nagy A, Semenkovich CF, Gordon JI (2004) The gut microbiota as an environmental factor that regulates fat storage. Proc Natl Acad Sci USA 101:15718–15723

Belanger AJ, Lu H, Date T, Liu LX, Vincent KA, Akita GY, Cheng SH, Gregory RJ, Jiang C (2002) Hypoxia up-regulates expression of peroxisome proliferator-activated receptor gamma angiopoietin-related gene (PGAR) in cardiomyocytes: role of hypoxia inducible factor 1alpha. J Mol Cell Cardiol 34:765–774

Bunger M, van den Bosch HM, van der Meijde J, Kersten S, Hooiveld GJ, Muller M (2007) Genome-wide analysis of PPARalpha activation in murine small intestine. Physiol Genomics 30:192–204

Cazes A, Galaup A, Chomel C, Bignon M, Brechot N, Le Jan S, Weber H, Corvol P, Muller L, Germain S, Monnot C (2006) Extracellular matrix-bound angiopoietin-like 4 inhibits endothelial cell adhesion, migration, and sprouting and alters actin cytoskeleton. Circ Res 99:1207–1215

Conklin D, Gilbertson D, Taft DW, Maurer MF, Whitmore TE, Smith DL, Walker KM, Chen LH, Wattler S, Nehls M, Lewis KB (1999) Identification of a mammalian angiopoietin-related protein expressed specifically in liver. Genomics 62:477–482

Desai U, Lee EC, Chung K, Gao C, Gay J, Key B, Hansen G, Machajewski D, Platt KA, Sands AT, et al (2007) Lipid-lowering effects of anti-angiopoietin-like 4 antibody recapitulate the lipid phenotype found in angiopoietin-like 4 knockout mice. Proc Natl Acad Sci USA 104:11766–11771

Dutton S, Trayhurn P (2008) Regulation of angiopoietin-like protein 4/fasting-induced adipose factor (Angptl4/FIAF) expression in mouse white adipose tissue and 3T3-L1 adipocytes. Br J Nutr 100:18–26

Fugier C, Tousaint JJ, Prieur X, Plateroti M, Samarut J, Delerive P (2006) The lipoprotein lipase inhibitor ANGPTL3 is negatively regulated by thyroid hormone. J Biol Chem 281:11553–11559

Fujimoto K, Koishi R, Shimizugawa T, Ando Y (2006) Angptl3-null mice show low plasma lipid concentrations by enhanced lipoprotein lipase activity. Exp Anim 55:27–34

Galaup A, Cazes A, Le Jan S, Philippe J, Connault E, Le Coz E, Mekid H, Mir LM, Opolon P, Corvol P, et al (2006) Angiopoietin-like 4 prevents metastasis through inhibition of vascular permeability and tumor cell motility and invasiveness. Proc Natl Acad Sci USA 103:18721–18726

Ge H, Yang G, Huang L, Motola DL, Pourbahrami T, Li C (2004a) Oligomerization and regulated proteolytic processing of angiopoietin-like protein 4. J Biol Chem 279:2038–2045

Ge H, Yang G, Yu X, Pourbahrami T, Li C (2004b) Oligomerization state-dependent hyperlipidemic effect of angiopoietin-like protein 4. J Lipid Res 45:2071–2079

Ge H, Cha JY, Gopal H, Harp C, Yu X, Repa JJ, Li C (2005) Differential regulation and properties of angiopoietin-like proteins 3 and 4. J Lipid Res 46:1484–1490

Gustavsson M, Mallard C, Vannucci SJ, Wilson MA, Johnston MV, Hagberg H (2007) Vascular response to hypoxic preconditioning in the immature brain. J Cereb Blood Flow Metab 27:928–938

Hatsuda S, Shoji T, Shinohara K, Kimoto E, Mori K, Fukumoto S, Koyama H, Emoto M, Nishizawa Y (2007) Association between plasma angiopoietin-like protein 3 and arterial wall thickness in healthy subjects. J Vasc Res 44:61–66

Hermann LM, Pinkerton M, Jennings K, Yang L, Grom A, Sowders D, Kersten S, Witte DP, Hirsch R, Thornton S (2005) Angiopoietin-like-4 is a potential angiogenic mediator in arthritis. Clin Immunol 115:93–101

Inaba T, Matsuda M, Shimamura M, Takei N, Terasaka N, Ando Y, Yasumo H, Koishi R, Makishima M, Shimomura I (2003) Angiopoietin-like protein 3 mediates hypertriglyceridemia induced by the liver X receptor. JBiolChem 278:21344–21351

Inukai K, Nakashima Y, Watanabe M, Kurihara S, Awata T, Katagiri H, Oka Y, Katayama S (2004) ANGPTL3 is increased in both insulin-deficient and -resistant diabetic states. Biochem Biophys Res Commun 317:1075–1079

Ito Y, Oike Y, Yasunaga K, Hamada K, Miyata K, Matsumoto S, Sugano S, Tanihara H, Masuho Y, Suda T (2003) Inhibition of angiogenesis and vascular leakiness by angiopoietin-related protein 4. Cancer Res 63:6651–6657

Jin W, Wang X, Millar JS, Quertermous T, Rothblat GH, Glick JM, Rader DJ (2007) Hepatic proprotein convertases modulate HDL metabolism. Cell Metab 6:129–136

Josephs T, Waugh H, Kokay I, Grattan D, Thompson M (2007) Fasting-induced adipose factor identified as a key adipokine that is up-regulated in white adipose tissue during pregnancy and lactation in the rat. J Endocrinol 194:305–312

Kaplan R, Zhang T, Hernandez M, Gan FX, Wright SD, Waters MG, Cai TQ (2003) Regulation of the angiopoietin-like protein 3 gene by LXR. J Lipid Res 44:136–143

Kathiresan S, Melander O, Guiducci C, Surti A, Burtt NP, Rieder MJ, Cooper GM, Roos C, Voight BF, Havulinna AS, et al (2008) Six new loci associated with blood low-density lipoprotein cholesterol, high-density lipoprotein cholesterol or triglycerides in humans. Nat Genet 40:189–197

Kersten S, Mandard S, Tan NS, Escher P, Metzger D, Chambon P, Gonzalez FJ, Desvergne B, Wahli W (2000) Characterization of the fasting-induced adipose factor FIAF, a novel peroxisome proliferator-activated receptor target gene. J Biol Chem 275:28488–28493

Kim I, Kim HG, Kim H, Kim HH, Park SK, Uhm CS, Lee ZH, Koh GY (2000) Hepatic expression, synthesis and secretion of a novel fibrinogen/angiopoietin-related protein that prevents endothelial-cell apoptosis. Biochem J 346:603–610

Koishi R, Ando Y, Ono M, Shimamura M, Yasumo H, Fujiwara T, Horikoshi H, Furukawa H (2002) Angptl3 regulates lipid metabolism in mice. Nat Genet 30:151–157

Koster A, Chao YB, Mosior M, Ford A, Gonzalez-DeWhitt PA, Hale JE, Li D, Qiu Y, Fraser CC, Yang DD, et al (2005) Transgenic angiopoietin-like (angptl)4 overexpression and targeted disruption of angptl4 and angptl3: regulation of triglyceride metabolism. Endocrinology 146:4943–4950

Le Jan S, Amy C, Cazes A, Monnot C, Lamande N, Favier J, Philippe J, Sibony M, Gasc JM, Corvol P, Germain S (2003) Angiopoietin-like 4 is a proangiogenic factor produced during ischemia and in conventional renal cell carcinoma. AmJPathol 162:1521–1528

Lichtenstein L, Berbee JF, van Dijk SJ, van Dijk KW, Bensadoun A, Kema IP, Voshol PJ, Muller M, Rensen PC, Kersten S (2007) Angptl4 upregulates cholesterol synthesis in liver via inhibition of LPL- and HL-dependent hepatic cholesterol uptake. Arterioscler Thromb Vasc Biol 27:2420–2427

Mandard S, Zandbergen F, Tan NS, Escher P, Patsouris D, Koenig W, Kleemann R, Bakker A, Veenman F, Wahli W, et al (2004) The direct peroxisome proliferator-activated receptor target fasting-induced adipose factor (FIAF/PGAR/ANGPTL4) is present in blood plasma as a truncated protein that is increased by fenofibrate treatment. J Biol Chem 279:34411–34420

Mandard S, Zandbergen F, van Straten E, Wahli W, Kuipers F, Muller M, Kersten S (2006) The fasting-induced adipose factor/angiopoietin-like protein 4 is physically associated with lipoproteins and governs plasma lipid levels and adiposity. J Biol Chem 281:934–944

Matsusue K, Miyoshi A, Yamano S, Gonzalez FJ (2006) Ligand-activated PPARbeta efficiently represses the induction of LXR-dependent promoter activity through competition with RXR. Mol Cell Endocrinol 256:23–33

Merkel M, Eckel RH, Goldberg IJ (2002) Lipoprotein lipase: genetics, lipid uptake, and regulation. J Lipid Res 43:1997–2006

Ono M, Shimizugawa T, Shimamura M, Yoshida K, Noji-Sakikawa C, Ando Y, Koishi R, Furukawa H (2003) Protein region important for regulation of lipid metabolism in ANGPTL3: ANGPTL3 is cleaved and activated in vivo. J Biol Chem

Padua D, Zhang XH, Wang Q, Nadal C, Gerald WL, Gomis RR, Massague J (2008) TGFbeta primes breast tumors for lung metastasis seeding through angiopoietin-like 4. Cell 133:66–77

Romeo S, Pennacchio LA, Fu Y, Boerwinkle E, Tybjaerg-Hansen A, Hobbs HH, Cohen JC (2007) Population-based resequencing of ANGPTL4 uncovers variations that reduce triglycerides and increase HDL. Nat Genet 39:513–516

Ruge T, Svensson M, Eriksson JW, Olivecrona G (2005) Tissue-specific regulation of lipoprotein lipase in humans: effects of fasting. Eur J Clin Invest 35:194–200

Schmuth M, Haqq CM, Cairns WJ, Holder JC, Dorsam S, Chang S, Lau P, Fowler AJ, Chuang G, Moser AH, et al (2004) Peroxisome proliferator-activated receptor (PPAR)-beta/delta stimulates differentiation and lipid accumulation in keratinocytes. J Invest Dermatol 122:971–983

Schug TT, Berry DC, Shaw NS, Travis SN, Noy N (2007) Opposing effects of retinoic acid on cell growth result from alternate activation of two different nuclear receptors. Cell 129:723–733

Shimamura M, Matsuda M, Kobayashi S, Ando Y, Ono M, Koishi R, Furukawa H, Makishima M, Shimomura I (2003) Angiopoietin-like protein 3, a hepatic secretory factor, activates lipolysis in adipocytes. Biochem Biophys Res Commun 301:604–609

Shimamura M, Matsuda M, Ando Y, Koishi R, Yasumo H, Furukawa H, Shimomura I (2004) Leptin and insulin down-regulate angiopoietin-like protein 3, a plasma triglyceride-increasing factor. Biochem Biophys Res Commun 322:1080–1085

Shimamura M, Matsuda M, Yasumo H, Okazaki M, Fujimoto K, Kono K, Shimizugawa T, Ando Y, Koishi R, Kohama T, et al (2007) Angiopoietin-like protein3 regulates plasma HDL cholesterol through suppression of endothelial lipase. Arterioscler Thromb Vasc Biol 27:366–372

Shimizugawa T, Ono M, Shimamura M, Yoshida K, Ando Y, Koishi R, Ueda K, Inaba T, Minekura H, Kohama T, Furukawa H (2002) ANGPTL3 decreases very low density lipoprotein triglyceride clearance by inhibition of lipoprotein lipase. J Biol Chem JID - 2985121R 277:33742–33748

Staiger H, Machicao F, Werner R, Guirguis A, Weisser M, Stefan N, Fritsche A, Haring HU (2008) Genetic variation within the ANGPTL4 gene is not associated with metabolic traits in white subjects at an increased risk for type 2 diabetes mellitus. Metabolism 57:637–643

Stejskal D, Karpisek M, Humenanska V, Solichova P, Stejskal P (2007) Angiopoietin-like protein 3: development, analytical characterization, and clinical testing of a new ELISA. Gen Physiol Biophys 26:230–233

Stejskal D, Karpisek M, Reutova H, Humenanska V, Petzel M, Kusnierova P, Vareka I, Varekova R, Stejskal P (2008) Angiopoietin-like protein 4: development, analytical characterization, and clinical testing of a new ELISA. Gen Physiol Biophys 27:59–63

Sukonina V, Lookene A, Olivecrona T, Olivecrona G (2006) Angiopoietin-like protein 4 converts lipoprotein lipase to inactive monomers and modulates lipase activity in adipose tissue. Proc Natl Acad Sci USA 103:17450–17455

Wang B, Wood IS, Trayhurn P (2007) Dysregulation of the expression and secretion of inflammation-related adipokines by hypoxia in human adipocytes. Pflugers Arch 455:479–492

Wiesner G, Morash BA, Ur E, Wilkinson M (2004) Food restriction regulates adipose-specific cytokines in pituitary gland but not in hypothalamus. J Endocrinol 180:R1–R6

Wiesner G, Brown RE, Robertson GS, Imran SA, Ur E, Wilkinson M (2006) Increased expression of the adipokine genes resistin and fasting-induced adipose factor in hypoxic/ischaemic mouse brain. Neuroreport 17:1195–1198

Willer CJ, Sanna S, Jackson AU, Scuteri A, Bonnycastle LL, Clarke R, Heath SC, Timpson NJ, Najjar SS, Stringham HM, et al (2008) Newly identified loci that influence lipid concentrations and risk of coronary artery disease. Nat Genet 40:161–169

Xu A, Lam MC, Chan KW, Wang Y, Zhang J, Hoo RL, Xu JY, Chen B, Chow WS, Tso AW, Lam KS (2005) Angiopoietin-like protein 4 decreases blood glucose and improves glucose tolerance but induces hyperlipidemia and hepatic steatosis in mice. Proc Natl Acad Sci USA 102:6086–6091

Yoon JC, Chickering TW, Rosen ED, Dussault B, Qin Y, Soukas A, Friedman JM, Holmes WE, Spiegelman BM (2000) Peroxisome proliferator-activated receptor gamma target gene encoding a novel angiopoietin-related protein associated with adipose differentiation. Mol Cell Biol 20:5343–5349

Yoshida K, Shimizugawa T, Ono M, Furukawa H (2002) Angiopoietin-like protein 4 is a potent hyperlipidemia-inducing factor in mice and inhibitor of lipoprotein lipase. J Lipid Res 43:1770–1772

Yu X, Burgess SC, Ge H, Wong KK, Nassem RH, Garry DJ, Sherry AD, Malloy CR, Berger JP, Li C (2005) Inhibition of cardiac lipoprotein utilization by transgenic overexpression of Angptl4 in the heart. Proc Natl Acad Sci USA 102:1767–1772

Zandbergen F, van Dijk S, Muller M, Kersten S (2006) Fasting-induced adipose factor/Angiopoietin-like protein 4: a potential target for dyslipidemia. Future Lipidol 1:227–236

Chapter 10
Thyroid Hormones and Lipid Metabolism: Thyromimetics as Anti-Atherosclerotic Agents?

Bernhard Föger, Andreas Wehinger, Josef R. Patsch, Ivan Tancevski, and Andreas Ritsch

B.F. and A.W. contributed mainly to Sect. 10.1 (Thyroid hormones, thyroid hormone-receptors and lipoprotein metabolism): J.R.P., I.T. and A.R. contributed mainly to Sect. 10.2 (Thyromimetics).

Abstract Overt hypothyroidism most often causes moderately elevated cholesterol and triglycerides; but, in genetically predisposed individuals, it can also trigger frank hyperlipidemia, e.g. type III hyperlipoproteinemia. In contrast, in subclinical hypothyroidism, T_4 lowers low-density lipoprotein cholesterol (LDL-C) by only 3.5–6.0 mg/dl LDL-C for each 1 mU/l thyroid-stimulating hormone (TSH) increase from baseline, a finding barely perceptible in individual patients. In humans, T_3 decreases apoB-containing IDL, LDL and chylomicron remnant particles by up-regulation of LDL receptors and despite a concomitant increase in HMGCoA reductase in the liver. In rodents, T3 also up-regulates CYP7A1, ABCG5 and ABCG8 which interact to increase bile acid synthesis and biliary elimination of cholesterol. Thyromimetics, selective for either thyroid hormone-receptor β or liver, lower both LDL-C in humans and HDL-C in mice, thereby promoting reverse cholesterol transport. Interestingly, the thyromimetic T-0681 protects cholesterol-fed rabbits from atherosclerosis. If safety concerns with TRβ-selective thyromimetics can be addressed (i.e. the observed pituitary feed-back inhibition leading to low TSH levels), this class of drugs may complement statins in tackling the burden of atherosclerosis in humans.

B. Föger and A. Wehinger
Department of Internal Medicine, Landeskrankenhaus Bregenz,
Carl-Pedenz-Strasse, 2, 6900 Bregenz, Austria

J.R. Patsch, I. Tancevski, and A. Ritsch
Department of Internal Medicine, Innsbruck Medical University,
Anichstrasse, 35, 6020 Innsbruck, Austria
e-mail: ivan.tancevski@i-med.ac.at

C. Ehnholm (ed.), *Cellular Lipid Metabolism*,
DOI 10.1007/978-3-642-00300-4_10, © Springer-Verlag Berlin Heidelberg 2009

10.1 Thyroid Hormones, Thyroid Hormone-Receptors and Lipoprotein Metabolism

10.1.1 Thyroid Hormone Signalling

Thyroid hormones (THs) play a key role in the regulation of development, growth, and metabolism of practically every cell and organ (for a review, see Oetting and Yen 2007). Two THs, thyroxine (T_4) and triiodothyronine (T_3), are synthesized by the thyroid gland, under the control of the hypothalamic-pituitary axis; and T_4 is the major secreted hormone, leading to a molar ratio of serum T_4:T_3 of ≈40. However, several naturally occuring TH metabolites (e.g. Triac, Tetrac) may retain some biological activity. T_4 binds to circulating TH-binding globuline (TBG), the main transporter in blood, and to prealbumin, albumin and also to lipoproteins, mainly high-density lipoproteins (HDL). In order to produce biological effects, sensitivity of the peripheral effector tissues is modulated at several levels. First, TH transporters located in the plasma membrane serve to control cellular uptake of TH; in the same vein, selective access to the nucleus mediated by the nuclear membrane has been postulated (Ichikawa et al. 2000). Binding of T_4 to low-density lipoprotein (LDL) generates LDL-T_4 complexes which are taken up by the LDL receptor (LDL-R), thereby affording T_4 an alternative mode of entry into cells (Benvenga and Robbins 1990). Second, T_3 (10-fold more potent of the two TH regarding TR-binding) is formed by 5' deiodination of the outer ring of T_4 by deiodinases. Type I deiodinase is present in peripheral tissues, e.g. liver, to convert circulating T_4; whereas type II deiodinase is present in pituitary gland and brain and contributes to both peripheral and intracellular conversion of T_4. Type III deiodinase (present mainly in brain) together with type I deiodinase serves to inactivate T_4 to *reverse* T_3, an enantiomer lacking relevant bioactivity. Third, and most importantly, signal transduction of T_3 is brought about by binding to thyroid hormone receptors (TRs), which act as ligand-regulatable transcription factors (Lazar 1993). TRs bind to DNA as monomers, homodimers, or heterodimers with other receptors from the steroid hormone receptor family, mainly with retinoic-X receptor-α (RXRα). Thus, T_3 binding to TR activates transcription of positively regulated target genes containing a TH response element (TRE), e.g. type I deiodinase. In humans, two genes (TRα, TRβ) are expressed in almost all tissues and encode several major TR isoforms which are formed as the result of alternative splicing: TRα1, TRα2, TRβ1, TRβ2. TRs are ≈500 amino acids long and contain a central DNA-binding domain, a C-terminal ligand-binding domain and a heterodimerization sequence with RXR, as well as interaction sites with co-activators and co-repressors. TRα1, TRβ1 and TRβ2 bind T_3 with similar affinity and display a comparable transcriptional activity. TRα2 does not bind T_3 and may act as a dominant inhibitor of TH action on some target genes. Major progress has been made in the past decade understanding the biological consequences of TR diversity. Expression of TRα and TRβ varies substantially between different tissues. In the heart, TRα makes up 75% of TR, whereas in the liver TRβ comprises 80% of TRs. However, genetic recombination studies in mice with targeted TR mutations have questioned in much greater detail whether TRα or TRβ serve redundant or specific functions. TRα1–/– mice show a normal anatomy and reproduction, however, they show decreased heart rate due to altered cardiac ion channel function, an effect which cannot be corrected by TRβ (Johansson et al. 1998; Gloss et al. 2001). In this context, it is very important to realize that a TR unoccupied by T_3 frequently acts as a dominant repressor of gene function, thus explaining why genetic ablation of a TR often displays a relatively mild phenotype compared to hypothyroidism. For example, mice with a total genetic ablation of all functional TR (TRα1–/– TRβ–/– mice) are viable, but display defects in pituitary–thyroid axis, growth and bone maturation (Göthe et al. 1999). To more fully recapitulate repression by the unliganded receptor, TRα knock-in mice were produced, with various defects in ligand binding. Dominant negative heterozygous TRα1-PV mice are dwarfs and show reduced adipose tissue (Ying et al. 2007); TRα1P398H show obesity combined with impaired catecholamine-stimulated lipolysis (Liu et al. 2003); and TRαR384C/TRβ–/– show central hypermetabolism, leading to resistance to diet-induced obesity, probably overriding direct effects of the mutated TRα in adipose (Sjögren et al. 2007). Synthesizing these intriguing data, TRα is crucial for heart rate, cardiac contractility, cardiac relaxation and adipogenesis in the mouse model. Human mutations in TRα have not been reported up to now. In contrast, more than 300 families with the syndrome of resistance to thyroid hormone have been found which are due to mutations in the TRβ gene disturbing T_3 binding. Studies of genetic ablation of TRβ, studies of knock-in mice for TRβ and experiments with TRβ-selective analogs have conclusively indicated that metabolic effects of TH on cholesterol and lipoprotein metabolism are predominantly mediated by TRβ. Many of the target genes are directly transcriptionally regulated by TRs, whereas others are indirectly transcriptionally regulated through intermediary genes, e.g. transcription factors. TRs can also regulate transcription through protein-protein interactions with other transcription-factors, alter mRNA levels by changing mRNA stability and directly induce rapid effects by engaging other cellular proteins, e.g. PI3-kinase (Storey et al. 2006).

10.1.2 Thyroid Function and Lipoprotein Metabolism

10.1.2.1 Overt Hypothyroidism and Lipoproteins

Overt hypothyroidism (OH) is both a relatively frequent condition (0.3% of the population) and a major cause of atherosclerotic disease, as evidenced both by current clinical studies (Nyirenda et al. 2005) and by autopsy studies performed already in the 1930s (for a review, see Cappola and Ladenson 2003). OH, due to primary thyroid disease, is defined as elevated TSH levels in the presence of decreased free T_4 and/or free T_3 values. Mechanistically, atherosclerosis this is caused by a combination of at least four factors: dyslipidemia (the topic of this Section), adverse changes in "novel" cardiovascular risk factors, increased peripheral vascular resistance with ensuing diastolic hypertension and direct adverse effects of T_3 deficiency on heart and vasculature, leading e.g. to diastolic dysfunction (Klein and Ojamaa 2001). Overt hypothyroidism is probably the most important cause of secondary hyperlipidemia and has to be ruled out in any patient evaluated at the outpatient lipid clinic (reviewed by Duntas 2002; Pearce 2004). Depending on the variation in lipid transport genes in individual patients, OH can present with either a moderately disturbed lipoprotein profile or, more rarely, with a frank hyperlipidemia of (respectively) the lipoprotein pattern type IIa, type IIb, type III, type IV or type V, according to Fredrickson. In other words, total cholesterol (TC) and/or triglycerides (TG) most often are moderately elevated, but can also be normal or dramatically increased. High-density lipoprotein cholesterol (HDL-C) often is moderately elevated, but can also be normal or low. Thus, the effect of OH on lipoproteins can not only induce atherosclerosis, but also the chylomicronemia syndrome, a life-threatening disorder with excessive TG-accumulation which may cause acute pancreatitis and/or multiple organ failure. All the above abnormalities are at least partly reversible with T_4 replacement, which in the setting of severe OH has a pronounced hypocholesterolemic effect of ≈100 mg/dl, similar in magnitude to standard doses of a statin. Although the reduction in LDL-C induced by T_4 in OH is quite variable, currently there are no genetic markers to predict this change (Wiseman et al. 1993; Diekman et al. 2000).

It is critical to initially treat the hyperlipidemia of OH with T_4 rather than statins not only because direct T_3 deficiency in numerous organs needs to be replaced, but also because OH leads to both myopathy and reduced statin clearance in the liver, two factors explaining the increased risk of rhabdomyolysis during statin treatment in the presence of OH. Usually, 4–6 weeks of maintenance therapy with an optimal dose of T_4 are necessary to fully correct the lipoprotein disorder due to OH. Combined treatment with T_3 and T_4 offers no clear advantages and is not advisable. It should be kept in mind, however, that T_4 replacement in OH also increases exercise, decreases food intake and, thus, induces weight loss of several kilograms, factors which are rarely controlled in clinical studies, but clearly have the potential to improve plasma lipids.

10.1.2.1.1 ApoB-Containing Lipoproteins in Overt Hypothyroidism

10.1.2.1.1.1 Lipids of apoB-Containing Lipoproteins in Overt Hypothyroidism

The typical lipid profile in pronounced OH [thyroid-stimulating hormone (TSH) >50 mU/l] is moderately disturbed: TC is clearly increased (ca. +35%), mainly due to increases in LDL-C (ca. +40%), intermediate-density lipoprotein cholesterol (IDL-C; ca. +100%) and HDL-C (ca. +25%), whereas VLDL-C is unchanged. Thus, in absolute terms, LDL-C is increased most (+80 mg/dl), followed by IDL-C (+10 mg/dl) and HDL_2-C (+10 mg/dl; Packard et al. 1993). However, normal or low HDL-C levels have also been reported (Agdeppa et al. 1979). Fasting TG tends to increase or shows moderate increases in OH (+20%). The postprandial TG response to a fatty meal, i.e. postprandial lipemia, is increased even in normotriglyceridemic OH (Weintraub et al. 1999). Interestingly, VLDL-TG, the major TG-containing lipoprotein in fasting plasma, is unchanged; and increases in TG in OH pertain mainly to LDL-TG and HDL-TG, suggesting increased core lipid exchange despite low CETP activity (Packard et al. 1993; Tan et al. 1998a, b). In line with these findings, the total mass (= lipid core + lipid surface components + apoproteins) of large $VLDL_1$, smaller $VLDL_2$ and HDL_3 are normal in OH, whereas IDL-, LDL- and HDL_2 mass are increased (Packard et al. 1993). Recently, NMR spectroscopy data complemented our understanding of lipid changes in OH (Pearce et al. 2008). The content of large VLDL particles was normal, but medium and small VLDL as well as IDL particles were increased two-fold. In line with these observations, remnant-like particles in fasting plasma, as defined by apolipoprotein composition (presumably overlapping with IDL), were increased in OH by 35% (Ito et al. 2007). Postprandial chylomicron remnant clearance, as measured by retinyl palmitate labelling, was delayed in OH (Weintraub et al. 1999). Importantly, only large LDL particles were increased by ≈100%, whereas medium and small LDL were normal (Pearce et al. 2008). As small LDL are considered to be especially atherogenic due to their propensity to be oxidized and to filter into the vessel wall, this may indicate that the increase in LDL-C in OH is confined to less dangerous LDL particle subsets (Pearce et al. 2008). It is not surprising that oxidized LDL is increased too in OH (Duntas et al. 2002). This can be explained both by an increased half-time of LDL in the circulation in OH and, alternatively, by the decrease in circulating LDL-bound T_4 itself, which may directly protect LDL from oxidation (Hanna et al. 1993).

10.1.2.1.1.2 Apolipoproteins of apoB-Containing Lipoproteins in Overt Hypothyroidism

Each VLDL-, IDL- and LDL-particle contains one molecule of apoB. **apoB** is consistently increased in OH, indicating an increased number of potentially pro-atherogenic particles (Tan et al. 1998a, b). **apoE** is also increased in OH, presumably as a component of smaller VLDL and IDL particles (Muls et al. 1985). apoB metabolism has been examined in some detail in OH (Packard et al. 1993). As expected from the lipid data above, VLDL-apoB is unchanged in OH, whereas IDL-apoB and LDL-apoB are increased. While VLDL-apoB kinetics were unchanged, the increased FCR of IDL-apoB was explained by an increased conversion of IDL-apoB to LDL-apoB, whereas direct liver uptake of IDL-apoB was unaltered. During OH, LDL-apoB removal was substantially delayed, explaining its increased plasma levels (Thompson et al. 1981), presumably due to down-regulation of LDL receptors.

Lipoprotein(a), Lp(a), is an LDL particle in which apoB is attached to a large plasminogen-like protein called apolipoprotein(a). Lp(a) values >30 mg/dl are strongly genetically determined, are widely regarded as a "novel" risk factor for atherosclerosis and respond poorly to treatment. Some reports suggest that Lp(a) decreases after T_4 replacement (de Bruin et al. 1993). However, currently, most papers report no change (for a review, see Duntas 2003).

10.1.2.1.2 HDL in Overt Hypothyroidism

10.1.2.1.2.1 HDL Lipids in Overt Hypothyroidism

In OH, HDL-C has most often been found to be high (Muls et al. 1984, 1985; Kuusi et al. 1988; Tan et al. 1998a, b; Pearce et al. 2008), however some reports also note normal values (Verdugo et al. 1987) or decreases (Agdeppa et al. 1979; Lithell et al. 1981). HDL_2-C is almost doubled compared to controls, whereas HDL_3-C is somewhat decreased (Tan et al. 1998a, b). Isopycnic ultracentrifugation confirmed a modified distribution of HDL subclasses in OH, with an increase in the lighter (and thus larger) HDL_{2b} particles ($d = 1.063–1.100$ g/ml; Muls et al. 1985). This has also been corroborated by NMR studies, which find large HDL to be increased in OH, whereas medium and small HDL are significantly decreased (Pearce et al. 2008).

10.1.2.1.2.2 HDL Apolipoproteins in Overt Hypothyroidism

In OH, apoA-I is increased, whereas apoA-II, the majority of which is associated with HDL_3, is unchanged (Verdugo et al. 1987). Correspondingly, HDL particles containing only apoA-I (LpA-I) are increased, whereas LpA-I:A-II are not (Tan et al. 1998a, b). Taken together, apoA-I levels during T_4 replacement are probably determined both by changes in production and catabolism, the latter in particular being responsive to the amount of TG-lowering afforded by T_4.

10.1.2.2 Subclinical Hypothyroidism and Lipoproteins

Subclinical hypothyroidism (SCH) is more than 10 times as prevalent as OH (4% vs 0.3% of the population; for a review, see Biondi and Cooper 2008). SCH is even more prevalent in women, the elderly and in populations with a high iodine intake. The etiology of SCH is similar to that of overt hypothyroidism. The main causes are autoimmune thyroiditis, other forms of thyroiditis, thyroid injury after thyroid resection, external radiotherapy, radioiodine-therapy or drugs impairing thyroid function, e.g. amiodarone, which act as a competitive antagonist of T_3 binding to TRβ, thereby not only explaining its antiarrhythmic property but also its ability to raise cholesterol (Bakker et al. 1994, 1998).

SCH is defined as the repeated measurement of increased TSH serum levels in the presence of normal free T_4 and free T_3 levels (Biondi and Cooper 2008). However, it is difficult to establish consensus as to what constitutes a normal TSH range because TSH levels in the population show no Gaussian distribution. The "upper tail" of TSH levels has been attributed either to occult autoimmunity against the thyroid or to the presence of TSH isoforms, e.g. due to variations in glycosylation. Thus, TSH cutoff values between 2.5 mU/l and 4.0 mU/l have been defined as the upper limit of normal. TSH values >2.5–4.0 mU/l in the presence of normal TH values correspond to mild SCH. TSH values >10 mU/l in the presence of normal TH values characterize a subgroup of patients with severe SCH (Biondi and Cooper 2008).

Whether SCH causes clinical symptoms of hypothyroidism (e.g. cold intolerance) is currently under debate. Of particular importance in the current context is the discussion whether or not SCH is risk factor for atherosclerosis and, if so, whether derangements in lipids are responsible. With regard to lipids, a large body of literature has examined mainly two questions (Biondi and Cooper 2008):

1. Do patients with SCH have increases in total and LDL-C?
2. Does replacement therapy with T_4 lower total and LDL-C in SCH patients?

10.1.2.2.1 Subclinical Hypothyroidism and Lipoproteins in Cross-Sectional Studies

Several large population-based studies have hitherto been unable to unequivocally establish increases in LDL-C in SCH especially when controlling for demographic variables and BMI. Some studies were negative, some showed even lower TC in SCH and some found the expected increase in TC, LDL-C and, more rarely, TG. HDL tends to be lower in SCH.

For example, in the study by Kanaya et al. (2002), in 2799 adults a TSH >5.5 mU/l was associated with an increase in TC of 10 mg/dl. In the study by Bauer et al. (1998), corrected LDL-C was 12% higher and HDL-C was 12% higher in women with TSH >5.5 mU/l. In the fifth Tromso study, using a nested case-control design, TC and LDL-C were higher and apoA-I was significantly lower than in the controls. Taken together, it remains biologically plausible that SCH leads to small increases in LDL-C, especially if there is more profound TSH elevation (>10 mU/l), and if background TC and BMI levels in the population are elevated. Heterogeneity regarding etiology, as well as extent and duration of disease may partly explain the variable results. The fact that the extrapolated increases in TC (3.5 mg/dl in men and 6.2 mg/dl in women for every 1 mU/l elevation in TSH) are quite small may help to explain why the noise of epidemiological, biological and analytical variation in LDL-C may have made it so difficult to clearly establish this link. However, if the above increases in LDL-C are correct, they alone may correspond to increases in 18–30% of CHD risk even within the cohort of mild SCH.

10.1.2.2.2 Lipoproteins in Subclinical Hypothyroidism after T_4 Replacement

A related, but actually quite different question is whether or not replacement of T_4 in SCH induces changes in lipoprotein metabolism. Even though the literature is also somewhat divided on this issue, the most probable answer is that T_4 reduces LDL-C (and with it TC) without changing HDL-C or Lp(a). Two meta-analyses summarize the individual studies. Tanis et al. (1996) analyzed 13 studies from 1976 to 1995 comprising 278 patients and found that T_4 decreased TC by 15 mg/dl. Danese et al. (2000) analyzed 13 studies from January 1966 to May 1999 comprising 247 patients and found that T_4 decreased TC by 8 mg/dl and LDL-C by 10 mg/dl with no changes in TG or HDL-C. Recently, three major RCTs resolved to some degree the above issue (Caraccio et al. 2002; Monzani et al. 2004; Razvi et al. 2007). LDL-C was reduced by –27.4%, –30.0% and –7.3%, without changes in HDL-C or Lp(a). Expectedly, apoB is also decreased by ≈16% (Arem and Patsch 1990). Importantly, one study also demonstrated that T_4 replacement reduced intima-media thickness by 10% after 6 months and that this decrease was directly related to improvements in lipids (Monzani et al. 2004). Taken together, there is substantial evidence for clinically relevant decreases in LDL-C resulting from T_4 replacement in SCH, a finding which may be especially relevant in SCH with TSH >10 mU/l, hypercholesterolemia and/or atherosclerotic vascular disease.

10.1.2.3 Overt Hyperthyroidism and Lipoproteins

In overt hyperthyroidism (OHyper), TC is decreased due to both decreases in LDL-C (by ≈50%) and HDL-C (by ≈35%); apoB and apoA-I are decreased in parallel (Tan et al. 1998a, b). Regarding HDL, mainly HDL_2 and LpA-I are decreased, whereas HDL_3 and LpA-I/A-II remain normal (Tan et al. 1998a, b). In OHyper, TG have been reported to be low (Abrams and Grundy 1981), normal (Tan et al. 1998a, b), or even high (Cachefo et al. 2001). High TG have been explained by increased TG hydrolysis in adipose, high FFA levels in plasma, an ensuing increase in hepatic lipogenesis (Cachefo et al. 2001), presumably leading to increased VLDL secretion.

10.1.2.4 Subclinical Hyperthyroidism and Lipoproteins

Subclinical hyperthyroidism (SCHyper) is a prevalent disorder affecting 0.7–3.2% of the population. Up to 10–30% of patients receiving replacement doses of T_4 for OH have exogenic SCHyper. Replacement doses of T_4 leading to supressed TSH, i.e. SCHyper, lower LDL-C more potently than doses resulting in normal TSH (Franklyn et al. 1993). Endogenic SCHyper significantly decreases TC by 12%, due to decreases of both LDL-C and HDL-C (both NS; Parle et al. 1992).

10.1.2.5 Thyroid Dysfunction in Rodents

Most of the animal data on thyroid dysfunction are derived from rodents. However, lipoprotein metabolism in rodents differs from humans in a number of aspects. In rodents on chow, cholesterol is mainly transported with HDL, with very little VLDL-C and LDL-C. Furthermore, rodents have extensive apoB mRNA-editing in the liver (Davidson 1988), monomeric apoA-II in plasma, hepatic lipase that circulates in plasma rather than binding to glucosaminoglycans in the liver; and, finally, they lack CETP. These changes contribute to make rodents notoriously resistant to atherosclerosis. Thus, extrapolation of lipoprotein changes in rodents with thyroid dysfunction to humans should be done with caution.

10.1.2.5.1 Hypothyroidism in Rodents

Hypothyroidism in **rats** increases TC, especially on a cholesterol-rich diet, due to the accumulation of apoE-rich large HDL, β-VLDL and LDL particles. In contrast, TG in hypothyroid rats are normal (Staels et al. 1990) or even decreased (Apostopoulos et al. 1990; Salter et al. 1991). In hypothyroid rats, **apoA-I** levels are unchanged, and apoB and apoE levels in plasma are moderately increased. Normal apoA-I levels result from a combination of normal apoA-I mRNA (Staels et al. 1990), a strong decrease in hepatic apoA-I secretion due to post-transcriptional effects of hypothyroidism (Wilcox et al. 1991) and, in turn, a reduced clearance of radiolabeled HDL proteins (Gross et al. 1987). Increased **apoB** levels in plasma are readily explained by increases in apoB mRNA in liver and intestine (Staels et al. 1990) and, also, by strongly decreased LDL-R mRNA, leading to a decreased FCR of LDL-apoB in hypothyroid rats (Gross et al. 1987). Hepatic **apoE** mRNA is unchanged; however, hepatic apoE synthesis is increased due to post-transcriptional effects (Davidson et al. 1988) which may, together with decreased HL and a strongly decreased LDL-R, explain the increased plasma levels of apoE in hypothyroid rats. Presumably, apoE in this model is associated primarily with IDL and large apoE-rich HDL.

In euthyroid **mice** fed normal chow or a cholesterol-rich diet, gene targeting of either TRβ or TRα does not substantially change serum cholesterol (Gullberg et al. 2000). No good data on apolipoproteins in TH-resistant mice are available.

10.1.2.5.2 Hyperthyroidism in Rodents

In **rats**, short-term (6–20 days) hyperthyroidism induced by T_3 injections does not change plasma TC levels. TG range from normal (Staels et al. 1990) to increases by 10-fold (Apostopoulos et al. 1990), presumably dependent on the degree of hyperthyroidism.

In euthyroid **mice** on chow, 14 days of T_3 injection resulting in severe hyperthyroidism reduces TC by 40%, whereas TG are unchanged (Tancevski et al. 2008). Reductions in TC are almost exclusively due to decreased HDL-C, with little apparent changes in VLDL and LDL. Metabolic studies show that HDL-cholesteryl ether clearance is unchanged, however, indicating that low HDL-C in hyperthyroid mice reflects decreased HDL-cholesteryl ester synthesis. In line with changes in HDL, serum apoA-I levels decrease by 50% (Tancevski et al. 2008).

Regulation of **apoA-I** by TH in rodents has been examined in some detail (Apostopoulos et al. 1990; Staels et al. 1990; Wilcox et al. 1991; Strobl et al. 1992; Lin-Lee et al. 1993, 1995; Soyal et al. 1995). Short-term treatment of euthyroid rats with T_3 increases transcription of apoA-I and apoA-II; however, the stimulation of apoA-I persists longer (Strobl et al. 1992). Chronic hyperthyroidism in rats increases the plasma levels of apoA-I by 2 to 3-fold, due to increased synthesis and secretion of apoA-I protein, as shown in perfused rat livers (Apostopolulos et al. 1990; Wilcox et al. 1991). Increased apoA-I is due to increases in apoA-I mRNA in the liver but not the intestine, which occur despite decreased transcription of apoA-I in chronic hyperthyroidism, indicating a post-transcriptional mechanism stabilizing apoA-I mRNA (Apostopoulos et al. 1990; Staels et al. 1990; Strobl et al. 1992).

Further analysis of the mechanism underlying decreased transcription of apoA-I in hyperthyroidism revealed that transcription initiation was similar to control rats, but that T_3-treated rats had a much higher rate of transcriptional arrest, explaining decreased overall transcription rates (Lin-Lee et al. 1995). Contradictory data have been reported regarding TM and apoA-I. In euthyroid rats, CGS 23425 afforded substantial increases in serum apoA-I (Taylor et al. 1997). In euthyroid mice on chow, in contrast, T-0681 decreased apoA-I (Tancevski et al. 2008).

Interestingly, **apoA-II**, the evolutionally related gene coding for the other major HDL structural protein, is clearly regulated differentially from apoA-I. In euthyroid rats, T_3 acutely increases apoA-II transcription (Strobl et al. 1992). In chronic hyperthyroidism, however, both apoA-II transcription and apoA-II mRNA levels decrease substantially (Staels et al. 1990; Strobl et al. 1992). Whether plasma levels of apoA-II are changed is unknown.

Acutely, T_3 increases **apoA-IV** transcription in euthyroid rat liver (Lin-Lee et al. 1993). Chronic hyperthyroidism decreases apoA-IV transcription by 55%, but increases apoA-IV mRNA 2.8-fold, respectively, indicating stabilization of apoA-IV mRNA in the cytoplasm which mechanistically differs from T_3 effects on apoA-I mRNA (Lin-Lee et al. 1993). Plasma levels of apoA-IV appear to be unchanged (Apostopoulos et al. 1990).

Short-term administration of T_3 increases **apoC-III** transcription in euthyroid rat liver (Lin-Lee et al. 1993). Chronic hyperthyroidism decreases apoC-III transcription by 72%, but decreases in apoC-III mRNA were less pronounced presumably due to more efficient maturation of apoC-III mRNA (Lin-Lee et al. 1993).

Hyperthyroidism does not change **apoB** and **apoE** mRNA (Staels et al. 1990), but increases transcription of LDL-R, which probably explains the substantial decrease in apoB and apoE plasma levels. However, T_3 enhances apoB mRNA-editing in rat liver, leading respectively to more hepatic production of apoB48 and less of apoB100 (Davidson et al. 1988).

Taken together, both transcriptional and post-transcriptional mechanisms make hepatic apolipoprotein synthesis (apoA-I, apoA-II, apoA-IV, apoC-III, apo-B) exquisitely sensitive to altered thyroid status. However, extrapolation to the human situation has to be done with caution.

10.1.2.6 Mechanistic Insight from Humans and Rodents

10.1.2.6.1 Factors Mainly Affecting apoB-Containing Lipoproteins

10.1.2.6.1.1 Cholesterol-7a Hydroxylase

Rodents, in contrast to man, are quite resistant to the development of hypercholesterolemia on a cholesterol-rich diet, at least in part because they are able to increase the formation and secretion of bile acids leading to increased removal of cholesterol from the body (Björkhem et al. 1997). However, both rats and mice accumulate lipoproteins in plasma when made hypothyroid, thus becoming susceptible to atherosclerosis. Because cholesterol-7a-hydroxylase (CYP7A1) is the rate-limiting enzyme for the conversion of cholesterol to bile acids in the liver via the "classical pathway", its regulation by TH has been thoroughly examined in rodents and man. Transcriptional, post-transcriptional and post-translational regulation of CYP7A1 activity has been documented (Pandak et al. 1997; Drover and Agellon 2004; Shin and Osborne 2006). However, transcriptional regulation appears to be the most important mechanism.

Early studies in hypophysectomized **rats** indicated that hepatic CYP7A1 mRNA and –activity respond rapidly to very low doses of T_3 reaching a level of 8-fold greater than control within 6 h (Ness et al. 1990). In these rats stimulation of biliary cholesterol secretion due to increased CYP7A1 preceded an increase in HMG-CoA reductase by 12 h, establishing that this effect is independent of the stimulation of cholesterol synthesis by TH (Day et al. 1989). T_3 induces CYP7A1 in hypophysectomized rats at the transcriptional level (Ness et al. 1990; Pandak et al. 1997).

In hypothyroid wild-type **mice**, T_3 substitution reduces cholesterol both on chow and on a cholesterol-rich diet, whereas no such effect of T_3 is seen in TRβ–/– mice, indicating that increased CYP7A1 transcription is unique to TRβ (Gullberg et al. 2000). Overexpression of TRα1 in the liver cannot overcome this defect (Gullberg et al. 2002). Paradoxically, however, T_3-deficient TRβ–/– mice show **augmented** CYP7A1 mRNA on a cholesterol-rich diet when compared to the wild type, making them **resistant** to dietary hypercholesterolemia despite overt hypothyroidism (Gullberg et al. 2000). This situation can presumably be explained by two mechanisms:

1. Unliganded TRβ potently suppresses CYP7A1 in hypothyroid wild-type mice, whereas in –/– mice this inhibition is relieved due to the absence of TRβ.
2. LXRα, an oxysterol-binding transcription factor directly activating the rodent CYP7A1 promotor is no longer competed by TRβ and gains the upper hand (Gullberg et al. 2000).

Regulation of CYP7A1 has also been analyzed in a model of human resistance to TH (Hashimoto et al. 2006a, b). On chow, CYP7A1 mRNA was undetectable in the livers of wild-type and mutant mice. T_3 substantially induced CYP7A1 in the wild type but, as expected with a ligand-binding defective TRβ, only marginally (<30%) in mutant mice. Paradoxically, on a high cholesterol diet, CYP7A1 mRNA rose to 3-fold the wild-type levels in mutant mice, presumably due to a lack of competition of the mutant TRβ with LXR-mediated promotor stimulation in this particular situation. Thus, cross-talk between TRβ and LXRα in the regulation of CYP7A1 during a hypercholesterolemic diet has also been reported in a TRβ knock-in mouse model (Hashimoto et al. 2006a, b). CYP7A1 gene transcription is inhibited by FXR by increasing the expression of small heterodimer partner (SHP), a non-DNA binding protein (Davis et al. 2002). Thus, in rodents, CYP7A1 gene transcription is, respectively, inhibited by FXR and unliganded TRβ and stimulated by liganded TRβ and liganded LXRα.

It has to be emphasized, however, that regulation of CYP7A1 regarding both TR and LXR is highly species-specific. Investigating the **human** CYP7A1 gene in cell culture, Drover et al. (2002) found that both TRα and TRβ bind to two regions in the proximal promoter. Upon addition of T_3 to cells, human CYP7A1 promotor activity is decreased by ≈50% (Drover et al. 2002). The study of CYP7A1 regulation by T_3 in humans in vivo has been hampered by the invasive nature of obtaining repeated liver samples. To overcome this limitation, transgenic mice were generated which expressed the human CYP7A1 gene under the control of its natural flanking sequences, but not the mouse CYP7A1 gene (Drover and Agellon 2004). Hypothyroidism did not alter human CYP7A1 mRNA levels in transgenic mice. In hyperthyroid male mice, expectedly, human CYP7A1 mRNA decreased; however, in hyperthyroid female mice human CYP7A1 mRNA remained unchanged (Drover and Agellon 2004). Thus, T_3 represses human CYP7A1 transcription in transgenic mice in a gender-dependent manner. Interestingly, however, despite lower CYP7A1 mRNA levels, CYP7A1 activity in liver microsomes was increased in hyperthyroid male transgenic mice, but not in female mice ≈2-fold. This finding suggests a major role of post-transcriptional mechanisms in the regulation of human CYP7A1 activity.

Regulation of CYP7A1 by LXRα also differs substantially between species. Whereas the rat/mouse CYP7A1 promotor contains LXR-binding sites that increase transcription in response to the generation of oxysterols induced by a cholesterol-rich diet, functionally active LXR-sites are absent in the human CYP7A1 (Chen et al. 2002; Menke et al. 2002). This, in part, explains why rodents are resistant and humans susceptible to diet-induced hypercholesterolemia.

Different thyromimetics (TM) stimulate bile acid synthesis, as assessed by monitoring serum levels of C4, an intermediary in bile acid synthesis, both in euthyroid mice (Johansson et al. 2005) and in euthyroid humans (Berkenstam et al. 2008). These data are corroborated by findings with the liver-specific TM T-0681, which increases liver CYP7A1 mRNA levels in mice (Tancevski et al. 2008) and by reduction of SHP mRNA levels in mice by ≈50%, which would be expected to deinhibit CYP7A1 expression (Davis et al. 2002). In humans with OHyper, cholic acid synthesis appears to be reduced (Pauletzki et al. 1989). Thus liver and TRβ-specific TM appear to stimulate bile acid synthesis both in rodents and in humans. However, robust data regarding the situation in humans are currently not available.

10.1.2.6.1.2 LDL Receptor

In humans, elegant LDL turnover studies have demonstrated quite early that LDL-R dependent-removal of LDL from the circulation is massively disturbed in OH; upon T_4 replacement LDL FCR in vivo increases 3.7-fold (Thompson et al. 1981). Chait et al. found that T_3 increases ^{131}I-LDL binding to human skin fibroblasts indicating upregulation of LDL-Rs (Chait et al. 1979). Fibroblasts that either lack the LDL-R or are from patients with the syndrome of resistance to TH do not respond to T_3 by taking up more LDL, indicating that both LDL-R and TRβ are indispensible for this effect. In hypophysectomized rats, hepatic LDL-R mRNA increased 20-fold 24 h after injection of T_3 (Ness et al. 1990); however, decreased growth hormone (GH) and cortisol levels may contribute to low baseline expression in this model. Similarly, in rats made hypothyroid by feeding propylthiouracil (PTU), hepatic mRNA for the LDL-R was significantly decreased (Staels et al. 1990; Salter et al. 1991), while there was no change in intestinal mRNA. In vivo, receptor-dependent clearance of ^{131}I-labelled LDL was decreased by 40% in hypothyroid rats (Gross et al. 1987). Conversely, hyperthyroid rats showed increased LDL-R mRNA levels (Staels et al. 1990). Co-transfection experiments of the human LDL-R promoter and TRβ1 in HepG2cells indicated a functional TRE in the proximal promoter and showed that constitutive activity is suppressed by unliganded TRβ1, while T_3 increases promoter activity up to 6-fold (Bakker et al. 1998). While the above data strongly support direct stimulation of LDL-R transcription by T_3, an indirect stimulation of transcription by T_3-induced increases in SREBP-2 was recently uncovered (Shin and Osborne 2003).

The contribution of TRs to the induction of LDL-R mRNA was clarified by Gullberg et al. (2000). T_3 increased LDL-R mRNA in wild-type mice; and both TRβ–/– and TRα–/– mice had higher mRNA levels than hypothyroid, but failed to increase LDL-R mRNA during T_3 substitution (Gullberg et al. 2000). Similar data were obtained in a TRβ knock-in model (Hashimoto et al. 2006a, b). Thus, stimulation of LDL-R transcription in vivo in mice cannot be attributed unequivocally to TRβ or TRα.

The liver-specific TM CGS23425 induced a 44% increase in ^{125}I-LDL binding to HepG2 cells, indicating an up-regulation of LDL-R number, an effect comparable to equimolar doses of T_3 (Taylor et al. 1997). Similarly, the liver-specific TM T-0681 increased LDL-R protein levels by 2.5-fold in rabbit liver (Heimberg et al. 1985; Tancevski et al. 2008).

Taken together, down-regulation of the LDL-R constitutes a major mechanism underlying the hyperlipidemia in OH. T_3 upregulates LDL-R transcription, directly by binding to a TRE and indirectly both by induction of SREBP-2 and, at least in rodents, by stimulation of bile acid synthesis, thereby lowering hepatic cholesterol levels.

10.1.2.6.1.3 HMGCoA Reductase

TH stimulate cholesterol synthesis in man, an effect alleviating both hypercholesterolemia in OH and hypocholesterolemia in OHyper (Abrams and Grundy 1981). HMGCoA reductase, the rate-limiting enzyme in cholesterol biosynthesis, is expressed in all tissues; however, contribution by the liver is crucial for whole body cholesterol homeostasis. If cholesterol accumulates in hepatocytes, HMGCoA reductase is transcriptionally down-regulated because the release of stimulatory membrane-bound transcription factors, i.e. SREBPs is blocked.

HMGCoA reductase is sensitive to T_3. In hypophysectomized **rats**, HMGCoA reductase mRNA, HMGCoA protein and HMGCoA activity are up-regulated after supplementation with T_3 (Ness et al. 1990). In rat liver nuclei, T_3 both increases

HMGCoA reductase transcription and stabilizes HMGCoA reductase mRNA (Simonet and Ness 1988). Studies in **mice** revealed that T_3 increases HMGCoA reductase mRNA only in the wild type, but not in TRβ–/– mice (Gullberg et al. 2000). In hypothyroid wild-type mice, expectedly, hepatic HMGCoA reductase mRNA was increased ≈10-fold by T_3 replacement on chow and on a high-cholesterol diet, whereas in TRβ knock-in mice the increase by T_3 was substantially blunted (Hashimoto et al. 2006a, b). Unexpectedly, in hypothyroid TRβ knock-in mice HMGCoA reductase mRNA was increased both on chow and on the atherogenic diet, suggesting activation of expression by unliganded TRβ (Hashimoto et al. 2006a, b). These findings support a critical role for TRβ in the regulation of cholesterol synthesis. Interestingly, TM GC-1 failed to upregulate HMGCoA reductase in euthyroid mice highlighting that effects of T_3 replacement and oversubstitution with TH may differ (Johansson et al. 2005).

Presumably, transcriptional, post-transcriptional and indirect mechanisms and indirect mechanisms (e.g. stimulation of SREBP-2) contribute to stimulation of HMGCoA reductase gene expression by T_3 in rodents in a TRβ-dependent fashion.

10.1.2.6.1.4 SREBP-1c, SREBP-2, LXRα and PPARγ

SREBP-1c preferentially increases the transcription of hepatic genes required for fatty acid synthesis. Recent reports indicate that T_3 lowers SREBP-1c mRNA levels in mouse liver, presumably by suppressing mouse SREBP-1c promoter activity in a mouse hepatocyte cell line in a TRβ-dependent manner (Hashimoto et al. 2006a, b). Because species differences in gene regulation by T_3 are frequent, the authors repeated these experiments examining human SREBP-1c promoter activity in HepG2 cells and obtained the same result (Hashimoto et al. 2006a, b). However, stimulation of human SREBP-1c mRNA by T_3 has also been reported (Kawai et al. 2004).

SREBP-2 preferentially increases the transcription of genes required for the maintenance of hepatocyte cholesterol levels, e.g. LDL-R and HMGCoA reductase. Recent reports indicate that in hypothyroid mice, decreased SREBP-2 mRNA and decreased nuclear protein levels of SREBP are present (Shin and Osborne 2003). Interestingly, the proximal promoter of SREBP-2 contains at least one TRE and is activated directly by heterodimers of TRα or TRβ with RXR in a T_3-dependent manner (Shin and Osborne 2003). This may in part explain down-regulation of LDL-R and low cholesterol synthesis in OH.

LXR-α is abundant in liver, whereas LXR-β is ubiquitously expressed. Although LXR and TR are structurally distinct and bind different ligands, they share some similarities, e.g. heterodimerization with RXR and a physiological role in cholesterol homeostasis. Both LXR and TR appear to regulate a series of target genes promoting reverse cholesterol transport (RCT; Naik et al. 2006), e.g. ABCA1, ABCG1, ABCG5/8, SREBP-1c, CETP and, in mice, CYP7A1. In this context, it is of interest that T_3 up-regulates LXRα, but not LXRβ mRNA expression in mouse liver by TR-β1 mediated binding to the –1240/+30 proximal promoter (Hashimoto et al.

2007). As a possible counterbalance to indirect activation of SREBP-1c by LXRα, T_3 directly down-regulates SREBP-1c mRNA expression in mice, which has been interpreted as fine tuning of this pathway. Thus, while enhancement of several facets of RCT due to TH is well documented, the relative contribution of direct TH-signalling as opposed to indirect signalling via LXR-α still needs to be resolved.

Cross-talk between TR and **PPARγ** regarding differentiation of adipose tissue has been reported (Ying et al. 2007).

10.1.2.6.1.5 ABCG5, ABCG8 and NPC1L1

ABCG5 and **ABCG8** are expressed in liver and intestine, where upon dimerization, they enhance biliary cholesterol secretion and, possibly, direct intestinal cholesterol loss, respectively (van der Velde et al. 2007). In hypophysectomised **rats**, ABCG5 and ABCG8 mRNA was almost totally suppressed (<20% of controls) and T_4, but not GH or cortisol increased their expression by >60-fold (Gälman et al. 2008). In line with these data, in hyperthyroid **mice**, hepatic ABCG5 and ABCG8 mRNA were substantially increased by 3.0- and 1.5-fold, respectively, compared to euthyroid controls (Tancevski et al. 2008). No effect on intestinal ABCG5 and ABCG8 mRNA levels was seen in hyperthyroid mice (Tancevski et al. 2008). Mice treated with T-0681 also showed moderate increases in liver ABCG5 and ABCG8 mRNA levels (Tancevski et al. 2008).

No major regulation of intestinal **NPC1L1** mRNA levels was observed in hypophysectomized rats during TH replacement (Gälman et al. 2008) and in mice on TM (Tancevski et al. 2008).

10.1.2.6.2 Factors Mainly Affecting HDL

10.1.2.6.2.1 Hepatic Lipase

HL hydrolyzes phospholipids (PL) and TG in IDL and large HDL particles, thereby facilitating IDL uptake by the liver and inducing HDL remodelling. In addition, non-catalytically active HL mediates uptake of apoB-containing lipoproteins in the liver (Amar et al. 1998) . HL consistently is decreased in OH. In human OH, postheparin HL increases by 40% (Kuusi et al. 1988) to 273% (Valdemarsson et al. 1982) after T_4 replacement and this increase is strongly indirectly related to the parallel decrease in HDL_2-C (Kuusi et al. 1988; Tan et al. 1998a, b). As HL is also regulated by sex steroids, the increase in SHBG due to T_4 was examined and found to be unrelated to the changes in lipids (Kuusi et al. 1988). Recently, HL was reported to be decreased by ≈30% in SCH (Brenta et al. 2007).

Only a slight decrease in HL mRNA in the liver was observed in hypothyroid rats; however, as HL activity was not examined, it remains possible that post-transcriptional stimulation of HL by T_4 was missed (Staels et al. 1990). No information regarding HL is available from studies targeting TR genes in mice.

In experimental OHyper in **humans**, HL increases by 46% (Hansson et al. 1983). In contradistinction, overdosing euthyroid **rats** with T_4 does not increase HL mRNA in the liver (Staels et al. 1990). Application of TM T-0681 for 4 weeks to euthyroid cholesterol-fed NZW **rabbits** failed to change postheparin HL (Tancevski et al. 2008). Thus, HL is low in OH, which in part explains distinctive features of the dyslipidemia in OH, i.e the accumulation of HL substrates, namely IDL, chylomicron remnants and HDL_2. The converse pertains to OHyper. The molecular regulation of HL by TH remains to be defined.

10.1.2.6.2.2 Scavenger Receptor Class B, Type I

Scavenger receptor class B, type I (SR-BI) functions as an HDL receptor in liver and steroid hormone-producing tissues. However, it also plays a major role in cholesterol efflux from cells, e.g. from macrophages. No data are available examining the role of SR-BI in **human** thyroid dysfunction. In euthyroid **mice** on chow, the liver and β-selective TM GC-1 and T_3 did not affect liver SR-BI mRNA levels. In these mice, surprisingly, GC-1 and equimolar amounts of T_3 increased SR-BI protein to levels 2.4-fold and 1.5-fold above baseline (Johansson et al. 2005). In euthyroid mice on a high-cholesterol diet, GC-1 and T_3 reduced hepatic SR-BI mRNA levels. Despite lower SR-BI mRNA, GC-1 (but not T_3) doubled hepatic SR-BI protein concentrations (Johansson et al. 2005). In line with these observations, in NZW **rabbits** fed a high-cholesterol diet, T-0681 increased liver SR-BI protein ≈2-fold (Tancevski et al. 2008). Up-regulation of SR-BI in the liver may be an important step in the turnover of HDL-C via the stimulation of RCT in humans, mice and rabbits. Presumably, this occurs by a post-transcriptional mechanism of action. It remains to be seen whether regulation in peripheral cells differs from the above paradigm.

10.1.2.6.2.3 ATP-Binding Cassette Transporter-AI

ATP-binding cassette transporter-AI (ABCA1) catalyzes cholesterol efflux from cells to lipid-poor apo A-I, the first step in RCT. Transcription of ABCA1 is regulated by promotor binding of heterodimers of LXRα/RXR and TRβ/RXR, which may compete for the same binding site (Huuskonen et al. 2004). No data on hypothyroidism and ABCA1 are available in humans and rodents. In severely hyperthyroid **mice**, both hepatic ABCA1 mRNA and hepatic ABCA1 protein were reduced by ≈50%, which may conceivably contribute to decreases in HDL-C (Tancevski et al. 2008). In contrast to liver, intestinal levels of ABCA1 mRNA were increased 4-fold (Tancevski et al. 2008). In mice treated with the liver-specific TM T-0681, hepatic ABCA1 mRNA levels were unchanged (Tancevski et al. 2008). Taken together, ABCA1 is clearly sensitive to regulation by T_3, however, important questions concerning tissue- and species-selective regulation still need to be worked out.

10.1.2.6.2.4 Cholesteryl Ester Transfer Protein, Lecithin:Cholesterol Acyl Transferase and Phospholipid Transfer Protein

Cholesteryl ester transfer protein (**CETP**) affords core neutral lipid exchange between lipoproteins. In fasting and/or postprandial hypertriglyceridemia (Föger et al. 1996), this leads to TG-enrichment of HDL and LDL, which thereby become a good substrate for intravascular remodelling by HL and phospholipid transfer protein (**PLTP**; Föger et al. 1997). CETP lowers HDL-C, which may be pro-atherogenic. However, CETP stimulates RCT via apoB-containing lipoproteins which may be anti-atherogenic. As rodents normally lack CETP, mainly **human** data contribute to our knowledge. In OH, CETP activity in plasma as measured independently of endogenous lipoproteins, is decreased by ≈35% (Tan et al. 1998a, b; Dullart et al. 1999). In contrast, OHyper increases CETP activity by ≈32% and correction of thyroid dysfunction normalizes CETP activity (Tan et al. 1998a, b). Thus, changes in CETP could help to explain high HDL-C in OH and low HDL in OHyper, respectively. However, Tan et al. (1998a, b) found no correlation between decreased CETP and increased HDL-C or HDL_2. While this may pertain to the lean, normo-triglyceridemic Chinese patients studied (Tan et al. 1998a, b) low CETP would be expected to increase HDL-C in obese and/or hypertriglyceridemic patients, respectively. Interestingly, CETP mass, not measured in the above work, was found to be normal in OH (Ritter et al. 1996), suggesting decreased specific activity in OH, e.g. due to increased inhibitory factors present in plasma. Cholesteryl ester (CE) net mass transfer from HDL to apoB-containing lipoproteins, influenced both by CETP activity and endogenous lipoproteins, was significantly decreased in OH (Ritter et al. 1996). This was due to qualitative changes in hypothyroid VLDL and LDL, which have been interpreted as potentially beneficial (Ritter et al. 1996). CETP activity in **rabbits** receiving the liver-specific TM T-0681 was unchanged (Tancevski et al. 2008). Thus, it is currently unclear at which level T_3 stimulates CETP-activity and also if this involves LXRα. While stimulation of CETP should be expected to lower HDL-C, this may in fact reflect stimulation of RCT.

Lecithin:cholesterol acyl transferase (**LCAT**) activity in plasma in **human** OH tends to be decreased; however, T_4 replacement did not result in increased LCAT activity (Valdemarsson et al. 1983). The same appears to be the case in MMI-treated hypothyroid **rats**. Hepatocyte cultures obtained from hypothyroid rats secrete substantially less LCAT into the medium (Ridgway and Dolphin 1985). Few data regarding regulation of LCAT are available indicating that few researchers feel LCAT plays an important role in the hyperlipoproteinemia in OH.

Human data on PLTP are lacking. PLTP activity was unchanged in severely hyperthyroid mice (Tancevski et al. 2008).

10.1.2.6.2.5 Lipoprotein Lipase

Lipoprotein lipase (**LPL**) is secreted from adipose and muscle to hydrolyze TG from large TG-rich lipoproteins, both postprandially and in the post-absorptive state, i.e. chylomicrons and VLDL. While some studies did not detect changes in

LPL during **human** OH (Krauss et al. 1974; Abrams et al. 1981; Packard et al. 1993; Tan et al. 1998a, b), others observed presumably decreased baseline values and an increase of up to 50% after T_4 replacement (Lithell et al. 1981; Valdemarsson et al. 1982; Kuusi et al. 1988). LPL in post-heparin plasma increases after T_4 for OH, varying in amount from marginal to +50% (Lithell et al. 1981). This is due to increases in LPL in adipose by 21% and in skeletal muscle by 42% (Lithell et al. 1981). Thus, in humans, LPL may be increased in both major tissues responsible for 75% of intravascular TG hydrolysis, leading to an increased clearance capacity after iv infusion of an artificial TG emulsion (Lithell et al. 1981). However, this effect is only observed in more profound OH with TSH >40 mU/l (Lithell et al. 1981). In line with these observations, after replacing T_4 in OH, the TG increase above fasting levels, as assessed by a fat tolerance test, was decreased by 20%; however, chylomicron RP-AUC was not decreased (Weintraub et al. 1999). In summary, a mild defect in LPL activity leading to TG intolerance in susceptible patients may be present in human OH, which is alleviated by T_4. The underlying molecular mechanisms remain to be clarified.

In ad-libitum fed hypothyroid **rats**, LPL mRNA was unchanged (Saffari et al. 1992). Nevertheless, LPL activity and mass in white adipose, cardiac and skeletal muscle was increased by 4.5-fold, 5.0-fold and 10.0-fold, respectively, indicating post-translational regulation (Saffari et al. 1992). Interestingly, pair-feeding hypothyroid rats with controls abolished the above differences, highlighting the often neglected role of alterations in motor activity and feeding behavior regarding lipids. No information regarding LPL is available from studies targeting TR genes in **mice**.

In OHyper, LPL was not found to be increased (Abrams et al. 1981, Tan et al. 1998a, b). Inducing experimental OHyper in healthy young volunteers does not increase LPL in post-heparin plasma (Hansson et al. 1983). Application of TM T-0681 for 4 weeks to euthyroid cholesterol-fed NZW **rabbits** failed to change postheparin LPL (Tancevski et al. 2008).

Thus, regulation of LPL in human and rat hypothyroidism is fundamentally different indicating species-specific responses. Hyperthyroidism does not appear to alter LPL.

10.2 Thyromimetics and Thyromimetic Compounds

10.2.1 Background

10.2.1.1 Desiccated Thyroid and Thyrotoxicosis

It has been known since 1930 that hyperthyroidism is associated with reduced plasma cholesterol levels (Mason et al. 1930) and since then many efforts have been made to exploit the ability of TH to lower cholesterol (Morkin et al. 2004; Moreno et al. 2008). In initial studies during the 1950s, desiccated thyroid was administered to a small group of patients, all of whom responded with a fall in cholesterol

(Strisower et al. 1954). A low dosage of dessicated thyroid led to a significant reduction in plasma cholesterol, "escape" occurred after 20–30 weeks of treatment (Strisower et al. 1955). Patients treated with high doses of desiccated thyroid were not refractory to treatment, but a large number presented with tachycardia, angina pectoris, diarrhea, weight loss and insomnia, in brief, with overt hyperthyroidism (Galioni et al. 1957). Thus, studies with thyroid preparations were stopped at that time and the search for synthetic analogs began.

10.2.1.2 First Synthetic TH Analogs and the "Twisted" D-T_4 Study

A large number of TH analogs were synthesized and tested in experimental animal models for their lipid-lowering activity (reviewed by Jorgensen 1979). 3,5,3'-triiodothyropropionic acid (Triprop) was one of the first analogs to be demonstrated to lower plasma cholesterol without affecting basal metabolic rate (Money et al. 1960). In addition to Triprop, tetraiodothyroformic acid, Tetrac, Triac, Diac and dextrothyroxine (D-T_4) were developed (Boyd and Oliver 1960; Fig. 10.1). Among all these analogs tested in animal studies, D-T_4 appeared to have the highest specific cholesterol-lowering action, without showing concomitant deleterious effects on the heart. In the late 1960s, a large clinical trial of D-T_4 therapy was conducted, as part of the **Coronary drug project** by the National Institutes of Health, which aimed to answer the question as to whether cholesterol reduction may prevent coronary heart disease (Coronary drug project research group 1972). The study was terminated after average follow-up of 36 months due to a higher proportion of deaths in the D-T_4-treated group, although this difference did not reach statistical significance. However, the design and performance of this study may have not been sufficient to elucidate the lipid-lowering effect of D-T_4 in humans. First, subsequent investigations revealed the preparations used in the D-T_4 study to be contaminated with as much as 0.5% L-T_4, equivalent to a dosage of 30 μg/day, which may have been the only active metabolite of the study (Young et al. 1984). Second, the deaths occurred in patients already carrying high cardiovascular risk factors at the initiation of the study, including angina pectoris, congestive heart failure and tachycardia.

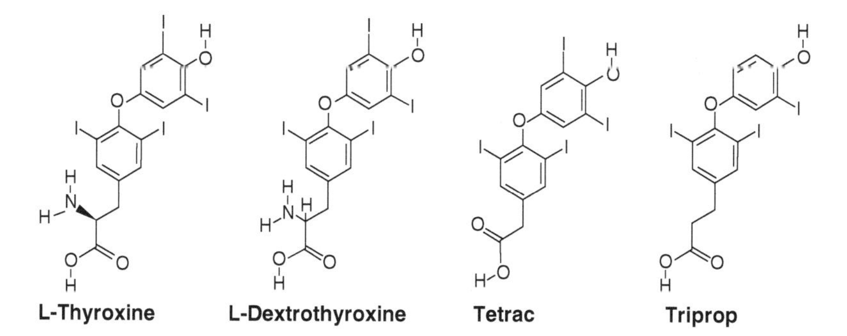

Fig. 10.1 Early synthetic TH analogues

After exclusion of high risk patients, the overall survival in the D-T_4-treated group was greater than with the controls (Baxter et al. 2001). The unfavorable recruitment of patients together with the accidental employment of preparations contaminated with the enantiomer of D-T_4 led to the discontinuation of clinical studies with TH analogs in the 1970s.

10.2.2 Selective Thyromimetic Compounds

With the introduction into clinical practice of 3-hydroxy-3-methyglutaryl coenzyme A reductase (HMG CoA reductase) inhibitors, usually known as "statins", to lower plasma cholesterol in the mid-1980s, efforts on the development of TH analogs slowed.

It was during this time period, however, that novel "selective" compounds mimicking the cholesterol-lowering actions of TH were developed (representative selective TH analogs are shown in Fig. 10.2a, b).

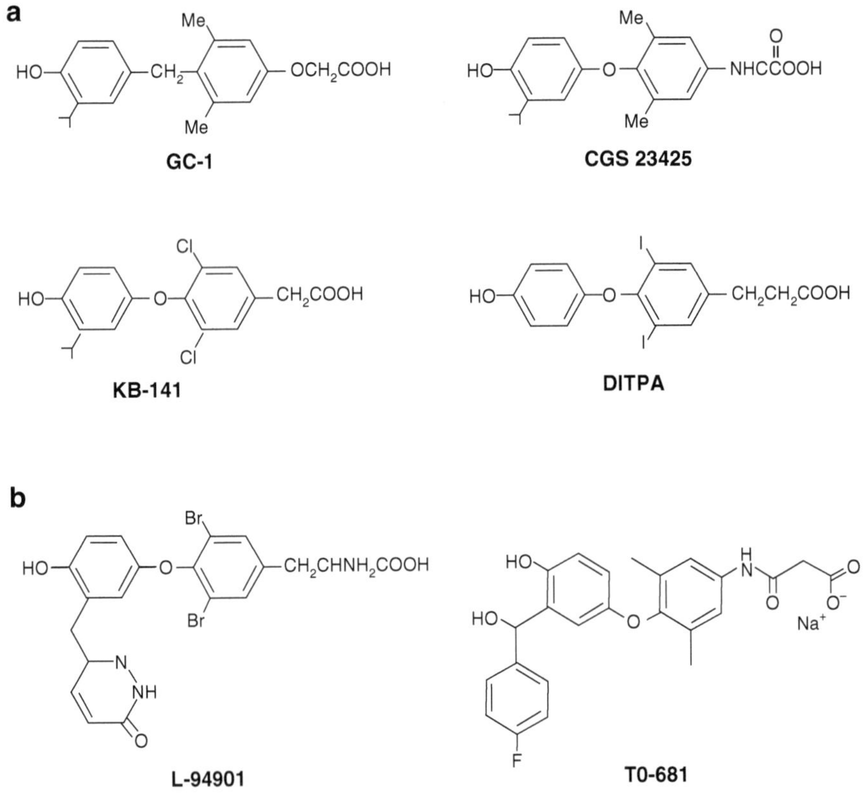

Fig. 10.2 Selective thyromimetic compounds. (**a**) TRβ-selective TH analogs. (**b**) Organ-selective TH analogs

10.2.2.1 Organ-Selective TH Analogs

The first described selective thyromimetic compound was 3,5-dibromo-3'-pyridazinone-L-thyronine (L-94901), which showed half of the binding affinity of T_3 to hepatic TRs, but only a minor affinity (1.3% of T_3) to cardiac TRs (Underwood et al. 1986; Leeson et al. 1988). This organ-selective thyromimetic has been reported to lower plasma cholesterol levels in experimental animals at doses that do not exhibit cardiotoxic side-effects. Very recently, another organ-selective compound with lipid-lowering properties was described, namely N-(4-{3-[(4-fluorophenyl) hydroxymethyl]-4-hydroxyphenoxy}-3,5-dimethylphenyl) malonamic acid sodium (T-0681, formerly KAT-681; Tancevski et al. 2007).

10.2.2.2 TRβ-Selective TH Analogs

In the late 1980s, different TR isoforms (TRα, TRβ) were cloned (Sap et al. 1986) and their tissue-specific expression characterized (reviewed by Apriletti et al. 1998; Lazar 1993). These findings led to the design of isoform-specific, TRβ1-selective thyromimetics, such as N-[3,5-dimethyl-4-(4'-hydroxy-3'-isopropylphenoxy)-phenyl] oxamic acid (CGS23425), 3,5-dimethyl-4[(4'-hydroxy-3'-isopropylbenzyl)-phenoxy] acetic acid (GC-1), 3,5-dichloro-4[(4-hydroxy-3-isopropylphenoxy)-phenyl] acetic acid (KB-141), 3-{3,5-dibromo-4-[4-hydroxy-3-(1-methylethyl)-phenoxy]-phenyl}-amino-3-oxopropanoic acid (KB2115) and 3,5-diiodothyropropionic acid (DITPA; for structures, please see Fig. 10.2a, b).

TRβ1-selectivity of CGS23425 and GC-1 seems to rely on the fact that both have methyl groups in place of iodides on the inner ring and that in the outer ring the single iodide of T_3 has been replaced by an isopropyl group (Moreno et al. 2008). It is of interest that the strong binding of GC-1 and CGS23425 to TRβ1 contradicts the usual substituent binding preference for the TR at the 3 and 5 positions on their inner ring (namely I>Br>methyl), according to which these compounds should bind with low affinity. However, as suggested by Moreno, the 3, 5 and 1 substituents of the inner ring should not be regarded independently, but as coupled moieties, when considering ligand-binding affinity (Moreno et al. 2008).

In summary, the past 20 years saw the development of either organ-selective or TRβ1-selective TH analogs, all of which lowered plasma cholesterol without deleterious effects on the heart.

10.2.3 Selective Thyromimetics as Hypolipidemic Drugs

10.2.3.1 Increased LDL Clearance

All of the above-mentioned thyromimetic compounds were demonstrated in preclinical studies to lower both plasma cholesterol and triglycerides (Underwood et al. 1986; Taylor et al. 1997; Grover et al. 2003, 2004; Tancevski et al. 2007). This effect is thought to be mainly brought about by an increased LDL-C plasma clearance through an increase in LDL-R expression in liver, similar to that described for TH action (Staels et al. 1990; Salter et al. 1991; Bakker et al. 1998). In cynomolgus monkeys, both GC-1 and KB-141 were reported to lower Lp(a) by up to 50% (Grover et al. 2003, 2004). First reports on clinical studies with DITPA (Morkin et al. 2002) and KB2115 (Berkenstam et al. 2008), respectively, showed TRβ1-selective thyromimetics to lower plasma LDL cholesterol and triglycerides also in humans, without untoward cardiac effects.

10.2.3.2 Promotion of Reverse Cholesterol Transport

Very recently, both GC-1 and T-0681 were reported to stimulate the expression of key players of reverse cholesterol transport (RCT; Johansson et al. 2005; Tancevski et al. 2007). They markedly induce hepatic SR-BI levels, stimulate the activity of cholesterol 7α-hydroxylase and induce the expression of hepatic cholesterol transporters ABCG5 and ABCG8. As a consequence, treated mice have a significant decrease in plasma HDL-C, with increased fecal excretion of bile acids and cholesterol. Interestingly, mice treated with T-0681 display reduced intestinal absorption of dietary sterols, most likely due to competition with sterols of biliary origin.

The hypothesis of promotion of RCT by a TM was tested by measuring RCT from macrophages to feces in mice treated with T-0681, according to the method developed by Rader and coworkers (Zhang et al. 2003). At 48 h after intraperitoneal injection of cholesterol-loaded, [^{3}H]-labelled J774 macrophages, T-0681-treated animals displayed a significant increase of both, fecal [^{3}H]-bile acids and [^{3}H]-cholesterol (Fig. 10.3; Tancevski et al. 2007). Thus, employment of TM may promote RCT from atherosclerotic plaque macrophages to the liver for fecal excretion.

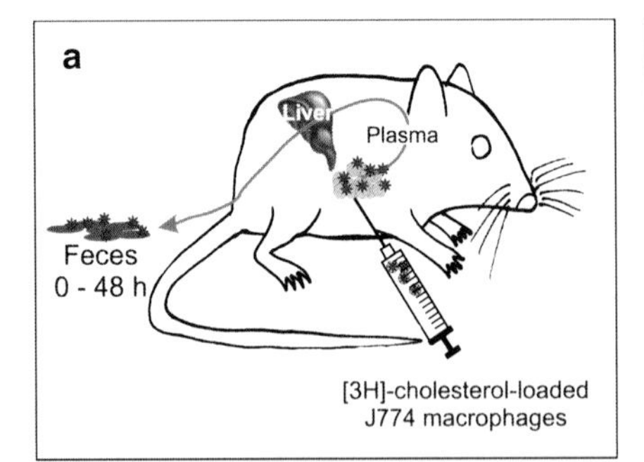

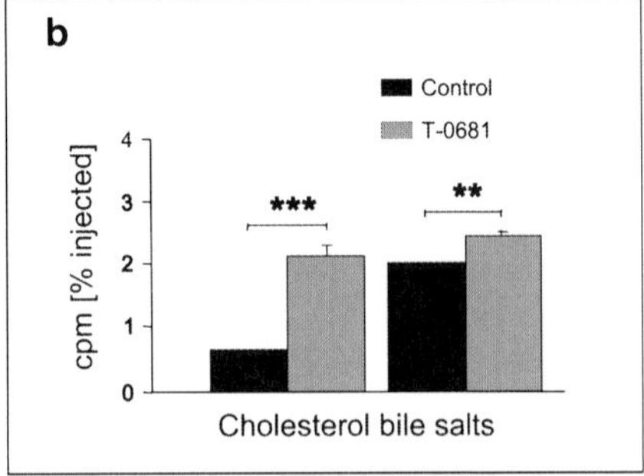

Fig. 10.3 Influence of thyromimetic T0-681 on in vivo macrophage reverse cholesterol transport in mice. (**a**) Schematic drawing of experimental approach. (**b**) Increased fecal [^{3}H]-bile acids and [^{3}H]-cholesterol after intraperitoneal injection of [^{3}H]-labeled J774 macrophages in control and T-0681 treated animals

RCT in humans is profoundly different from that found in rodents in that cholesterol from plaque macrophages can be transported to the liver either directly via HDL particles, or – after transfer to VLDL and LDL mediated by CETP – via apoB-containing lipoproteins (Ritsch et al. 2003). Interestingly, CETP-transgenic mice require hepatic expression of LDL-R to counterbalance accumulation of apoB-containing lipoproteins in plasma. In line with these findings, adenoviral over-expression of SR-BI in rabbits, which naturally express CETP in plasma, led to accumulation of VLDL and LDL cholesterol (Tancevski et al. 2005). Thus, hepatic stimulation of SR-BI expression necessitates a concomitant, appropriate clearance of apoB-lipoproteins to guarantee maintenance of RCT in CETP-expressing species, like humans. Simultaneous upregulation of hepatic SR-BI and LDL-R may represent a rational approach to direct excessive cholesterol from the periphery to the liver in humans; and selective TM may prove useful to promote this mechanism.

10.2.3.3 Prevention of Atherosclerosis

First evidence that accelerated clearance of LDL cholesterol and promotion of RCT by a TM may constitute a powerful approach to prevent the development of atherosclerosis came from studies in cholesterol-fed NZW rabbits treated with T-0681 (Tancevski et al. 2007). T-0681 reportedly decreased plasma cholesterol by 60%, triglyceride levels by more than 80% and hepatic expression of SR-BI and LDL-R were found increased 2-fold. Lipid staining of rabbit aortas revealed a 60% decrease in atherosclerotic lesion area, when compared to placebo-treated controls.

In summary, current data heavily suggest that selective thyromimetic compounds have great clinical potential as agents to treat hyperlipidemia and to protect from atherosclerosis and its clinical sequelae.

10.2.4 Potential Additional Applications

10.2.4.1 Obesity

Besides their lipid-lowering properties, both GC-1 and KB-141 were shown to increase the metabolic rate, to increase oxygen consumption and to reduce body-weight in primates (Grover et al. 2003, 2004). These effects were not accompanied by deleterious effects on heart, skeletal muscle and bone mass. TRβ1-selective compounds may therefore have potential to be used as therapeutic agents for the treatment of obesity. However, first studies with KB2115 in humans failed to influence the body-weight of probands (Berkenstam et al. 2008), probably due to the smaller doses employed when compared to previous animal studies (Grover et al. 2003, 2004; Berkenstam et al. 2008).

10.2.4.2 Angiogenesis

GC-1 was shown to promote angiogenesis in the chick chorioallantoic membrane model (Mousa et al. 2005); and DITPA was reported to be angiogenic in the postinfarction rat heart (Tomanek et al. 1998), suggesting the employment of a TRβ1-selective thyromimetic as proangiogenic agents in coronary artery disease, e.g. by administration via the coating of a coronary stent (Mousa et al. 2005).

10.2.4.3 Congestive Heart Failure

In hypothyroid rats, DITPA increased cardiac performance with approximately half the chronotropic effect and less metabolic stimulation than L-T_4 (Barker et al. 1951). DITPA also improved left ventricular performance in rabbit and rat post-infarction heart failure models (Mahaffey et al. 1995). In primates, DITPA enhanced the in vivo force–frequency and relaxation–frequency relationships of the heart without increasing heart rate (Khoury et al. 1996; Hoit et al. 1997). In a phase I clinical study, DITPA did not affect the heart rate or blood pressure of healthy volunteers, while total plasma cholesterol and triglycerides were decreased significantly (Morkin et al. 2002). In heart-failure patients receiving this drug for 4 weeks, cardiac index was increased on an average to 17%, systemic vascular resistance index was decreased and diastolic function was improved. Total serum cholesterol and triglycerides were significantly decreased. These results indicated that DITPA could represent a useful agent to treat congestive heart failure and dyslipidemia.

10.2.4.4 Chemoprevention for Hepatocellular Carcinoma

Interestingly, both T-0681 and GC-1 were found to stimulate rat hepatocyte proliferation in vivo (Hayashi et al. 2004; Hayashi et al. 2005; Columbano et al. 2006). Of note, T-0681 inhibited the development of hepatocellular carcinoma in rats induced by 2-acetylaminofluorene and partial hepatectomy after diethylnitrosamine initiation (Hayashi et al. 2004, 2005). Together, these data suggest the potential for TM as chemopreventive agents for hepatocarcinogenesis.

10.2.5 Off-Target Toxicity of Selective Thyromimetics

There are at least four main untoward effects of selective TM to be mentioned: (i) **relative hypothyroidism**, (ii) elevation of liver enzymes, (iii) cardiotoxicity and (iv) disturbed bone metabolism. Ideally, a selective TR agonist would cause modest increase in metabolic rate without tachycardia but would not reduce TSH and/or T_4, as observed with all of the selective analogs at therapeutic doses. TRβ-selective TM lead to feedback-inhibition at the hypothalamus/pituitary, which reduces TSH and T_4 levels, thereby causing a paradoxical hypothyroidism in some tissues. However, so far there are no data available on the consequences of such a relative hypothyroidism.

Both KB2115 and T-0681 were reported to induce an **elevation of liver enzymes**, when administered in high doses (Tancevski et al. 2007; Berkenstam et al. 2008). Either through TRβ1-selectivity of KB2115 or through organ-selectivity of T-0681, both kinds of compounds may accumulate in hepatocytes and lead to the observed hepatotoxicity at high dosages.

Another important side-effect, observed solely at highest doses, is **tachycardia**, which may be explained by a loss of selectivity in such oversaturated settings. However, the use of TRβ1-selective thyromimetics is likely to be safe, as GC-1 was reported to have a relative selectivity for cholesterol-lowering versus tachycardia of 18-fold (Grover et al. 2003) when compared to T_3. KB-141 was shown to be approximately 27-fold more selective for cholesterol-lowering when normalized for T_3 (Grover et al. 2004).

TRβ is active in bone, and OHyper is well known to induce **osteoporosis**. Thus, skeletal safety of TRβ-selective TM needs to be demonstrated.

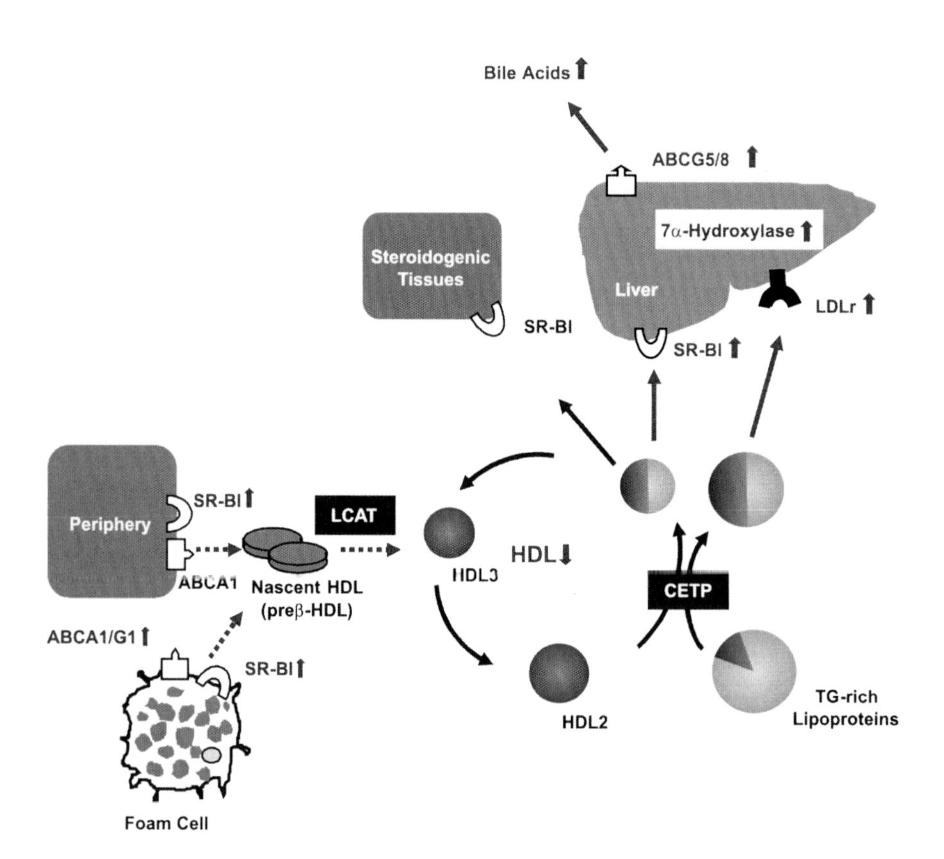

Fig. 10.4 Influence of thyromimetic T-0681 on reverse cholesterol transport

In conclusion, efforts spanning more than 50 years led to the development of selective thyromimetic compounds, a new drug class for the treatment of hypercholesterolemia and for prevention of atherosclerosis. The influence of thyromimetic compounds on lipid metabolism, i.e. on the reverse cholesterol transport, is illustrated in Fig. 10.4. Drugs of this type may be useful in treating patients who are intolerant to statins, or who do not respond to this medication. They also may be used to treat obesity or congestive heart failure. However, current studies do not provide sufficient data on the central question as to whether the observed relative hypothyroidism may have deleterious effects on the human organism. Thus, further studies on the mechanism of negative feedback on the thyroid axis induced by both TRβ1-selective and organ-selective TH analogs are awaited to increase the potential of this novel drug class for the prevention of cardiovascular diseases.

References

Abrams JJ, Grundy SM (1981) Cholesterol metabolism in hypothyroidism and hyperthyroidism in man. J Lipid Res 22:323–338

Abrams JJ, Grundy SM, Ginsberg H (1981) Metabolism of plasma triglycerides in hypothyroidism and hyperthyroidism in man. J Lipid Res 22:307–322

Agdeppa D, Macaron C, Mallik T, Schnuda ND (1979) Plasma high density lipoprotein cholesterol in thyroid disease. J Clin Endocrinol Metab 49:726–729

Amar MJ, Dugi KA, Haudenschild CC, Shamburek RD, Föger B, Chase M, Bensadoun A, Hoyt RF Jr, Brewer HB Jr, Santamarina-Fojo S (1998) Hepatic lipase facilitates the selective uptake of cholesteryl esters from remnant lipoproteins in apoE-deficient mice. J Lipid Res 39:2436–2442

Apostolopoulos JJ, Marshall JF, Howlett GJ (1990) Triiodothyronine increases rat apolipoprotein A-I synthesis and alters high-density lipoprotein composition in vivo. Eur J Biochem 194:147–154

Apriletti JW, Ribeiro RC, Wagner RL, Feng W, Webb P, Kushner PJ, West BL, Nilsson S, Scanlan TS, Fletterick RJ, Baxter JD (1998) Molecular and structural biology of thyroid hormone receptors. Clin Exp Pharmacol Physiol Suppl 25:S2–S11

Arem R, Patsch W (1990) Lipoprotein and apolipoprotein levels in subclinical hypothyroidism. Effect of levothyroxine therapy. Arch Intern Med 150:2097–2100

Bakker O, van Beeren HC, Wiersinga WM (1994) Desethylamiodarone is a noncompetitive inhibitor of the binding of thyroid hormone to the thyroid hormone beta 1-receptor protein. Endocrinology 134:1665–1670

Bakker O, Hudig F, Meijssen S, Wiersinga WM (1998) Effects of triiodothyronine and amiodarone on the promoter of the human LDL receptor gene. Biochem Biophys Res Commun 249:517–521

Barker SB, Kiely CE, Jr., Klitgarrd HM, Dirks HB, Jr., Wang SC, Wawzonek S (1951) Metabolic effects of some halogenated acrylic acid analogues of thyroxine. Endocrinology 48:70–74

Bauer DC, Ettinger B, Browner WS (1998) Thyroid functions and serum lipids in older women: a population-based study. Am J Med 104:546–551

Baxter JD, Dillmann WH, West BL, Huber R, Furlow JD, Fletterick RJ, Webb P, Apriletti JW, Scanlan TS (2001) Selective modulation of thyroid hormone receptor action. J Steroid Biochem Mol Biol 76:31–42

Benvenga S, Robbins J (1990) Enhancement of thyroxine entry into low density lipoprotein(LDL) receptor-competent fibroblasts by LDL: an additional mode of entry of thyroxine into cells. Endocrinology 126:933–941

Berkenstam A, Kristensen J, Mellstrom K, Carlsson B, Malm J, Rehnmark S, Garg N, Andersson CM, Rudling M, Sjoberg F, Angelin B, Baxter JD (2008) The thyroid hormone mimetic compound KB2115 lowers plasma LDL cholesterol and stimulates bile acid synthesis without cardiac effects in humans. Proc Natl Acad Sci USA 105:663–667

Biondi B, Cooper DS (2008) The clinical significance of subclinical thyroid dysfunction. Endocr Rev 29:76–131

Bjorkhem I, Lund E, Rudling M (1997) Coordinate regulation of cholesterol 7 alpha-hydroxylase and HMG-CoA reductase in the liver. Subcell Biochem 28:23–55

Boyd GS, Oliver MF (1960) Thyroid hormones and plasma lipids. Br Med Bull 16:138–142

Brenta G, Berg G, Arias P, Zago V, Schnitman M, Muzzio ML, Sinay I, Schreier L (2007) Lipoprotein alterations, hepatic lipase activity, and insulin sensitivity in subclinical hypothyroidism: response to L-t(4) treatment. Thyroid 17:453–460

Cachefo A, Boucher P, Vidon C, Dusserre E, Diraison F, Beylot M (2001) Hepatic lipogenesis and cholesterol synthesis in hyperthyroid patients. J Clin Endocrinol Metab 86:5353–5357

Cappola AR, Ladenson PW (2003) Hypothyroidism and atherosclerosis. J Clin Endocrinol Metab 88:2438–2444

Caraccio N, Ferrannini E, Monzani F (2002) Lipoprotein profile in subclinical hypothyroidism: response to levothyroxine replacement, a randomized placebo-controlled study. J Clin Endocrinol Metab 87:1533–1538

Chait A, Bierman EL, Albers JJ (1979) Regulatory role of triiodothyronine in the degradation of low density lipoprotein by cultured human skin fibroblasts. J Clin Endocrinol Metab 48:887–889

Chen JY, Levy-Wilson B, Goodart S, Cooper AD (2002) Mice expressing the human CYP7A1 gene in the mouse CYP7A1 knock-out background lack induction of CYP7A1 expression by cholesterol feeding and have increased hypercholesterolemia when fed a high fat diet. J Biol Chem 277:42588–42595

Columbano A, Pibiri M, Deidda M, Cossu C, Scanlan TS, Chiellini G, Muntoni S, Ledda-Columbano GM (2006) The thyroid hormone receptor-beta agonist GC-1 induces cell proliferation in rat liver and pancreas. Endocrinology 147:3211–3218

Coronary drug project research group (1972) The coronary drug project. Findings leading to further modifications of its protocol with respect to dextrothyroxine. JAMA 220:996–1008

Danese MD, Ladenson PW, Meinert CL, Powe NR (2000) Clinical review 115:effect of thyroxine therapy on serum lipoproteins in patients with mild thyroid failure: a quantitative review of the literature. J Clin Endocrinol Metab 85:2993–3001

Davidson NO, Carlos RC, Drewek MJ, Parmer TG (1988) Apolipoprotein gene expression in the rat is regulated in a tissue-specific manner by thyroid hormone. J Lipid Res 29:1511–1522

Davis RA, Miyake JH, Hui TY, Spann NJ (2002) Regulation of cholesterol-7alpha-hydroxylase: BAREly missing a SHP. J Lipid Res 43:533–543

Day R, Gebhard RL, Schwartz HL, Strait KA, Duane WC, Stone BG, Oppenheimer JH (1989) Time course of hepatic 3-hydroxy-3-methylglutaryl coenzyme A reductase activity and messenger ribonucleic acid, biliary lipid secretion, and hepatic cholesterol content in methimazole-treated hypothyroid and hypophysectomized rats after triiodothyronine administration: possible linkage of cholesterol synthesis to biliary secretion. Endocrinology 125:459–468

de Bruin TW, van Barlingen H, van Linde-Sibenius Trip M, van Vuurst de Vries AR, Akveld MJ, Erkelens DW (1993) Lipoprotein(a) and apolipoprotein B plasma concentrations in hypothyroid, euthyroid, and hyperthyroid subjects. J Clin Endocrinol Metab 76:121–126

Diekman MJ, Anghelescu N, Endert E, Bakker O, Wiersinga WM (2000) Changes in plasma low-density lipoprotein (LDL)- and high-density lipoprotein cholesterol in hypo- and hyperthyroid patients are related to changes in free thyroxine, not to polymorphisms in LDL receptor or cholesterol ester transfer protein genes. J Clin Endocrinol Metab 85:1857–1862

Drover VA, Agellon LB (2004) Regulation of the human cholesterol 7alpha-hydroxylase gene (CYP7A1) by thyroid hormone in transgenic mice. Endocrinology 145:574–581

Drover VA, Wong NC, Agellon LB (2002) A distinct thyroid hormone response element mediates repression of the human cholesterol 7alpha-hydroxylase. (CYP7A1) gene promoter. Mol Endocrinol 16:14–23

Dullaart RP, Hoogenberg K, Groener JEM, Dikkeschei LD, Erkelens DW, Doorenbos H (1990) The activity of CETP is decreased in hypothyroidism: a possible contribution to alterations in HDL. Eur J Clin Invest 20:581–587
Duntas LH (2002) Thyroid disease and lipids. Thyroid 12:287–293
Duntas LH (2003) Lipoprotein. (a) and apoliprotein. (a) isoform size in thyroid disease: the quest for the golden fleece. Thyroid 13:345–346
Duntas LH, Mantzou E, Koutras DA (2002) Circulating levels of oxidized low-density lipoprotein in overt and mild hypothyroidism. Thyroid 12:1003–1007
Föger B, Ritsch A, Doblinger A, Wessels H, Patsch JR (1996) Relationship of plasma cholesteryl ester transfer protein to HDL cholesterol. Studies in normotriglyceridemia and moderate hypertriglyceridemia. Arterioscler Thromb Vasc Biol 16:1430–1436
Föger B, Santamarina-Fojo S, Shamburek RD, Parrot CL, Talley GD, Brewer HB Jr (1997) Plasma phospholipid transfer protein. Adenovirus-mediated overexpression in mice leads to decreased plasma high density lipoprotein. (HDL) and enhanced hepatic uptake of phospholipids and cholesteryl esters from HDL. J Biol Chem 272:27393–27400
Franklyn JA, Daykin J, Betteridge J, Hughes EA, Holder R, Jones SR, Sheppard MC (1993) Thyroxine replacement therapy and circulating lipid concentrations. Clin Endocrinol. (Oxf) 38:453–459
Galioni EF, Gofman JW, Guzvich P, Pouteau J, Rubinger JH, Strisower B (1957) Long-term effect of dried thyroid on serum-lipoprotein and serum-cholesterol levels. Lancet 272:120–123
Galman C, Bonde Y, Matasconi M, Angelin B, Rudling M (2008) Dramatically increased intestinal absorption of cholesterol following hypophysectomy is normalized by thyroid hormone. Gastroenterology 134:1127–1136
Gloss B, Trost S, Bluhm W, Swanson E, Clark R, Winkfein R, Janzen K, Giles W, Chassande O, Samarut J, Dillmann W (2001) Cardiac ion channel expression and contractile function in mice with deletion of thyroid hormone receptor alpha or beta. Endocrinology 142:544–550
Göthe S, Wang Z, Ng L, Kindblom JM, Barros AC, Ohlsson C, Vennstrom B, Forrest D (1999) Mice devoid of all known thyroid hormone receptors are viable but exhibit disorders of the pituitary-thyroid axis, growth, and bone maturation. Genes Dev 13:1329–1341
Gross G, Sykes M, Arellano R, Fong B, Angel A (1987) HDL clearance and receptor-mediated catabolism of LDL are reduced in hypothyroid rats. Atherosclerosis 66:269–275
Grover GJ, Mellstrom K, Ye L, Malm J, Li YL, Bladh LG, Sleph PG, Smith MA, George R, Vennstrom B, Mookhtiar K, Horvath R, Speelman J, Egan D, Baxter JD (2003) Selective thyroid hormone receptor-beta activation: a strategy for reduction of weight, cholesterol, and lipoprotein. (a) with reduced cardiovascular liability. Proc Natl Acad Sci USA 100:10067–10072
Grover GJ, Egan DM, Sleph PG, Beehler BC, Chiellini G, Nguyen NH, Baxter JD, Scanlan TS (2004) Effects of the thyroid hormone receptor agonist GC-1 on metabolic rate and cholesterol in rats and primates: selective actions relative to 3,5,3'-triiodo-L-thyronine. Endocrinology 145:1656–1661
Gullberg H, Rudling M, Forrest D, Angelin B, Vennstrom B (2000) Thyroid hormone receptor beta-deficient mice show complete loss of the normal cholesterol 7alpha-hydroxylase. (CYP7A) response to thyroid hormone but display enhanced resistance to dietary cholesterol. Mol Endocrinol 14:1739–1749
Gullberg H, Rudling M, Salto C, Forrest D, Angelin B, Vennstrom B (2002) Requirement for thyroid hormone receptor beta in T3 regulation of cholesterol metabolism in mice. Mol Endocrinol 16:1767–1777
Hanna AN, Feller DR, Witiak DT, Newman HA (1993) Inhibition of low density lipoprotein oxidation by thyronines and probucol. Biochem Pharmacol 45:753–762
Hansson P, Valdemarsson S, Nilsson-Ehle P (1983) Experimental hyperthyroidism in man: effects on plasma lipoproteins, lipoprotein lipase and hepatic lipase. Horm Metab Res 15:449–452
Hashimoto K, Cohen RN, Yamada M, Markan KR, Monden T, Satoh T, Mori M, Wondisford FE (2006a) Cross-talk between thyroid hormone receptor and liver X receptor regulatory pathways is revealed in a thyroid hormone resistance mouse model. J Biol Chem 281:295–302

Hashimoto K, Yamada M, Matsumoto S, Monden T, Satoh T, Mori M (2006b) Mouse sterol response element binding protein-1c gene expression is negatively regulated by thyroid hormone. Endocrinology 147:4292–4302

Hashimoto K, Matsumoto S, Yamada M, Satoh T, Mori M (2007) Liver X receptor-alpha gene expression is positively regulated by thyroid hormone. Endocrinology 148:4667–4675

Hayashi M, Ohnota H, Tamura T, Kuroda J, Shibata N, Akahane M, Moriwaki H, Machida N, Mitsumori K (2004) Inhibitory effects of KAT-681, a liver-selective thyromimetic, on development of hepatocellular proliferative lesions in rats induced by 2-acetylaminofluorene and partial hepatectomy after diethylnitrosamine initiation. Arch Toxicol 78:460–466

Heimberg M, Olubadewo JO, Wilcox HG (1985) Plasma lipoproteins and regulation of hepatic metabolism of fatty acids in altered thyroid states. Endocr Rev 6:590–607

Hoit BD, Pawloski-Dahm CM, Shao Y, Gabel M, Walsh RA (1997) The effects of a thyroid hormone analog on left ventricular performance and contractile and calcium cycling proteins in the baboon. Proc Assoc Am Phys 109:136–145

Huuskonen J, Vishnu M, Pullinger CR, Fielding PE, Fielding CJ (2004) Regulation of ATP-binding cassette transporter A1 transcription by thyroid hormone receptor. Biochemistry 43:1626–1632

Ichikawa K, Miyamoto T, Kakizawa T, Suzuki S, Kaneko A, Mori J, Hara M, Kumagai M, Takeda T, Hashizume K (2000) Mechanism of liver-selective thyromimetic activity of SK&F L-94901: evidence for the presence of a cell-type-specific nuclear iodothyronine transport process. J Endocrinol 165:391–397

Ito M, Arishima T, Kudo T, Nishihara E, Ohye H, Kubota S, Fukata S, Amino N, Kuma K, Sasaki I, Hiraiwa T, Hanafusa T, Takamatsu J, Miyauchi A (2007) Effect of levo-thyroxine replacement on non-high-density lipoprotein cholesterol in hypothyroid patients. J Clin Endocrinol Metab 92:608–611

Johansson C, Vennstrom B, Thoren P (1998) Evidence that decreased heart rate in thyroid hormone receptor-alpha1-deficient mice is an intrinsic defect. Am J Physiol 275:R640–R646

Johansson L, Rudling M, Scanlan TS, Lundasen T, Webb P, Baxter J, Angelin B, Parini P (2005) Selective thyroid receptor modulation by GC-1 reduces serum lipids and stimulates steps of reverse cholesterol transport in euthyroid mice. Proc Natl Acad Sci USA 102:10297–10302

Jorgensen EC (1979) Thyromimetics and antithyroid drugs. In: Wolff ME. (ed) Burger's medicinal chemistry, part III, 4th edn. Wiley, New York, pp 103–145

Kanaya AM, Harris F, Volpato S, Pérez-Stable EJ, Harris T, Bauer DC (2002) Association between thyroid dysfunction and total cholesterol level in an older biracial population: the health, aging and body composition study. Arch Intern Med. 162:773–779

Kawai K, Sasaki S, Morita H, Ito T, Suzuki S, Misawa H, Nakamura H (2004) Unliganded thyroid hormone receptor-β represses liver X receptor α/oxysterol-dependent transactivation. Endocrinology 145:5515–5524

Khoury SF, Hoit BD, Dave V, Pawloski-Dahm CM, Shao Y, Gabel M, Periasamy M, Walsh RA (1996) Effects of thyroid hormone on left ventricular performance and regulation of contractile and Ca(2+)-cycling proteins in the baboon. Implications for the force-frequency and relaxation-frequency relationships. Circ Res 79:727–735

Klein I, Ojamaa K (2001) Thyroid hormone and the cardiovascular system. N Engl J Med 344:501–509

Krauss RM, Levy RI, Fredrickson DS (1974) Selective measurement of two lipase activities in postheparin plasma from normal subjects and patients with hyperlipoproteinemia. J Clin Invest 54:1107–1124

Kuusi T, Taskinen MR, Nikkila EA (1988) Lipoproteins, lipolytic enzymes, and hormonal status in hypothyroid women at different levels of substitution. J Clin Endocrinol Metab 66:51–56

Lazar MA (1990) Sodium butyrate selectively alters thyroid hormone receptor gene expression in GH3 cells. J Biol Chem 265:17474–17477

Lazar MA (1993) Thyroid hormone receptors: multiple forms, multiple possibilities. Endocr Rev 14:184–193

Leeson PD, Ellis D, Emmett JC, Shah VP, Showell GA, Underwood AH (1988) Thyroid hormone analogues. Synthesis of 3'-substituted 3,5-diiodo-L-thyronines and quantitative structure-activity studies of in vitro and in vivo thyromimetic activities in rat liver and heart. J Med Chem 31:37–54
Lin-Lee YC, Strobl W, Soyal S, Radosavljevic M, Song M, Gotto AM Jr, Patsch W (1993) Role of thyroid hormone in the expression of apolipoprotein A-IV and C-III genes in rat liver. J Lipid Res 34:249–259
Lin-Lee YC, Soyal SM, Surguchov A, Sanders S, Strobl W, Patsch W (1995) Thyroid hormone influences conditional transcript elongation of the apolipoprotein A-I gene in rat liver. J Lipid Res 36:1586–1594
Lithell H, Boberg J, Hellsing K, Ljunghall S, Lundqvist G, Vessby B, Wide L (1981) Serum lipoprotein and apolipoprotein concentrations and tissue lipoprotein-lipase activity in overt and subclinical hypothyroidism: the effect of substitution therapy. Eur J Clin Invest 11:3–10
Liu YY, Schultz JJ, Brent GA (2003) A thyroid hormone receptor alpha gene mutation (P398H) is associated with visceral adiposity and impaired catecholamine-stimulated lipolysis in mice. J Biol Chem 278:38913–38920
Mahaffey KW, Raya TE, Pennock GD, Morkin E, Goldman S (1995) Left ventricular performance and remodeling in rabbits after myocardial infarction. Effects of a thyroid hormone analogue. Circulation 91:794–801
Mason RL, Hunt HM, Hurxthal L (1930) Blood cholesterol values in hyperthyroidism and hypothyroidism. N Engl J Med 203:1273–1278
Menke JG, Macnaul KL, Hayes NS, Baffic J, Chao YS, Elbrecht A, Kelly LJ, Lam MH, Schmidt A, Sahoo S, Wang J, Wright SD, Xin P, Zhou G, Moller DE, Sparrow CP (2002) A novel liver X receptor agonist establishes species differences in the regulation of cholesterol 7alpha-hydroxylase (CYP7a). Endocrinology 143:2548–2558
Money WL, Kumaoka S, Rawson RW, Kroc RL (1960) Comparative effects of thyroxine analogues in experimental animals. Ann NY Acad Sci 86:512–544
Monzani F, Caraccio N, Kozakowa M, Dardano A, Vittone F, Virdis A, Taddei S, Palombo C, Ferrannini E (2004) Effect of levothyroxine replacement on lipid profile and intima-media thickness in subclinical hypothyroidism: a double-blind, placebo-controlled study. J Clin Endocrinol Metab 89:2099–2106
Moreno M, de Lange P, Lombardi A, Silvestri E, Lanni A, Goglia F (2008) Metabolic effects of thyroid hormone derivatives. Thyroid 18:239–253
Morkin E, Ladenson P, Goldman S, Adamson C (2004) Thyroid hormone analogs for treatment of hypercholesterolemia and heart failure: past, present and future prospects. J Mol Cell Cardiol 37:1137–1146
Mousa SA, O'Connor LJ, Bergh JJ, Davis FB, Scanlan TS, Davis PJ (2005) The proangiogenic action of thyroid hormone analogue GC-1 is initiated at an integrin. J Cardiovasc Pharmacol 46:356–360
Muls E, Rosseneu M, Blaton V, Lesaffre E, Lamberigts G, de Moor P (1984) Serum lipids and apolipoproteins A-I, A-II and B in primary hypothyroidism before and during treatment. Eur J Clin Invest 14:12–15
Muls E, Rosseneu M, Lamberigts G, De Moor P (1985) Changes in the distribution and composition of high-density lipoproteins in primary hypothyroidism. Metabolism 34:345–353
Naik SU, Wang X, Da Silva JS, Jaye M, Macphee CH, Reilly MP, Billheimer JT, Rothblat GH, Rader DJ (2006) Pharmacological activation of liver X receptors promotes reverse cholesterol transport in vivo. Circulation 113:90–97
Ness GC, Pendleton LC, Li YC, Chiang JY (1990) Effect of thyroid hormone on hepatic cholesterol 7 alpha hydroxylase, LDL receptor, HMG-CoA reductase, farnesyl pyrophosphate synthetase and apolipoprotein A-I mRNA levels in hypophysectomized rats. Biochem Biophys Res Commun 172:1150–1156
Nyirenda MJ, Clark DN, Finlayson AR, Read J, Elders A, Bain M, Fox KA, Toft AD (2005) Thyroid disease and increased cardiovascular risk. Thyroid 15:718–724
Oetting A, Yen PM (2007) New insights into thyroid hormone action. Best Pract Res Clin Endocrinol Metab 21:193–208

Packard CJ, Shepherd J, Lindsay GM, Gaw A, Taskinen MR (1993) Thyroid replacement therapy and its influence on postheparin plasma lipases and apolipoprotein-B metabolism in hypothyroidism. J Clin Endocrinol Metab 76:1209–1216
Pandak WM, Heuman DM, Redford K, Stravitz RT, Chiang JY, Hylemon PB, Vlahcevic ZR (1997) Hormonal regulation of cholesterol 7alpha-hydroxylase specific activity, mRNA levels, and transcriptional activity in vivo in the rat. J Lipid Res 38:2483–2491
Parle JV, Franklyn JA, Cross KW, Jones SR, Sheppard MC (1992) Circulating lipids and minor abnormalities of thyroid function. Clin Endocrinol. (Oxf) 37:411–414
Pauletzki J, Stellaard F, Paumgartner G (1989) Bile acid metabolism in human hyperthyroidism. Hepatology 9:852–855
Pearce EN (2004) Hypothyroidism and dyslipidemia: modern concepts and approaches. Curr Cardiol Rep 6:451–456
Pearce EN, Wilson PW, Yang Q, Vasan RS, Braverman LE (2008) Thyroid function and lipid subparticle sizes in patients with short-term hypothyroidism and a population-based cohort. J Clin Endocrinol Metab 93:888–894
Razvi S, Ingoe L, Keeka G, Oates C, McMillan C, Weaver JU (2007) The beneficial effect of L-thyroxine on cardiovascular risk factors, endothelial function, and quality of life in subclinical hypothyroidism: randomized, crossover trial. J Clin Endocrinol Metab 92:1715–1723
Ridgway ND, Dolphin PJ (1985) Serum activity and hepatic secretion of lecithin:cholesterol acyltransferase in experimental hypothyroidism and hypercholesterolemia. J Lipid Res 26:1300–1313
Ritsch A, Patsch JR (2003) Cholesteryl ester transfer protein: gathering momentum as a genetic marker and as drug target. Curr Opin Lipidol 14:173–179
Ritter MC, Kannan CR, Bagdade JD (1996) The effects of hypothyroidism and replacement therapy on cholesteryl ester transfer. J Clin Endocrinol Metab 81:797–800
Saffari B, Ong JM, Kern PA (1992) Regulation of adipose tissue lipoprotein lipase gene expression by thyroid hormone in rats. J Lipid Res 33:241–249
Salter AM, Hayashi R, al-Seeni M, Brown NF, Bruce J, Sorensen O, Atkinson EA, Middleton B, Bleackley RC, Brindley DN (1991) Effects of hypothyroidism and high-fat feeding on mRNA concentrations for the low-density-lipoprotein receptor and on acyl-CoA:cholesterol acyltransferase activities in rat liver. Biochem J 276:825–832
Shin DJ, Osborne TF (2003) Thyroid hormone regulation and cholesterol metabolism are connected through sterol regulatory element-binding protein-2. (SREBP-2). J Biol Chem 278:34114–34118
Simonet WS, Ness GC (1988) Transcriptional and posttranscriptional regulation of rat hepatic 3-hydroxy-3-methylglutaryl-coenzyme A reductase by thyroid hormones. J Biol Chem 263:12448–12453
Sjögren M, Alkemade A, Mittag J, Nordstrom K, Katz A, Rozell B, Westerblad H, Arner A, Vennstrom B (2007) Hypermetabolism in mice caused by the central action of an unliganded thyroid hormone receptor alpha1. EMBO J 26:4535–4545
Soyal SM, Seelos C, Lin-Lee YC, Sanders S, Gotto AM, Jr., Hachey DL, Patsch W (1995) Thyroid hormone influences the maturation of apolipoprotein A-I messenger RNA in rat liver. J Biol Chem 270:3996–4004
Staels B, Van Tol A, Chan L, Will H, Verhoeven G, Auwerx J (1990) Alterations in thyroid status modulate apolipoprotein, hepatic triglyceride lipase, and low density lipoprotein receptor in rats. Endocrinology 127:1144–1152
Storey NM, Gentile S, Ullah H, Russo A, Muessel M, Erxleben C, Armstrong DL (2006) Rapid signaling at the plasma membrane by a nuclear receptor for thyroid hormone. Proc Natl Acad Sci USA 103:5197–5201
Strisower B, Gofman JW, Galioni EF, Almada AA, Simon A (1954) Effect of thyroid extract on serum lipoproteins and serum cholesterol. Metabolism 3:218–227
Strisower B, Gofman JW, Galioni E, Rubinger JH, O'Brien GW, Simon A (1955) Effect of long-term administration of desiccated thyroid on serum lipoprotein and cholesterol levels. J Clin Endocrinol Metab 15:73–80
Strobl W, Chan L, Patsch W (1992) Differential regulation of hepatic apolipoprotein A-I and A-II gene expression by thyroid hormone in rat liver. Atherosclerosis 97:161–170

Tan KC, Shiu SW, Kung AW (1998a) Effect of thyroid dysfunction on high-density lipoprotein subfraction metabolism: roles of hepatic lipase and cholesteryl ester transfer protein. J Clin Endocrinol Metab 83:2921–2924

Tan KC, Shiu SW, Kung AW (1998b) Plasma cholesteryl ester transfer protein activity in hyper- and hypothyroidism. J Clin Endocrinol Metab 83:140–143

Tancevski I, Frank S, Massoner P, Stanzl U, Schgoer W, Wehinger A, Fievet C, Eller P, Patsch JR, Ritsch A (2005) Increased plasma levels of LDL cholesterol in rabbits after adenoviral overexpression of human scavenger receptor class B type I. J Mol Med 83:927–932

Tancevski I, Wehinger A, Demetz E, Duwensee K, Eller P, Schgoer W, Fievet C, Stellaard F, Rudling M, Patsch JR, Ritsch A (2007) The novel thyromimetic KAT-681 promotes macrophage reverse cholesterol transport (RCT) in vivo and protects from atherosclerosis. Circulation Suppl 116:II-298

Tancevski I, Wehinger A, Demetz E, Eller P, Duwensee K, Huber J, Hochegger K, Schgoer W, Fievet C, Stellaard F, Rudling M, Patsch JR, Ritsch A (2008) Reduced plasma high-density lipoprotein cholesterol in hyperthyroid mice coincides with decreased hepatic adenosine 5'-triphosphate-binding cassette transporter 1 expression. Endocrinology 149:3708–3712

Tanis BC, Westendorp GJ, Smelt HM (1996) Effect of thyroid substitution on hypercholesterolaemia in patients with subclinical hypothyroidism: a reanalysis of intervention studies. Clin Endocrinol. (Oxf) 44:643–649

Taylor AH, Stephan ZF, Steele RE, Wong NC (1997) Beneficial effects of a novel thyromimetic on lipoprotein metabolism. Mol Pharmacol 52:542–547

Thompson GR, Soutar AK, Spengel FA, Jadhav A, Gavigan SJ, Myant NB (1981) Defects of receptor-mediated low density lipoprotein catabolism in homozygous familial hypercholesterolemia and hypothyroidism in vivo. Proc Natl Acad Sci USA 78:2591–2595

Tomanek RJ, Zimmerman MB, Suvarna PR, Morkin E, Pennock GD, Goldman S (1998) A thyroid hormone analog stimulates angiogenesis in the post-infarcted rat heart. J Mol Cell Cardiol 30:923–932

Underwood AH, Emmett JC, Ellis D, Flynn SB, Leeson PD, Benson GM, Novelli R, Pearce NJ, Shah VP (1986) A thyromimetic that decreases plasma cholesterol levels without increasing cardiac activity. Nature 324:425–429

Valdemarsson S (1983) Plasma lipoprotein alterations in thyroid dysfunction. Roles of lipoprotein lipase, hepatic lipase and LCAT. Acta Endocrinol Suppl. (Copenh) 255:1–52

Valdemarsson S, Hedner P, Nilsson-Ehle P (1982) Reversal of decreased hepatic lipase and lipoprotein lipase activities after treatment of hypothyroidism. Eur J Clin Invest 12:423–428

van der Velde AE, Vrins CL, van den Oever K, Kunne C, Oude Elferink RP, Kuipers F, Groen AK (2007) Direct intestinal cholesterol secretion contributes significantly to total fecal neutral sterol excretion in mice. Gastroenterology 133:967–975

Verdugo C, Perrot L, Ponsin G, Valentin C, Berthezene F (1987) Time-course of alterations of high density lipoproteins (HDL) during thyroxine administration to hypothyroid women. Eur J Clin Invest 17:313–316

Weintraub M, Grosskopf I, Trostanesky Y, Charach G, Rubinstein A, Stern N (1999) Thyroxine replacement therapy enhances clearance of chylomicron remnants in patients with hypothyroidism. J Clin Endocrinol Metab 84:2532–2536

Wilcox HG, Frank RA, Heimberg M (1991) Effects of thyroid status and fasting on hepatic metabolism of apolipoprotein A-I. J Lipid Res 32:395–405

Wiseman SA, Powell JT, Humphries SE, Press M (1993) The magnitude of the hypercholesterolemia of hypothyroidism is associated with variation in the low density lipoprotein receptor gene. J Clin Endocrinol Metab 77:108–112

Ying H, Araki O, Furuya F, Kato Y, Cheng SY (2007) Impaired adipogenesis caused by a mutated thyroid hormone alpha1 receptor. Mol Cell Biol 27:2359–2371

Young WF Jr, Gorman CA, Jiang NS, Machacek D, Hay ID (1984) L-Thyroxine contamination of pharmaceutical D-thyroxine: probable cause of therapeutic effect. Clin Pharmacol Ther 36:781–787

Zhang Y, Zanotti I, Reilly MP, Glick JM, Rothblat GH, Rader DJ (2003) Overexpression of apolipoprotein A-I promotes reverse transport of cholesterol from macrophages to feces in vivo. Circulation 108:661–663

Chapter 11
Adipokines: Regulators of Lipid Metabolism

Oreste Gualillo and Francisca Lago

Abstract Obesity has reached epidemic dimensions, affecting millions of individuals; and it is the major contributor to type 2 diabetes, cardiovascular diseases and certain cancer types. The discovery in the mid-1990s of leptin, the first paradigmatic adipokine, renewed dynamic interest in the research of "forgotten" adipose tissue, until then believed to be only the body energy reservoir. During recent years several hormones, most of them with cytokine-like activity, have been isolated and characterized from white adipose tissue. Adipokines have been widely viewed, both in the context of obesity and its associated metabolic syndrome and insulin resistance, but also as modulators of the inflammatory state and immune response. However, the secretory endocrine function of adipose tissue cannot be separated from intracellular lipid storage and handling; hence, the goal of this chapter is to give the reader an overview of the signals evoked by the most relevant adipokines in the context of adipose tissue primordial physiology (i.e. lipid metabolism).

11.1 Introduction

A right upkeep of energy balance is a mandatory condition for the survival of all species, including man. Indeed, neither deficient nor massive nutrient intake is correct for energy homeostasis and this in turn results respectively in chronic starvation or obesity.

O. Gualillo
NEIRID (Neuroendocrine Interactions in Rheumatic and Inflammatory Diseases) Laboratory, University Clinical Hospital, Travesía Choupana s/n, 15706 Santiago de Compostela, Spain
e-mail: oreste.gualillo@usc.es

F. Lago
Cellular and Molecular Cardiology Research Laboratory, University Clinical Hospital, Travesía Choupana s/n, 15706 Santiago de Compostela, Spain
e-mail: mfrancisca.lago@usc.es

C. Ehnholm (ed.), *Cellular Lipid Metabolism*,
DOI 10.1007/978-3-642-00300-4_11, © Springer-Verlag Berlin Heidelberg 2009

It is generally conceived that starvation is a quickly detrimental process and, when it is prolonged up to a depletion of more than 30% (most of which is due to loss of fat stores) of body weight in a short timescale, it results in death. So, because of the fatal consequences of chronic starvation, human beings evolved along centuries in order to face various periods of famine and scarcity, by developing a greater capacity for storing energy as fat. Nowadays, this key characteristic for survival makes individuals more susceptible to obesity and its complications during times of abundance.

Obesity in adult age is a worldwide alarming situation, but more alarming is the increasing rate of obesity in children. Childhood obesity is particularly troubling because the extra pounds often start kids on the path to health problems that were once confined to adults, such as diabetes, high blood pressure and high cholesterol.

While the prevalence of obesity increases wordwide, the knowledge about its pathogenesis and metabolic consequences markedly advances.

Energy homeostasis requires a fine regulation of food intake which must be matched by an appropriate equilibrium among nutrient absorption, energy storage and fuel expenditure.

Appetite represents a key regulator of food intake which precedes intestinal carbohydrate and lipid uptake, whose consequent metabolic changeover to triacylglycerides provides the energy substrate that is stored within fat depots. The integration of these signals occurs mostly in the central nervous system, which orchestrates afferent signals and their transduction into homeostatic adjustments and balancing body economy.

Although the initial suggestion of the existence of a soluble factor circulating in blood that reflected the amount of fat store was made in the 1950s (Kennedy 1953), after a silence of more than 35 years, a renaissance in the interest about white adipose tissue occurred in the late 1980s when studies using obese mutant mice in parabiotic experiments gave evidence for the relevance of this hypothetical but still elusive circulating substance. The mystery was solved, at least in terms of compound identification, in 1994 when Zhang and collaborators identified the product of the gene *ob* and named it leptin (Greek, *leptos* = thin) due to its anorexigenic action. After the discoveryof leptin, the past 16 years witnessed a revolution regarding the role of white adipose tissue. Indeed, more than 50 adipocyte-derived products, most of them with cytokine-like activity (which correctly have been named adipokines) have been identified.

This chapter focuses on the metabolic activities of the most relevant adipokines, with a special emphasis on lipid metabolism.

11.2 Regulation of Lipid Metabolism by Adipokines

11.2.1 Leptin

Leptin is a 16-kDa non-glycosylated peptide hormone encoded by the gene *obese* (*ob*), the mouse homologue of the human gene *LEP* (Zhang et al. 1994). Structurally, it belongs to the class I cytokine superfamily, consisting of a bundle of four α-helices. It is mainly produced by adipocytes; and circulating leptin levels are directly correlated with white adipose tissue (WAT) mass. It decreases food intake and increases energy consumption by acting on hypothalamic cell populations (Ahima et al. 1996; Chan et al. 2003), inducing anorexigenic factors (cocaine- and amphetamine-related transcript, proopiomelanocortin) and inhibiting orexigenic neuropeptides (neuropeptide Y, Agouti-related protein, orexin); and leptin levels are correlated negatively with glucocorticoids (Zakrzewska et al. 1997) and positively with insulin (Boden et al. 1997). Its own synthesis is mainly regulated by food intake and eating-related hormones, but also depends on energy status, sex hormones (being inhibited by testosterone and increased by ovarian sex steroids) and a wide range of inflammation mediators (Sarraf et al. 1997; Gualillo et al. 2000), being increased or suppressed by pro-inflammatory cytokines depending on whether their action is acute or chronic. Through the mediation of these latter agents, leptin synthesis is increased by acute infection and sepsis. As a result of the effects of sex hormones, leptin levels are higher in women than in men even when adjusted for body mass index (BMI), which may be relevant to the influence of sex on the development or frequency of certain diseases (Blum et al. 1997). Thus, leptin appears to act not only as an adipostatin, the function in relation to which it was discovered, but also as a general signal of energy reserves (Otero et al. 2005) that is involved in a wide variety of other functions, including glucose metabolism, synthesis of glucocorticoids, proliferation of CD4+ T lymphocytes, cytokine secretion, phagocytosis, regulation of the hypothalamic–pituitary–adrenal axis, reproduction and angiogenesis (Otero et al. 2006). It can accordingly be described as a cytokine-like hormone with pleiotropic actions. For pathological conditions to which dysregulated leptin activity contributes (see below), the blockade/activation of this activity in specific target tissues may be a useful therapeutic strategy in the near future. Leptin exerts its biological actions by binding to its receptors. These are encoded by the gene *diabetes* (*db*) and belong to the class I cytokine receptor superfamily, which includes receptors for IL6, LIF, CNTF, OSM, G-CSF and gp130. Alternative splicings of *db* give rise to six receptor isoforms:

A. The soluble form Ob-Re, which lacks a cytoplasmic domain,
B. Four forms with short cytoplasmic domains (Ob-Ra, Ob-Rc, Ob-Rd, Ob-Rf),
C. The long form Ob-Rb, which is found in almost all tissues and appears to be the only form capable of transducing the leptin signal (Frühbeck 2006).

As in the case of other class I cytokine receptors, the main routes by which Ob-Rb appears to transmit the extracellular signal it receives are JAK-STAT pathways (Frühbeck 2006), which involve JAK2 phosphorylating tyrosines in the cytoplasmic domain of the receptor. In particular, mutation of the intracellular tyrosine Y1138 of mouse Ob-Rb prevents STAT3 activation and results in hyperphagia, obesity and impaired thermoregulation; and replacing Y1138 with a serine residue likewise causes pronounced obesity in knock-in mice. However, since Y1138S knock-in mice do not exhibit other defects of *db*/*db* mice, such as infertility, the role of leptin in the processes that are disrupted in these latter conditions must be independent of STAT3 (Bates et al. 2003). Indeed, the other two cytoplasmic tyrosines of mouse Ob-Rb (Y985, Y1077) have been shown to bind other intracellular signalling molecules (Gualillo et al. 2002; Prodi and Obici 2006). The early studies of leptin focused on its anorexigenic action. Both in humans and rodents, leptin levels are closely correlated with BMI; and defects of the genes encoding for leptin and its receptors give rise to severe obesity and diabetes (Lago et al. 2007). Treating leptin-deficient mice with leptin induces a reduction in food intake, accompanied by an increase in metabolic rate and weight loss. Mutations of these genes in humans appear to be rare, but the cases that are known have occurred in families with a high prevalence of morbid obesity; again, leptin administration has ameliorated all the problems associated with leptin deficiency (Lago et al. 2007). As noted in previous sections, leptin participates in the control of food intake by acting on an intricate neuronal circuit involving hypothalamic and brainstem nuclei, where it integrates a variety of different orexigenic and anorexigenic signals (Ahima and Lazar. 2008). Leptin therapy is not an effective treatment for morbid obesity that is not due to congenital deficiency of leptin or leptin receptors. In these non-congenital types of obesity, leptin concentrations are already high as a consequence of increased fat mass. The persistence of obesity in spite of high leptin levels suggests that high leptin levels can induce leptin resistance. This may occur due to a leptin-induced increase of SOCS3, which blocks intracellular transmission of the leptin signal (Bjorbaek et al. 1999), but our understanding of leptin resistance is still limited.

Lipid metabolism is the balance between lipid synthesis and degradation that in turns determines fat mass. Almost the totality of energy stores is accumulated in adipocytes as triacylglycerol that can be hydrolyzed (lipolysis) following specific stimulations to release free fatty acids. Fatty acids have two possible fates: β-oxidation to produce ATP or reesterification back into triacylglycerol. A number of experiments have shown that leptin has a direct autocrine or paracrine mode of action on the rates of synthesis and degradation of lipids; however, before describing these effects it is important to keep in mind that these experiments primarily used in vitro cell or tissue preparations. Therefore, caution should be taken when extrapolating these results to in vivo conditions, where many other regulating factors exist. An auto- or paracrine mode of action was initially demonstrated by measuring leptin-induced changes in the lipolytic rates of cultured adipose tissue (Frühbeck et al. 1997). Isolated lean wild-type mouse adipocytes had, respectively, a 34% and 40% higher rate of lipolysis than *ob/ob* and *db/db* mice, even before leptin treatment. When adipocytes from lean mice were treated with leptin, the lipolytic rate increased by 28%, whereas the *ob/ob* mouse lipolytic rate increased to 123%. No change occurred in *db/db* mouse adipocytes that lacked a functional leptin receptor. Since leptin-driven increase in lipolytic rate was measured on an in vitro preparation, it is likely that the lipolytic effect of leptin on adipocytes is not dependent upon hypothalamic, neural or adrenergic control, although these components certainly modify in vivo leptin effects. A similar study was carried out to measure the lipolytic rates of ex vivo adipocytes of wild-type and *fa/fa* rats with or without leptin treatment (*fa/fa* rats do not possess a functional leptin receptor). Similarly to mice, adipocytes of the wild-type rats increased their lipolytic rate upon leptin stimulation, whereas no change was observed in the adipocytes lacking a functional receptor (Siegrist-Kaiser et al. 1997). In another study, lean, *ob/ob* and *db/db* mice were treated with three different doses of leptin before isolating adipocytes and measuring lipolytic rates. As expected, no effect of leptin treatment was observed in the *db/db* mice. In addition, the two highest doses induced a change in the *ob/ob* mice and only the highest dose caused a significant increase in lipolysis in the wild-type mice (Frühbeck et al. 1998). The fact that the lipolytic rates of the *ob/ob* adipocytes were much more strongly affected by leptin than the wild-type adipocytes in both their experiments, suggests that, in the absence of a functional leptin protein as is the case in *ob/ob* mice, the receptor is strongly up-regulated.

Leptin likely mediates fatty acid metabolism by changing specific enzyme mRNA levels. For instance, the presence of leptin inhibits the expression of acetyl CoA carboxylase in adipocytes (Bai et al. 1996). This is the rate-limiting enzyme for long-chain fatty acid synthesis and it is essential for the conversion of carbohydrate to fatty acids and caloric storage as triacylglycerol. Under basal conditions (serum-free starved cells), cultured differentiating adipocytes expressing the *ob* gene have lower acetyl CoA carboxylase and fatty acid and triacylglycerol synthesis compared with cells that do not express the leptin gene. Long-term treatment of wild-type mice with elevated doses of leptin increased mRNA expression of hormone-sensitive lipase, the key lipolytic enzyme, while causing a decrease in mRNA expression of the lipogenic enzyme, fatty acid synthase (Sarmiento et al. 1997). Therefore, it is most likely that leptin actually regulates lipolysis by controlling the activity of hormone-sensitive lipase. This is an enzyme which is controlled by cellular levels of cyclic AMP (cAMP), and although the regulation of lipolysis by leptin at the molecular level has not yet been fully described, preliminary evidence suggests that leptin, like glucagon and catecholamines, stimulates lipolysis by increasing cAMP concentrations (Takekoshi et al. 1999). Leptin may also work via the central nervous system to reduce adipose tissue mass by inducing adipocyte apoptosis. Due to the fact that leptin treatment induced a rapid loss of fat mass with a very slow reversal of this process when treatment was terminated, it has been hypothesized that the delayed recovery was due to the deletion of some adipocytes (Qian et al. 1998). Indeed, histological analysis of adipose tissue and quantification of DNA fragmentation of the adipocytes from leptin-treated rats compared with the controls provided some evidence that apoptotic events occurred following leptin injection. Further research is certainly needed to confirm whether this interesting effect is a normal occurrence that takes place under physiological conditions. Leptin-driven modifications in lipid metabolism have also been measured in other tissues that store triacylglycerol. Leptin treatment of isolated pancreatic islets of rats causes an increase in fatty acid oxidation and a decrease in esterification, leading to a reduction in intracellular triacylglycerol content (Shimabukuro et al. 1997). Modifications in triacylglycerol content, fatty acid oxidation or esterification has been reported in rats with the *fa/fa* mutation (Heo et al. 2002). Pancreatic islets of *fa/fa* rats contain as much as 20× the amount of triacylglycerol found in lean rats (Lee et al. 1997). Rats lacking a functional leptin receptor exhibit significantly higher expression of acyl-CoA synthetase and glycerol-3-PO_4 acyltransferase (two enzymes required for lipogenesis) and reduced expression of acyl CoA oxidase and carnitine palmitoyl transferase I (two enzymes involved in fatty acid oxidation). Because of these differences in enzyme expression, as well as the much higher lipid content of *fa/fa* non-adipocytes, it has been hypothesized that one of the functions of leptin is to keep the triacylglycerol content of non-adipocyte cells low, essentially limiting triacylglycerol storage to adipocytes (Walder et al. 1997). A physiologic role of the hyperleptinemia of caloric excess could be to protect non-adipocytes from steatosis and lipotoxicity by preventing the up-regulation of lipogenesis and increasing fatty acid oxidation (Lee et al. 2001). Leptin-induced increases in insulin sensitivity are well established and may be related to the effects of leptin on lipid metabolism: insulin resistance induced by an acute lipid infusion is prevented by hyperleptinemia (Dube et al. 2007). Leptin also mediates lipid metabolism indirectly by reducing the lipogenic effects of insulin. Addition of insulin to cultured leptin-deficient adipocytes induces an increase in synthesis of acetyl CoA carboxylase, fatty acids and triacylglycerol than in adipocytes that do produce leptin (Bai et al. 1996). This may be due in part because leptin might inhibit binding of insulin to adipocytes (Walder et al. 1997). In skeletal muscle the situation is almost similar; indeed, it has been demonstrated that both the rates of fatty acid oxidation and fatty acid incorporation in triacylglycerol are modulated upon stimulation of leptin, insulin, or both into isolated mouse skeletal muscle (Muoio et al. 1997). They found that fatty acid oxidation was increased in the presence of leptin. Conversely, insulin had the opposite effect, depressing fatty acid oxidation and increasing triacylglycerol synthesis. However, when insulin and leptin were administered together, leptin decreased insulin's effects on lipid oxidation and synthesis. The inhibitory effect of leptin on insulin action in both adipocytes and muscle suggest that high leptin levels in vivo probably depress the temporary postprandial lipogenic effect of insulin. This is relevant because it means that, when circulating levels of leptin are high (i.e. obesity), triacylglycerol synthesis is not fully activated, even when lipogenic hormones are present. Also, even though leptin is able to greatly decrease insulin-induced changes in lipid metabolism, it appears that leptin does not modulate the effects of insulin on glycogen synthesis, glucose oxidation or lactate production in non-adipocytes (Muoio et al. 1997). Therefore, although leptin can prevent full activation of insulin-induced lipogenesis, it probably does not interfere with insulin's primary function: the reduction of high circulating glucose levels by upregulation of muscle and liver glycogenesis. In adipocytes, however, leptin has been shown to impair many aspects of insulin's effects on glucose metabolism, including stimulation of glucose transport and glycogen synthase activity (Müller et al. 1997). A number of feedback loops exist among leptin and other hormones involved in lipid metabolism. Indeed, leptin has been implicated in promoting the production of T_3 (Legradi et al. 1997), whereas high T_3 levels have been shown to reduce circulating leptin levels (Escobar Morreale et al. 1997). Growth hormone (GH) also appears to have a negative feedback loop with leptin. Indeed, obese subjects exhibit a marked decrease in plasma GH levels (Luke and Kineman 2006). However, the mechanisms by which increased adiposity leads to an impairment of GH secretion are poorly understood. It has been suggested that the adipose tissue can markedly influence GH secretion via two different signals, namely free fatty acids (FFA) and leptin (Diéguez et al. 2000). FFA appear to inhibit GH secretion mainly by acting directly at the pituitary level. Interestingly, reduction in circulating FFA levels in obese subjects leads to a marked increase in GH responses to different GH secretagogues. This indicates that FFAs exert a tonic inhibitory effect that contributes to blunted GH secretion in obese subjects. Recent data have shown that leptin is a metabolic signal that regulates GH secretion, since the administration of leptin antiserum to adult rats leads to a marked decrease in spontaneous GH secretion. However, leptin prevents the inhibitory effect exerted by fasting on plasma GH levels. The effect of leptin in adult rats appears to be exerted at hypothalamic level by regulating growth hormone releasing hormone (GHRH), somatostatin and neuropeptide Y (NPY)-producing neurones. In addition, during foetal life or following the development of pituitary tumors, leptin can also act directly at the anterior pituitary (Diéguez et al. 2000). It is notable that prolactin, another lactogenic hormone, is able to stimulate leptin secretion by white adipose tissue. Intriguingly, the effect of prolactin on leptin expression is likely gender-dependent (Gualillo et al. 1999).

Tumor necrosis factor alpha (TNF-α), a classic cytokine with lipolytic activity, also has leptin-modulating properties. Leptin and TNF-α are both secreted by adipocytes and play a role in controlling lipolysis. In vitro treatment of rodent and human adipocytes with TNF has been shown to inhibit leptin secretion and mRNA expression (Yamaguchi et al. 1998) whereas in vivo treatment causes an increase (Kirchgessner et al. 1997). Supporting this in vivo effect is the fact that TNF-deficient mice have circulating leptin levels that are lower than those found in the wild types. Finally, it appears that leptin mRNA expression is also regulated by adipocyte transcription factors involved in cell differentiation. These transcription factors are peroxisome proliferator-activated receptors (PPARs) and CCAAT/enhancer binding proteins (C/EBPs). PPAR-α has been involved in the regulation of fatty acid oxidation enzymes (Leone et al. 1999) and UCP3 expression (Brun et al. 1999), while both PPAR-α and PPAR-γ upregulate fatty acid-binding protein expression. Leptin has been found to up-regulate mRNA expression of both PPAR-α and PPAR-γ (Ahima et al. 1996). It is still not clear, however, whether PPARs are able to regulate leptin gene expression, since not all studies reached a clear consensus about this topic. Finally, the relationship between leptin and lipid metabolism during fasting is particularly interesting. Independent of changes in weight loss, circulating leptin levels actually decrease with fasting or caloric restriction (Ahima et al. 1996), while rates of lipolysis and fatty oxidation increase (Klein et al. 1993). It is not yet known what triggers the decrease in leptin that is observed during fasting; however, a decrease in plasma insulin concentration and increases in FFA and corticosterone levels precede the drop in leptin. In addition, it appears that this drop is initially caused by a decrease in leptin secretion rather than mRNA expression (Dallman et al. 1999). The decrease in leptin concentration observed with fasting obviously does not act to depress lipolysis and fatty acid oxidation; therefore, other hormones or factors involved in controlling lipid metabolism must take control. Another interesting aspect of fasting metabolism is the decline in energy expenditure that is usually concurrently observed with a fast or caloric restriction. This decline in metabolic rate is most likely mediated through the fall in T_3 levels that occurs under these circumstances. It is possible that the decrease in leptin concentration observed with fasting contributes to this decrease in energy expenditure through its effects on thyroid function and uncoupling proteins (Legradi et al. 1997). However, this has still not been entirely clarified.

11.2.2 Adiponectin

Adiponectin [also called gelatin-binding protein 28 (GBP28), adipose most abundant gene transcript 1(apM1) and 30-kDa adipocyte complement-related protein (Acrp30, AdipoQ)] is a 244-residue protein that, as far as is known, is produced prevalently by white adipose tissue (WAT). It increases fatty acid oxidation and reduces the synthesis of glucose in the liver (Berg and Scherer 2005). Ablation of the adiponectin gene has no dramatic effect on knock-out mice on a normal diet, but when placed on a high-fat, high-sucrose diet, they develop severe insulin resistance and exhibit lipid accumulation in muscles (Whitehead et al. 2006). Circulating adiponectin levels tend to be low in morbidly obese patients and increase both with weight loss and with the use of thiazolidinediones, which enhance sensitivity to insulin (Maeda et al. 2001). Adiponectin acts mainly via two recently described receptors, one (AdipoR1) found mainly in skeletal muscle and the other (AdipoR2) in liver (for a third route, see Sect. 11.2.3). Transduction of the adiponectin signal by AdipoR1 and AdipoR2 involves the activation of AMPK, PPAR (both α and γ) and presumably other signalling molecules also. Adiponectin exhibits structural homology with collagen VIII and X and complement factor C1q and circulates in the blood in relatively large amounts in oligomeric forms (mainly trimers and hexamers, but also a 12- to 18-mer form; Kadowaki et al. 2006), constituting about 0.01% of total plasma protein. It is somewhat controversial whether the various oligomers have different activities, which would make the effect of adiponectin controllable through its oligomerization state; and this may depend on target cell type. Although authors working with myocytes reported that trimers activated AMP-activated protein kinase (AMPK) whereas higher oligomers activated NFκβ, it has also been reported that 12- to 18-mers promote AMPK in hepatocytes (Lago et al. 2007).

Increasing evidences showed that adiponectin is involved in the regulation of both lipid and carbohydrate metabolism. Adiponectin likely has direct and indirect actions that would be related to a protective effect against cardiovascular disease (Gualillo et al. 2007). Indeed, it has been hypothesized that reduced adiponectin concentrations observed in obese patients (Arita et al. 1999) are involved in the development of atherosclerosis and cardiovascular diseases (Funahashi et al. 1999; Matsuzawa et al. 1999). Decreased adiponectin levels have been linked to small dense LDL and high apoB and triglyceride levels (Kazumi et al. 2002). Some reports have shown that adiponectin has direct actions on vascular endothelium that would protect against cardiovascular disease in part by suppressing lipid accumulation in human macrophages (Okamoto et al. 2000; Ouchi et al. 2001; Tian et al. 2008). More evidence comes from studies with adiponectin knock-out mice which showed that these animals exhibited an increase in inflammatory response to vascular injury (Kubota et al. 2002). In addition, the fact that adiponectin administration prevented atherosclerosis in apoE-deficient mice (Okamoto et al. 2002; Yamauchi et al. 2003) provides further support to the concept that adiponectin protects against cardiovascular disease. Several genes linked to circulating adiponectin levels have pleiotropic genetic effects on serum HDL and triglyceride levels (Havel 2004). In addition, data from two large cross-sectional studies indicate that, after adjusting for gender and body adiposity, circulating adiponectin concentrations are negatively correlated with triglyceride levels and strongly positively correlated with plasma HDL concentrations (Cnop et al. 2003; Tschritter et al. 2003).

11.2.3 Other Relevant Adipokines Contributing to Lipid Metabolism

11.2.3.1 Tumor Necrosis Factor Alpha and Other Pro-Inflammatory Cytokines

White adipose tissue secrete a multitude of inflammatory cytokines including TNF-α, and several interleukins (e.g. IL-2, IL-6) which adversely affect both glucose and lipid metabolism. TNF-α is a paradigmatic proinflammatory cytokine produced by numerous cells, but mainly macrophages and lymphocytes. Adipocytes significantly contribute to synthesize TNF-α in rodents, and in lower amounts in humans. In rodents, TNF-α is involved in the pathophysiology of insulin resistance (Hotamisligil et al. 1993). However, as above-mentioned, TNF-α is poorly expressed in human adipose tissue and its role in alterations of dyslipidemia in humans is not completely understood. It is likely that the TNF contribution to alterations in lipid metabolism comes from an active cross-talk between invading stromal mononuclear cells and adipocytes. Interestingly, the infiltrating macrophages constitute the major source of inflammatory mediators within the adipose tissue and likely act synergistically with adipocytes amplifying local and systemic inflammation. TNF-α acts at several levels on adipocyte lipid metabolism. First, TNF-α inhibits the uptake of free fatty acids through a mechanism that likely involves down-regulation of fatty acid transport protein (FATP), fatty acid translocase (FAT) and fatty acid-binding protein FABP4/aP2. TNF-α also regulates lipoprotein lipase (LPL) expression, but studies investigating the role of TNF-α on LPL in human adipose tissue have reported conflicting results (Cawthorn et al. 2008). In addition to suppressing gene expression of key proteins of fatty acid uptake, TNF-α reduces the transcript levels and expression of many proteins involved in glyceroneogenesis, de novo fatty acid synthesis and esterification. This leads to impaired triglyceride storage in adipose tissue. Notably, most of these genes are regulated by PPARγ activity, so it could mediate those effects primarily through the inhibition of PPARγ activity and expression. TNF-α also induces lypolisis by a complex mechanism in absence of insulin, suggesting the importance of nutritional state in this process. Although the precise mechanism by which TNF-α promotes lypolysis is far to be completely defined, it is clear that activation of TNF receptor type 1 by TNF binding involves downstream signals such as ERK1/2, JNK, AMPK, IKK and PKA. In addition, other transient mechanisms (i.e. controlled at trancriptional regulation level) are at play. The past 10 years has unveiled many novel aspects of TNF-α besides its classic effects as oncolytic and cachectic factor. Indeed, TNF-α significantly alters lipid storage and oxidative capacity of WAT. (For a review, see Cawthorn et al. 2008).

11.2.3.2 Acylation Stimulating Protein

Acylation-simulating protein (ASP) is a complement-derived protein which is an adipokine that both increases triglycerides clearance and triacylglycerol storage and enhances insulin sensitivity by increasing glucose transport into adipocytes. ASP also inhibits lipolysis by decreasing the hormone-sensitive lipase (Van Harmelen et al. 1999) and is increased in human obesity, being a marker for this condition much the same as is leptin (Cianflone et al. 2003).

11.2.3.3 Apelin

Apelin is a recently discovered adipokine widely increased during adipocyte differentiation and whose levels are elevated in obesity, particularly when it is associated with insulin resistance (Boucher et al. 2005). It is also a mediator of endothelial vasodilation that exerts beneficial cardiovascular effects and lowers blood pressure by a nitric oxide stimulation-dependent mechanism (Ashley et al. 2005). It is an antagonist to angiotensin II since it decreases vascular tone and angiogenesis as an anti-inflammatory secretagogue (Kasai et al. 2004). Apelin appears to regulate adiposity and lipid metabolism in both lean and obese mice. Additionally, apelin treatment increased mRNA expression of uncoupling protein 1 (UCP1; a marker of peripheral energy expenditure in brown adipose tissue; BAT) and of UCP3, a regulator of fatty acid export, in skeletal muscle (Higuchi et al. 2007).

Other secreted proteins indicated to be involved in energy metabolism (Sethi and Vidal-Puig 2007) are rather obscure regarding their source, function and relevance to humans.

11.2.3.4 Visfatin

Visfatin is a recently intensively studied new "adipokine," because it was shown to mimic insulin effects (Kasai et al. 2004) and as such might have linked obesity and insulin resistance. However, subsequent studies in human subjects reported conflicting results regarding its relation with adiposity, with subcutaneous or visceral fat distribution and with insulin resistance (Berndt et al. 2005; Chen et al. 2006; Pagano et al. 2006), suggesting that the role of this protein in the development of obesity and insulin resistance is unclear. In addition, other authors could not repeat the insulin-mimic effect in vivo and in vitro (Revollo et al. 2007). Visfatin is a rediscovered cytokine which is identical to Pre-B-cell colony-enhancing factor 1 and is functionally characterized as a cytosolic nicotinamide phosphoribosyltransferase involved in the biosynthesis of nicotinamide adenine dinucleotide (NAD), a vital factor in cell survival and in the regulation of insulin secretion in β-cells. Although visfatin has been regarded as an adipose tissue-related hormone, it is ubiquitously expressed in the body. Among different human tissues, the maximum gene expression was found in liver and peripheral blood leukocytes. Apparently, adipocytes and adipose tissue are not the major contributor to the high concentration of visfatin in human circulation (10–40 ng/ml). The lack of taking accpunt of other tissues and organs, such as liver and immune system, may explain the inconsistent conclusions with respect to the relation of circulating visfatin to insulin resistance and adiposity. Recent studies found that circulating visfatin is associated with HDL-cholesterol in humans (Smith et al. 2006; Wang et al. 2007). The relation between visfatin and lipid metabolism may be explained by its function in the biosynthesis of NAD, because one of the NAD precursors, nicotinic acid, is able to increase HDL-cholesterol considerably (Carlson 2005). This suggests that visfatin is a novel modulator of HDL metabolism. Interestingly, early studies showed that visfatin is strongly induced in white blood cells by cytokines and lipopolysaccharides in experimental inflammation and clinical sepsis. It is known that lipopolysaccharides and sepsis affect a wide range of apolipoproteins, plasma enzymes, lipid transfer factors and receptors that are involved in HDL metabolism.

11.2.3.5 Omentin

Omentin is another recently rediscovered protein which was found to be related to the visceral adipose depot and was suggested to regulate insulin action (Yang et al. 2006). It has been identified before as intelectin 1, a protein that is involved in body defense by binding to galactofuranoses on bacteria (Tsuji et al. 2001). Although omentin is related to adipose tissue, it is not secreted from adipocytes, but from stromal vascular cells. Also, omentin has been found expressed abundantly in human tissues other than adipose tissue, such as vasculature, small intestine, colon, heart and thymus. Recent studies focused on its role in the response to infections and only a few studies investigated its role in obesity and insulin action. Although it is neither clear nor confirmed whether this protein contributes to the development of obesity and insulin resistance, it remains of interest to understand how omentin contributes to the physiological difference between visceral and subcutaneous adipose tissue and, very probably, to the regulation of lipid metabolism.

11.2.3.6 Chemerin

Chemerin is an adipokine identified in 2007 (Bozaoglu et al. 2007; Goralski et al. 2007) whose plasma levels are strongly associated with several key aspects of metabolic syndrome. Chemerin also influences adipose cell function and, importantly, it is able to induce lipolysis en adipocytes (Roh et al. 2007) acting through its receptor. The chemerin receptor, now termed chemerinR, was isolated and found to be an orphan G protein-coupled receptor known to stimulate intracellular calcium release, phosphorylate extracellular signal-regulated kinase-1 and -2 (ERK 1/2) and to inhibit cAMP accumulation through its binding to Gi-coupled heterotrimeric G proteins. Now, it has also been demonstrated also that chemerin enhances insulin signaling and potentiates insulin-stimulated glucose uptake in adipocytes (Takahashi et al. 2008) and probably in the near future the role played by this adipokine in lipid metabolism regulation will be clarified.

11.3 Conclusions

A large number of other adipocyte-derived factors have also been recently identified, including among these vaspin or hepcidin; however, they have as yet not been studied in terms of their ability to modulate lipid metabolism in either health or disease. The discovery of leptin nearly 15 years ago enlightened a new epoch in research as it was eventually demonstrated that this adipocyte-derived peptide not only regulates energy production but also exerts peripheral effects on a large number of tissues. The relevance of adipokines to lipid metabolic function in either physiology or pathology is still emerging; and studies addressing this question represent an exciting area of research not only with respect to obesity, since adipokine production is generally related to adiposity, but also with respect to the central issue of the potential role of adipose tissue as an endocrine organ regulating overall metabolic function in health and disease. There are numerous challenges facing researchers in this field. Important among these is the fundamental question of precisely how leptin, adiponectin and other adipokines affect metabolic diseases. This task will undoubtedly be made easier and hastened by the development of new pharmacological tools targeting specific adipokine systems. A second major challenge is to understand how the various adipokines interact with each other since several adipokines with diverse biological properties can be released simultaneously and the net effect of increased adipokine production may not reflect the actions of a single individual substance. This remains a challenge for future investigations which are important, not only for a full understanding of the role of adipokines in the metabolic regulation of lipids, but also in terms of the potential to develop novel therapeutic targets.

References

Ahima RS, Lazar MA (2008) Adipokines and the peripheral and neural control of energy balance. Mol Endocrinol 22:1023–1031

Ahima RS, Prabakaran D, Mantzoros C, Qu D, Lowell B Maratos-Flier E, Flier JS (1996) Role of leptin in the neuroendocrine response to fasting. Nature 382:250–252

Arita Y, Kihara S, Ouchi N, Takahashi M, Maeda K, Miyagawa J, Hotta K, Shimomura I, Nakamura T, Miyaoka K, Kuriyama H, Nishida M, Yamashita S, Okubo K, Matsubara K, Muraguchi M, Ohmoto Y, Funahashi T, Matsuzawa Y (1999) Paradoxical decrease of an adipose-specific protein, adiponectin, in obesity. Biochem Biophys Res Commun 257:79–83

Ashley EA, Powers J, Chen M, Kundu R, Finsterbach T, Caffarelli A, Deng A, Eichhorn J, Mahajan R, Agrawal R, Greve J, Robbins R., Patterson AJ, Bernstein D, Quertermous T (2005) The endogenous peptide apelin potently improves cardiac contractility and reduces cardiac loading in vivo. Cardiovasc Res 65:73–82

Bai Y, Zhang S, Kim KS, Lee JK, Kim KH (1996) Obese gene expression alters the ability of 30A5 preadipocytes to respond to lipogenic hormones. J Biol Chem 271:13939–13942

Bates SH, Stearns WH, Dundon TA, Schubert M, Tso AW, Wang Y, Banks AS, Lavery HJ, Haq AK, Maratos-Flier E, Neel BG, Schwartz MW, Myers MG Jr (2003) STAT3 signalling is required for leptin regulation of energy balance but not reproduction. Nature 421:856–859

Berg AH, Scherer PE (2005) Adipose tissue, inflammation, and cardiovascular disease. Circ Res 96:939–949

Berndt J, Kloting N, Kralisch S, Kovacs P, Fasshauer M, Schon MR, Stumvoll M, Bluher M (2005) Plasma visfatin concentrations and fat depot-specific mRNA expression in humans. Diabetes 54:2911–2916

Bjorbaek C, Elmquist JK, Frantz JD, Flier JS (1999) The role of SOCS3 in leptin signalling and leptin resistance. J Biol Chem 274:30059–30065

Blum WF, Englaro P, Hanitsch S, Juul A, Hertel NT, Muller J, Skakkebaek NE, Heiman ML, Birkett M, Attanasio AM, Kiess W, Rascher W (1997) Plasma leptin levels in healthy children and adolescents: dependence on body mass index, body fat mass, gender, pubertal stage, and testosterone. Clin Endocrinol Metab 82:2904–2910

Boden G, Chen X, Kolaczynski JW, Polansky M (1997) Effects of prolonged hyperinsulinemia on serum leptin in normal human subjects. J Clin Inv 100:1107–1113

Boucher J, Masri B, Daviaud D, Gesta S, Guigné C, Mazzucotelli A, Castan-Laurell I, Tack I, Knibiehler B, Carpéné C, Audigier Y, Saulnier-Blache JS, Valet P (2005) Apelin, a newly identified adipokine up-regulated by insulin and obesity. Endocrinology 146:1764–1771

Bozaoglu K, Bolton K, McMillan J, Zimmet P, Jowett J, Collier G, Walder K, Segal D (2007) Chemerin is a novel adipokine associated with obesity and metabolic syndrome. Endocrinology 148:4687–4694

Brun S,Carmona MC, Mampel T, Vinas O, Giralt M, Iglesias R, Villarroya F (1999) Activators of peroxisome proliferator-activated receptor-alpha induce the expression of the uncoupling protein-3 gene in skeletal muscle: a potential mechanism for the lipid intake-dependent activation of uncoupling protein-3 gene expression at birth. Diabetes 48:1217–1222

Carlson LA (2005) Nicotinic acid: The broad-spectrum lipid drug. A 50th anniversary review. J Intern Med 258: 94–114

Cawthorn WP, Sethi JK (2008) TNF-alpha and adipocyte biology. FEBS Lett 582:117–131

Chan JL, Heist K, DePaoli AM, Veldhuis JD, Mantzoros CS (2003) The role of falling leptin levels in the neuroendocrine and metabolic adaptation to short-term starvation in healthy men. J Clin Invest 111:1409–1421

Chen MP, Chung FM, Chang DM, Tsai JC, Huang HF, Shin SJ, Lee YJ (2006) Elevated plasma level of visfatin/pre-B cell colony-enhancing factor in patients with type 2 diabetes mellitus. J Clin Endocrinol Metab 91:295–299

Cianflone K, Xia Z, Chen LY (2003) Critical review of acylation-stimulating protein physiology in humans and rodents. Biochim Biophys Acta 1609:127–43

Cnop M, Havel PJ, Utzschneider KM, Carr DB, Sinha MK, Boyko EJ, Retzlaff BM, Knopp RH, Brunzell JD, Kahn SE (2003) Relationship of adiponectin to body fat distribution, insulin sensitivity and plasma lipoproteins: evidence for independent roles of age and sex. Diabetologia 46:459–469

Dallman MFAkana SF, Bhatnagar S, Bell ME, Choi S, Chu A, Horsley C, Levin N, Meijer O, Soriano LR, Strack AM, Viau V (1999) Starvation: early signals, sensors, and sequelae. Endocrinology 140:4015–4023

Dieguez C, Carro E, Seoane LM, Garcia M, Camina JP, Senaris R, Popovic V, Casanueva FF (2000) Regulation of somatotroph cell function by the adipose tissue. Int J Obes Relat Metab Disord 24:S100–S103

Dube JJ, Bhatt BA, Dedousis N, Bonen A, O'Doherty RM (2007) Leptin, skeletal muscle lipids, and lipid-induced insulin resistance. Am J Physiol Regul Integr Comp Physiol 293:R642–R650

Escobar-Morreale HF, Escobar del Rey F, Morreale de Escobar G (1997) Thyroid hormones influence serum leptin concentrations in the rat. Endocrinology 138:4485–4488

Fruhbeck G (2006) Intracellular signalling pathways activated by leptin. Biochem J 393:7–20

Frühbeck GAguado M, Martinez JA (1997) In vitro lipolytic effect of leptin on mouse adipocytes: evidence for a possible autocrine/paracrine role of leptin. Biochem Biophys Res Commun 240:590–594

Frühbeck GAguado M, Gomez-Ambrosi J, Martinez JA (1998) Lipolytic effect of in vivo leptin administration on adipocytes of lean and ob/ob mice, but not db/db mice. Biochem Biophys Res Commun 250:99–102

Fukuhara A, Matsuda M, Nishizawa M, Segawa K, Tanaka M, Kishimoto K, Matsuki Y, Murakami M, Ichisaka T, Murakami H, Watanabe E, Takagi T, Akiyoshi M, Ohtsubo T, Kihara S, Yamashita S, Makishima M, Funahashi T, Yamanaka S, Hiramatsu R, Matsuzawa Y, Shimomura I (2005) Visfatin: a protein secreted by visceral fat that mimics the effects of insulin. Science 307:426–430

Funahashi T, Nakamura T, Shimomura I, Maeda K, Kuriyama H, Takahashi M, Arita Y, Kihara S, Matsuzawa Y (1999) Role of adipocytokines on the pathogenesis of atherosclerosis in visceral obesity. Intern Med 38:202–206

Goralski KB, McCarthy TC, Hanniman EA, Zabel BA, Butcher EC, Parlee SD, Muruganandan S, Sinal CJ (2007) Chemerin a novel adipokine that regulates adipogenesis and adipocyte metabolism. J Biol Chem 282:28175–28188

Gualillo O, Lago F, García M, Menéndez C, Señarís R, Casanueva FF, Diéguez C (1999) Prolactin stimulates leptin secretion by rat white adipose tissue. Endocrinology 140:5149–5153

Gualillo O, Eiras S, Lago F, Dieguez C, Casanueva FF (2000) Elevated serum leptin concentrations induced by experimental acute inflammation. Life Sci 67:2433–2441

Gualillo O, Eiras S, White DW, Dieguez C, Casanueva FF (2002) Leptin promotes the tyrosine phosphorylation of SHC proteins and SHC association with GRB2. Mol Cell Endocrinol 190:83–89

Gualillo O, González-Juanatey JR, Lago F (2007) The emerging role of adipokines as mediators of cardiovascular function:physiologic and clinical perspectives. Trends Cardiovasc Med 17:275–283

Havel PJ (2002) Control of energy homeostasis and insulin action by adipocyte hormones: leptin, acylation stimulating protein, and adiponectin. Curr Opin Lipidol 13:51–59

Havel PJ (2004) Regulation of energy balance and carbohydrate metabolism. Diabetes 53:S143–S151

Heo YR, Claycombe K, Jones BH, Wright P, Truett GE, Zemel M, Banz W, Maher M, Moustaid-Moussa N (2002) Effects of fatty (fa) allele and high-fat diet on adipose tissue leptin and lipid metabolism. Horm Metab Res 34:686–690

Higuchi K, Masaki T, Gotoh K, Chiba S, Katsuragi I, Tanaka K, Kakuma T,Yoshimatsu H (2007) Apelin, an APJ receptor ligand, regulates body adiposity and favors the messenger ribonucleic acid expression of uncoupling proteins in mice. Endocrinology 148:2690–2697

Hotamisligil GS, Shargill NS, Spiegelman BM (1993) Adipose expression of tumor necrosis factor-alpha: direct role in obesity-linked insulin resistance. Science 259:87–91
Kadowaki T, Yamauchi T, Kubota N, Hara K, Ueki K, Tobe K(2006)Adiponectin and adiponectin receptors in insulin resistance diabetes, and the metabolic syndrome. J Clin Invest 116:1784–1792
Kasai A, Shintani N, Oda M, Kakuda M, Hashimoto H, Matsuda T, Hinuma S, Baba A (2004) Apelin is a novel angiogenic factor in retinal endothelial cells. Biochem Biophys Res Commun. 325:395–400
Kazumi T, Kawaguchi A, Sakai K, Hirano T, Yoshino G (2002) Young men with high-normal blood pressure have lower serum adiponectin smaller LDL size and higher elevated heart rate than those with optimal blood pressure. Diabetes Care 25:971–976
Kennedy GC (1953) The role of depot fat in the hypothalamic control of food intake in the rat. Proc R Soc Lond B Biol Sci 140:578–596
Kirchgessner TGUysal KT, Wiesbrock SM, Marino MW, Hotamisligil GS (1997) Tumor necrosis factor-alpha contributes to obesity-related hyperleptinemia by regulating leptin release from adipocytes. J Clin Invest 100:2777–2782
Klein SSakurai Y, Romijn JA, Carroll RM (1993) Progressive alterations in lipid and glucose metabolism during short-term fasting in young adult men. Am J Physiol 265:E801–E806
Kubota N, Terauchi Y, Yamauchi T, Kubota T, Moroi M, Matsui J, Eto K, Yamashita T, Kamon J, Satoh H, Yano W, Froguel P, Nagai R, Kimura S, Kadowaki T, Noda T (2002) Disruption of adiponectin causes insulin resistance and neointimal formation. J Biol Chem 277:25863–25866
Lago F, Dieguez C, Gomez-Reino J, Gualillo O (2007) The emerging role of adipokines as mediators of inflammation and immune responses. Cytokine Growth Factor Rev 18:313–325
Lee Y, Hirose H, Zhou YT, Esser V, McGarry JD, Unger RH (1997) Increased lipogenic capacity of the islets of obese rats: a role in the pathogenesis of NIDDM. Diabetes 46:408–413
Lee Y, Wang MY, Kakuma T, Wang ZW, Babcock E, McCorkle K, Higa M, Zhou YT, Unger RH (2001) Liporegulation in diet-induced obesity. The antisteatotic role of hyperleptinemia. J Biol Chem 276:5629–5235
Legradi GEmerson CH, Ahima RS, Flier JS, Lechan RM (1997) Leptin prevents fasting-induced suppression of prothyrotropin-releasing hormone messenger ribonucleic acid in neurons of the hypothalamic paraventricular nucleus. Endocrinology 138:2569–2576
Leone TC,Weinheimer CJ, Kelly DP (1999) A critical role for the peroxisome proliferator-activated receptor alpha (PPARalpha) in the cellular fasting response: the PPARalpha-null mouse as a model of fatty acid oxidation disorders. Proc Natl Acad Sci USA 96:7473–7478
Luque RM, Kineman RD (2006) Impact of obesity on the growth hormone axis: evidence for a direct inhibitory effect of hyperinsulinemia on pituitary function. Endocrinology 147:2754–2763
Maeda N, Takahashi M, Funahashi T, Kihara S, Nishizawa H, Kishida K Nagaretani H, Matsuda M, Komuro R, Ouchi N, Kuriyama H, Hotta K, Nakamura T, Shimomura I,Matsuzawa Y (2001) PPARgamma ligands increase expression and plasma concentrations of adiponectin, an adipose-derived protein. Diabetes 50:2094–2099
Matsuzawa Y, Funahashi T, Nakamura T (1999) Molecular mechanism of metabolic syndrome X contribution of adipocytokines adipocyte-derived bioactive substances. Ann NY Acad Sci 892:146–154
Müller GErtl J, Gerl M, Preibisch G (1997) Leptin impairs metabolic actions of insulin in isolated rat adipocytes. J Biol Chem 272:10585–10593
Muoio DM,Dohn GL, Fiedorek FT Jr, Tapscott EB, Coleman RA (1997) Leptin directly alters lipid partitioning in skeletal muscle. Diabetes 46:1360–1363
Okamoto Y, Arita Y, Nishida M, Muraguchi M, Ouchi N, Takahashi M, Igura T, Inui Y, Kihara S, Nakamura T, Yamashita S, Miyagawa J, Funahashi T, Matsuzawa Y (2000) An adipocyte-derived plasma protein, adiponectin, adheres to injured vascular walls. Horm Metab Res 32:47–50
Okamoto Y, Kihara S, Ouchi N, Nishida M, Arita Y, Kumada M, Ohashi K, Sakai N, Shimomura I, Kobayashi H, Terasaka N, Inaba T, Funahashi T, Matsuzawa Y (2002) Adiponectin reduces atherosclerosis in apolipoprotein E-deficient mice. Circulation 106:2767–2770

Otero M, Lago R, Lago F, Casanueva FF, Dieguez C, Gomez-Reino JJ, Gualillo O (2005) Leptin, from fat to inflammation: old questions and new insights. FEBS Lett 579:295–301

Otero M, Lago R, Gomez R, Dieguez C, Lago F, Gomez-Reino JJ, Gualillo O (2006) Towards a pro-inflammatory and immunomodulatory emerging role of leptin. Rheumatology (Oxf) 45:944–950

Ouchi N, Kihara S, Arita Y, Nishida M, Matsuyama A, Okamoto Y, Ishigami M, Kuriyama H, Kishida K, Nishizawa H, Hotta K, Muraguchi M, Ohmoto Y, Yamashita S, Funahashi T, Matsuzawa Y (2001) Adipocyte-derived plasma protein, adiponectin, suppresses lipid accumulation and class A scavenger receptor expression in human monocyte-derived macrophages. Circulation 103:1057–1063

Pagano C, Pilon C, Olivieri M, Mason P, Fabris R, Serra R, Milan G, Rossato M, Federspil G, Vettor R (2006) Reduced plasma visfatin/pre-B cell colony-enhancing factor in obesity is not related to insulin resistance in humans. J Clin Endocrinol Metab 91:3165–3170

Prodi F, Obici S (2006) The brain as a molecular target for diabetic therapy. Endocrinology 147:2664–2669

Qian H,Azain MJ, Compton MM, Hartzell DL, Hausman GJ, Baile CA (1998) Brain administration of leptin causes deletion of adipocytes by apoptosis. Endocrinology 139:791–794

Revollo JR, Korner A, Mills KF, Satoh A, Wang T, Garten A, Dasgupta B, Sasaki Y, Wolberger C, Townsend RR, Milbrandt J, Kiess W, Imai S-I (2007) Nampt/PBEF/Visfatin regulates insulin secretion in [beta] cells as a systemic NAD biosynthetic enzyme. Cell Metab 6:363–375

Roh SG, Song SH, Choi KC, Katoh K, Wittamer V, Parmentier M, Sasaki S (2007) Chemerin - a new adipokine that modulates adipogenesis via its own receptor. Biochem Biophys Res Commun 362:1013–1018

Sarmiento UBenson B, Kaufman S, Ross L, Qi M, Scully S, DiPalma C (1997) Morphologic and molecular changes induced by recombinant human leptin in the white and brown adipose tissues of C57BL/6 mice. Lab Invest 77:243–255

Sarraf P, Frederich RC, Turner EM, Ma G, Jaskowiak NT, Rivet DJ 3rd, Flier JS, Lowell BB, Fraker DL, Alexander HL (1997) Multiple cytokines and acute inflammation raise mouse leptin levels: potential role in inflammatory anorexia. J Exp Med 185:171–175

Sethi JK, Vidal-Puig AJ (2007) Thematic review series: adipocyte biology. Adipose tissue function and plasticity orchestrate nutritional adaptation. J Lipid Res 48:1253–1262

Shimabukuro MKoyama K, Chen G, Wang M-Y, Trieu F, Lee Y, Newgard CB, Unger RH (1997) Direct antidiabetic effect of leptin through triglyceride depletion of tissues. Proc Natl Acad Sci USA 94:4637–4641

Siegrist-Kaiser CAPauli VJuge-Aubry CE, Boss O, Pernin A, Chin WW, Cusin I, Rohner-Jeanrenaud F, Burger AG, Zapf J, Meier CA (1997) Direct effects of leptin on brown and white adipose tissue. J Clin Inves 100:2858–2864

Smith J, Al-Amri M, Sniderman A, Cianflone K (2006) Visfatin concentration in Asian Indians is correlated with high density lipoprotein cholesterol and apolipoprotein A1. Clin Endocrinol (Oxf) 65:667–672

Stefan N, Stumvoll M (2002) Adiponectin: its role in metabolism and beyond. Horm Metab Res 34:469–474

Takahashi M, Takahashi Y, Takahashi K, Zolotaryov FN, Hong KS, Kitazawa R, Iida K, Okimura Y, Kaji H, Kitazawa S, Kasuga M, Chihara K (2008) Chemerin enhances insulin signaling and potentiates insulin-stimulated glucose uptake in 3T3-L1 adipocytes. FEBS Lett 582:573–578

Takekoshi KMotooka M, Isobe K, Nomura F, Manmoku T, Ishii K, Nakai T (1999) Leptin directly stimulates catecholamine secretion and synthesis in cultured porcine adrenal medullary chromaffin cells. Biochem Biophys ResCommun 261:426–431

Tian L, Luo N, Klein RL, Chung BH, Garvey WT, Fu Y (2008) Adiponectin reduces lipid accumulation in macrophage foam cells. Atherosclerosis. doi:10.1016/j.atherosclerosis.2008.04-011

Tschritter O, Fritsche A, Thamer C, Haap M, Shirkavand F, Rahe S, Staiger H Maerker E, Haring H, Stumvoll M (2003) Plasma adiponectin concentrations predict insulin sensitivity of both glucose and lipid metabolism. Diabetes 52:239–243

Tsuji S, Uehori J, Matsumoto M, Suzuki Y, Matsuhisa A, Toyoshima K, Seya T (2001) Human intelectin is a novel soluble lectin that recognizes galactofuranose in carbohydrate chains of bacterial cell wall. J Biol Chem 276:23456–23463

Unger RH,Zhou Y-T, Orci L (1999) Regulation of fatty acid homeostasis in cells: novel role of leptin. Proc Natl Acad Sci USA 96:2327–2332

Van Harmelen V, Reynisdottir S, Cianflone K, Degerman E, Hoffstedt J, Nilsell K, Sniderman A, Arner P (1999) Mechanisms involved in the regulation of free fatty acid release from isolated human fat cells by acylationstimulating protein and insulin. J Biol Chem 274:18243–18251

Walder KFilippis A, Clark S, Zimmet P, Collier GR (1997) Leptin inhibits insulin binding in isolated rat adipocytes. J Endocrinol 155:R5–R7

Wang P, van Greevenbroek MM, Bouwman FG, Brouwers MC, van der Kallen CJ, Smit E, Keijer J, Mariman EC (2007) The circulating PBEF/NAMPT/visfatin level is associated with a beneficial blood lipid profile. Pflugers Arch 454:971–976

Whitehead JP, Richards AA, Hickman IJ, MacDonald GA, Prins JB (2006) Adiponectin: a key adipokine in the metabolic syndrome. Diab Obes Metab 8:264–280

Yamaguchi MMurakami T, Tomimatsu T, Nishio Y, Mitsuda N, Kanzaki T, Kurachi H, Shima K, Aono T, Murata Y (1998) Autocrine inhibition of leptin production by tumor necrosis factor-alpha (TNF-alpha) through TNF-alpha type-I receptor in vitro. Biochem Biophys Res Commun 244:30–34

Yamauchi T, Kamon J, Waki H, Imai Y, Shimozawa N, Hioki K, Uchida S, Ito Y, Takakuwa K, Matsui J, Takata M, Eto K, Terauchi Y, Komeda K, Tsunoda M, Murakami K, Ohnishi Y, Naitoh T, Yamamura K, Ueyama Y, Froguel P, Kimura S, Nagai R, Kadowaki T (2003) Globular adiponectin protected ob/ob mice from diabetes and ApoE-deficient mice from atherosclerosis. J Biol Chem 278:2461–2468

Yang RZ, Lee MJ, Hu H, Pray J, Wu HB, Hansen BC, Shuldiner AR, Fried SK, McLenithan JC, Gong DW (2006) Identification of omentin as a novel depot-specific adipokine in human adipose tissue: Possible role in modulating insulin action. Am J Physiol Endocrinol Metab 290:E1253–E1261

Zakrzewska KE, Cusin I, Sainsbury A, Rohner-Jeanrenaud F, Jeanrenaud B (1997) Glucocorticoids as counterregulatory hormones of leptin: toward an understanding of leptin resistance. Diabetes 46:717–719

Zhang Y, Proenca R, Maffei M, Barone M, Leopold L, Friedman JM (1994) Positional cloning of the mouse obese gene and its human homologue. Nature 372:425–432

Zhou Y-T, Shimabukuro M, Wang M-Y, Lee Y, Higa M, Milburn JL, Newgard CB, Unger RH (1998) Role of peroxisome proliferator-activated receptor alpha in disease of pancreatic beta cells. Proc Natl Acad Sci USA 95:8898–8903

Chapter 12
Cellular Cholesterol Transport – Microdomains, Molecular Acceptors and Mechanisms

Christopher J. Fielding

Abstract "Reverse" cholesterol transport (RCT) from peripheral tissues to the liver is believed to play a major role in preventing accumulation of this lipid locally. Lipid-poor (prebeta-migrating) high-density lipoprotein (prebeta-HDL) plays a key and probably rate-limiting role in RCT, even though only a small proportion of the cholesterol content of circulating HDL originates from RCT. Normal RCT is explained here on the basis of a two-compartment recycling model. Prebeta-HDL is lipidated in interstitial fluid and lymph by ATP-dependent lipid transporters. These particles are then passed to the plasma compartment, where they become lipid-filled under the influence of the lecithin:cholesterol acyltransferase (LCAT) reaction without further input from transporters. In atherosclerosis, where activated macrophages are uniquely in contact with plasma, lipid transporters can directly stimulate RCT driven by LCAT.

12.1 Overview

The cholesterol content of peripheral tissues is maintained by the balance between influx and efflux pathways. Preformed cholesterol is transported out of plasma to the peripheral tissues, mainly in the form of the low-density lipoproteins (LDL) that originate as triglyceride (TG)-rich particles from the liver. Some cholesterol is made locally. This synthesis is suppressed by LDL. The sum of these two fluxes is balanced: (i) in gonadal and adrenal tissues by the conversion of some cholesterol to steroid hormones, (ii) in skin via the secretion of cholesterol from the sebaceous glands by direct loss of cholesterol directly from the skin (Capponi 2002; Fielding and Fielding 2008; Nikkari et al. 1975) and (iii) mainly by centripetal or "reverse" cholesterol transport (RCT) that drives cholesterol net transport from peripheral tissues to the liver for catabolism (Fielding and Fielding 2007; Lewis and Rader 2005).

C.J. Fielding
Cardiovascular Research Institute, Box 0130, University of California
San Francisco, 4th & Parnassus, San Francisco, CA 94143, USA
e-mail: christopher.fielding@ucsf.edu

C. Ehnholm (ed.), *Cellular Lipid Metabolism*,
DOI 10.1007/978-3-642-00300-4_12,

RCT begins when a peripheral cell transfers "excess" cholesterol to acceptors in interstitial fluid (IF). Most of these cells divide rarely or not at all, and they have little use for any additional cholesterol. From there, via the lymphatic collecting ducts, the complex of cholesterol with its protein carrier moves into the plasma compartment. This complex, together with lipids transferred from circulating lipoprotein particles, transfers cholesterol to the liver which is then secreted into bile, partly as cholesterol itself and partly, after catabolism, as bile acid. In both IF and plasma, RCT uses high-density lipoprotein (HDL) for its major carrier. In IF, a lipid-poor HDL fraction with a prebeta-electrophoretic migration rate is also present (Nanjee et al. 2000). It contains about 10% w/w lipid, particularly phospholipids [PL; 2– 3 mol mol^{-1} apolipoprotein (apo)-A-I; Lee et al. 2004]. Prebeta-HDL makes up ~ 5– 10% of HDL particles in normal plasma. The balance of HDL is present in alpha-migrating particles. Alpha-HDL contains ~ 50% of lipid. Apo-A-I is the major and diagnostic protein of all HDL, and is the only protein of its prebeta-HDL fraction.

Two origins for lipid-poor HDL are now recognized. Prebeta-HDL are generated at the surface of hepatocytes that secrete lipid-free apo-A-I. This combines extracellularly with small amounts of PL, mostly lecithin, to generate an efficient acceptor of cellular cholesterol. A little apo-A-I originates from intestinal cells (Krimbou et al. 2006). However, the apo-A-I in most circulating prebeta-HDL is not newly synthesized. It has been detached from larger, more lipid-rich HDL during remodeling of the particle surface in the plasma compartment. This process, still poorly understood, takes place when large HDL interact with plasma lipases and lipid transfer proteins (Hennessy et al. 1993). It is not known whether this "recycled" prebeta-HDL is released from alpha-HDL as lipid-free apo-A-I, or as a lipid-poor complex. The practical effect of apo-A-I recycling between plasma and IF is that each molecule can function multiple times to promote RCT. In this, apo A-I differs from other lipoprotein structural proteins. The apo-A-I polypeptide is exceptionally flexible and can refold into several different conformations. This is probably how apo-A-I adapts to changes in HDL volume. Nevertheless changes in the surface lipid composition of alpha-HDL, as well as HDL diameter, occur at the same time. It is not clear which of these properties is the major driving force in apo-A-I recycling.

Almost all cholesterol in peripheral tissues is free (unesterified). Free cholesterol (FC) transfers spontaneously by simple diffusion through the aqueous phase, or by collision between lipid surfaces (cell membranes, lipoprotein particles). The unproductive exchange of FC between surfaces is much more rapid than observed rates of RCT. This means that the chemical potential of FC in all lipid surfaces would normally be at an equilibrium; and without input of metabolic energy, RCT would be zero.

Two different mechanisms have been identified to drive RCT. The effective concentration of FC in membranes is strongly influenced by their FC/PL molar ratio. Active transport of PL out of cells, driven by an ATP-dependent PL transporter, could promote the exodus of FC. This would continue until the chemical potential of FC in IF once more reached that of the donor membrane. ATP-dependent PL transporters are expressed by most peripheral and hepatic parenchymal cells (O'Connell et al. 2004). They are inactive, or nearly so, in vascular endothelial cells and the formed elements of blood (Hassan et al. 2006; O'Connell et al. 2004). The role of transmembrane or extracellular proteins is emphasized by the observation that RCT to prebeta-HDL is blocked if living peripheral cells (fibroblasts, macrophages, vascular smooth muscle cells) pretreated with an extracellular protease (Kawano et al. 1993).

In plasma (but not IF or lymph), a second mechanism can drive RCT. FC exchanges readily between lipid surfaces while cholesteryl ester (CE) does not. Lecithin:cholesterol acyltransferase (LCAT) converts FC and lecithin (the major PL of both cell membranes and lipoproteins) into CE and lysolecithin. LCAT activity is restricted almost entirely to the plasma compartment (Clark and Norum, 1977). As a result of plasma LCAT activity, most (>75%) of the total plasma HDL cholesterol is esterified. Apo-A-I is cofactor for LCAT activity with HDL.

These data define a two-compartment model (Fig. 12.1). In the first (peripheral cells, IF, lymph) movement of FC out of cells is driven by lipid transporters. These generate prebeta-HDL which contain FC and PL but no CE. The second includes only the plasma. Lipid-poor HDL react with LCAT to form lipid-enriched, alpha-migrating particles. As part of this process additional apo-A-I molecules are recruited to the

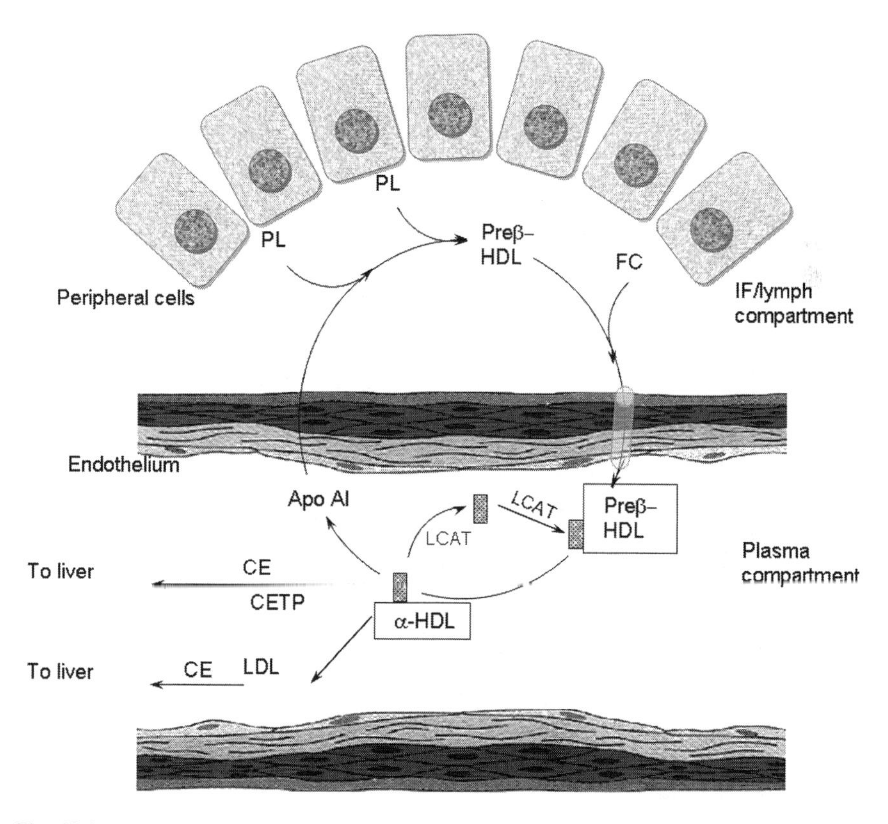

Fig. 12.1 Two-step model of RCT illustrating the lipidation of prebeta-HDL, the intravascular metabolism of prebeta-HDL and alpha-HDL formation via the LCAT reaction and recycling of apo-A-I as part of the surface remodeling of alpha-HDL. *CETP* Cholesteryl ester transfer protein

surface of each enlarging HDL (Nakamura et al. 2004). The detailed organization of apo-A-I at the HDL surface changes as they grow. This is indicated by modification in the pattern of apo-A-I epitopes recognized by monoclonal antibodies (Curtiss et al. 2000). This process involves changes in the composition of both HDL "core" (CE, triglyceride; TG) and surface (mainly FC, PL, apo-A-I). If CE is removed by selective transfer into the liver, a reduction of HDL particle size can lead to the release of apo-A-I. Part of this is filtered through the endothelium back to the IF.

Much of the current interest in RCT has been driven by the hope that a better understanding of the steps involved could lead to strategies that would reduce or prevent the cholesterol accumulation in the vascular bed which leads to atherosclerosis. It might be thought that details of the apo-A-I cycle would be by now well established. However, this is far from the case. Many aspects of the RCT reaction sequence and its regulation are poorly understood, in particular the mechanism of apo-A-I recycling. No assay is yet available to measure apo-A-I recycling in plasma. The rate-limiting step of the apo-A-I cycle is unknown. It is not clear whether the rate of apo-A-I recycling (a major contributor to the availability of prebeta-HDL for RCT) can be modified by drugs or diet. The sections that follow review recent research on each step of this complex reaction.

12.2 Structure and Properties of the Cell Surface

RCT begins at the plasma membrane of quiescent peripheral cells. Like most cells in vivo, these retain little newly synthesized sterol, or FC from internalized LDL. This situation contrasts with that seen in transformed and other continuous cell lines. Here much of the new FC helps form new membrane. The choice of an appropriate cell model to study RCT has bedeviled research in this field.

Quiescent cells have more cholesterol than transformed cells in surface FC-rich microdomains. Much of this cholesterol is in caveolae. These are membrane invaginations stabilized by the structural protein caveolin (Sens and Turner 2004). When primary cells do divide, much of their plasma membrane FC is withdrawn from caveolae and FC efflux is decreased. Caveolin at the cell surface is transferred to intracellular pools (Fielding et al. 1999). Synthesis of new caveolin is inhibited at the transcriptional level. Multiprotein complexes at the cell surface associated with FC microdomains dissociate (Zeidan et al. 2003). Each of these processes is reversed in a co-ordinated way when the cell re-enters the quiescent stage.

There is controversy as to whether FC-rich (caveolar) or FC-poor microdomains of the peripheral cell surface contribute most to the FC that is lost from peripheral cells by RCT. It has been argued, based on the structure of synthetic phospholipid vesicles and in particular from its condensing effect on PL acyl chains, that less energy would be needed to transfer FC from lipid-poor areas than from FC-rich domains of the plasma membrane (Rothblat et al. 1992). However it was recently shown that lipid packing is looser, not tighter, in caveolae compared to FC-poor domains in the same membranes (Hill et al. 2005). This probably reflects that distorting effects on caveolin on local lipid packing in caveolae.

When the FC content of caveolae was monitored with cholesterol oxidase, this pool was selectively depleted when the cells were incubated either with β-methyl cyclodextrin, a synthetic FC sequestrant (Parpal et al. 2001), or with prebeta-HDL in normal plasma (Fielding and Fielding 1995). It has been reported that, in detergent-treated membranes, PL transporters and caveolin appear in different fractions (Mendez et al. 2001; Sdrobnik et al. 2002). However it has since been recognized that this treatment can modify the composition of membranes (Ortegren et al. 2004). Most recently, confocal microscopy of intact plasma membranes has shown that lipid transporters and caveolae may be in the same microdomain in some intact cells (Lin et al. 2007).

There is also disagreement about the contributions of different molecular mechanisms to RCT. Passive FC exchange, not RCT, was found to predominate in some continuous cell lines (Adorni et al. 2007). In contrast, receptor-based mechanisms depending on lipid-poor HDL appeared to be most important in cultured primary cells, such as those from human aortic endothelium and smooth muscle (Fielding and Fielding 2001). These arguments emphasize the relevance of primary cells as models of RCT.

12.3 Role of Cell-Surface Lipid Transporters in RCT

In IF, LCAT is almost absent (Clark and Norum 1977). A major role for plasma membrane PL transporters there is indicated. Contributions by several members of the ATP-binding cassette (ABC) family of lipid transporters have been suggested. One of these, ABCA1, is now broadly accepted to play a key role (Singaraja et al. 2006). In human Tangier disease, where ABCA1 is genetically defective, no normal HDL are formed, despite normal synthesis and secretion rates for apo-A-I. The low circulating concentration of apo-A-I in the plasma of Tangier patients (1– 2% of normal) is secondary, reflecting an inability of Tangier cells to transfer PL to apo-A-I to initiate RCT. Also, the lipid-free polypeptide is rapidly filtered by the kidney. A high proportion of apo-A-I in Tangier plasma circulates as a proprotein that retains a 6-amino-acid propeptide. Proapo-A-I in Tangier disease plasma reflects the high proportion of newly secreted protein in this fraction. Newly secreted proapo-A-I bound PL poorly compared to the mature protein (Chau et al. 2006). We recently showed proapo-A-I to be cleaved extracellularly by a metallic proteinase, bone morphogenetic protein-1 (BMP-1) (Chau et al. 2007). These data confirm that ABCA1 plays a key role in the early synthesis of normal HDL for RCT.

Other studies suggest that ABCA1 plays additional roles. Transport of proteins through the Golgi secretion pathway is inhibited in ABCA1 –/– cells (Zha et al. 2003). The transporter may play a role in regulating the PL composition of Golgi transport vesicles. A second activity supported by ABCA1 is that of a molecular chaperone, refolding apo-A-I at the surface of hepatocytes, prior to its PL lipidation by ABCA1. Transfer of PL mediated by ABCA1 follows, and depends on, this earlier refolding step. The chaperone and PL transporter roles of ABCA1 can be distinguished by the sensitivity of only the latter function to the ATPase inhibitor glyburide (Chau et al. 2006).

There has been controversy as to whether FC is carried to apo-A-I by primary active transport of ABCA1. Because of the complex three-dimensional structure of this transporter and a lack of information about its mechanism, protein binding sites that might be directly involved in PL and FC transport have not yet been identified. However, several studies throw additional light on this question. PL transfer to the exofacial leaflet of the membrane bilayer by ABCA1 is not dependent on the presence of apo-A-I or other extracellular acceptors. Cells expressing ABCA1 that are incubated in the absence of apo-A-I accumulate "blebs" on their exofacial surface that represent "excess" PL (Wang et al. 2000). Some Tangier cell lines with mutations within the extracellular loops of ABCA1 lack PL transfer activity, though they retain the ability to bind apo-A-I (Singaraja et al. 2006). Transfer of PL to apo-A-I does not depend on co-transport of FC, since this lipid can be transferred to preformed prebeta-HDL by cells without functional ABCA1 (Fielding et al. 2000). These data argue that direct (primary) FC transport is not a necessary, and possibly, not even a significant function of this transporter.

ABCA1 may play one additional role, by mediating PL transfer from a cell surface to the plasma phospholipid transfer protein (PLTP; Oram et al. 2003). PLTP actively redistributes PL between lipoprotein and PL vesicles. PLTP deficiency reduced FC efflux promoted by ABCA1 in macrophage foam cells (Lee-Rueckert et al. 2006). Part of the PL leaving the cell under the influence of ABCA1 may be directed to lipoproteins other than prebeta-HDL (alpha-HDL is not an acceptor for ABCA1-derived lipids]. It is not known whether either lipid-free apo-A-I or prebeta-HDL could accept PL directly from PLTP.

12.4 Cholesterol Efflux and the LCAT Reaction

Newly synthesized apo-A-I of hepatic origin and recycled apo-A-I can both be converted to lipid-poor, prebeta-migrating HDL that initiates RCT. It was widely accepted until recently that prebeta-HDL could grow in IF and/or lymph by additional of further PL and FC from surrounding cells, in the absence of LCAT, to generate a discoidal product resembling that formed in vitro from isolated lipid-free apo-A-I and PL. These discoidal HDL could then be converted to spherical alpha-HDL by the action of LCAT. This model seemed to be supported by the observation that in congenital and acquired LCAT deficiency, discoidal lipoproteins were a prominent feature of the lipoprotein spectrum in plasma and lymph.

More recently, it seemed likely that this otherwise attractive scenario does not fit experimental observations on the properties and composition of lymph HDL and on the substrate specificity of LCAT. Discoidal HDL are almost absent from normal human lymph (Kujiraoka et al. 2003). Discoidal HDL identified in plasma and lymph in human LCAT deficiency were found to contain bound apo-E, but very little apo-A-I. In fact, prebeta-HDL of apparently normal properties were prominent there, in co-existence with the apo-E discs (Mitchell et al. 1980). Further, in cultures of human hepatocytes, even extended incubation of prebeta-HDL did not lead to the appearance of apo-A-I discoidal HDL or other PL-rich forms larger than prebeta-HDL (Chau et al. 2006, 2007). These data suggest that discoidal HDL might not, after all, be an obligatory intermediate in the conversion of prebeta- to alpha-HDL.

By studying synthetic lipid-poor recombinants of apo-A-I, PL and FC with a composition similar to that of prebeta-HDL, Sparks et al. (1999) showed that LCAT was very reactive with apo-A-I monomers containing as few as 2– 3 moles of PL and FC. The V_{max} of LCAT with these lipid-poor apo-A-I particles was greater than with native alpha-HDL. These observations were extended by our laboratory. We studied the metabolism of lipid-poor (prebeta-)HDL in native plasma (Nakamura et al. 2004). These particles not only efficiently bound ^{3}H-FC newly transferred from equilibrium-labeled cultured cell monolayers. They also reacted more efficiently with LCAT than the rest of HDL in plasma. The significance of this observation is that a pathway by which cell-derived FC was a preferential substrate for LCAT would significantly increase the efficiency of RCT. CE accumulated in the labeled alpha-HDL particle. Of interest, these particles became enlarged by fusion with additional apo-A-I units. These data strongly support a model in which prebeta-HDL, released into the plasma from lymph, is directly a substrate for LCAT. No evidence was found for an intermediate discoidal HDL. As the lipid content of unreacted prebeta-HDL is very small, we were unable to determine whether lipid-free apo-A-I or prebeta-HDL bound to the enlarging complex initiated by prebeta-HDL and LCAT. These data suggest a model in which prebeta-HDL, first formed in IF by binding cell-derived PL and FC, is converted to alpha-HDL in plasma by LCAT. It is possible that apo-A-I recycles from alpha-HDL with a small amount of PL. The evidence is not yet available.

There is an important consequence of these findings in RCT in vivo. It is that most of the FC supplied to alpha-HDL over its life-time must originate not by RCT from peripheral cells, but by spontaneous transfer from other lipoproteins (and perhaps from red blood cells) in the plasma compartment. This means that the rate of LCAT activity and that of other parameters reflecting cholesterol metabolism in the plasma compartment such as HDL cholesterol (HDL-C) levels, need not reflect the rate of RCT. The rate-limiting step of RCT is likely to be determined by events in the IF/lymphatic pool – the activity of PL transporter ABCA1, or the availability of its PL acceptor (newly secreted or recycled apo-A-I).

LCAT converts PL and FC in equal numbers to CE. About 30% of the CE formed on the reaction product of LCAT and prebeta-HDL was transferred away to other plasma lipoprotein particles (Nakamura et al. 2004). Some CE is likely lost to the liver via receptor CLA-1 (the human analog of SR-BI, though the proportions of HDL-CE uptake by this and endocytosis pathways are not yet established and its expression in human liver is lower than in steroid-forming tissues; Cao et al. 1997). We can first estimate how many molecules of FC and PL are needed to convert prebeta-HDL in IF to alpha-HDL in plasma (Fig. 12.2).

Each prebeta-HDL would contribute 1– 2 mol of FC, or a maximum of 3– 6 mol for an alpha-HDL particle containing three apo-A-I. The CE content of the alpha-HDL can be shown to contain ~ 38 mol of CE, representing 70% of the CE formed there (54 mol of CE) formed by LCAT from the same number of FC. The HDL contains 10 mol of FC. Formation of an alpha-HDL particle from three prebeta-HDL requires roughly 64 FC, of which no more than three to six would have originated, by RCT, from peripheral cell membranes. The balance (58– 61 FC) would have been transferred by simple diffusion of collision from other plasma lipoprotein particles, mostly LDL. The LCAT activity in plasma is ~ 20 μ;g FC esterified ml^{-1} plasma h^{-1} or about 2.4 g day^{-1}

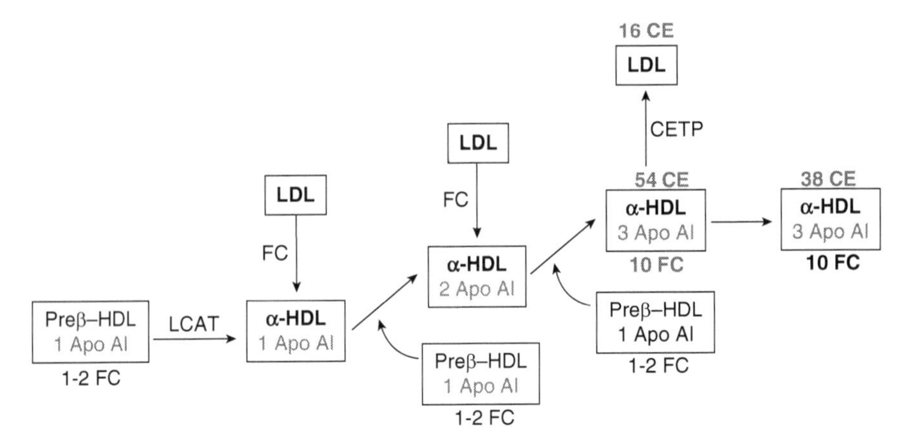

Fig. 12.2 Phospholipid (*PL*), free cholesterol (*FC*) and cholesteryl ester (*CE*) fluxes on HDL during RCT

for a human subject of 75 kg and a plasma volume of 5 l. Applying the ratio calculated above, this suggests that RCT in this subject is 0.12– 0.24 g FC day^{-1}. If RCT contributes only a small proportion of FC required for LCAT activity, this could explain why plasma parameters such as LCAT and CETP activities, and HDL cholesterol itself, are normally not sensitive indicators of RCT (Dietschy and Turley 2002).

12.5 Significance of ABCG1

A key role in RCT has been ascribed to a second cell surface transporter, ABCG1 (Cavelier et al. 2006). This promotes FC from cell surfaces to alpha-migrating HDL and PL vesicles, but not to lipid-free apo-A-I. In normal mice in which ABCG1 had been knocked out, the plasma cholesterol level, including that of HDL cholesterol, remained within normal limits. The HDL in these animals also had a normal composition. The main pathological finding was an accumulation of PL and FC in lung macrophages. Although ABCG1 was widely expressed, levels of FC and CE were not increased in most cells in ABCG1 – /– mice (Kennedy et al. 2005). Homozygous ABCG1 mutation in human subjects has not been reported, so whether this would be linked to disease is not known. ABCG1, though present, was inactive in endothelium in promoting FC efflux to HDL (O'Connell et al. 2004). In view of the large contribution of plasma FC and PL to the enlargement of alpha-HDL, it would be surprising if ABCG1 played an obligatory role in normal HDL metabolism.

12.6 Recycling of apo-A-I

It is widely agreed that the stability of the hydrophobic core of plasma lipoprotein particles depends on the presence of a monomolecular film of polar lipids and protein. Alpha-HDL are subject to a number of different reactions that affect their core/surface ratio. CE is transferred from the core of HDL to acceptor tissues (adrenals, gonads, liver) via scavenger receptor B-I. Independently, part of HDL-CE is exchanged for TG on other lipoprotein particles via the activity of CETP. Unlike the reaction with SR-BI, this has little direct effect on HDL core volume, though a TG molecule is somewhat larger than a CE. However the exchange of CE for TG makes these HDL a substrate by plasma lipases, whose product, unesterified fatty acids, is transferred away to albumin.

A net decrease in core can lead to one of two effects: further CE formation and storage under the influence of LCAT, or loss of "surface" components (PL, FC, protein, especially apo-A-I). The equilibrium between these processes appears to be one major factor that stabilizes the distribution of HDL species in the plasma compartment. Other changes affect the composition of HDL without changing the ratio of surface to volume. For example, another protein could binding to HDL, displacing apo-A-I and promoting recycling and RCT. It was recently found that when a newly identified HDL protein, apolipoprotein M (apo-M) was knocked out in mice, prebeta-HDL entirely disappeared from plasma. If apo-M was over-expressed, prebeta-HDL levels were increased (Wolfrum et al. 2005). These data suggest that an equilibrium between apo-M and apo-A-I could determine recycling. However other interpretations are possible. When LCAT activity is blocked in plasma, prebeta-HDL levels increase. When LCAT acts in plasma in the absence of nucleated cells, prebeta-HDL disappears (Kawano et al. 1993). Apo-M could mediate its observed effects not by promoting apo-A-I recycling, but by regulating LCAT activity. Apo-M could stimulate PLTP activity to promote prebeta-HDL production indirectly. A systematic study of the effects of apo-M on the metabolic rates of lipid factors in native plasma will be needed to unravel this skein of connected reactions.

A second unresolved question is the composition of the apo-A-I that dissociates from alpha-HDL. Is this lipid-poor HDL with the composition of prebeta-HDL in plasma? Or is it lipid-free apo-A-I? If so, can this apo-A-I immediately bind PL from plasma lipoproteins, or does it need to interact first in IF with ABCA1? This is an important issue, because it frames whether ABCA1 has a continuing role to play in activating apo-A-1 for RCT, or this is limited to the activation of newly secreted apo-A-I. We recently presented data indicating that new apo-A-I was monomeric not dimeric. The same seems likely for recycled apo-A-I. Otherwise, co-existence of alpha-HDL with 2, 3 and 4 apo-A-1/particle would be difficult to explain.

12.7 RCT from Activated Macrophages

RCT from extrahepatic tissues is important in determining the FC content of their cells. The ideal response of cells to increased cholesterol uptake local synthesis of FC, would be to increase RCT via the two-compartment pathway described above by increasing the activity of ABCA1 or concentration of prebeta-HDL in IF. However another mechanism may come into play in atherosclerosis. In this condition, the physical separation of the earlier and later steps of RCT breaks down. Vascular lesions (plaques) containing lipid-filled activated macrophages (foam cells) are exposed at the endothelial surface of the large vessels. These cells express high levels of ABCA1 and ABCG1 (Out et al. 2008) and these activities are directly

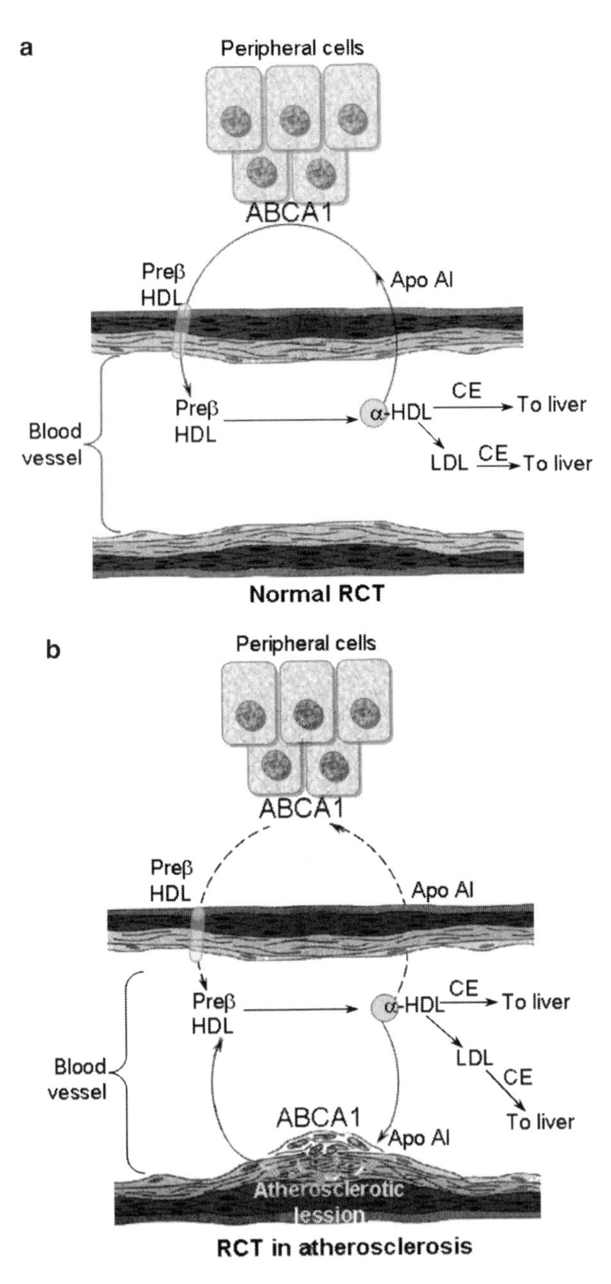

Fig. 12.3 Comparison of RCT between normal peripheral cells (**a**) and vascular lesion-activated macrophages (**b**)

accessible to both prebeta-HDL and alpha-HDL in the plasma (Fig. 12.3). If this model is correct, the expression of lipid transporters in macrophage/foam cells may represent a defense (though ultimately not successful) against increased FC levels in the vascular bed, using increased levels of apo-A-I recycling, equilibrium prebeta-HDL and RCT.

References

Adorni MP, Zimetti F, Billheimer JT, Wang N, Rader DJ, Phillips MC, Rothblat GH (2007) The roles of different pathways in the release of cholesterol from macrophages. J Lipid Res 48:2453–2462

Cao G, Garcia CK, Wyne KL, Scultz RA, Parker KL, Hobbs HH (1997) Structure and localization of the human gene encoding SR-BI/CLA-1. Evidence for transcriptional control by steroidogenic factor-1. J Biol Chem 272:33068–33076

Capponi AM (2002) Regulation of cholesterol supply for mineralocorticoid biosynthesis. Trends Encrocrin Metab 13:118–121

Cavelier C, Lorenzi I, Rohrer L, von Eckardstein A (2006) Lipid efflux by the ATP-binding cassette transporters ABCA1 and ABCG1. Biochim Biophys Acta 1761:655–666

Chau P, Nakamura Y, Fielding CJ, Fielding PE (2006) Mechanism of prebeta HDL formation and activation. Biochemistry 45:3981–3987

Chau P, Fielding PE, Fielding CJ (2007) Bone morphogenetic protein-1 (BMP-1) cleaves human proapolipoprotein A1 and regulates its activation for lipid binding. Biochemistry 46:8445–8450

Clark SB, Norum KR (1977) The lecithin-cholesterol acyltrasnferase activity of lymph. J Lipid Res 18:293–300

Curtiss LK, Bonnet DJ, Rye KA (2000) The conformation of apolipoprotein A-I in high density lipoproteins is influenced by lipid composition and particle size: a surface plasmon resonance study. Biochemistry 39:5712–5721

Dietschy JM, Turley SD (2002) Control of cholesterol turnover in the mouse. J Biol Chem 277:3801–3804

Fielding CJ, Fielding PE (2001) Cellular cholesterol efflux. Biochim Biophys Acta 1533:175–189

Fielding CJ, Fielding PE (2007) Reverse cholesterol transport - new roles for prebeta1-HDL and lecithin:cholesterol acyltransferase. In: Fielding CJ (ed) High density lipoproteins. Wiley-VCH, Weinheim, pp 143–161

Fielding CJ, Fielding PE (2008) In: Vance D, Vance J(eds) Lipids, lipoproteins and membranes, 5th edn. Elsevier, London, pp 533–553

Fielding PE, Fielding CJ (1995) Plasma membrane caveolae mediate the efflux of cellular free cholesterol. Biochemistry 34:14288–14292

Fielding CJ, Bist A, Fielding PE (1999) Intracellular cholesterol transport in synchronized human skin fibroblasts. Biochemistry 38:2506–2513

Fielding PE, Nagao K, Hakamata H, Chimini G, Fielding CJ (2000) A two-step mechanism for free cholesterol and phospholipid efflux from human vascular cells to apolipoprotein A-I. Biochemistry 39:14113–14120

Hassan HH, Denis M, Krimbou L, Marcil M, Genest J (2006) Cellular cholesterol homeostasis in vascular endothjelial cells. Can J Cardiol 22:35B–40B

Hennessy LK, Kunitake ST, Kane JP (1993) Apolipoprotein-AI-containing lipoproteins, with or without apolipoprotein A-II, as progenitors of prebeta high density lipoprotein particles. Biochemistry 32:5759–5765

Hill WG, Almasri E, Ruiz WG, Apodaca G, Zeidel ML (2005) Water and solute permeability of rat lung caveolae: high permeabilities explained by acyl chain unsaturation. Am J Physiol Cell Physiol 289:C33–C41

Kawano M, Miida T, Fielding CJ, Fielding PE (1993) Quantitation of prebeta-HDL-dependent and nonspecific components of the total efflux of cellular cholesterol and phospholipids. Biochemistry 32:5025–5028

Kennedy MA, Barrera GC, Nakamura K, Baldan A, Tarr P, Fishbein MC, Frank J Francone OL, Edwards PA (2005) ABCG1 has a critical role in mediating cholesterol efflux to HDL and preventing cellular lipid accumulation. Cell Metab 1:121–131

Krimbou L, Marcil M, Genest J (2006) New insights into the biogenesis of human high-density lipoproteins. Curr Opin Lipidol 17:258–267

Kujiraoka T, Nanjee MN, Oka T, Ito M, Nagano M, Cooke CJ, Takahashi S, Olszewski WL, Wong JS, Stepanova IP, Hamilton RL, Egashira T, Hattori H, Miller NE (2003) Effects of intravenous apolipoprotein A-I/phosphatidylcholine discs on LCAT, PLTP and CETP in plasma and peripheral lymph in humans. Arterioscler Thromb Vasc Biol 23:1653–1659

Lee JY, Lanningham-Foster L, Boudyguina EY, Smith TL, Young ER, Colvin PL, Thomas MJ, Parks JS (2004) Prebeta high density lipoprotein has two metabolic fates in human apolipoprotein A-I transgenic mice. J Lipid Res 45:716–728

Lee-Rueckert M, Vikstedt R, Metso J, Ehnholm C, Kovanen PT, Jauhiainen M (2006) Absence of endogeneous phospholipid transfer protein impairs ABCA1-dependent efflux of cholesterol from macrophage foam cells. J Lipid Res 47:1725–1732

Lewis GF, Rader DJ (2005) New insights into the regulation of HDL metabolism and reverse cholesterol transport. Circ Res 96:1221–1232

Lin YC, Ma C, Hsu WC, Lo HF, Yang VC (2007) Molecular interaction between caveolin-1 and ABCA1 on high density lipoprotein-mediated cholesterol efflux in aortic endothelial cells. Cardiovascular Res 75:575–583

Mendez AJ, Lin G, Wade DP, Lawn RM, Oram JF (2001) Membrane domains distinct from cholesterol/sphingomyelin-rich rafts are involved in the ABCA1-mediated lipid secretory pathway. J Biol Chem 276:3158–3166

Mitchell CD, King WC, Applegate KR, Forte T, Glomset JA, Norum KR, Gjone E (1980) Characterization of apolipoprotein E-rich high density lipoproteins in familial lecithin:cholesterol acyltransferase deficiency. J Lipid Res 21:625–634

Nakamura Y, Kotite L, Gan Y, Spencer TA, Fielding CJ, Fielding PE (2004) Molecular mechanism of reverse cholesterol transport: reaction of prebeta-migrating high density lipoprotein with plasma lecithin:cholesterol acyltransferase. Biochemistry 43:14811–14820

Nanjee MN, Cooke CJ, Olszewski WL, Miller NE (2000) Concentrations of electrophoretic and size subclasses of apolipoprotein AI-containing particles in human peripheral lymph. Arterioscler Thromb Vasc Biol 20:2148–2155

Nikkari T, Schreibman PH, Ahrens EH (1975) Isotope kinetics of human skin cholesterol secretion. J Exp Med 141:620–634

O'Connell BJ, Denis M, Genest J (2004) Cellular physiology of cholesterol efflux in vascular endothelial cells. Circulation 110:2881–2888

Oram JF, Wolfbauer G, Vaughan AM, Tang C, Albers JJ (2003) Phospholipid transfer protein interacts with and stabilizes ATP-binding cassette transporter A1 and enhances cholesterol efflux from cells. J Biol Chem 278:52379–52385

Ortegren U, Karlsson M, Blazic N, Blomqvist M, Nystrom FH, Gustavsson J, Fredman P, Stralfors P (2004) Lipids and glycosphingolipids in caveolae and surrounding plasma membrane of primary rat adipocytes. Eur J Biochem 271:2028–2036

Out R, Hoekstra M, Habets K, Meurs I, de Waard V, Hildebrand RB, Wang Y, Chimini G, Kuiper J, Van Berkel TJ, Van Eck M (2008) Combined deletion of macrophage ABCA1 and ABCG1 leads to massive lipid accumulation in tissue macrophages and distinct atherosclerosis at relatively low plasma cholesterol levels. Arterioscler Thromb Vasc Biol 28:258–264

Parpal S, Karlsson M, Thorn H, Stralfors P (2001) Cholesterol depletion disrupts caveolae and insulin receptor signaling for metabolic control via insulin receptor substrate-1 but not for mitogen-activated protein kinase control. J Biol Chem 276:9670–9678

Rothblat GH, Mahlberg FH, Johnson WJ, Phillips MC (1992) Apolipoproteins, membrane cholesterol domains, and the regulation of cholesterol efflux. J Lipid Res 33:1091–1097

Sdrobnik W, Borsukova H, Bottcher A, Pfeiffer A, Liebisch G, Schutz GJ, Schindler H, Schmitz G (2002) Apo AI/ABCA1-dependent and HDL3-mediated lipid efflux from compositionally distinct cholesterol-based microdomains. Traffic 3:268–278

Sens P, Turner MS (2004) Theoretical model for the formation of caveolae and similar membrane invaginations. Biophys J 86:2049–2057

Singaraja RR, Visscher H, James ER, Chroni A, Coutinho JM, Brunham LR, Kang MH, Zannis VI, Chimini G, Hayden MR (2006) Specific mutations in ABCA1 have discrete effects on ABCA1 function and lipid phenotypes both in vivo and in vitro. Circ Res 99:389–397

Sparks DL, Frank PG, Braschi S, Neville TA, Marcel YL (1999) Effect of apolipoprotein A-I lipidation on the formation and function of prebeta- and alpha-migrating LpA-I particles. Biochemistry 38:1727–1735

Wang N, Silver DL, Costet P, Tall AR (2000) Specific binding of apo A-I, enhanced cholesterol efflux and altered plasma membrane morphology in cells expressing ABCA1. J Biol Chem 275:33053–33058

Wolfrum C, Poy MN, Stoffel M (2005) Apolipoprotein M is required for prebeta-HDL formation and cholesterol efflux to HDL and protects against atherosclerosis. Nat Med 11:418–422

Zeidan A, Broman J, Hellstrand P, Sward K (2003) Cholesterol dependence of vascular ERK1/2 activation and growth in response to stretch: role of endothelin-1. Arterioscler Thromb Vasc Biol 23:1528–1534

Zha X, Gauthier A, Genest J, McPherson R (2003) Secretory vesicular transport from the Golgi is altered during ATP-binding cassette protein A1 (ABCA1)-mediated cholesterol efflux. J Biol Chem 278:10002–10005

Chapter 13
The Ins and Outs of Adipose Tissue

Thomas Olivecrona and Gunilla Olivecrona

Abstract The aim of this chapter is to discuss what mechanisms are available to rapidly modulate fatty acid uptake/mobilization in adipose tissue. The major pathway for net uptake is lipoprotein lipase (LPL)-mediated hydrolysis of lipoprotein lipids. There are several mechanisms for control and they all serve to suppress LPL activity on a time-scale of hours in the setting of essentially unchanged LPL mRNA and mass. A protein complex that specifically binds to LPL mRNA can block synthesis of new enzyme. The Ca^{2+} milieu, and perhaps other conditions in the ER, can partition more of the enzyme towards intracellular degradation and less for export. After secretion from the adipocytes, active LPL can be converted into inactive monomers through interaction with angiopoietin-like proteins. At the vascular endothelium, product control may balance LPL action. If fatty acids accumulate at sites of lipolysis they eliminate the effect of apolipoprotein CII, which is a necessary activator for LPL. Intracellular lipolysis is initiated by adipose tissue triglyceride lipase (ATGL) which hydrolyzes triglycerides to diglycerides. These can either be re-esterified by a diacylglycerol acyl transferase (DGAT) enzyme or further hydrolyzed by hormone-sensitive lipase (HSL). The system is controlled by phosphorylation mediated by protein kinase A, and perhaps other protein kinases, as well as protein phosphatases. The prime target is perilipin, a lipid droplet protein which in its unphosphorylated form suppresses the activity of both ATGL and HSL. The two lipase systems are modulated by different mechanisms and on different time-scales. Both systems seem to operate at levels that generate an excess of fatty acids. The overriding control of how much gets deposited in the tissue as triglyceride and how much spills over into blood as albumin-bound fatty acids (NEFA) is exerted by the rate of glyceride synthesis. Recent studies show that glycerol-3-phosphate for this is generated mainly through glyceroneogenesis from citric acid cycle intermediates.

T. Olivecrona and G. Olivecrona
Department of Medical Biosciences/Physiological Chemistry, Bldg 6M,
3rd Floor, Umeå University, SE-90187 Umeå, Sweden
e-mail: thomas.olivecrona@medbio.umu.se
e-mail: gunilla.olivecrona@medbio.umu.se

C. Ehnholm (ed.), *Cellular Lipid Metabolism*,
DOI 10.1007/ 978-3-642-00300-4_13,

Abbreviations Angptl, angiopoietin-like protein; Apo, apolipoprotein; AKAP, A kinase-binding protein; AQP, aquaporin; ATGL, Adipose triglyceride lipase; CETP, cholesteryl ester transfer protein; CGI-58, comparative gene identification 58; cld, combined lipase deficiency; CoA, coenzyme A; DGAT, diacylglycerol acyl transferase; DG, diglyceride; EM, electron microscopy; ER, endoplasmatic reticulum; FATP, fatty acid transport protein; GPI, glycerolphosphatidylinositol; GPIHBP, glycerolphosphatidylinositol-linked high-density binding protein; HDL, high-density lipoprotein; HSL, hormone-sensitive lipase; HSPG, heparan sulfate proteoglycan; LDL, low-density lipoprotein; Lmf, lipase maturation factor; LPL, lipoprotein lipase; LPS, lipopolysaccharide; LRP, low-density lipoprotein receptor-related protein; MAGH, monoacylglycerol hydrolase; MG, monoglyceride; MGAT, monoacylglycerol acyl transferase; NEFA, non-esterified fatty acids (also called albumin-bound free fatty acids); PTH, parathyroid hormone; PKA, protein kinase A; RAP, receptor-associated protein; SR-B1, scavenger receptor type B1; SREBP, steroid regulatory element binding protein; TG, triglyceride; UTR, untranslated region; VLDL, very low-density lipoprotein

13.1 Introduction

The major chemical form in which the body stores energy substrate is the triglyceride. This is a smart molecule for the purpose. Triglycerides can be stored in very compact form as lipid droplets (Chap. 1). When mobilized during starvation, the triglyceride molecule provides fatty acids as a direct energy source for most tissues and a substrate for production of ketone bodies which, together with glucose produced by gluconeogenesis from the glycerol released, can provide an energy substrate for the nervous system. On a day to day basis, the main function of adipose tissue is to serve as a buffer for daily lipid flux (Frayn 2002), with net uptake of fatty acids in the postprandial state and net release between meals. The tissue also sends out a number of signal molecules that participate in the regulation of energy metabolism, substrate selection and appetite (Chap. 12).

The triglyceride molecule cannot be efficiently moved across cell membranes and extracellular spaces. For this it has to be split into fatty acids and monoglycerides which move more readily. Therefore, lipases are central for both uptake and release of lipids from adipose tissue. This chapter focuses on the properties of these lipases and what mechanisms are available to control their action in harmony with overall energy metabolism in the body. Our objective is not to cover the physiology, endocrinology or pathology of these processes. There are several excellent recent reviews on these subjects (Arner and Langin 2007; Brasaemle 2007; Granneman and Moore 2008; Jaworski et al. 2007; Mead et al. 2002; Merkel et al. 2002a; Otarod and Goldberg 2004; Preiss-Landl et al. 2002; Pulinilkunnil and Rodrigues 2006; Stein and Stein 2003; Watt and Steinberg 2008; Zechner et al. 2000, 2005, 2008).

A word of caution is in place before we begin our discussion. Most of the data we discuss come from experiments on rodents and are not necessarily directly applicable

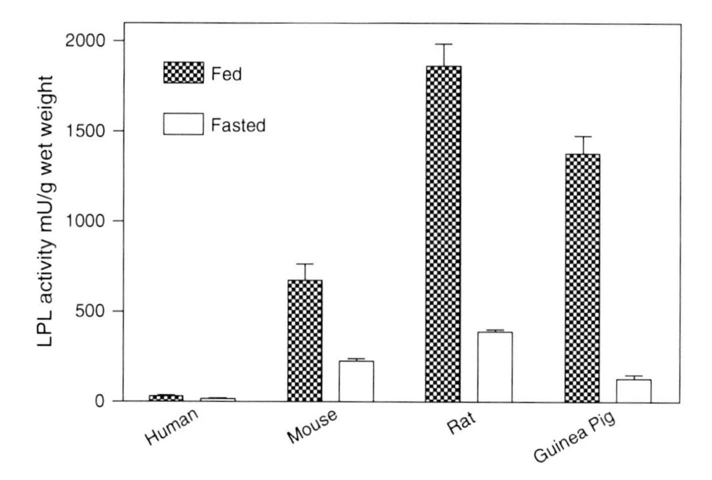

Fig. 13.1 Changes of adipose tissue LPL activity with nutritional state. The data are from previous publications from our laboratory (Bergo et al. 1996; Ruge et al. 2004, 2005; Semb and Olivecrona 1986) and represent total LPL activity per gram tissue before and after fasting overnight. For rats it is known that the activities change with age and body weight. The same is probably true for guinea pigs and mice, so that the activities in the fed state in these three species should be considered as in the same range. The ability to down-regulate LPL activity on fasting appears to be more pronounced in guinea pigs than in rats or mice. The most marked species difference is, however, the much lower adipose tissue LPL activity in humans compared to the other three species, both fed and fasted

to human physiology. On the contrary, we know that there are major differences in the lipid transport systems. Compared to humans, rats and mice have low LDL and high HDL, they have no CETP in their blood and they produce chylomicron-like apoB48-lipoproteins in their livers. In mice the hepatic lipase circulates in plasma, whereas in humans (and rats) it is confined to the liver. Important to the current topic is the large difference in adipose tissue lipoprotein lipase (LPL) activity between rodents and humans (Fig. 13.1). Also important for interpretation of the data are the regional differences between different localities of adipose tissue, subcutaneous, visceral, pericardial, mammary, etc (Santosa and Jensen 2008). A further complexity is the differences between small adipocytes from young/lean animals compared to enlarged adipocytes in older/obese animals (Hartman 1977). We do not address these variations between species and between anatomic locations in any systematic manner since the objective of our review is to discuss the mechanisms involved in lipid uptake and release from a biochemical/molecular biological perspective.

13.2 Sources of Lipids for Deposition in Adipose Tissue

In principle, fatty acids can be deposited in adipose tissue by net uptake from blood or by local synthesis in the tissue (Fig. 13.2). The adipocytes have the enzymes required for fatty acid synthesis, but under most conditions de novo fatty acid synthesis in

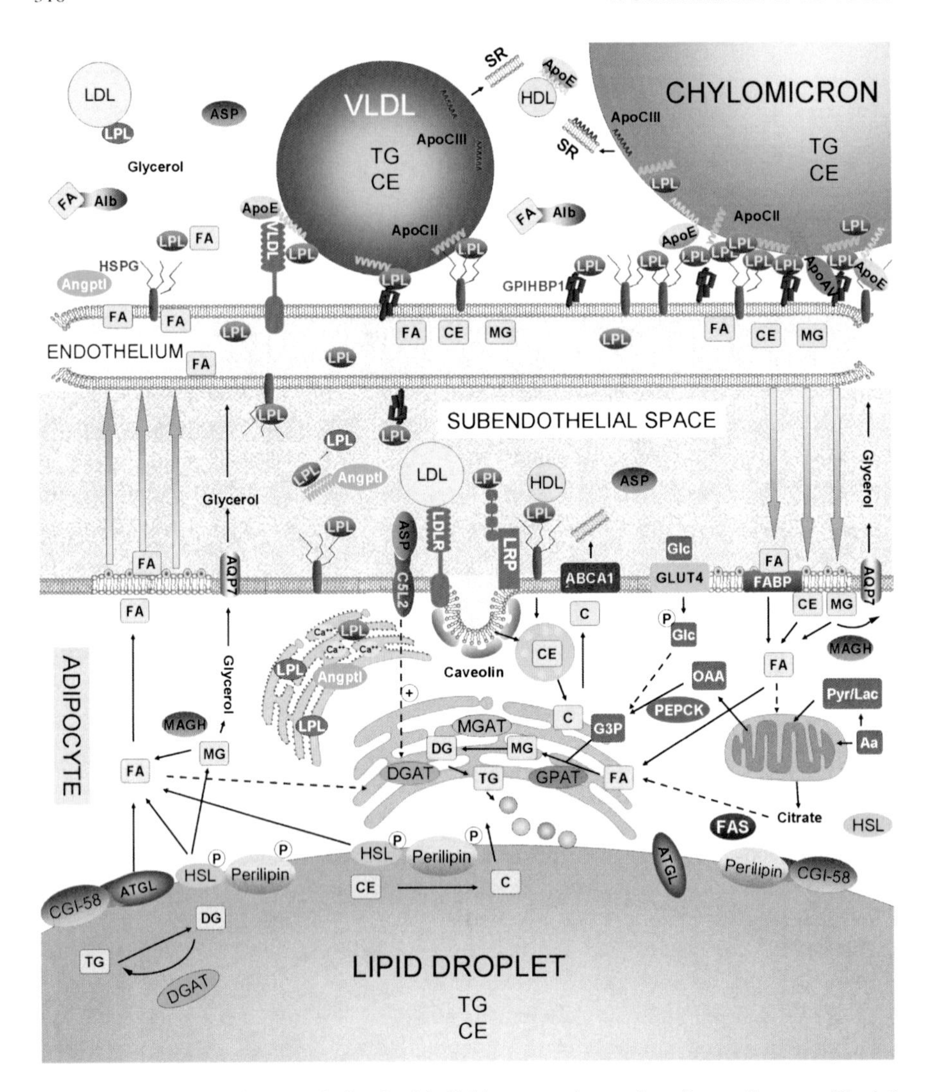

Fig. 13.2 The metabolic networks involved in lipid transport into and out from adipocytes. The *left* part shows pathways involved in mobilization. Perilipin is phosphorylated and has brought HSL (which is also phosphorylated) to the lipid droplet. ATGL forms a complex with its activator (CGI-58) and hydrolyzes TG to DG which are either re-esterified to TG by DGAT, or further hydrolyzed to MG by HSL and to glycerol by MAGH. The glycerol moves out from the adipocytes through the AQP7 channel. FA from all three steps of hydrolysis (also from hydrolysis of CE) probably mix and move out from the adipocyte, perhaps primarily through raft areas of the plasma membrane. After crossing (or moving around in the membrane of) the endothelial cells the fatty acids bind to albumin and leave the tissue. LPL is synthesized in RER (still inactive, *gray* symbol) and matures with the aid of calcium ions (Ca^{2+}), into active dimers (*red* symbol) which are secreted and bind to heparan sulfate proteoglycans (HSPG), or other LPL-binding molecules, at the surface of the adipocytes. In the fasting state, production of angptl4 in the adipocytes is up-regulated and the protein converts most of the LPL dimers to inactive monomers, probably outside the adipocyte. Some LPL molecules escape this and move along cell surfaces to reach the luminal surface of blood vessels. High affinity binding to GPIHBP1 may cause a gradient (and/or in vesicles through the endothelial cells) that drives the movement of active LPL.

the tissue is of small magnitude (Chong et al. 2007). The main source of fatty acids is the lipoproteins in blood. This is true in man as well as in rats and mice. One should be aware that the conditions may be different in other animal species. For instance, in ruminants, a major pathway for lipid deposition is de novo synthesis from acetate.

There are two potential sources of fatty acids for uptake from blood, plasma non-esterified fatty acids (NEFA) and fatty acids released by LPL-mediated hydrolysis of lipoprotein-triglycerides. Bragdon and Gordon (1958) compared the uptake of labeled fatty acids into adipose tissue from chylomicrons and from NEFA and found that a much larger fraction was taken up from chylomicrons. This observation has been repeated many times in animals. Recently, Bickerton et al. (2007)

Fig. 13.2 (continued) At the endothelium the enzyme can engage VLDL particles and hydrolyze their TG. Several other molecules take part in formation of such a "binding-lipolysis" site, apoCII which activates LPL, the VLDL receptor which can bind LPL and perhaps escort the enzyme (and lipoproteins) across the endothelial cells. The VLDL receptor may also bind apolipoproteins B (not shown) and E on the lipoprotein. ApoAV promotes the lipolysis in some yet undefined way. The FA generated by LPL bind to albumin and join the FA from intracellular lipolysis into the circulating blood. The MG move into the adipocytes, perhaps mix with MG from intracellular lipolysis and are hydrolyzed by MAGH or possibly re-esterified by MGAT. When the particle loses core volume by triglyceride hydrolysis, excess surface is shed as "surface remnants" (*SR*) in the form of cholesterol (*C*)-phospholipid-apolipoprotein discs which leave the tissue and VLDL are remodeled into HDL. Excess FA may displace some LPL from the cell surface into blood. Some of the LPL rendered inactive by interaction with angptl4 also leaves the tissue with blood to be taken up and degraded in the liver. In the fed state (*right* part of the figure) there are chylomicrons in blood and insulin suppresses production of angptl4. Most of the LPL molecules now make it to the vascular endothelium in active form. Large "binding-lipolysis" sites are generated where many LPL molecules act simultaneously on the chylomicron and cause the very rapid delipidation characteristic for chylomicron metabolism. Some of the FA generated spill over into blood, but most of the FA move into the tissue, driven by a gradient generated by active synthesis of TG. The MG also move into the adipocyte and are hydrolyzed by MAGH or perhaps are directly esterified by MGAT to DG and then by DGAT to TG. De novo synthesis of TG is stimulated by ASP which binds to its receptor CSL2 on the adipocyte membrane. The glycerol-3-phosphate (*G3P*) needed for GPAT may come in part from glycolysis, but most is probably generated by glyceroneogenesis by way of oxaloacetic acid (*OOA*) from citric acid cycle intermediates and from pyruvate/lactate (*Pyr/Lac*) that the adipocyte may have taken up from blood. Some of the glycerol from hydrolysis of MG may be phosphorylated to G3P but most of the glycerol returns to blood by way of the AQP7 channel. Glucose (*Glc*), transported into the adipocyte by GLUT4, may also serve as a substrate for de novo sythesis of FA (not shown in detail). In this nutritional state perilipin and HSL are dephosphorylated. Perilipin now captures CGI-58 and ATGL displays low activity as it is deprived of its activator. Some basal lipolysis continues and generates DG but most of these are re-esterified by DGAT since HSL has been displaced from the droplets into the cytoplasm. In both nutritional states (*center* of figure) lipoproteins can bind to receptors (LDL-receptor, *LRP*) on the adipocytes, be endocytosed and delivered to lysosomes (blue circle with yellow border). Some CE may move into the adipocyte by selective transfer of core lipids from lipoproteins mediated by the bridging function of LPL. The cholesterol may either be esterified into CE that are deposited in the lipid droplets, be included in intracellular membranes, or be transferred to the plasma membrane and pumped out of the cell by the ABCA1 transporters which use precursor HDL discs as acceptors. Several important steps are omitted so as not to overcrowd the figure, for instance lmf1 and other chaperons needed for correct folding of LPL in the ER, conversion of FA to CoA esters, etc addressed the question of what the sources are for uptake of lipids into adipose tissue in humans. They studied arteriovenous differences in subjects given a combination of a test meal containing [^{13}C]palmitate to label chylomicrons and an intravenous infusion of [^{2}H]palmitate to label plasma NEFA, and from them VLDL. The results showed a clear preference for uptake of fatty acids from lipoprotein triglycerides compared to from plasma NEFA. In addition, the study showed a greater fractional extraction of fatty acids from chylomicrons than from VLDL (Bickerton et al. 2007).

LPL is necessary for normal lipoprotein metabolism. Deficiency of the enzyme or its activator, apolipoprotein (apo)CII, leads to massive hypertriglyceridemia. Mouse models in which the endogenous LPL gene was inactivated and a construct was inserted expressing human LPL demonstrated that expression of LPL in a single tissue is enough to rescue nearly normal lipoprotein metabolism. This is true for expression in cardiac muscle (Levak-Frank et al. 1999) skeletal muscle (Levak-Frank et al. 1997) and even in the liver (Merkel et al. 1998), a tissue which usually does not express LPL. Growth and body mass composition was similar among all groups, including essentially normal adipose tissue mass. In line with this, humans (Brun et al. 1989; Ullrich et al. 2001), cats (Veltri et al. 2006) and mink (Christophersen et al. 1997) with LPL deficiency are not pathologically lean. However, ob/ob mice rendered deficient in adipose tissue LPL could not store lipid as rapidly as normal ob/ob mice (Weinstock et al. 1997).

The lipid composition of adipose tissue in LPL deficiency shows a marked decrease in polyunsaturated fatty acids in humans (Ullrich et al. 2001), cats (Veltri et al. 2006) and in mouse genetic models (Levak-Frank et al. 1999; Levak-Frank et al. 1997). There appears to be an SREBP-1 driven induction of a number of enzymes involved in fatty acid and triglyceride biosynthesis (Wagner et al. 2004). Furthermore, adipose tissue from mice that lacked LPL expressed large amounts of mRNA for endothelial lipase, which is not expressed by adipocytes in wild-type mice (Kratky et al. 2005). This suggests that endothelial lipase, which acts primarily on phospholipids in HDL and is structurally closely related to LPL, provided a source of long-chain polyunsaturated fatty acids for the adipose tissue. Another source of essential fatty acids may be net uptake from circulating NEFA. Hence, even though LPL hydrolysis of triglycerides in plasma lipoproteins is the main source of fatty acids for the adipose tissue under normal conditions, there are compensatory mechanisms when LPL-mediated fatty acid delivery does not keep up.

It was early recognized that the nutritional state had a large effect on the uptake of lipids into adipose tissue (Bragdon and Gordon 1958). More chylomicron fatty acids were taken up in fed compared to fasted rats. In contrast, the fraction of fatty acids administered as NEFA that was taken up into adipose tissue did not differ much with the nutritional state. Robinson and his coworkers demonstrated an almost linear correlation over a more than tenfold range between the LPL activity in adipose tissue and the uptake of labeled fatty acids from injected chylomicrons (Cryer et al. 1976). This shows that the LPL activity has a strong, directive effect on the uptake of lipoprotein-derived fatty acids into adipose tissue (Fielding and Frayn 1998; Hartman 1977). This concept is supported by recent studies in mouse models where uptake of lipids into different tissues correlated with the expression of LPL (van Bennekum et al. 1999). The concept is also illustrated by a number of studies on the relation between LPL expression in heart and heart function (Lee and Goldberg 2007; Park et al. 2007). Knock-out of the enzyme in cardiomyocytes leads to a compensatory increase of glucose utilization and cardiac dysfunction with age (Augustus et al. 2004), while overexpression leads to increased fatty acid oxidation, signs of lipotoxicity and cardiomyopathy (Yokoyama et al. 2004).

13.3 Lipoprotein Lipase

13.3.1 Molecular Properties

Active LPL is a non-covalent dimer of two identical subunits. The human LPL subunit has 448 amino acid residues and two asparagine-linked oligosaccharide chains giving a total molecular mass of about 55,000 (Mead et al. 2002; Merkel et al. 2002a; Olivecrona and Olivecrona 1999; Stein and Stein 2003; Zechner et al. 2000). The subunits are probably arranged in a head to tail orientation (Wong et al. 1997). The enzyme is member of a gene family that also comprises endothelial lipase (Rader and Jaye 2000), hepatic lipase and pancreatic lipase (Kirchgessner et al. 1987). Modelling on the basis of X-ray structures of pancreatic lipase suggests that the subunit in LPL has a similar structure, with two independently folded domains (van Tilbeurgh et al. 1994). This is supported by biochemical evidence from limited proteolysis (Bengtsson-Olivecrona et al. 1986) and by the fact that functional chimeric enzymes have been produced by genetic engineering joining the N-terminal domain of LPL with the C-terminal domain of the related hepatic lipase and vice versa (Davis et al. 1992). The larger N-terminal domain contains the active site and the site for interaction with apoCII, whereas the C-terminal domain contains sites for interaction with lipoproteins, receptors and heparin (Kobayashi et al. 2002; Lookene et al. 2000).

It is a common experience that LPL is unstable under physiological conditions (pH, ionic concentrations, temperature). Loss of activity was traced to a slight change in structure as evidenced by the CD spectrum, and was associated with dissociation of active LPL dimers into monomers (Osborne et al. 1985). Further studies showed that there is rapid subunit exchange between the active dimers, but the intermediate active monomer is prone to change structure with loss of catalytic activity (Lookene et al. 2004). It is concluded that the enzyme is spring-loaded and endowed with a built-in mechanism to self-destruct (Osborne et al. 1985). This is an important property for an enzyme that acts extracellularly but whose activity needs to be rapidly regulated.

13.3.1.1 Mode of Action

There are two steps in the action of a lipase (Verger 1976). First the enzyme must adsorb to the lipid–water interface of its insoluble substrate and be properly aligned at the surface so that single substrate molecules can enter the active site. The second event is the actual chemical steps of catalysis. Because of this sequence of events, the kinetic parameters associated with "substrate affinity" take on a different meaning than for enzymes acting on water-soluble substrates. The dependency of the overall reaction on substrate concentration reflects the propensity of the lipase to bind to the lipid–water interface. LPL can hydrolyze lipids in all types of lipoproteins, chylomicrons, VLDL, LDL and HDL, but when presented with a mixture, as in plasma, the enzyme hydrolyzes almost exclusively lipids in the triglyceride-rich lipoproteins. This is because it prefers to bind to that type of lipid structure. In contrast, hepatic lipase (Bengtsson and Olivecrona 1980a) and endothelial lipase (Rader and Jaye 2000) prefer HDL.

The active site is that of a typical serine hydrolase (Faustinella et al. 1992). The first catalytic event is that a fatty acid transfers from the substrate (e.g. a triglyceride) to the active site serine with formation of an acyl–enzyme intermediate. This is then cleaved by water with release of the fatty acid. The active site is located at the bottom of a hydrophobic pocket in the lipase molecule. A loop of 22 amino acids covers the opening to this pocket (Faustinella et al. 1992). Analogy to pancreatic lipase suggests that, when the lid is closed, the enzyme exposes a hydrophilic surface towards the surrounding aqueous phase but when the lipase adsorbs to a lipid interface, for instance on a lipoprotein, the loop swings around a hinge formed by a disulfide bond. What used to be the inner surface of the loop now becomes part of a hydrophobic lipid-binding site, and the entrance to the active site opens for substrate molecules to enter (Santamarina-Fojo and Dugi 1994).

Rojas et al. (1991) questioned how substrate molecules enter and products leave the active site of LPL. To approach this they compared hydrolysis of triglycerides in liposomes and in emulsion droplets, made up of the same phospholipids and triglycerides. This approach was based on data of Hamilton and Small, who showed that about 3 mol% triolein can be accommodated in phosphatidylcholine bilayers and suggested, on the basis of thermodynamic considerations, that the same applies to the surface structure of a phospholipid–triglyceride emulsion droplet and that of lipoproteins (Hamilton et al. 1983). Rojas et al. (1991) found that the relation between the rate of hydrolysis and the surface area was similar for liposomes containing 3% triolein and for emulsion droplets. Hence the lipid binding site(s) on the lipase did not detect any marked difference between the surfaces of the two types of particles. If each hydrolytic event was followed by dissociation of the lipase into the aqueous phase, one would expect to see the same ratio of phospholipid to triglyceride hydrolysis and similar kinetics with liposomes and emulsion droplets. The ratios were, however, quite different. With liposomes LPL hydrolyzed more phospholipid than triglyceride, whereas with emulsion droplets triglyceride hydrolysis was much faster. Furthermore, the maximal rate of triglyceride hydrolysis was more than 40× higher with the emulsion droplets than with the liposomes. These results show that, when LPL binds to the substrate particle, it stays for several rounds of lipolysis, i.e. the action is processive. Furthermore, the results suggest that when the products leave the active site it is easier for a new substrate molecule to come in from below, i.e. the core of the particle, than from the side, i.e. the surface layer. This seems to fit the proposed three-dimensional structure of LPL, which shows the active site at the bottom of a hydrophobic pocket (van Tilbeurgh et al. 1994), after the lid has been turned away as presumably happens on binding to a lipid–water interface.

The main initial product formed by LPL action on a triglyceride is a *sn*-2,3-diglyceride (Morley and Kuksis 1972), which is then hydrolyzed further to a 2-monoglyceride. Diglycerides do not accumulate during the hydrolysis of triglycerides (Nilsson-Ehle et al. 1973; Scow and Olivecrona 1977). This is in striking contrast to what is observed with the intracellular adipose tissue triglyceride lipase (ATGL; Zechner et al. 2005) and indicates that diglycerides are a preferred substrate for LPL. It is likely that, when a triglyceride enters the active site of LPL from the core of the lipid particle (see above), it is hydrolyzed to a diglyceride, which tends to stay in the active site and be further hydrolyzed to a monoglyceride before a new triglyceride can enter.

13.3.1.1.1 ApoCII

ApoCII is a necessary activator for LPL (LaRosa et al. 1970; Shen et al. 2002). Individuals with a genetic deficiency of apoCII have massive hypertriglyceridemia and the same clinical symptoms as patients with LPL deficiency (Fojo and Brewer 1992). When LPL is added to chylomicrons from a CII-deficient individual, no hydrolysis occurs. When apoCII is added, hydrolysis immediately starts (Olivecrona and Beisiegel 1997). The detailed mechanism of the activation is not known. LPL displays high activity without apoCII against some model substrates, e.g. the short-chain triglyceride tributyrin (Rapp and Olivecrona 1978). It is only with long-chain triglycerides, phospholipids and lipoproteins as substrate that the enzyme requires its activator (Bengtsson and Olivecrona 1980c).

It was early shown that peptide fragments corresponding to the C-terminal one-third of apoCII could activate LPL in some model systems (Kinnunen et al. 1977). The N-terminal two-thirds were assumed to be responsible for binding the apolipoprotein to the lipoprotein surface. With chylomicrons from an apoCII-deficient individual, the lipid-binding N-terminal part of apoCII was required; a C-terminal fragment that could activate LPL in model systems had no effect with natural lipoproteins (Olivecrona and Beisiegel 1997).

Comparative studies of apoCII sequences from widely different vertebrates (fish, birds, a number of mammals) showed that, in the C-terminal part of the molecule, seven amino acids are strictly conserved (Shen et al. 2000). Mutagenesis confirmed the functional importance of these residues (Shen et al. 2002). Combining this with the 3-D structure (Zdunek et al. 2003) revealed that some of these amino acids are in a helical segment and form a common surface. This is presumably the structure in apoCII that interacts with LPL. In contrast to the strict conservation of structure in this part of the molecule, the N-terminal part showed widely different structures (Shen et al. 2000). In human apoCII this part of the molecule forms two amphipathic helixes that anchor the protein in the lipoprotein surface (Zdunek et al. 2003).

It is not entirely clear what property apoCII imparts to the LPL reaction. ApoCII is provided as an integral part of the substrate lipoprotein. There is no evidence for LPL–apoCII complexes at the endothelium, in the absence of lipoproteins. It should be noted, however, that apoCII is expressed in adipose tissue (Gonzales and Orlando 2007). There is also no evidence that the amount of apoCII is limiting for LPL action in vivo. The related pancreatic lipase needs a small protein cofactor, colipase, to allow the enzyme to bind to lipid droplets in the presence of bile salts which remove most proteins from the droplets (Borgstrom et al. 1979; Lowe 2002). This is a problem that LPL does not have to overcome; and there is no sequence or structural similarity whatsoever between apoCII and pancreatic colipase. ApoCII does not seem to be important for the binding of LPL to lipoproteins (Olivecrona and Beisiegel 1997) or model substrate droplets (Bengtsson and Olivecrona 1980c), but probably orients the enzyme in the correct manner at the interface (Bengtsson and Olivecrona 1980c; Shirai et al. 1983). The role of apoCII in the product control of LPL action is revisited in Sect. 13.3.1.2.

13.3.1.1.2 Other Apolipoproteins

An important parameter in lipase action is the "quality of the interface" (Verger 1976) which is a function of the lipid and protein composition of the lipid–water interface. LPL action can be severely inhibited by a variety of synthetic peptides that have no sequence homology to native apolipoproteins (Chung et al. 1996). Hence, when one considers reports on effects of apolipoproteins on the LPL reaction, one has to bear in mind that the effect may be relatively non-specific.

13.3.1.1.3 Apolipoprotein CIII

Apolipoprotein CIII is the most abundant C-apolipoprotein in humans. There are many reports that apoCIII inhibits LPL, both with model substrates and with lipoproteins (Jong et al. 1999). ApoCIII has also been shown to enhance binding of triglyceride-rich lipoproteins to SR-B1 (Zannis et al. 2006) and impede binding to receptors in the LDL-R family (Narayanaswami et al. 2004). One mechanism by which apoCIII may do this is by displacing apoE from the lipoprotein surface (Narayanaswami et al. 2004). Likewise, apoCIII may displace apoCII from the surface, explaining in part its inhibition of LPL action (Jong et al. 1999). Overexpression of apoCIII in mice causes hypertriglyceridemia, which appears to be due to impeded lipolysis of triglyceride-rich lipoproteins (Ito et al. 1990). The expression of apoCIII in the liver is decreased by fibrates and other PPAR-α agonists and this is accompanied by decreased plasma triglyceride levels (Kolovou et al. 2008). There is a correlation between apoCIII expression and triglyceride levels in plasma (Fredenrich 1998). Mutations in the apoCIII gene are associated with deranged triglyceride metabolism (Talmud et al. 2002). Hence, there is strong evidence that apoCIII impedes LPL action in vivo but the mechanism has not been revealed. The 3-D structure for apoCIII at a lipid–water interface was recently reported (Gangabadage et al. 2008). ApoCIII is arranged in the interface as six amphipathic helices (each approx. ten residues long), which are connected via semiflexible hinges. Several structural motifs along the solvent-exposed face of apoCIII are highly conserved among mammals. The authors propose molecular mechanisms for some of the multifaceted actions of apoCIII in lipoprotein metabolism (Gangabadage et al. 2008).

13.3.1.1.4 Apolipoprotein CI

Apolipoprotein CI is less abundant in human plasma than apoCIII or apoCII. It has been suggested that a major function of apoCI is to aid in the innate immune defense system (Berbee et al. 2006). The protein contains a highly conserved consensus KVKEKLK binding motif for lipopolysaccharide (LPS), an outer-membrane component of gram-negative bacteria. ApoCI may bind LPS and present it to macrophages. It was recently shown that HDL from patients with sepsis are almost depleted from apoCI, indicating that the apolipoprotein has been consumed in the defense reaction.

Mice overexpressing human apoCI display gross hypertriglyceridemia (Berbee et al. 2005). This goes along with a function in the innate immune system, since triglyceride-rich lipoproteins bind LPS. The hypertriglyceridemia is not due to interference with binding to hepatic receptors (LDL-R, VLDL-R or LRP) and cannot be due to interference of apoCI with the functions of apoCIII or apoE, since it is seen also in mice deficient in these apolipoproteins (van der Hoogt et al. 2006). In vitro apoCI directly inhibits LPL, indicating that this is a factor in the development of hypertriglyceridemia. These effects may also explain why mice overexpressing apoCI are protected against obesity (Berbee et al. 2005).

13.3.1.1.5 Apolipoprotein AV

Apolipoprotein AV is present in very low amounts in plasma compared to the other apolipoproteins. ApoAV was therefore not discovered until the year 2001, although the gene is located in the *APOA1/C3/A4* gene complex (Kluger et al. 2008; Talmud 2007; Wong and Ryan 2007). ApoAV is a 40-kDa protein with low solubility due to long stretches of hydrophobic and amphipatic sequences (Beckstead et al. 2003). ApoAV is produced mainly (or only) in the liver, but in chickens it is also found in brain, kidney and ovarian follicles (Dichlberger et al. 2007). When overexpressed in hepatoma cells (Hep3B) it accumulates around intracellular lipid droplets but does not affect secretion of apoB-containing lipoproteins (Shu et al. 2007). In plasma, apoAV is found predominantly on triglyceride-rich lipoproteins and on HDL, just like apoCIII, but the levels of apoCIII are about 300-fold higher. Mice lacking apoAV have fourfold increased plasma triglyceride levels (Pennacchio et al. 2001), while overexpression of apoAV results in decreased plasma triglycerides due to faster catabolism of chylomicrons and VLDL (Kluger et al. 2008). From experiments by several groups the current view is that apoAV does not stimulate LPL activity in vitro, but there is strong evidence that LPL action is markedly stimulated in vivo

(Kluger et al. 2008). Stimulation of LPL activity against VLDL was demonstrated with LPL attached to heparan sulfate-covered wells of microtiter plates (Dorfmeister et al. 2008; Merkel et al. 2005). ApoAV has high affinity for heparin and was shown to mediate binding of chylomicrons to heparin-covered surfaces (Lookene et al. 2005). ApoAV also binds to receptors in the LDL-R family (LRP, SorLA; Nilsson et al. 2007) and in the Vsp10p domain families (SorLA and sortilin; Nilsson et al. 2008). Binding of apoAV to these receptors leads to endocytosis of the complex (Nilsson et al. 2008). Similar data were reported with chicken apoAV and the major avian LDL receptor (Dichlberger et al. 2007). A puzzling fact is that the levels of apoAV in human plasma are positively correlated with plasma triglyceride levels (just like apoCIII; Kluger et al. 2008; Talmud 2007). From detailed studies on obese and non-obese subjects, Wolfgang Patch's group concluded that the expression of *APOA5,* and the plasma levels of apoAV, are intimately linked to hepatic lipid metabolism (Hahne et al. 2008). At this moment the mechanism for stimulating triglyceride clearance by apoAV remains in the dark.

13.3.1.2 Fatty Acids – an Unusual Mechanism for Product Control

For the LPL hydrolysis of triglycerides to proceed, a fatty acid acceptor must be present. Otherwise the reaction comes to near stop when only a few percent of the triglycerides have been hydrolyzed (Bengtsson and Olivecrona 1980b). Inhibition by fatty acids is seen with other lipases, but is much more pronounced with LPL. The related pancreatic lipase is even activated by fatty acids (van Kuiken and Behnke 1994). Hence, some special mechanism must be at play to cause the very strong inhibition of LPL. Detailed studies of this revealed that there are three factors that contribute (Bengtsson and Olivecrona 1980b).

One mechanism is general for lipases with the serine hydrolase type of reaction mechanisms. The fatty acids and monoglycerides formed on hydrolysis locate at the lipid–water interface and, since they are substrates for the reaction, act as competitive inhibitors of triglyceride hydrolysis. This "reverse reaction" is widely used by the chemical industry, where lipases are used to synthesize various types of esters from fatty acids in water-free or low-water conditions. In the LPL system, however, this type of mechanism explains only a small part of the fatty acid inhibition.

Another, more important mechanism is that LPL forms complexes with fatty acids (Bengtsson and Olivecrona 1979a, b, 1980b). Therefore, as fatty acids accumulate in the system, the enzyme is sequestered into enzyme–fatty acid complexes. This impedes further lipolysis and can break the binding of LPL to heparan sulfate (Saxena and Goldberg 1990) and cause the lipase to dissociate from the capillary wall into the blood. Albumin can prevent the formation of such complexes since it has a higher affinity than the enzyme for fatty acids. The cut-off point when albumin can no longer prevent the formation of LPL–fatty acid complexes is when the fatty acid to albumin ratio exceeds 7–8 (Bengtsson and Olivecrona 1980b; Scow and Olivecrona 1977).

The third and most important component to the inhibition is that, when fatty acids accumulate, apoCII no longer activates the lipase (Bengtsson and Olivecrona 1979a, 1980b; Saxena and Goldberg 1990). This may be why evolution has endowed LPL with a strict dependence on an activator: to be able to stop the reaction when products start to accumulate. The question on what role this may have for control of the LPL system in vivo is tackled in Sect. 13.3.4.

13.3.1.3 Action on Lipoproteins

LPL is an efficient enzyme, with a turnover number for triglyceride hydrolysis of about 1000 s^{-1} (Scow and Olivecrona 1977). A typical chylomicron contains millions of triglyceride molecules. Hydrolysis of most of these to fatty acids and monoglycerides would require several million hydrolytic events and it would take a single LPL molecule hours to accomplish this. The metabolism of chylomicrons is much quicker. It follows that several lipase molecules must act simultaneously on the particle. We return to this concept when we discuss LPL at the endothelium (Sect. 13.3.4).

LPL hydrolyzes both triglycerides and phospholipids. Deckelbaum and associates (1992) studied the relation between these processes by measuring the hydrolysis of [^{14}C]triolein and [^{3}H]dipalmitoylphosphatidylcholine incorporated into lipoproteins by lipid transfer processes. The relation between phospholipid and triglyceride hydrolysis was generally linear until at least half of the particle triglyceride had been hydrolyzed. Phospholipid hydrolysis, relative to triglyceride hydrolysis, was most efficient in VLDL, but could not fully account for the loss of surface phospholipid that accompanies triglyceride hydrolysis and decreasing core volume. Thus, shedding phospholipid molecules from the VLDL particle, presumably as phospholipid–apolipoprotein discs, must be a major mechanism for losing excess surface as large lipoprotein particles are converted to smaller particles. The discs are precursors for HDL formation, by remodeling processes in blood (Chap. 9). This creates a link between efficient LPL action and HDL levels in blood (Patsch et al. 1978).

13.3.1.4 Polyanion Binding Site

LPL was early found to bind to heparin and it was suggested that this is how the enzyme is anchored to the vascular endothelium via heparan sulfate proteoglycans (Olivecrona et al. 1977). This is an attractive model since it positions the enzyme at a distance from the membrane, in the periphery of the glycocalyx. This should facilitate the capture of and action on lipoproteins from the circulating blood. The model also explains how heparin injection releases the enzyme into the circulating blood. Studies in model systems showed that LPL is catalytically active when bound to heparin (Bengtsson and Olivecrona 1981).

Several segments in the LPL molecule have been shown to contribute to heparin binding. A main heparin-binding site is probably in a cleft between the N- and

C-terminal folding units (van Tilbeurgh et al. 1994). Within this segment (residues 260–306 in human LPL) there are 14 positively charged residues. Structures in the C-terminal domain and at the C-terminal end of the molecule also contribute to heparin binding (Lookene et al. 2000; Sendak and Bensadoun 1998).

Lookene et al. (1996) studied the interaction of LPL with heparan sulfate and with size-fractionated fragments of heparin by several approaches. Heparin decasaccharides formed a 1:1 complex with dimeric LPL and were the shortest heparin fragments which could completely satisfy the heparin-binding regions. Equimolar concentrations of octasaccharides also stabilized dimeric LPL, while shorter fragments (hexa-, tetrasaccharides) were less efficient. The number of ionic interactions between LPL and high-affinity decasaccharides was estimated to be ten. Binding of heparin did not induce major rearrangements in the conformation of LPL, indicating that the heparin-binding region is preformed in the native structure. Using a model system with immobilized heparan sulfate bound to a biosensor chip, it was shown that LPL moves rapidly between individual heparan sulfate chains within the layer (Lookene et al. 1996). The overall dissociation constant is very low. This creates a high concentration of LPL along the surface layer of heparan sulfate chains, but the lipase is free to move within the layer.

Early studies showed that several structurally different linear polyanions, e.g. dextrane sulfate and polynucleotides bind LPL avidly, so it was clear that a relatively non-specific electrostatic interaction is enough for binding of LPL (Olivecrona and Bengtsson 1978). There have been several searches for a structure in heparan sulfate specific for binding of LPL, but most of these studies have returned the result that the more sulfated the polysaccharide chain is the higher is the affinity for LPL (Bengtsson et al. 1980; Larnkjaer et al. 1995; Parthasarathy et al. 1994). The most recent study came to the conclusion that LPL, "containing several clusters of positive charges on each subunit, may constitute an ideal structure for a protein that needs to bind with reasonable affinity to a variety of modestly sulfated sequences of the type that is abundant in heparan sulfate chains" (Spillmann et al. 2006).

Heparan sulfate is a component of the outer glycocalyx on virtually all cells (Bishop et al. 2007). The high-affinity binding of LPL to heparan sulfate can explain why the enzyme avidly binds to cells, but is hardly specific enough to fully explain why the enzyme gets concentrated on the luminal surface of endothelial cells.

13.3.1.5 LPL as a Ligand – Binding to Receptors and Cell Surfaces

In experiments on the binding of lipoproteins to cultured cells, for instance to study binding to the LDL receptor, it is usually only a small fraction of the lipoproteins that bind. If LPL is added, binding increases many-fold (Eisenberg et al. 1992). This demonstrates the so-called bridging function of the lipase. Dimeric LPL presumably binds with one of its subunits to the lipoprotein and with the other subunit to the cell surface. This binding can be to cell surface heparan sulfate but it can also be to receptors. That LPL is not only an enzyme, but also a receptor ligand was first suggested by Felts et al. (1975). Experimental proof for this hypothesis came in 1991 when Beisiegel and her collaborators demonstrated that LPL binds with high affinity to LRP (the low-density lipoprotein receptor-related protein; Beisiegel et al. 1991). This has been extended to most members of the LDL receptor family (Gliemann 1998) and to the SorLA (Jacobsen et al. 2001) and Sortilin receptors (Nielsen et al. 1999). The binding site for the receptors has been localized to the surface loop on the C-terminal domain in LPL (Nykjaer et al. 1993, 1994). This is close to and perhaps partly overlapping an important site for binding to lipid–water interfaces and to lipoproteins. Therefore one LPL subunit can only bind to either a lipoprotein or a receptor. Bridging of lipoproteins to receptors requires dimeric LPL.

A common feature for all the receptors we have mentioned is that their interactions with ligands are blocked by the receptor associated protein (RAP). This is true also for their interaction with LPL. RAP is a 39-kDa resident ER protein which is thought to act as a chaperone/escort protein (Willnow 1998). RAP binds tightly to LDL receptor family members in the endoplasmic reticulum (ER). After escorting them to the Golgi, RAP dissociates from the receptors (Lee et al. 2006). It has been suggested that the function of RAP is to prevent premature binding of ligands to the receptors during their passage through the secretory pathway.

RAP is thought to block binding of ligands, e.g. lipoproteins, by binding to the receptors. Surprisingly, RAP also binds directly to LPL forming a stable complex between monomeric LPL and RAP (Page et al. 2006; van Vlijmen et al. 1999). Overexpression of RAP in the liver of mice that lack both LRP and the LDL receptor caused a marked hypertriglyceridemia in addition to the pre-existing hypercholesterolemia in these animals and also caused a sevenfold increase of circulating, but largely inactive, LPL (van Vlijmen et al. 1999). RAP-deficient adipocytes secreted mostly inactive LPL (Page et al. 2006). These studies suggest that RAP may act as a chaperone/escort protein also for LPL, as for some of the receptors. This may prevent premature interaction of LPL in the secretory pathway with other molecules that it could potentially bind, e.g. LRP and other receptors, angiopoietin-like proteins and heparan sulfate proteoglycans.

13.3.2 Synthesis, Maturation and Transport of LPL

LPL is expressed in many cell types (Mead et al. 2002; Preiss-Landl et al. 2002). A common theme is that the enzyme is synthesized in parenchymal, not in endothelial cells. Therefore LPL must be secreted and transported to the luminal side of the vascular endothelium where it can engage lipoproteins from blood and hydrolyze their triglycerides. In adipose tissue the enzyme is produced in adipocytes and preadipocytes. It is also produced by macrophages, but in the adipose tissue this makes only a small contribution to overall tissue LPL activity.

The LPL transcript starts with a signal sequence which targets it for vectorial synthesis into the ER. The enzyme has been visualized by electron microscopy

(EM) in vesicles of the secretory pathway (Blanchette-Mackie et al. 1989). There are reports that LPL is localized in a special type of secretory vesicles, different from those used by leptin in adipocytes (Roh et al. 2001) or by insulin in beta cells (Cruz et al. 2005).

13.3.2.1 Folding into Active LPL

It has been debated at what stage of assembly and processing LPL becomes catalytically active. Zhang et al. (2005) studied the folding of LPL in vitro. The first step was rapid and resulted in an inactive monomer with a completely folded C-terminal domain, whereas the N-terminal domain was in a molten globule state. The second step converted the monomers to a more tightly folded state that rapidly formed dimers. The second step was slow and it appears that proline isomerization (rather than dimerization as such) is rate-limiting.

Interestingly, the second step was dependent on Ca^{2+} (Zhang et al. 2005). There are several other pieces of evidence that Ca^{2+} may affect LPL activity. In an early study it was shown that incubation of rat adipocytes in the absence of extracellular calcium produced a rapid decline of LPL activity in the cells (Soma et al. 1989). The enzyme could be rapidly reactivated by the addition of calcium. The degree of reactivation was proportional to the concentration of extracellular calcium. Others reported that parathyroid hormone (PTH), which causes a rise in cytosolic calcium, suppressed LPL activity in rat adipocytes by a post-translational mechanism, and this could be prevented by addition of the calcium channel blocker verapamil (Querfeld et al. 1999). The Ca^{2+}-dependence of LPL folding and secretion is so pronounced that Osibow et al. (2006) used the LPL activity state to monitor the impact of ER Ca^{2+} dynamics on protein folding. These studies raise the possibility that Ca^{2+}-dependent control of LPL dimerization might be involved in post-translational regulation of LPL activity.

In studies of folding of transfected human LPL in CHO cells, it was found that most of the cellular LPL was in the ER, while only a small fraction passed to or through the Golgi, as evidenced by the pattern of glycosylation (Ben-Zeev et al. 2002). In the ER there were two fractions of LPL: (a) properly folded, active dimers and (b) disulfide-liked aggregated LPL proteins. The aggregates were formed directly during synthesis. They could not be recruited into active LPL, nor were they formed from the active dimers. LPL molecules in the Golgi and released into the medium were in the form of active dimers. Hence, it appears that only a fraction of newly synthesized LPL subunits fold correctly are approved by ER quality control mechanisms and are passed to the Golgi for further transfer to the cell surface and, in tissues, to the vascular endothelium. A relatively large fraction of newly synthesized LPL subunits do not attain a proper folding but form large, irreversible aggregates which are retained in the ER and degraded there. It had previously been noted that a large fraction of LPL in 3T3-L1 cells is inactive and mostly insoluble (Olivecrona et al. 1987). In rat adipocytes, isolated after collagenase digestion, only about one-third of the LPL protein was in the active, dimeric form (Wu et al. 2003).

Two-thirds was catalytically inactive and probably corresponded to the LPL aggregates characterized by Ben-Zeev et al. (2002). The pattern was the same in adipocytes from both fed and fasted rats (Wu et al. 2003).

Folding of LPL requires ER molecular chaperones. This was evidenced in attempts to express LPL in large scale using *Baculovirus*-infected insect cells (Zhang et al. 2003). The cells expressed and secreted LPL protein, but virtually no LPL activity. Co-infection of the insect cells with mammalian molecular chaperones, in particular with calreticulin, endowed them with the ability to form increased amounts of active LPL dimers (Zhang et al. 2003). Ben-Zeev et al. (2002) demonstrated interaction of LPL with several chaperones by co-immunoprecipitation, mainly with calnexin and BiP. Another molecule that may have a role in the maturation of LPL is RAP, which has been shown to bind tightly to LPL (Sect. 13.3.1.5).

Mice homozygous for the combined lipase deficiency (cld/cld) mutation show impaired maturation of LPL and hepatic lipase in the ER. Tissues and cells from mutant mice express only a few percent of normal lipase activity, even though they produce relatively normal amounts of lipase protein (Masuno et al. 1990; Olivecrona et al. 1985). The defective lipase molecules are retained in the ER (Masuno et al. 1990) The gene containing the cld mutation was recently identified and given the name lipase maturation factor 1, Lmf1 (Peterfy et al. 2007). It encodes a transmembrane protein which localizes in the ER. Lmf1 consists of an N-terminal part that is predicted to have five transmembrane segments which presumably anchor the protein in the ER membrane. The C-terminal domain is conserved among species and the cld mutation is located in this part. In further support of the importance of Lmf1, a patient homozygous for a mutation causing loss of part of the C-terminal domain had severe hypertriglyceridemia (Peterfy et al. 2007). Lmf1 is expressed in virtually all tissues in the mouse, including all major LPL-producing tissues. However, the highest abundance of Lmf1 mRNA is in kidney and testes, tissues that have low LPL. This suggests that Lmf1 may have additional functions.

Taken together, these studies demonstrate that the folding of LPL into its active form is not a simple, spontaneous process, but is a major bottle-neck that requires molecular chaperones, a special lipase maturation factor and a suitable Ca^{2+} concentration within the ER. This is not surprising since the lipase has to fold into a metastable conformation to form active dimers that are spring-loaded and prone to reorganize into a more stable, catalytically inactive monomeric state (Zhang et al. 2005). What role, if any, the folding process has in tissue-specific regulation of LPL activity is not clear.

13.3.2.2 The Role of Glycosylation

During the 1980s it was noted that, when glycosylation was blocked by tunicamycin, cells synthesized a non-glycosylated, inactive form of LPL that was not secreted (Chajek-Shaul et al. 1985; Olivecrona et al. 1987; Ong and Kern 1989b). Along the same line, when adipocytes were cultured in a medium devoid of glucose, more than 90% of the LPL was unglycosylated, inactive and not secreted (Ong and Kern 1989b).

A glucose concentration of 1 mg ml^{-1} in the medium, much below the physiological range, was enough to restore glycosylation, catalytic activity and secretion. LPL from most animal species has two oligosaccharide chains, but only one of these, attached to Asn43 in the human sequence, must be glycosylated for the enzyme to fold correctly (Ben-Zeev et al. 1994). It was debated whether core glycosylation was enough for LPL to fold into its active form or whether the oligosaccharide chains must be processed from their initial high mannose form into more complex oligosaccharide structures (Chajek-Shaul et al. 1985; Masuno et al. 1991; Vannier and Ailhaud 1989). Isolation of LPL containing only high-mannose oligosaccharides by lectin affinity chromatography from homogenates of guinea-pig adipocytes showed that these LPL molecules were catalytically active and could be secreted (Semb and Olivecrona 1989a). Others reached the same conclusion (Ben-Zeev et al. 1992, 2002). It thus appears that processing of the oligosaccharide chains indicate that the enzyme subunits have been approved by the quality control mechanisms in the ER (Ellgaard and Helenius 2003) and have been passed on to the Golgi, but trimming and processing as such is not necessary for transport or for activity (Ben-Zeev et al. 2002).

13.3.2.3 Secretion

The intracellular movement and metabolism of LPL has been traced in pulse-chase experiments by several groups. The enzyme begins to appear in the medium within 30 min, and secretion is largely complete by 1 h. These times are compatible with unrestricted movement through the default secretory pathway. When collagenase-isolated adipocytes are incubated in medium without heparin, only a relatively small fraction appears in the medium and much of the lipase is degraded. Less than 20% of pulse-labeled LPL was secreted into the medium with guinea pig adipocytes and nearly 40% was degraded within 1 h (Semb and Olivecrona 1987). In experiments with cultured 3T3-F442 cells, less than 20% of the pulse-labeled LPL was secreted and about 70% was degraded in 70 min (Vannier and Ailhaud 1989). Only 2% of the labeled lipase was secreted and about 50% was degraded in 2 h with adipocytes from fasted rats (Lee et al. 1998). Similar figures were obtained in experiments where protein synthesis was blocked by cycloheximide (Ben-Zeev et al. 2002; Semb and Olivecrona 1987). Hence, without heparin in the medium, the predominant fate of newly synthesized LPL is intracellular degradation. When heparin is added, much more lipase is secreted. With guinea pig adipocytes the secretion increased from 20% to 50% (Semb and Olivecrona 1987); with 3T3-F442 cells secretion increased from less than 20% to 90% (Vannier and Ailhaud 1989). The increased secretion was balanced by decreased degradation in these studies. The amount of labeled lipase remaining in the cells was about the same. These data can be reinterpreted in light of more recent data on the maturation of the lipase (Ben-Zeev et al. 2002). If a lipase molecule folds properly to active dimers, it traverses the secretory path and arrives at the cell surface where it initially binds, perhaps mainly to heparan sulfate chains. If a high-affinity ligand is available (e.g. heparin) the lipase detaches from the cell surface. In the absence of a ligand, the lipase recycles into the cells and is degraded.

Perhaps it goes through several rounds of recycling as has been observed with LPL-transfected CHO cells (Berryman and Bensadoun 1995).

13.3.2.4 Transport

It is not known in detail how LPL finds its way to the vascular endothelium. A picture of this was obtained in a study where LPL was immunolocalized by EM in mouse hearts (Blanchette-Mackie et al. 1989). Within the myocytes, LPL was found in ER, Golgi and in secretory vesicles. In the extracellular space, the enzyme was found near the orifice of secretory vesicles of the myocytes, along all cell surfaces and crossing the endothelium in vesicles or intracellular channels. This is in concert with the observations by Goldberg and his associates that LPL undergoes bidirectional transport over layers of cultured endothelial cells (Obunike et al. 2001) and that the VLDL receptor appears to be involved in the transcytosis of active LPL across endothelial cells (Obunike et al. 2001). The EM study showed that the endothelial surface had processes that extended into the lumen of the capillaries. The density of immunogold particles over these projections, which comprised about 10% of the total luminal surface, was 2.4× that over the rest of the luminal surface. In fasted mice, the amount of immunoreactive LPL at the luminal projections increased by a factor of five (Blanchette-Mackie et al. 1989).

This pattern of LPL suggests that the enzyme moves in the tissue by an essentially two-dimensional route along cell surfaces. In this process, the enzyme may move from one heparan sulfate chain to the next (Lookene et al. 1996; Obunike et al. 2001). Interaction with such chains may also help the enzyme traverse the basal lamina. To impart directionality to the movement of the lipase, there must be higher-affinity binding sites at the luminal surface of the endothelial cells, particularly at the projections. An interesting possibility is that such high-affinity binding sites are provided by the newly discovered GPIHBP1 protein (Young et al. 2007). This protein, which binds to both LPL and chylomicrons, is expressed on the luminal surface of capillaries in several tissues, notably adipose tissue, heart and skeletal muscle. Mice lacking GPIHBP1 manifest chylomicronemia, demonstrating that the protein is of crucial importance for normal lipoprotein catabolism. The mice display an abnormal pattern of LPL release after injection of heparin. The rapid component is lacking, indicating that the amount of LPL at the endothelial surface is much reduced, supporting the view that GPIHBP1 is required for binding LPL at the endothelium (Weinstein et al. 2008). The N-terminal domain of GPIHBP1 has a highly negatively charged stretch of residues (17 of 25 residues in the mouse sequence and 21 of 25 in the human are glutamates and aspartates). The C-terminal domain shows homology to the lymphocyte antigen 6 (Ly-6) motif, which is defined by a distinct disulfide-bonding pattern between eight or ten cysteine residues (Ploug et al. 1993). Many members of the large Ly-6 gene family are receptors or have other protein–protein binding properties and, like GPIHBP1, have a glycosyl-phosphatidylinositol anchor that tethers them to membranes. It seems likely that the acidic stretch of residues in the N-terminal domain engages the heparin-binding sites on the lipase molecule, whereas the C-terminal Ly-6 domain interacts in a more specific way with some other part of the lipase molecule. The expression of GPIHBP1 is regulated by dietary factors and by PPAR-γ (Davies et al. 2008).

In the EM study it was estimated that, in mouse heart, about 80% of immunoreactive LPL protein was in myocytes and most of the remainder was associated with endothelial cells (Blanchette-Mackie et al. 1989). In adipose tissue much more of the enzyme is extracellular. Less than one-third of the LPL protein was recovered with adipocytes after digestion of rat adipose tissue with collagenase (Wu et al. 2003). The rest was presumably extracellular. This includes both active and inactive forms of the lipase. Considering only active LPL, about 10% was released from the adipose tissue of fed rats within 5 min after an intravenous injection of heparin (Wu et al. 2003). This presumably represents easily accessible LPL at or close to the endothelial surface. About 30% of the LPL activity was recovered with the adipocytes. The rest, about 60%, must have been in extracellular locations other than the endothelial surface. Hence, LPL is not restricted to the lipase-producing cells and the luminal aspect of the endothelial cells, but there is LPL throughout the extracellular spaces of the tissue.

13.3.2.5 Turnover of Extracellular LPL

LPL turns over rapidly. Several studies in which synthesis of new protein has been blocked by puromycin (Schotz and Garfinkel 1965) or cycloheximide (Semb and Olivecrona 1987; Wing et al. 1967; Wu et al. 2003) have indicated that the half-life measured as LPL activity is 2 h or less. Inactive LPL protein turns over at a similar rate (Wu et al. 2003) or even faster (Ben-Zeev et al. 2002). How does this occur? Binding, endocytosis and degradation of LPL has been demonstrated with several cell types, including adipocytes (Berryman and Bensadoun 1995; Obunike et al. 1996; Olivecrona et al. 1987). Thus, degradation by cells within the tissue is a clear possibility. Friedman et al. (1982) compared the ability of different cell types to bind, internalize and degrade the enzyme. Fibroblasts and endothelial cells, which do not synthesize LPL, showed a low ratio of degradation to surface binding. In contrast, cardiomyocytes and preadipocytes, which synthesize LPL, showed a much higher ratio of degradation to binding. The authors speculated that LPL-producing cells have a prominent role in re-uptake and degradation of extracellular LPL. Another possibility is extracellular degradation by matrix metalloproteinases, as has been demonstrated in explants of adipose tissue (Wu et al. 2005).

A third, interesting, possibility is that the enzyme is released into the blood, followed by uptake and degradation in the liver (Vilaro et al. 1988). There are both active and inactive LPL in blood. A typical value for LPL activity in human plasma is 1.7±1.1 mU ml^{-1} (1 mU is 1 nmol min^{-1} of fatty acid released; Tornvall et al. 1995). This corresponds to less than 1% of the activity in post-heparin plasma and could account for hydrolysis of less than 5% of all plasma triglycerides per hour. Active LPL in plasma is bound mainly to triglyceride-rich lipoproteins, VLDL and chylomicrons (Zambon et al. 1996) and it has been suggested that the enzyme enhances hepatic binding and uptake of the lipoproteins, particularly chylomicron remnants (Heeren et al. 2002). In line with this, the amount of active LPL in plasma increases after a lipid meal (Karpe et al. 1998) or after infusion of a fat emulsion (Hultin et al. 1992; Nordenstrom et al. 2006; Peterson et al. 1990). It has been shown that the addition of triglyceride-rich lipoproteins causes release of LPL from the surface of endothelial cells (Sasaki and Goldberg 1992; Saxena et al. 1989).

Active LPL is rapidly cleared from the blood and more than half is taken up by the liver (Vilaro et al. 1988), where it initially localizes along sinusoids (Neuger et al. 2004). With time, the immunostaining for LPL shifts to the hepatocytes, becomes granular and then fades, indicating internalization and degradation. Hepatocytes in livers of adult rats do not express LPL (Peinado-Onsurbe et al. 1992). Nonetheless, the liver contains substantial amounts of LPL, most of which is inactive (Neuger et al. 2004), in accord with the hypothesis that there is continuous transport of LPL from adipose and other tissues to the liver, where the enzyme is degraded. In further support of this, Chajek-Shaul et al. (1988) perfused rat livers with LPL-containing media and found that the enzyme lost its catalytic activity soon after binding/uptake. It is not known whether active LPL binds to heparan sulfate or to receptors in the liver, or perhaps first to heparan sulfate and then transfers to receptors; but the binding is impeded by injection of heparin (Neuger et al. 2004). Overall clearance of the lipase becomes much slower and the initial immunostaining shifts from the sinusoids to the Kupffer cells (Neuger et al. 2004).

There is much more inactive than active LPL in plasma (Tornvall et al. 1995). The nature and metabolic significance of inactive LPL is not known. Western blot analysis shows that it is full length and not bound in disulphide-linked aggregates (Olivecrona et al. 1995). On heparin-agarose it elutes in the position expected for monomeric LPL (Olivecrona et al. 1995). It is bound to lipoproteins, mainly LDL, corresponding to about one lipase molecule for every thousand LDL particles in human plasma (Vilella et al. 1993). Inactive LPL enhances the binding of LDL to macrophages, but is much less efficient than active LPL (Makoveichuk et al. 2004). When labeled LPL monomers were injected to rats, they disappeared even more rapidly from plasma than active LPL does (Neuger et al. 2004). Virtually all label ended up in the liver. In contrast to active LPL, most of the inactive lipase was taken up by Kupffer cells (Neuger et al. 2004).

Karpe and his associates have studied the arteriovenous gradients of plasma LPL activity and mass across adipose tissue and skeletal muscle in humans (Karpe et al. 1998). Adipose tissue releases relatively large quantities of inactive LPL, whilst at the same time extracting small amounts of active LPL. In contrast, skeletal muscle releases less LPL overall, and this is mostly active LPL. From these data, and some assumptions, it was estimated that transport away in blood can account for a large fraction of LPL turnover (Karpe et al. 1998). In a more recent paper, a group of patients with diabetes type 2 were treated for 12 weeks with rosiglitazone (Tan et al. 2006). The release of LPL protein from adipose tissue increased after treatment and this correlated to an overall increase of inactive LPL in plasma. The authors conclude that the adipose tissue is a major source of the inactive LPL in plasma. An observation that goes along with this was made on a group of patients who underwent major elective surgery. Three days after this trauma, which would be expected to decrease adipose tissue LPL production, LPL mass in plasma had decreased by about 50% (Thorne et al. 2005).

A study of pre- and post-heparin LPL in survivors of myocardial infarction and controls showed no correlation between LPL mass in pre- and postheparin plasma, indicating that these parameters reflect different aspects of the LPL system. In contrast, there was a rather strong correlation between LPL mass in pre-heparin plasma and HDL cholesterol (Tornvall et al. 1995). This interesting finding was followed up in a number of studies, mainly by Japanese groups (Saiki et al. 2007). These studies confirmed the relation between LPL mass and HDL cholesterol and extended the correlation of LPL mass to a number of parameters characteristic for the metabolic syndrome, such as fasting plasma insulin and homeostasis model assessment of insulin resistance (HOMA-R). It thus appears that the level of inactive LPL mass in plasma is a marker of insulin resistance (Saiki et al. 2007). Taken together with information on the nature and turnover of inactive LPL in plasma, this suggests that the underlying correlation is between insulin resistance and LPL production rate in adipose tissue.

13.3.3 LPL at the Endothelium

This is where LPL exerts its well known physiological function (Fig. 13.2). Examples of other enzymatic reactions that occur at the endothelium are blood coagulation, fibrinolysis and complement activation. All are complex processes regulated by fine-tuned protein–protein interactions, rather than by gene expression or phosphorylation, as most intracellular processes are. LPL is extracellular and the substrate molecules are contained in large macromolecular lipoproteins. The products of the reaction are surface-active and, if allowed to accumulate, threaten to disrupt membrane organization in adjacent cells.

The notion that the lipase is anchored at the endothelium solely by electrostatic interaction with heparan sulfate chains (Olivecrona et al. 1977) is widely accepted but may be too simplistic. Esko and his collaborators studied mice in which the gene for an enzyme responsible for addition of sulfate to heparan sulfate chains (Ndst1) was inactivated either in hepatocytes (MacArthur et al. 2007) or on endothelial cells (Wang et al. 2005). Hepatocyte Ndst1 null mice developed hypertriglyceridemia due to accumulation of VLDL-like particles, while, surprisingly, endothelial Ndst1 null mice showed no effects on plasma triglyceride levels, suggesting that heparan sulfate proteoglycans present on extrahepatic cells may not interact significantly with lipoproteins or with LPL (Bishop et al. 2008; MacArthur et al. 2007). This argues against the hypothesis that it is only the interaction with the sulfate groups on endothelial heparan sulfate proteoglycans that anchors LPL to the vessel wall. An apo-B related protein was reported to play a role in the binding of LPL (Sivaram et al. 1994), but this has not been followed up. The VLDL receptor is likely to participate in transport of LPL to the endothelium and its action there

(Goudriaan et al. 2004; Obunike et al. 2001). The newly discovered GPIHBP1 protein probably has an important role (Sect. 13.3.2.4).

EM of heart (Blanchette-Mackie et al. 1989) and lactating mammary gland (Schoefl and French 1967) show chylomicrons attached to the luminal surface of the capillary endothelium. The chylomicrons are often partially enveloped by the endothelium, but they are never seen inside cells or penetrating the basal lamina. As discussed in Sect. 13.3.1.3, several LPL molecules must act simultaneously on a chylomicron to account for the rapid delipidation observed in vivo. In one of the EM studies, antibodies demonstrated clusters of LPL molecules over chylomicrons at endothelial projections (Blanchette-Mackie et al. 1989). A likely sequence of events is that chylomicrons initially are attracted to the endothelial surface through electrostatic interactions between cell surface heparan sulfate (and perhaps other polyanions) and the clusters of positively charged amino acid residues on apolipoproteins B and E on the particle. If the particle encounters a binding molecule, LPL, apoAV and/or GPIHBP1, or some other yet unknown molecule, the interaction is enhanced and the particle stays a little longer. LPL molecules move in by lateral diffusion along the endothelium, the interaction is further enhanced and the particle is retained at this "endothelial binding-lipolysis site" (Fig. 13.2). The picture that emerges is that of a large lipolytic complex. Several LPL molecules act on the lipoprotein and are themselves tethered to the endothelium via heparan sulfate chains and/or by interaction with lipase-binding proteins such as GPIHBP1. Important components of the lipolytic complex are apoCII (which binds to and activates the lipase), apoAV (which facilitates hydrolysis in a yet undefined manner) and apo-B and E (which help bind the particle to the cell surface by interacting with heparan sulfate and receptors). ApoCI and apoCIII may also have roles in modulating lipolysis (Sect. 13.3.1.1). There may be more, as yet undiscovered actors that participate in the lipolytic process. It is important to realize that lipolysis is not a single LPL molecule that acts in isolation on the lipoprotein; the process is much more complex. Lipolysis generates several different products (fatty acids, monoglycerides, lysophospholipids), all of which have different fates, as we discuss in Sect. 13.3.3.1. In addition, excess surface components (phospholipids, cholesterol, apolipoproteins) are shed from the lipoprotein and return to blood as a substrate for remodeling reactions that produce HDL (Chap. 10).

In such a lipolytic complex there are a multitude of interactions. The lipoprotein must stay at the same site for many rounds of lipolysis. There is not time for dissociation, movement to a new site and alignment of all necessary interactions, at least not during the rapid lipolysis of large chylomicron particles. One must then ask what terminates the lipolysis and causes the complex to dissociate. This may be a purely stochastic phenomenon, but an interesting possibility is that lipolysis goes on until the tissue and/or the available albumin gets saturated so that fatty acids start to accumulate at the site of lipolysis. As discussed in Sect. 13.3.1.2, experiments in model systems and with cultured cells show that the accumulation of fatty acids causes lipolysis to stop and lipase–lipoprotein as well as lipase–heparan sulfate interactions to dissociate (Bengtsson and Olivecrona 1980b; Saxena and Goldberg 1990; Saxena et al. 1989). This would make good sense in relation to control of energy metabolism, but would also guard against dangerously high local concentrations of fatty acids that could endanger membrane integrity. In vitro studies have shown that albumin binds fatty acids more strongly than the lipase does (Bengtsson and Olivecrona 1980b). It is only when the fatty acid:albumin ratio exceeds 7–8 that lipase–fatty acid complexes are formed. These ratios are never reached in the circulating blood. It follows that modulation of lipase action by fatty acids is not caused by high amounts of NEFA in the blood, but is a local phenomenon at the endothelial lipolysis sites.

The model predicts that at any given moment a substantial fraction of chylomicron particles are "marginalized" at endothelial "binding-lipolysis sites". Direct evidence for this was presented in experiments where humans were injected with retinyl palmitate-labeled chylomicrons (collected the day before from the same individual; Karpe et al. 1997). Shortly thereafter, a large bolus dose of a triglyceride emulsion was injected. Retinyl palmitate-containing lipoproteins in the Sf > 400 fraction rapidly appeared in plasma, indicating that marginalized chylomicrons had returned to the circulating blood as a consequence of being displaced from endothelial sites by the emulsion droplets.

Triglyceride-rich lipoproteins and their remnants have been observed to bind to endothelial cells in culture. Interestingly, this binding appears to occur predominantly in lipid-raft-type membrane microdomains (Wang et al. 2008). This would be in accord with GPIHBP1 having a role in the binding, since GPI-linked proteins tend to localize in raft areas. Addition of LPL with ensuing release of lipolysis products has a number of effects on endothelial cells in culture. Membrane lipids are re-organized via aggregation of raft microdomains (Eiselein et al. 2007). There are signs of endothelial cell inflammation (Wang et al. 2008) and apoptosis (Reinbold et al. 2008). Endothelial permeability increases (Rutledge et al. 1997). Monocyte adhesion increases (Saxena et al. 1992; Williams et al. 2004). A large number of genes are up- or down-regulated (Williams et al. 2004). It is apparent that lipolysis at the endothelium in vivo must be kept under control so that products do not accumulate in sufficient amounts to trigger some of these deleterious effects. Most of the experiments cited above were with cultured endothelial cells. In such systems there is no sink that can absorb the products of lipolysis. In vivo, utilization of fatty acids in the underlying cells and binding by albumin provides such a sink. A safety valve (in case the capacity of the sink is overwhelmed) is provided by fatty acid inhibition of LPL action.

13.3.3.1 Fate of the Products of Lipolysis

The main products of the LPL reaction are fatty acids and 2-monoglycerides. Scow and his associates perfused rat parametrial adipose tissue with chylomicrons labeled either with glycerol or fatty acids (Scow 1977). Of the chylomicron triglycerides perfused, 1.1% was removed and more than 95% of that taken up by the tissue was hydrolyzed to fatty acids and glycerol. Using tissue from fed rats about 55% of the fatty acids liberated from triglycerides were retained in the tissue as newly synthesized triglycerides, about 45% returned to the perfusion medium as albumin-bound fatty acids. Several other studies also show that a large fraction of the fatty acids return to blood as albumin-bound free fatty acids (NEFA). Frayn and his associates have shown this directly, by measuring the arteriovenous difference over human subcutaneous adipose tissue (Bickerton et al. 2007). Mathematical modeling of data from experiments where labeled fat has been given as a fat meal or by injection of chylomicrons also show that a large fraction of the fatty acids returns to the blood in the NEFA pool (Hultin et al. 1996; Teusink et al. 2003).

In contrast to the escape of fatty acids into the effluent when chylomicrons were perfused through adipose tissue, all of the glycerol released returned to the perfusion medium. Retention of glycerol in the tissue was negligible (Scow 1977). Fatty acid release began immediately, while there was a time-lag of a minute before glycerol appeared. When perfusion was switched to medium without chylomicrons, release of fatty acids dropped quickly, whereas release of labeled glycerol continued for several minutes. The authors concluded that the chylomicron triglycerides were hydrolyzed by LPL at sites in near contact with the medium, whereas the monoglycerides probably moved into the adipocytes where they were hydrolyzed. In accord with this, plasma concentrations of mono- or diglycerides are low (Fielding et al. 1993) and in studies on rats only small amounts of labeled di- or monoglycerides appeared in the circulating blood during the clearing of injected labeled chylomicrons (Belfrage et al. 1965; Fielding et al. 1993; Morley et al. 1977).

LPL can hydrolyze monoglycerides, but this reaction is far too slow to keep pace with triglyceride hydrolysis (Scow and Olivecrona 1977). Hence, some other enzyme must step in to complete the hydrolysis. The obvious candidate is monoacylglycerol hydrolase (MAGH), the same enzyme that catalyzes the final step in hydrolysis of intracellular triglycerides (Karlsson et al. 1997). This is an intracellular enzyme present in many tissues and known to be present in adipocytes. It is not known if the enzyme is also present in endothelial cells. It would seem likely that the monoglycerides move to and are hydrolyzed in the adipocytes and that glycerol returns to the blood (Fig. 13.2). An interesting corollary is that, in situations where there is net flux of lipids from blood into adipose tissue, more than half of the fatty acids that are retained originate from monoglycerides. Using the data of Scow (1977), 45% of the chylomicron fatty acids return to the blood as albumin-bound NEFA, 30% move into the adipocytes as monoglycerides and only 25% move as fatty acids.

13.3.3.2 Transport of Fatty Acids and Monoglycerides Generated by LPL

The molecular mechanism for transport of fatty acids in tissues has been much debated. The driving force for their movement is disposal by metabolic reactions within the cells. In adipocytes the main pathway is synthesis of triglycerides. This creates a sink that generates a gradient along which the fatty acids flow, as long as the sink can accommodate them. If/when the capacity of the sink is overwhelmed by the rate of LPL-mediated lipolysis, fatty acids and other products of lipolysis start to accumulate at the endothelial sites and shut off the reaction, as discussed in

Sect. 13.3.1.2. It seems somewhat unlikely that the net direction of fatty acid flow would be determined by the transport process as such.

The fatty acids (and the monoglycerides) that are generated by LPL in the capillaries need to move across or around the endothelial cells and then move through the basal lamina before they can enter the adipocytes. Scow, Blanchette-Mackie and their coworkers suggested that the fatty acids and the monoglycerides move mainly by lateral diffusion in the endothelial cell plasma membrane (Scow and Blanchette-Mackie 1985). They also argued that there are enough membrane continuities that penetrate the basal lamina so that here too the fatty acids and monoglycerides can move in membranes rather than through aqueous spaces. From studies in model systems, it seems likely that these processes are rapid and not rate-limiting. The next step for the fatty acids and monoglycerides is to move across the adipocyte plasma membrane. How this occurs is the subject of much current debate (Hamilton 2007; Schaffer 2002). On the basis of physical studies, all three necessary steps for transmembrane movement of a fatty acid (adsorption of a fatty acid anion into the outer leaflet of a lipid bilayer, flip-flop over the membrane as a protonated fatty acid, desorption from the inner leaflet in the form of a fatty acid anion) are rapid enough so that they would not be rate-limiting for overall metabolism. Other investigators argue that transmembrane movement of fatty acids depends on specific transport proteins, analogous to the transport of other nutrients, such as glucose and amino acids. There are a number of proteins which have been suggested to facilitate the uptake of fatty acids by cells (Hamilton 2007; Schaffer 2002). Unfortunately, these studies do not consider how monoglycerides move across the membrane.

Fatty acid transport proteins (FATP) form a family of six related membrane proteins that enhance fatty acid import when expressed in cultured cells (Gimeno 2007). Members of this family are present in all tissues engaged in active fatty acid uptake and metabolism. All of the FATPs display a structural motif characteristic of ATP-dependent enzymes and several of the FATPs have been shown to catalyze the first step in the cellular metabolism of fatty acids, the formation of fatty acyl-CoA esters. The FATP that has received most interest in terms of adipose tissue function is FATP1 (Gimeno 2007). There is evidence that insulin causes FATP1 to translocate from intracellular membranes to the plasma membrane in adipocytes, analogously to the glucose transporter GLUT4. FATP1 knockout mice show no obvious abnormality on a normal diet, but are less prone to develop insulin resistance and obesity on a high-fat diet (Gimeno 2007). At least one study demonstrates that FATP1 enhances fatty acid uptake even when localized in intracellular membranes (Garcia-Martinez et al. 2005), suggesting that the protein may act primarily by catalyzing the formation of fatty acyl-CoA derivatives and thereby accelerate further metabolism of the fatty acid, rather than acting on the transmembrane movement as such.

FABPpm (plasma membrane fatty acid binding protein) was identified in brush border plasma membranes of the jejunum (Stremmel 1988). It has the same amino acid sequence as an aspartate aminotransferase (Stump et al. 1993). This is a protein that is synthesized on free polysomes in the cytosol and gets imported into mitochondria by virtue of an N-terminal targeting sequence. Overexpression of this protein has been reported to increase fatty acid uptake (Isola et al. 1995). What role, if any, this has in adipocyte lipid uptake is unclear.

Caveolin 1 is a peripheral membrane protein that associates with the cytoplasmic side of raft domains in the plasma membrane and causes bending and formation of invaginations of the membrane, named caveolae. Adipocytes are very rich in these structures, which have been suggested to have important roles in fatty acid flux across the plasma membrane (Pilch et al. 2007). Caveolin-1 null mice show hypertriglyceridemia and are resistant to diet-induced obesity (Razani et al. 2002). Hamilton (2007) has suggested a model for how caveolin 1 may facilitate fatty acid uptake and utilization. He notes that there are 14 basic residues in caveolin 1 which are close to the inner leaflet of the phospholipid bilayer. Fatty acids that flip-flop over the membrane and form anions on the cytosolic side may be sequestered by these positive charges on caveolin and favorably organized for further metabolism. In this respect it is of interest that triglyceride synthesis in adipocytes appears to be localized to the caveolae (Ost et al. 2005).

CD36 is a multifunctional membrane glycoprotein (Ibrahimi and Abumrad 2002). It has been described as a class B scavenger receptor involved in angiogenesis, atherosclerosis, inflammation and lipid metabolism. CD36 binds a diverse list of ligands that includes, in addition to long-chain fatty acids, oxidized low-density lipoproteins, anionic phospholipids, as well as a number of proteins (e.g. thrombospondin, some collagens) and apoptotic cells. CD36 is expressed in many tissues, with high expression in cells with active fatty acid transport such as intestinal fat-absorbing cells, adipocytes and oxidative fiber myocytes. In adipocytes the expression is modulated by a number of factors that go along with a function in lipid assimilation, including PPAR-γ agonists (Ibrahimi and Abumrad 2002). CD36 null mice show increased levels of triglycerides and NEFA in blood (Goudriaan et al. 2005) and defects in fatty acid uptake/utilization in heart (Park et al. 2007). It has been suggested that the increased plasma triglyceride levels are caused by decreased LPL-mediated hydrolysis of triglyceride-rich lipoproteins resulting from fatty acid inhibition of LPL action (Goudriaan et al. 2005).

A mutation that introduces a premature stop in the CD36 gene in Asian and African populations is thought to cause some forms of heart hypertrophy (Ibrahimi and Abumrad 2002). CD36 was also shown to be the defective gene in a strain of rats with spontaneous hypertension (SHR rats; Ibrahimi and Abumrad 2002). These rats display hypertriglyceridemia and insulin resistance which is supposed to be secondary to deranged fatty acid metabolism. Interestingly, CD36 has been shown to be preferentially located in caveolae (Zhang et al. 2008). Perhaps this is a clue to how the protein facilitates fatty acid uptake, as discussed above for caveolin 1.

13.3.3.3 LPL Facilitates the Movement of Cholesterol and Other Lipids into Cells

Scow and his associates found that, when adipose tissue or mammary gland was perfused with chylomicrons, not only fatty acids but also cholesterol and other lipids moved into the tissue (Scow 1977). This uptake was related to the LPL activity in the tissue. The mammary gland of a normal, lactating rat had high LPL activity and removed about 15% of the chylomicron cholesterol infused, whereas the tissue of a hypophysectomized rat had low LPL activity and removed less than 1%. Others found that rat hearts took up substantial amounts of cholesteryl esters when perfused with chylomicrons (Fielding 1978) and that LPL catalyzed the transfer of tocopherol from lipoproteins to fibroblasts (Traber et al. 1985). The Steins and their collaborators demonstrated that addition of LPL markedly enhanced transfer of cholesteryl esters from chylomicrons to cultured cells (Friedman et al. 1981). Release of surface-bound endogenous or exogenous LPL by heparin was accompanied by almost complete elimination of uptake of cholesteryl esters, even though rapid hydrolysis of triglycerides continued in the medium (Chajek-Shaul et al. 1982). They concluded that LPL must bind to the cell surface to catalyze the uptake of cholesteryl esters. Transfer of cholesteryl esters did not require that they were presented in lipoproteins, LPL-catalyzed transfer was also observed with phospholipid liposomes. The transfer did not require any lipid hydrolysis, but occurred with non-hydrolyzable diether phospholipid analogues and cholesteryl ethers. A likely mechanism for the transfer is that the lipase binds on the one hand to the lipoprotein particle/liposome and on the other hand to the cell surface, thereby creating an apposition that favors exchange and/or net transfer of lipids.

Goldberg, Deckelbaum and their associates showed that, in a system where labeled LDL was incubated with LDL receptor-negative fibroblasts, LPL mediated a greater uptake of [^{3}H]-cholesteryloleoyl ether than of [^{125}I]-LDL protein (Seo et al. 2000). This result indicated selective lipid uptake, as in the experiments from the Steins' laboratory, described above. Selective lipid uptake was not affected by tetrahydrolipstatin (which inhibits LPL hydrolysis) but was nearly abolished by heparin, monoclonal anti-LPL antibodies and by chlorate treatment of cells (which inhibits sulfation of proteoglycans) and was not found using CHO cells deficient in heparan sulfate proteoglycans. Similar results were obtained with HDL and it appears likely that LPL can facilitate selective uptake of lipids from all types of lipoproteins and model lipid particles, such as liposomes and emulsion droplets. In continued experiments, this group used a mutated form of LPL where exchange of one of the critical residues in the active site rendered the LPL catalytically inactive (Merkel et al. 2002b). Transgenic expression of this mutated inactive LPL enhanced selective transfer of cholesteryl esters both in cell systems and in vivo.

13.3.3.4 LPL Mediates Particle Uptake

LPL enhances the binding of lipoproteins to cells. This is a necessary feature for its function as an enzyme that hydrolyzes lipoprotein triglycerides at the surface of endothelial cells. In this case, the binding is reversible and the particles return to the circulating blood after a round of lipolysis. In other cases, the binding can result in endocytosis (Casaroli-Marano et al. 1998; Eisenberg et al. 1992; Fernandez-Borja et al. 1996). For this function, the lipase can bind to proteoglycans in the same way as when it functions as an enzyme, but LPL also binds to several cell surface receptors (Sect. 13.3.1.5). It has been suggested that chylomicron remnants bring some LPL with them when they leave the sites of lipolysis in adipose and other tissues and that this facilitates the binding and uptake of remnants in the liver (Felts et al. 1975; Heeren et al. 2002). It is of interest to note that the liver does not normally produce LPL. This may be a case when the non-enzymatic ligand function of the lipase dominates. It seems clear that most chylomicron and VLDL particles are metabolized by sequential binding-lipolysis in extrahepatic tissues, followed by binding-endocytosis in the liver, but some of the particles are cleared in extrahepatic tissues (Fielding 1978; Hultin et al. 1996; Merkel et al. 2002b) and this occurs mainly in LPL-rich tissues (van Bennekum et al. 1999; Yokoyama et al. 2007). It has also been suggested that LPL-mediated uptake of lipoproteins by macrophages can cause excessive lipid load on the cells and accelerate athrosclerosis (Babaev et al. 2000). Bridging by LPL can also contribute to retention of lipoproteins in the extracellular matrix (Edwards et al. 1993; Neuger et al. 2001; Pentikainen et al. 2002), a process that may also accelerate atherosclerosis (Mead et al. 1999; Stein and Stein 2003).

13.3.4 Regulation/Modulation of Tissue LPL Activity

LPL activities change rapidly and profoundly in a tissue-specific manner (Scow et al. 1977). Several-fold changes over one or a few hours have been reported both for heart (Pedersen and Schotz 1980; Pulinilkunnil and Rodrigues 2006) and for adipose tissue LPL activity (Semb and Olivecrona 1986). Thus, LPL activity is one of the most, if not the most, dynamic parameters in lipoprotein metabolism. The LPL reaction is the first step in lipoprotein catabolism and it is a common feature of biochemical pathways that the initial step is the main site of control. We discuss the regulation of LPL action from an adipose tissue perspective and we focus on the nature of the mechanisms involved.

13.3.4.1 LPL Turns Over Rapidly

A number of studies in vivo, in tissue explants or in adipocytes have shown that LPL decays rapidly when protein synthesis is blocked (Bergo et al. 2002; Cryer et al. 1973). The turnover time for the lipase protein, calculated from these data is in the order of 2–3 h. All information available indicates that modulation of tissue LPL activity occurs by active or inactive LPL molecules flowing through the system, or by the conversion of active LPL to inactive. There is no evidence for reversible shifts of long-lived lipase molecules between active and inactive states.

13.3.4.2 Transcriptional Control is Important But Too Slow to Explain the Rapid Modulation of LPL Activity

There are several examples of major changes in LPL expression (in rats or mice): (a) increase in mammary gland at the onset of lactation (Jensen et al. 1994) with concomitant decrease of LPL expression in adipose tissue (Jensen et al. 1994; Ling et al. 2003; Martin-Hidalgo et al. 1994), (b) increase in brown adipose tissue on cold exposure (Carneheim et al. 1988), (c) induction of expression on development of preadipocytes into adipocytes (Dani et al. 1990), (d) suppression of expression in liver in the neonatal period (Peinado-Onsurbe et al. 1992), (e) induction of LPL expression on activation of monocytes to macrophages (Auwerx et al. 1988). Insulin (Ong et al. 1988), cortisol (Ong et al. 1992b), cathecolamines, cellular Ca^{2+} signaling (Querfeld et al. 1999), PPARγ agonists (Laplante et al. 2008), statins (Saiki et al. 2005) and other hormones/agents have been reported to change LPL mRNA abundance in adipocytes, but results vary between laboratories. In general these effects are slow with time-scales of days, indicating that they mediate long-term adaptations to the environment, rather than rapid modulation in response to meals, exercise and other rapid/transient events. This is not to say that gene expression is not an important aspect of the LPL system (Enerback and Gimble 1993). On the contrary, LPL gene expression has been suggested to be a risk factor for development of obesity (Chen et al. 2008), insulin resistance (Holzl et al. 2002) and dyslipidemia (Sprecher et al. 1996). Most impressive are two recent large studies that used advanced genetic techniques to identify networks of genes of importance for metabolic diseases (Chen et al. 2008), which identified LPL as a member of such a network with impact on obesity. In-depth discussions of the LPL gene, the potential regulatory sites that it contains and the hormones/agents that regulate its expression can be found in recent reviews (Auwerx et al. 1996; Enerback and Gimble 1993; Mead et al. 2002; Merkel et al. 2002a).

13.3.4.3 Rapid Modulation of LPL Activity is (Mainly) Post-Transcriptional

Adipose tissue LPL activity changes rapidly (within one or a few hours) in response to changes in nutritional state (Doolittle et al. 1990; Semb and Olivecrona 1989b) and other changes, e.g. physical activity versus inactivity (Bey and Hamilton 2003). In contrast, LPL mRNA in (rat) adipose tissue turns over relatively slowly; a half-life of 17 h has been reported for white adipose tissue (Bergo et al. 2002) and 20–30 h in brown adipose tissue (Carneheim et al. 1988). This is far too slow to account for the rapid changes in LPL activity. In further support of post-transcriptional mechanisms, the changes in LPL activity are much larger then the changes in LPL mass (Bergo et al. 1996; Ong and Kern 1989a), at least in experimental animals. In humans, LPL transcription may be modulated on a relatively rapid time-scale in parallel with LPL activity and mass (Ruge et al. 2005).

13.3.4.4 Translation of LPL mRNA into Protein can be Controlled

A highly specific mechanism to suppress LPL translation, without effects on translation of other proteins, has been described. Stimulation of the PKA system in cultured adipocytes causes a decrease in LPL synthesis (Ball et al. 1986; Ong et al. 1992a). Kern and his associates demonstrated that this is mediated by a protein complex that binds to the 3' UTR of LPL mRNA and blocks its translation into LPL protein. The inhibiting protein complex consists of the catalytic subunit of PKA and the A kinase anchoring protein (AKAP) (Ranganathan et al. 2005). This protein, which is abundant in white adipocytes, is expressed at low levels or not at all in myocytes or brown adipocytes. Hence, this mechanism for control of LPL translation appears to be specific for white adipose tissue. Kern and his associates reported that, while the rate of LPL synthesis was decreased, the rate at which cell-associated LPL is degraded was also decreased (Ong et al. 1992a) so that LPL protein mass would not change much. Ball and her coworkers, in contrast, reported that the rate of LPL degradation was increased by β-adrenergic stimulation (Ball et al. 1986).

It was noted in several studies on the synthesis of LPL that the specific effects are compounded by a general anabolic effect of insulin on overall protein synthesis in adipocytes. In the system (rat fat pads) used by Ashby and Robinson (1980), insulin more than doubled the rate of incorporation of labeled amino acids into proteins. Others found a 40–50% decrease of protein synthesis in adipocytes of rats starved for 1 day (Lee et al. 1998) or found that methionine incorporation was only 30–60% in adipocytes from starved compared to fed guinea pigs (Semb and Olivecrona 1989b). Ong and Kern found that LPL synthetic rate in rat adipocytes was reduced by 80% in the absence of glucose in the medium (Ong and Kern 1989b). Synthesis increased directly with glucose concentration. Insulin doubled the synthetic rate at all glucose concentrations (Ong and Kern 1989b). Hence, studies on the effects of nutritional state on adipose LPL activity are likely to be confounded by this general effect on protein synthesis.

13.3.4.5 Extracellular LPL Activity Changes more than Intracellular

This was first shown for rat heart. The technique used was to perfuse the heart with heparin-containing medium. LPL appears immediately, i.e. the first drop of perfusate that contains heparin also contains LPL. Within one or a few minutes a peak of LPL activity appears. This is interpreted to be the lipase that was at or near the vascular endothelium (Borensztajn and Robinson 1970; Pedersen and Schotz 1980). This peak is several-fold higher with hearts from fasted compared to fed rats, and the change can occur rapidly. In one early experiment, fasted rats were given glucose and this caused the heparin-releasable LPL activity in heart to decrease by 85% within 1 h (Pedersen and Schotz 1980). In contrast to this rapid change in heparin-releasable LPL, there is only a small change of total tissue activity, often statistically not significant. The amount of LPL released is typically in the order of 10–25% of tissue total so that even a large change in the endothelium-localized

LPL causes only a small change in the total. Rodrigues and his associates have shown directly that LPL activity within the cardiomyocytes does not change significantly (An et al. 2005). This protocol with brief heparin release of heart LPL has been used in several laboratories to study LPL turnover (Liu and Olivecrona 1991) and regulation (Pulinilkunnil and Rodrigues 2006).

The same type of observation has been made for adipose tissue. The technique used to separate intra- and extracellular LPL has been to digest the tissue with collagenase and isolate the adipocytes. When adipose tissue from fed rats was digested with collagenase, only about 20% of the LPL activity was recovered (Cunningham and Robinson 1969). With tissue from fasted rats, virtually all the LPL activity was recovered with the adipocytes. Moreover, the LPL activity in the adipocytes was the same irrespective of whether they came from fed or fasted animals. The overall conclusion is the same for adipose tissue as for heart: the rapid modulation of LPL activity in response to changes in nutrition engages only the amount of active extracellular LPL; the activity in the lipase-producing cells changes much less, or not at all.

Additional insight has come from studies of LPL mass. When adipose tissue of fed and fasted rats were compared, the LPL activity was 4.8× higher in adipose tissue from the fed rats, but there was no difference in LPL mass (Bergo et al. 1996). Separation on heparin-agarose confirmed that the difference lay in the proportion of inactive to active LPL. With tissue from fasted rats most of the lipase protein eluted in the position of inactive lipase, whereas with tissue from fed rats most of the lipase protein was in the position of active lipase. Hence it was not a change in the activity state of all lipase molecules (as could for instance have been brought about by a phosphorylation/dephosphorylation event), but it was the fraction of lipase molecules that were in the active state that changed. The specific activity (activity/mass) was the same for the active species whether it came from fed or fasted rats. Further studies showed that within the adipocytes there was no difference in LPL mRNA, protein mass or activity between fed and fasted rats. Separation on heparin-agarose showed the same distribution between active and inactive forms of the lipase (Wu et al. 2003). Hence, the difference lay in the activity state of lipase molecules located extracellularly. Studies with transcription blockers (actinomycin D, α-amanitin) demonstrated that for this mechanism to operate in going from the fed to the fasted state, another gene(s), separate from the LPL gene has to be expressed to suppress LPL activity. When transcription is blocked in fasted rats, LPL activity rapidly increases in a manner similar to when the animals are given food (Bergo et al. 2002). Thus, a rapidly turning-over LPL-controller protein was implicated.

13.3.4.6 Angiopoietin-Like Proteins can Convert Active LPL into Inactive Monomers

A recently discovered family of proteins, the angiopoietin-like proteins (Angptl; Chap. 10), has at least three members (angptl 3, 4, 6) that inactivate LPL (Li 2006). Work in several laboratories showed that deficiency of either angptl3 or angptl4 was associated with high LPL activity and low plasma triglyceride levels. Conversely, overexpression resulted in low LPL activity and hypertriglyceridemia (Chap. 10). In vitro model studies showed that the C-terminal coiled-coil domain of angptl4 binds to active LPL dimers and converts them into inactive monomers (Sukonina et al. 2006). The turnover of the mRNA for angptl4 in rat adipose tissue was shown to be rapid enough to modulate LPL activity in response to feeding/fasting and the LPL activity levels in adipose tissue correlated to the angptl4 mRNA levels (Sukonina et al. 2006). Hence, the angptl proteins are likely candidates for the LPL-controlling protein inferred from the studies on effect of transcription blockade (Bergo et al. 2002). Angptl4 has affinity both for heparin (Sukonina et al. 2006) and components in the extracellular matrix (Cazes et al. 2006) and can thus be positioned to inactivate LPL on its travel from the adipocyte surface to the luminal side of the endothelium.

13.4 Intracellular Lipases

For a long time it was believed that hormone-sensitive lipase (HSL) was the only lipase hydrolyzing tri- and diglycerides in adipocytes and that regulation of its activity by phosphorylation of the lipase itself was the main rate-determining step in the mobilization of fatty acids. It was discovered that: (a) perilipin is a lipid droplet protein whose state of phosphorylation has a decisive influence on the rate of lipolysis (Greenberg et al. 1991) and then (b) Adipose triglyceride lipase (ATGL) is the enzyme that catalyzes the first step in overall lipolysis (Zechner et al. 2008); and these discoveries have generated an intense interest in intracellular lipolysis. We limit our discussion here to the properties of the main proteins that participate in intracellular lipolysis and the molecular mechanisms by which they are regulated as a background for a discussion of how their action is integrated with that of the LPL system and with triglyceride synthesis to enable a fine-tuned regulation of lipid deposition/mobilization over the entire spectrum of circumstances from starvation to caloric excess. For more detailed discussions of aspects of intracellular lipolysis, the reader is referred to several excellent reviews (Arner and Langin 2007; Brasaemle 2007; Duncan et al. 2007; Granneman and Moore 2008; Watt and Steinberg 2008; Zechner et al. 2008).

13.4.1 Adipose Triglyceride Lipase

This enzyme catalyzes the first step in the breakdown of stored triglycerides: their hydrolysis to diglycerides (Zechner et al. 2008). ATGL was discovered only recently, in 2004. This is surprising in view of the fundamental role the enzyme has. Apparently, its activity did not show up under the assay conditions used by the many groups that studied HSL or LPL in tissue extracts. ATGL is a 486-amino-acid protein. Related enzymes are found in all eucaryotes; vertebrates, flies, fungi and plants (Zechner et al. 2008). Mammalian ATGL belongs to a gene family that is characterized by the presence of a patatain-like domain. This is named after the most abundant protein in potato tuber, which has lipase activity. There are several other members of this family in the human and/or mouse genomes. Some of these are expressed in adipocytes and some have lipase and/or phospholipase activity (Lake et al. 2005). The physiological roles of these proteins are unknown and we will not discuss them further. A comprehensive list of all the 23 putative lipases in mammalian tissues is given by Watt and Steinberg (2008). The combined action of ATGL and HSL accounts for more than 95% of adipose tissue lipolysis (Schweiger et al. 2006). ATGL shows a strong preference for triglycerides as substrate (Zimmermann et al. 2004). It hydrolyzes diglycerides and phospholipids only slowly, and does not hydrolyze cholesteryl esters or monoglycerides at all.

Most cell types in the body can produce lipid droplets as (a) a local buffer that ensures uninterrupted access to fatty acids for metabolism and (b) as a way to detoxify a temporary surplus of fatty acids. All these cells contain ATGL (Zechner et al. 2008), which is distributed between the cytosol and the surface of the lipid droplets. ATGL knockout mice show a severe phenotype, indicating gross disturbance of overall energy metabolism and decreased ability to respond to metabolic stress, e.g. cold acclimatization (Haemmerle et al. 2006). These mice accumulate triglycerides in virtually all organs. When fasted, they have low plasma NEFA, ketone bodies and triglycerides, indicating defective mobilization of fatty acids from the adipose tissue.

The catalytic activity of ATGL is regulated through interaction with proteins on the lipid droplets. Comparative gene identification 58 (CGI-58), also known as α/β hydrolase fold domain 5 (ABHD5), acts as a specific activator. It is a 349-amino-acid protein that interacts with and enhances the activity of (mouse) ATGL about 20-fold (Lass et al. 2006). Mutations in CGI-58 cause the rare Chanarian–Dorfman syndrome characterized by ichtyosis and accumulation of triglycerides in most tissues (Lass et al. 2006).

13.4.2 *Hormone-Sensitive Lipase*

Hormone-sensitive lipase (HSL) was for a long time considered to be the only enzyme responsible for hydrolysis of triglycerides and cholesteryl esters in lipid droplets. It came as a surprise that HSL knockout mice displayed a rather mild phenotype (Haemmerle et al. 2002; Osuga et al. 2000; Wang et al. 2001). The animals had an apparently normal energy metabolism. In particular, they were not overweight. HSL-deficient adipocytes responded to stimulation by catecholamines and released fatty acids at an almost normal rate. A main difference compared to wild-type mice was that the HSL knockout mice accumulated diglycerides in several tissues, indicting that HSL was rate-limiting for diglyceride hydrolysis.

HSL has a complex gene structure with ten exons. Alternative exon usage generates mRNA transcripts and proteins of different sizes in a tissue-specific manner. Modeling of the 3-D structure indicates that the molecule is organized in two major structural domains, encoded by exons 1–4 and 5–9, respectively (Osterlund et al. 1999). The

C-terminal domain contains the active site and a loop region (amino acids 521–669) where all known phosphorylation sites are located. The enzyme hydrolyzes a wide variety of substrates in model systems. Of the substrates relevant in vivo, diglycerides are the best and triglycerides are the worst (Zechner et al. 2008).

HSL is present throughout the body. In most tissues its main role is probably, as in adipose tissue, to act in a co-ordinated manner with ATGL to hydrolyze tri- and diglycerides for metabolic use. While HSL knockout mice appear to be able to mobilize lipid at a sufficient rate during basal conditions, the enzyme is needed for accelerated lipolysis under stressed conditions, e.g. submaximal exercise (Fernandez et al. 2008a). HSL hydrolyzes not only tri- and diglycerides but also cholesteryl esters, a reaction that ATGL cannot carry out. The importance of this reaction is readily demonstrated in steroidogenic tissues, where HSL is needed for hormone synthesis both from stored cholesteryl esters and from cholesteryl esters taken up from lipoproteins by so-called selective transfer mediated by the SR-B1 receptor (Kraemer 2007). In a similar manner, HSL is probably needed to split cholesteryl esters in many cell types so that the cholesterol can be used by the cell or pumped out via the ABCA-1 transporter. Therefore, it is not surprising that HSL knockout mice show signs of disturbed cholesterol homeostasis (Fernandez et al. 2008b).

Under basal conditions, HSL is a cytoplasmic protein. The enzyme has several sites for phosphorylation by PKA and some other protein kinases (Zechner et al. 2008). Phosphorylation stimulates the enzyme activity modestly (twofold or less) in model systems, whereas adrenergic stimulation of adipocytes can increase the release of fatty acids by up to 100-fold. This dramatic stimulation is mediated by perilipin, as we discuss below.

13.4.3 Monoacylglycerol Hydrolase

The enzyme monoacylglycerol hydrolase (MAGH), which is a 302-amino-acid protein, is present in most tissues (Watt and Steinberg 2008). It catalyzes the last step in the hydrolysis of stored triglycerides in the adipocyte (Fig. 13.2). It is presumably the same enzyme that hydrolyzes the monoglycerides that emanate from LPL-mediated action, although it is not known if these monoglycerides have to move into the adipocytes or if they can be hydrolyzed already in the endothelial cells (Sect. 13.3.3.1). Another possibility is that some of the monoglycerides, from intracellular lipolysis or from LPL action, are re-esterified by the monoacylglycerol acyl transferase (MGAT), which has been demonstrated in adipose tissue (Schultz et al. 1971). The transmembrane movement and metabolism of monoglycerides must be efficient. There are no reports of conditions where these molecules accumulate.

MAGH has no activity against tri- or diglycerides but it has been shown to participate in the degradation of some endocannabinoids (Saario and Laitinen 2007). There is a large literature dealing with pharmacological inhibition of MAGH to target endocannabinoid metabolism (Viso et al. 2008). An interesting connection to our present topic is that one of the major endocannabinoids is 2-arachidonoylglycerol, a molecule that could derive from lipase action on triglycerides.

13.4.4 *Perilipin and the Orchestration of Lipolysis*

Perilipin is a lipid droplet associated protein that has important functions in modulating lipolysis in mammalian adipocytes (Fig. 13.2) (Brasaemle 2007; Ducharme and Bickel 2008). There are three isoforms of perilipin that arise from differential splicing but perilipin A, the largest protein (517 amino acids in mice), is the most abundant in adipocytes. There are six phosphorylation sites in perilipin. One of these, serine 517 appears to be a master switch (Miyoshi et al. 2007). What role phosphorylation of HSL itself has is at present not clear.

While perilipin has a pivotal role in regulating lipolysis, the details are not yet known. A current model suggests the following sequence of events (Granneman and Moore 2008). Under basal conditions perilipin binds CGI-58 at the surface of the lipid droplets and thereby denies ATGL access to its activator. HSL is mainly in the cytosol and overall lipolysis is slow. When the adipocytes are stimulated, for instance by a β-adrenergic agonist, a signaling relay causes protein kinase A (PKA) to phosphorylate perilipin. This abolishes the binding between perilipin and CGI-58 which now binds to and activates ATGL. At the same time the phosphorylated perilipin recruits (phosphorylated) HSL from the cytosol to the lipid droplets (Granneman and Moore 2008). Now the stage is set for rapid hydrolysis of triglycerides to diglycerides by ATGL and further hydrolysis of these to monoglycerides by HSL. MAGH can then finish the job and release a fatty acid and glycerol. Another effect of β-adrenergic stimulation of adipocytes is that the lipid droplets show a dramatic remodeling: a few large centrally located lipid droplets fragment into a myriad of tiny microlipid droplets that scatter throughout the cytoplasm (Brasaemle 2007), thus enlarging the surface at which the lipases can act (Chap. 1).

13.5 Triglyceride Synthesis

Net release of fatty acids is not a simple function of the rate of lipolysis in adipocytes, but reflects a balance between lipolysis and re-esterification, sometimes referred to as the triglyceride–fatty acid cycle. Current thinking is that lipolysis generates an excess of fatty acids and that it is the rate of re-esterification that determines the net outflow of fatty acids from the adipocytes (Fielding and Frayn 1998). The fatty acids from the LPL reaction must also be factored-in. During periods of nutrient excess there is net uptake of fatty acids from lipoprotein triglycerides. During starvation there is net release of fatty acids from the tissue. These fatty acids come both from intracellular lipolysis and from LPL-mediated hydrolysis of lipoprotein triglycerides (Fielding and Frayn 1998). If we assume that fatty acids from the two sources mix, then (re-)esterification becomes the overriding control of their fate (Fielding and Frayn 1998).

Glyceride synthesis requires that the fatty acids are presented as their coenzyme A (CoA) esters. Such esters can not pass over the cell membrane. Therefore conversion to an acyl-CoA ester traps the fatty acid in the cell and favors triglyceride synthesis (Soupene and Kuypers 2008). In this respect it is of interest to note that some of the putative fatty acid transport proteins have acyl-CoA synthetase activity (Sect. 13.3.3.2). These proteins have been shown to enhance transport of fatty acids into cells, but the mechanism by which they do this is not entirely clear. Perhaps trapping is the way some of them work.

13.5.1 A Triglyceride–Diglyceride Cycle?

There are two diacylglycerol acyl transferase (DGAT) enzymes in adipocytes (Yen et al. 2008). They are located in the ER and perhaps also on the lipid droplets. In the adipocytes, diglycerides are continuously generated by ATGL action. Some of these can be further hydrolyzed by ATGL, but most are probably re-esterified by DGAT. When adipocytes are stimulated, ATGL action accelerates and HSL becomes active. In this situation there will be a competition for the diglycerides between re-esterification into triglycerides by DGAT and further hydrolysis by HSL. Diglycerides also arise as intermediates in de novo synthesis of triglycerides. The balance between HSL and DGAT activity probably determines whether diglycerides are hydrolyzed or esterified (Fig. 13.2).

There is much evidence to support that DGAT enzymes are regulated and have important roles for lipid deposition in adipose tissue. In cultured adipocytes, DGAT1 mRNA is increased by glucose and insulin (Meegalla et al. 2002) and by PPAR-γ agonists (Payne et al. 2007). Overexpression of DGAT enzymes leads to triglyceride accumulation in cells (Chen et al. 2002). DGAT1-deficient mice are viable and lean, even on a high-fat diet (Smith et al. 2000). A synthetic small molecule inhibitor to DGAT1 reduced weight gain in mice on a high fat diet (Zhao et al. 2008).

13.5.2 Reacylation of Monoglycerides

Monoglycerides are generated both in intracellular lipolysis by HSL (Sect. 13.4.1.2) and in the LPL reaction (Sect. 13.3.1.1). We know that the monoglycerides from LPL must move into the tissue, since they do not appear in appreciable amounts in the blood (Belfrage et al. 1965; Fielding et al. 1993; Scow 1977). Whether monoglycerides from the two sources mix is not known. There is monoacylglycerol acyl transferase (MGAT) activity in adipocytes (Schultz et al. 1971) and one of the DGAT enzymes can use monoglycerides as substrate (Yen et al. 2008). Hence, the fate of monoglycerides must be determined by the relative activities of MAGH, leading to breakdown to fatty acids and glycerol, and of MGAT enzymes, leading to synthesis of triglycerides. How this balance is set under different conditions is not known, but reacylation of monoglycerides is probably a minor pathway.

13.5.3 De Novo Synthesis of Triglycerides

It was thought for a long time that the only source of glycerol-3-phosphate for triglyceride synthesis was glucose that the adipocyte took from blood by the insulin-regulated GLUT-4 transporter. This led to the view that insulin suppresses the net release of fatty acids both by decreasing lipolysis and by providing glycerol-3-phosphate for re-esterification. This view is now changing. The main source of glycerol-3-phosphate for triglyceride synthesis in adipocytes appears to be glyceroneogenesis from lactate/pyruvate or amino acids via the citric acid cycle, with glycolysis from glucose being less important (Nye et al. 2008a, b). The key regulatory enzyme for glyceroneogenesis is phosphoenolpyruvate kinase (PEPCK). Knockout of this enzyme from adipocytes resulted in lean mice, with 25% of the animals displaying lipodystrophy (Olswang et al. 2002). Overexpression resulted in obese mice (Franckhauser et al. 2002). In contrast, when GLUT-4 (which mediates insulin-stimulated glucose uptake in adipocytes) was knocked out, the mice had normal fat mass (Abel et al. 2001).

There is a glycerol channel (aquaporin 7; AQP7) in adipocytes (Fain et al. 2008) and in endothelial cells in adipose tissue (Skowronski et al. 2007). The expression of AQP7 is up-regulated in response to starvation and down-regulated by insulin (Kishida et al. 2001). Knockout mice showed impaired glycerol release from adipose tissue (Skowronski et al. 2007); and they developed obesity and insulin resistance, even at a young age (Hibuse et al. 2005). Studies with adipocytes from these mice showed that AQP7 disruption lead to elevated adipose glycerol kinase activity and accelerated triglyceride synthesis (Hibuse et al. 2005). These studies demonstrate that glycerol can be used for triglyceride synthesis in adipocytes. Under most conditions glycerol is, however, efficiently released through the AQP7 channel to serve as a substrate for gluconeogenesis in the liver. In accord with this view, AQP7 knockout mice displayed severe fasting-induced hypoglycemia (Hibuse et al. 2005).

13.5.4 Acylation-Stimulating Protein

Acylation-stimulating protein (ASP) is often described as a lipogenic hormone (Kalant and Cianflone 2004). It is produced in adipose tissue from complement C3, primarily through the alternative pathway of complement activation (Paglialunga et al. 2008). ASP acts by binding to a G protein-coupled membrane receptor, C5L2, and has been shown to increase glucose transport and triglyceride synthesis in adipocytes (Kalant et al. 2005). C3 knockout mice (which lack the precursor and are therefore deficient in ASP) are lean, yet hyperphagic (Paglialunga et al. 2008). The increased energy intake appears to be balanced by increased fatty acid oxidation in muscle. ASP levels in blood are increased in obesity, both in humans (Kalant and Cianflone 2004) and in mice (Paglialunga et al. 2008). The proposed mechanism for ASP is that it enhances triglyceride synthesis by an effect on the DGAT enzyme.

In accord with this, studies of arteriovenous differences of triglycerides and fatty acids over adipose tissue in lean and obese individuals after a meal showed a strong correlation between fatty acid trapping in the tissue and ASP levels in blood (Kalant et al. 2000).

13.6 Conclusion: an Integrated View of the Lipase Systems in Adipose Tissue

The intracellular lipase system with ATGL and HSL, as well as the LPL system, are regulated at two levels. Relatively slow changes in gene transcription set the potential activity of the systems in accord with the environment, but for rapid hour to hour responses to changes in nutritional state and other factors, both systems rely on post-transcriptional mechanisms. In this review we do not discuss the long-term adaptations; our focus is the rapid changes that occur against a background of essentially stable mRNA and protein levels.

It is often stated that intracellular lipolysis and LPL action are regulated in a reciprocal manner. This would seem expedient to ensure a smooth transition when net release of fatty acids during fasting changes to uptake after a meal. There is however, as yet, no established molecular mechanism that couples the two processes directly to each other. On the contrary, they operate on different time-scales. β-Adrenergic stimulation, insulin and several other hormones can change the rate of intracellular lipolysis within minutes by phosphorylation/dephosphorylation of relatively long-lived proteins in the lipase system, perilipin, HSL and perhaps other. Changes in LPL activity at its site of action, the vascular endothelium, are accomplished by changes in the amount of short-lived LPL molecules by several different mechanisms. Two intracellular mechanisms are: (a) synthesis of LPL can be specifically blocked by a protein complex that binds to and blocks translation of LPL mRNA and (b) changes in the ER milieu can lead to changes in the partitioning between export of active lipase molecules and degradation of misfolded lipase molecules. After the enzyme has been secreted from the adipocytes, angiopoietin-like proteins can convert active LPL dimers into inactive LPL monomers (Sect. 13.3.4.6).

Even though these mechanisms are separate and different, they all depend on the synthesis and secretion of new LPL molecules. Therefore the changes in LPL activity take a few hours to be fully developed. This is in accord with what is observed in experiments on rats (Bergo et al. 2002; Picard et al. 1999). It is also in accord with the results of studies of arteriovenous differences over adipose tissue in man. When fasted individuals are given a meal it takes 2 h before LPL-mediated hydrolysis increases appreciably (Fielding and Frayn 1998). Hence, the mechanisms by which intracellular lipolysis and LPL activity changes occur by different mechanisms and act on different time-scales.

An additional level of control that acts on LPL but not directly on the ATGL-HSL system is exerted at the vascular endothelium. When fatty acids and other products of lipolysis accumulate, they shut off LPL action by eliminating the effect of apoCII, which is a necessary activator for LPL.

This is not to say that there is no co-ordination between intracellular lipolysis and the LPL reaction at the vascular endothelium, but such co-ordination is not brought about by common mechanisms for regulation, or by direct crosstalk between the lipases. The lipases probably feed their products, fatty acids and monoglycerides into common pools and the ultimate result, net deposition or mobilization, is determined by the balance between rates of lipolysis and rates of esterification.

References

Abel ED, Peroni O, Kim JK, Kim YB, Boss O, Hadro E, Minnemann T, Shulman GI, Kahn BB (2001) Adipose-selective targeting of the GLUT4 gene impairs insulin action in muscle and liver. Nature 409:729–733

An D, Pulinilkunnil T, Qi D, Ghosh S, Abrahani A, Rodrigues B (2005) The metabolic "switch" AMPK regulates cardiac heparin-releasable lipoprotein lipase. Am J Physiol Endocrinol Metab 288:E246–E253

Arner P, Langin D (2007) The role of neutral lipases in human adipose tissue lipolysis. Curr Opin Lipidol 18:246–250

Ashby P, Robinson DS (1980) Effects of insulin, glucocorticoids and adrenaline on the activity of rat adipose-tissue lipoprotein lipids. Biochem J 188:185–192

Augustus A, Yagyu H, Haemmerle G, Bensadoun A, Vikramadithyan RK, Park SY, Kim JK, Zechner R, Goldberg IJ (2004) Cardiac-specific knock-out of lipoprotein lipase alters plasma lipoprotein triglyceride metabolism and cardiac gene expression. J Biol Chem 279:25050–25057

Auwerx J, Deeb S, Brunzell JD, Peng R, Chait A (1988) Transcriptional activation of the lipoprotein lipase and apolipoprotein E genes accompanies differentiation in some human macrophage-like cell lines. Biochemistry 27:2651–2655

Auwerx J, Schoonjans K, Fruchart JC, Staels B (1996) Transcriptional control of triglyceride metabolism: fibrates and fatty acids change the expression of the LPL and apo C-III genes by activating the nuclear receptor PPAR. Atherosclerosis 124[Suppl]:S29–S37

Babaev VR, Patel MB, Semenkovich CF, Fazio S, Linton MF (2000) Macrophage lipoprotein lipase promotes foam cell formation and atherosclerosis in low density lipoprotein receptor-deficient mice. J Biol Chem 275:26293–26299

Ball KL, Speake BK, Robinson DS (1986) Effects of adrenaline on the turnover of lipoprotein lipase in rat adipose tissue. Biochim Biophys Acta 877:399–405

Beckstead JA, Oda MN, Martin DD, Forte TM, Bielicki JK, Berger T, Luty R, Kay CM, Ryan RO (2003) Structure–function studies of human apolipoprotein A-V: a regulator of plasma lipid homeostasis. Biochemistry 42:9416–9423

Beisiegel U, Weber W, Bengtsson-Olivecrona G (1991) Lipoprotein lipase enhances the binding of chylomicrons to low density lipoprotein receptor-related protein. Proc Natl Acad Sci USA 88:8342–8346

Belfrage P, Elovson J, Olivecrona T (1965) Radioactivity in blood and liver partial glycerides, and liver phospholipids after intravenous administration to carbohydrate-fed rats of chyle containing double-labeled triglycerides. Biochim Biophys Acta 106:45–55

Ben-Zeev O, Doolittle MH, Davis RC, Elovson J, Schotz MC (1992) Maturation of lipoprotein lipase. Expression of full catalytic activity requires glucose trimming but not translocation to the cis-Golgi compartment. J Biol Chem 267:6219–6227

Ben-Zeev O, Stahnke G, Liu G, Davis RC, Doolittle MH (1994) Lipoprotein lipase and hepatic lipase: the role of asparagine-linked glycosylation in the expression of a functional enzyme. J Lipid Res 35:1511–1523
Ben-Zeev O, Mao HZ, Doolittle MH (2002) Maturation of lipoprotein lipase in the endoplasmic reticulum. Concurrent formation of functional dimers and inactive aggregates. J Biol Chem 277:10727–10738
Bengtsson G, Olivecrona T (1979a) Apolipoprotein CII enhances hydrolysis of monoglycerides by lipoprotein lipase, but the effect is abolished by fatty acids. FEBS Lett 106:345–348
Bengtsson G, Olivecrona T (1979b) Binding of deoxycholate to lipoprotein lipase. Biochim Biophys Acta 575:471–474
Bengtsson G, Olivecrona T (1980a) The hepatic heparin releasable lipase binds to high density lipoproteins. FEBS Lett 119:290–292
Bengtsson G, Olivecrona T (1980b) Lipoprotein lipase. Mechanism of product inhibition. Eur J Biochem 106:557–562
Bengtsson G, Olivecrona T (1980c) Lipoprotein lipase: some effects of activator proteins. Eur J Biochem 106:549–555
Bengtsson G, Olivecrona T (1981) Heparin-bound lipoprotein lipase is catalytically active and can be stimulated by apolipoprotein CII. FEBS Lett 128:9–12
Bengtsson G, Olivecrona T, Hook M, Riesenfeld J, Lindahl U (1980) Interaction of lipoprotein lipase with native and modified heparin-like polysaccharides. Biochem J 189:625–633
Bengtsson-Olivecrona G, Olivecrona T, Jornvall H (1986) Lipoprotein lipases from cow, guinea-pig and man. Structural characterization and identification of protease-sensitive internal regions. Eur J Biochem 161:281–288
Berbee JF, van der Hoogt CC, Sundararaman D, Havekes LM, Rensen PC (2005) Severe hypertriglyceridemia in human APOC1 transgenic mice is caused by apoC-I-induced inhibition of LPL. J Lipid Res 46:297–306
Berbee JF, van der Hoogt CC, Kleemann R, Schippers EF, Kitchens RL, van Dissel JT, Bakker-Woudenberg IA, Havekes LM, Rensen PC (2006) Apolipoprotein CI stimulates the response to lipopolysaccharide and reduces mortality in gram-negative sepsis. Faseb J 20:2162–2164
Bergo M, Olivecrona G, Olivecrona T (1996) Forms of lipoprotein lipase in rat tissues: in adipose tissue the proportion of inactive lipase increases on fasting. Biochem J 313:893–898
Bergo M, Wu G, Ruge T, Olivecrona T (2002) Down-regulation of adipose tissue lipoprotein lipase during fasting requires that a gene, separate from the lipase gene, is switched on. J Biol Chem 277:11927–11932
Berryman DE, Bensadoun A (1995) Heparan sulfate proteoglycans are primarily responsible for the maintenance of enzyme activity, binding, and degradation of lipoprotein lipase in Chinese hamster ovary cells. J Biol Chem 270:24525–24531
Bey L, Hamilton MT (2003) Suppression of skeletal muscle lipoprotein lipase activity during physical inactivity: a molecular reason to maintain daily low-intensity activity. J Physiol 551:673–682
Bickerton AS, Roberts R, Fielding BA, Hodson L, Blaak EE, Wagenmakers AJ, Gilbert M, Karpe F, Frayn KN (2007) Preferential uptake of dietary Fatty acids in adipose tissue and muscle in the postprandial period. Diabetes 56:168–176
Bishop JR, Schuksz M, Esko JD (2007) Heparan sulphate proteoglycans fine-tune mammalian physiology. Nature 446:1030–1037
Bishop JR, Stanford KI, Esko JD (2008) Heparan sulfate proteoglycans and triglyceride-rich lipoprotein metabolism. Curr Opin Lipidol 19:307–313
Blanchette-Mackie EJ, Masuno H, Dwyer NK, Olivecrona T, Scow RO (1989) Lipoprotein lipase in myocytes and capillary endothelium of heart: immunocytochemical study. Am J Physiol 256:E818–E828
Borensztajn J, Robinson DS (1970) The effect of fasting on the utilization of chylomicron triglyceride fatty acids in relation to clearing factor lipase. (lipoprotein lipase) releasable by heparin in the perfused rat heart. J Lipid Res 11:111–117
Borgstrom B, Erlanson-Albertsson C, Wieloch T (1979) Pancreatic colipase: chemistry and physiology. J Lipid Res 20:805–816

Bragdon JH, Gordon RS Jr (1958) Tissue distribution of C14 after the intravenous injection of labeled chylomicrons and unesterified fatty acids in the rat. J Clin Invest 37:574–578

Brasaemle DL (2007) Thematic review series: adipocyte biology. The perilipin family of structural lipid droplet proteins: stabilization of lipid droplets and control of lipolysis. J Lipid Res 48:2547–2559

Brun LD, Gagne C, Julien P, Tremblay A, Moorjani S, Bouchard C, Lupien PJ (1989) Familial lipoprotein lipase-activity deficiency: study of total body fatness and subcutaneous fat tissue distribution. Metabolism 38:1005–1009

Carneheim C, Nedergaard J, Cannon B (1988) Cold-induced beta-adrenergic recruitment of lipoprotein lipase in brown fat is due to increased transcription. Am J Physiol 254:E155–E161

Casaroli-Marano RP, Garcia R, Vilella E, Olivecrona G, Reina M, Vilaro S (1998) Binding and intracellular trafficking of lipoprotein lipase and triacylglycerol-rich lipoproteins by liver cells. J Lipid Res 39:789–806

Cazes A, Galaup A, Chomel C, Bignon M, Brechot N, Le Jan S, Weber H, Corvol P, Muller L, Germain S, Monnot C (2006) Extracellular matrix-bound angiopoietin-like 4 inhibits endothelial cell adhesion, migration, and sprouting and alters actin cytoskeleton. Circ Res 99:1207–1215

Chajek-Shaul T, Friedman G, Stein O, Olivecrona T, Stein Y (1982) Binding of lipoprotein lipase to the cell surface is essential for the transmembrane transport of chylomicron cholesteryl ester. Biochim Biophys Acta 712:200–210

Chajek-Shaul T, Friedman G, Knobler H, Stein O, Etienne J, Stein Y (1985) Importance of the different steps of glycosylation for the activity and secretion of lipoprotein lipase in rat preadipocytes studied with monensin and tunicamycin. Biochim Biophys Acta 837:123–134

Chajek-Shaul T, Friedman G, Ziv E, Bar-On H, Bengtsson-Olivecrona G (1988) Fate of lipoprotein lipase taken up by the rat liver. Evidence for a conformational change with loss of catalytic activity. Biochim Biophys Acta 963:183–191

Chen HC, Stone SJ, Zhou P, Buhman KK, Farese RV Jr (2002) Dissociation of obesity and impaired glucose disposal in mice overexpressing acyl coenzyme a:diacylglycerol acyltransferase 1 in white adipose tissue. Diabetes 51:3189–3195

Chen Y, Zhu J, Lum PY, Yang X, Pinto S, MacNeil DJ, Zhang C, Lamb J, Edwards S, Sieberts SK, Leonardson A, Castellini LW, Wang S, Champy MF, Zhang B, Emilsson V, Doss S, Ghazalpour A, Horvath S, Drake TA, Lusis AJ, Schadt EE (2008) Variations in DNA elucidate molecular networks that cause disease. Nature 452:429–435

Chong MF, Fielding BA, Frayn KN (2007) Metabolic interaction of dietary sugars and plasma lipids with a focus on mechanisms and de novo lipogenesis. Proc Nutr Soc 66:52–59

Christophersen B, Nordstoga K, Shen Y, Olivecrona T, Olivecrona G (1997) Lipoprotein lipase deficiency with pancreatitis in mink: biochemical characterization and pathology. J Lipid Res 38:837–846

Chung BH, Palgunachari MN, Mishra VK, Chang CH, Segrest JP, Anantharamaiah GM (1996) Probing structure and function of VLDL by synthetic amphipathic helical peptides. J Lipid Res 37:1099–1112

Cruz WS, Kwon G, Marshall CA, McDaniel ML, Semenkovich CF (2001) Glucose and insulin stimulate heparin-releasable lipoprotein lipase activity in mouse islets and INS-1 cells. A potential link between insulin resistance and beta-cell dysfunction. J Biol Chem 276:12162–12168

Cryer A, Foster B, Wing DR, Robinson DS (1973) The effect of cycloheximide on adipose-tissue clearing-factor lipase. Biochem J 132:833–836

Cryer A, Riley SE, Williams ER, Robinson DS (1976) Effect of nutritional status on rat adipose tissue, muscle and post-heparin plasma clearing factor lipase activities: their relationship to triglyceride fatty acid uptake by fat-cells and to plasma insulin concentrations. Clin Sci Mol Med 50:213–221

Cunningham VJ, Robinson DS (1969) Clearing-factor lipase in adipose tissue. Distinction of different states of the enzyme and the possible role of the fat cell in the maintenance of tissue activity. Biochem J 112:203–209

Dani C, Amri EZ, Bertrand B, Enerback S, Bjursell G, Grimaldi P, Ailhaud G (1990) Expression and regulation of pOb24 and lipoprotein lipase genes during adipose conversion. J Cell Biochem 43:103–110

Davies BS, Waki H, Beigneux AP, Farber E, Weinstein MM, Wilpitz DC, Tai LJ, Evans RM, Fong LG, Tontonoz P, Young SG (2008) The expression of GPIHBP1, an endothelial cell binding site for lipoprotein lipase and chylomicrons, is induced by PPARγ. Mol Endocrinol 22:2496–2504

Davis RC, Wong H, Nikazy J, Wang K, Han Q, Schotz MC (1992) Chimeras of hepatic lipase and lipoprotein lipase. Domain localization of enzyme-specific properties. J Biol Chem 267:21499–21504

Deckelbaum RJ, Ramakrishnan R, Eisenberg S, Olivecrona T, Bengtsson-Olivecrona G (1992) Triacylglycerol and phospholipid hydrolysis in human plasma lipoproteins: role of lipoprotein and hepatic lipase. Biochemistry 31:8544–8551

Dichlberger A, Cogburn LA, Nimpf J, Schneider WJ (2007) Avian apolipoprotein A-V binds to LDL receptor gene family members. J Lipid Res 48:1451–1456

Doolittle MH, Ben-Zeev O, Elovson J, Martin D, Kirchgessner TG (1990) The response of lipoprotein lipase to feeding and fasting. Evidence for posttranslational regulation. J Biol Chem 265:4570–4577

Dorfmeister B, Zeng WW, Dichlberger A, Nilsson SK, Schaap FG, Hubacek JA, Merkel M, Cooper JA, Lookene A, Putt W, Whittall R, Lee PJ, Lins L, Delsaux N, Nierman M, Kuivenhoven JA, Kastelein JJ, Vrablik M, Olivecrona G, Schneider WJ, Heeren J, Humphries SE, Talmud PJ (2008) Effects of six APOA5 variants, identified in patients with severe hypertriglyceridemia, on in vitro lipoprotein lipase activity and receptor binding. Arterioscler Thromb Vasc Biol 28:1866–1871

Ducharme NA, Bickel PE (2008) Lipid droplets in lipogenesis and lipolysis. Endocrinology 149:942–949

Duncan RE, Ahmadian M, Jaworski K, Sarkadi-Nagy E, Sul HS (2007) Regulation of lipolysis in adipocytes. Annu Rev Nutr 27:79–101

Edwards IJ, Goldberg IJ, Parks JS, Xu H, Wagner WD (1993) Lipoprotein lipase enhances the interaction of low density lipoproteins with artery-derived extracellular matrix proteoglycans. J Lipid Res 34:1155–1163

Eiselein L, Wilson DW, Lame MW, Rutledge JC (2007) Lipolysis products from triglyceride-rich lipoproteins increase endothelial permeability, perturb zonula occludens-1 and F-actin, and induce apoptosis. Am J Physiol Heart Circ Physiol 292:H2745–H2753

Eisenberg S, Sehayek E, Olivecrona T, Vlodavsky I (1992) Lipoprotein lipase enhances binding of lipoproteins to heparan sulfate on cell surfaces and extracellular matrix. J Clin Invest 90:2013–2021

Ellgaard L, Helenius A (2003) Quality control in the endoplasmic reticulum. Nat Rev Mol Cell Biol 4:181–191

Enerback S, Gimble JM (1993) Lipoprotein lipase gene expression: physiological regulators at the transcriptional and post-transcriptional level. Biochim Biophys Acta 1169:107–125

Fain JN, Buehrer B, Bahouth SW, Tichansky DS, Madan AK (2008) Comparison of messenger RNA distribution for 60 proteins in fat cells vs the nonfat cells of human omental adipose tissue. Metabolism 57:1005–1015

Faustinella F, Smith LC, Chan L (1992) Functional topology of a surface loop shielding the catalytic center in lipoprotein lipase. Biochemistry 31:7219–7223

Felts JM, Itakura H, Crane RT (1975) The mechanism of assimilation of constituents of chylomicrons, very low density lipoproteins and remnants – a new theory. Biochem Biophys Res Commun 66:1467–1475

Fernandez C, Hansson O, Nevsten P, Holm C, Klint C (2008a) Hormone-sensitive lipase is necessary for normal mobilization of lipids during submaximal exercise. Am J Physiol Endocrinol Metab 295:E179–E186

Fernandez C, Lindholm M, Krogh M, Lucas S, Larsson S, Osmark P, Berger K, Boren J, Fielding B, Frayn K, Holm C (2008b) Disturbed cholesterol homeostasis in hormone-sensitive lipase-null mice. Am J Physiol Endocrinol Metab 295:E820–E831

Fernandez-Borja M, Bellido D, Vilella E, Olivecrona G, Vilaro S (1996) Lipoprotein lipase-mediated uptake of lipoprotein in human fibroblasts: evidence for an LDL receptor-independent internalization pathway. J Lipid Res 37:464–481

Fielding BA, Frayn KN (1998) Lipoprotein lipase and the disposition of dietary fatty acids. Br J Nutr 80:495–502

Fielding BA, Humphreys SM, Allman RF, Frayn KN (1993) Mono-, di- and triacylglycerol concentrations in human plasma: effects of heparin injection and of a high-fat meal. Clin Chim Acta 216:167–173

Fielding CJ (1978) Metabolism of cholesterol-rich chylomicroms. Mechanism of binding and uptake of cholesteryl esters by the vascular bed of the perfused rat heart. J Clin Invest 62:141–151

Fojo SS, Brewer HB (1992) Hypertriglyceridaemia due to genetic defects in lipoprotein lipase and apolipoprotein C-II. J Intern Med 231:669–677

Franckhauser S, Munoz S, Pujol A, Casellas A, Riu E, Otaegui P, Su B, Bosch F (2002) Increased fatty acid re-esterification by PEPCK overexpression in adipose tissue leads to obesity without insulin resistance. Diabetes 51:624–630

Frayn KN (2002) Adipose tissue as a buffer for daily lipid flux. Diabetologia 45:1201–1210

Fredenrich A (1998) Role of apolipoprotein CIII in triglyceride-rich lipoprotein metabolism. Diabetes Metab 24:490–495

Friedman G, Chajek-Shaul T, Stein O, Olivecrona T, Stein Y (1981) The role of lipoprotein lipase in the assimilation of cholesteryl linoleyl ether by cultured cells incubated with labeled chylomicrons. Biochim Biophys Acta 666:156–164

Friedman G, Chajek-Shaul T, Olivecrona T, Stein O, Stein Y (1982) Fate of milk 125I-labelled lipoprotein lipase in cells in culture. Comparison of lipoprotein lipase- and non-lipoprotein lipase-synthesizing cells. Biochim Biophys Acta 711:114–122

Gangabadage CS, Zdunek J, Tessari M, Nilsson S, Olivecrona G, Wijmenga SS (2008) Structure and dynamics of human apolipoprotein CIII. J Biol Chem 283:17416–17427

Garcia-Martinez C, Marotta M, Moore-Carrasco R, Guitart M, Camps M, Busquets S, Montell E, Gomez-Foix AM (2005) Impact on fatty acid metabolism and differential localization of FATP1 and FAT/CD36 proteins delivered in cultured human muscle cells. Am J Physiol Cell Physiol 288:C1264–C1272

Gimeno RE (2007) Fatty acid transport proteins. Curr Opin Lipidol 18:271–276

Gliemann J (1998) Receptors of the low density lipoprotein (LDL) receptor family in man. Multiple functions of the large family members via interaction with complex ligands. Biol Chem 379:951–964

Gonzales AM, Orlando RA (2007) Role of adipocyte-derived lipoprotein lipase in adipocyte hypertrophy. Nutr Metab (Lond) 4:22

Goudriaan JR, Espirito Santo SM, Voshol PJ, Teusink B, van Dijk KW, van Vlijmen BJ, Romijn JA, Havekes LM, Rensen PC (2004) The VLDL receptor plays a major role in chylomicron metabolism by enhancing LPL-mediated triglyceride hydrolysis. J Lipid Res 45:1475–1481

Goudriaan JR, den Boer MA, Rensen PC, Febbraio M, Kuipers F, Romijn JA, Havekes LM, Voshol PJ (2005) CD36 deficiency in mice impairs lipoprotein lipase-mediated triglyceride clearance. J Lipid Res 46:2175–2181

Granneman JG, Moore HP (2008) Location, location: protein trafficking and lipolysis in adipocytes. Trends Endocrinol Metab 19:3–9

Greenberg AS, Egan JJ, Wek SA, Garty NB, Blanchette-Mackie EJ, Londos C (1991) Perilipin, a major hormonally regulated adipocyte-specific phosphoprotein associated with the periphery of lipid storage droplets. J Biol Chem 266:11341–11346

Haemmerle G, Zimmermann R, Strauss JG, Kratky D, Riederer M, Knipping G, Zechner R (2002) Hormone-sensitive lipase deficiency in mice changes the plasma lipid profile by affecting the tissue-specific expression pattern of lipoprotein lipase in adipose tissue and muscle. J Biol Chem 277:12946–12952

Haemmerle G, Lass A, Zimmermann R, Gorkiewicz G, Meyer C, Rozman J, Heldmaier G, Maier R, Theussl C, Eder S, Kratky D, Wagner EF, Klingenspor M, Hoefler G, Zechner R

(2006) Defective lipolysis and altered energy metabolism in mice lacking adipose triglyceride lipase. Science 312:734–737
Hahne P, Krempler F, Schaap FG, Soyal SM, Hoffinger H, Miller K, Oberkofler H, Strobl W, Patsch W (2008) Determinants of plasma apolipoprotein A-V and APOA5 gene transcripts in humans. J Intern Med 264:452–462
Hamilton JA (2007) New insights into the roles of proteins and lipids in membrane transport of fatty acids. Prostaglandins Leukot Essent Fatty Acids 77:355–361
Hamilton JA, Miller KW, Small DM (1983) Solubilization of triolein and cholesteryl oleate in egg phosphatidylcholine vesicles. J Biol Chem 258:12821–12826
Hartman AD (1977) Lipoprotein lipase distribution in rat adipose tissues: effect on chylomicron uptake. Am J Physiol 232:E316–E323
Heeren J, Niemeier A, Merkel M, Beisiegel U (2002) Endothelial-derived lipoprotein lipase is bound to postprandial triglyceride-rich lipoproteins and mediates their hepatic clearance in vivo. J Mol Med 80:576–584
Hibuse T, Maeda N, Funahashi T, Yamamoto K, Nagasawa A, Mizunoya W, Kishida K, Inoue K, Kuriyama H, Nakamura T, Fushiki T, Kihara S, Shimomura I (2005) Aquaporin 7 deficiency is associated with development of obesity through activation of adipose glycerol kinase. Proc Natl Acad Sci USA 102:10993–10998
Holzl B, Iglseder B, Sandhofer A, Malaimare L, Lang J, Paulweber B, Sandhofer F (2002) Insulin sensitivity is impaired in heterozygous carriers of lipoprotein lipase deficiency. Diabetologia 45:378–384
Hultin M, Bengtsson-Olivecrona G, Olivecrona T (1992) Release of lipoprotein lipase to plasma by triacylglycerol emulsions. Comparison to the effect of heparin. Biochim Biophys Acta 1125:97–103
Hultin M, Savonen R, Olivecrona T (1996) Chylomicron metabolism in rats: lipolysis, recirculation of triglyceride-derived fatty acids in plasma FFA, and fate of core lipids as analyzed by compartmental modelling. J Lipid Res 37:1022–1036
Ibrahimi A, Abumrad NA (2002) Role of CD36 in membrane transport of long-chain fatty acids. Curr Opin Clin Nutr Metab Care 5:139–145
Isola LM, Zhou SL, Kiang CL, Stump DD, Bradbury MW, Berk PD (1995) 3T3 fibroblasts transfected with a cDNA for mitochondrial aspartate aminotransferase express plasma membrane fatty acid-binding protein and saturable fatty acid uptake. Proc Natl Acad Sci USA 92:9866–9870
Ito Y, Azrolan N, O'Connell A, Walsh A, Breslow JL (1990) Hypertriglyceridemia as a result of human apo CIII gene expression in transgenic mice. Science 249:790–793
Jacobsen L, Madsen P, Jacobsen C, Nielsen MS, Gliemann J, Petersen CM (2001) Activation and functional characterization of the mosaic receptor SorLA/LR11. J Biol Chem 276:22788–22796
Jaworski K, Sarkadi-Nagy E, Duncan RE, Ahmadian M, Sul HS (2007) Regulation of triglyceride metabolism. IV. Hormonal regulation of lipolysis in adipose tissue. Am J Physiol Gastrointest Liver Physiol 293:G1–G4
Jensen DR, Gavigan S, Sawicki V, Witsell DL, Eckel RH, Neville MC (1994) Regulation of lipoprotein lipase activity and mRNA in the mammary gland of the lactating mouse. Biochem J 298:321–327
Jong MC, Hofker MH, Havekes LM (1999) Role of ApoCs in lipoprotein metabolism: functional differences between ApoC1, ApoC2, and ApoC3. Arterioscler Thromb Vasc Biol 19:472–484
Kalant D, Cianflone K (2004) Regulation of fatty acid transport. Curr Opin Lipidol 15:309-314.
Kalant D., Phelis S., Fielding B.A., Frayn K.N., Cianflone K., Sniderman A.D (2000) Increased postprandial fatty acid trapping in subcutaneous adipose tissue in obese women. J Lipid Res 41:1963-1968.
Kalant D, MacLaren R, Cui W, Samanta R, Monk PN, Laporte SA, Cianflone K (2005) C5L2 is a functional receptor for acylation stimulating protein. J Biol Chem 280:23936–23944
Karlsson M, Contreras JA, Hellman U, Tornqvist H, Holm C (1997) cDNA cloning, tissue distribution, and identification of the catalytic triad of monoglyceride lipase. Evolutionary relationship to esterases, lysophospholipases, and haloperoxidases. J Biol Chem 272:27218–27223

Karpe F, Olivecrona T, Hamsten A, Hultin M (1997) Chylomicron/chylomicron remnant turnover in humans: evidence for margination of chylomicrons and poor conversion of larger to smaller chylomicron remnants. J Lipid Res 38:949–961
Karpe F, Olivecrona T, Olivecrona G, Samra JS, Summers LK, Humphreys SM, Frayn KN (1998) Lipoprotein lipase transport in plasma: role of muscle and adipose tissues in regulation of plasma lipoprotein lipase concentrations. J Lipid Res 39:2387–2393
Kinnunen PK, Jackson RL, Smith LC, Gotto AM Jr, Sparrow JT (1977) Activation of lipoprotein lipase by native and synthetic fragments of human plasma apolipoprotein C-II. Proc Natl Acad Sci USA 74:4848–4851
Kirchgessner TG, Svenson KL, Lusis AJ, Schotz MC (1987) The sequence of cDNA encoding lipoprotein lipase. A member of a lipase gene family. J Biol Chem 262:8463–8466
Kishida K, Shimomura I, Kondo H, Kuriyama H, Makino Y, Nishizawa H, Maeda N, Matsuda M, Ouchi N, Kihara S, Kurachi Y, Funahashi T, Matsuzawa Y (2001) Genomic structure and insulin-mediated repression of the aquaporin adipose. (AQPap), adipose-specific glycerol channel. J Biol Chem 276:36251–36260
Kluger M, Heeren J, Merkel M (2008) Apoprotein A-V: an important regulator of triglyceride metabolism. J Inherit Metab Dis 31:281–288
Kobayashi Y, Nakajima T, Inoue I (2002) Molecular modeling of the dimeric structure of human lipoprotein lipase and functional studies of the carboxyl-terminal domain. Eur J Biochem 269:4701–4710
Kolovou GD, Kostakou PM, Anagnostopoulou KK, Cokkinos DV (2008) Therapeutic effects of fibrates in postprandial lipemia. Am J Cardiovasc Drugs 8:243–255
Kraemer FB (2007) Adrenal cholesterol utilization. Mol Cell Endocrinol 265/266:42–45
Kratky D, Zimmermann R, Wagner EM, Strauss JG, Jin W, Kostner GM, Haemmerle G, Rader DJ, Zechner R (2005) Endothelial lipase provides an alternative pathway for FFA uptake in lipoprotein lipase-deficient mouse adipose tissue. J Clin Invest 115:161–167
Lake AC, Sun Y, Li JL, Kim JE, Johnson JW, Li D, Revett T, Shih HH, Liu W, Paulsen JE, Gimeno RE (2005) Expression, regulation, and triglyceride hydrolase activity of Adiponutrin family members. J Lipid Res 46:2477–2487
Laplante M, Festuccia WT, Soucy G, Blanchard PG, Renaud A, Berger JP, Olivecrona G, Deshaies Y. (2008) Tissue-specific postprandial clearance is the major determinant of PPARγ-induced triglyceride lowering in the rat. Am J Physiol Regul Integr Comp Physiol. DOI:10.1152/ajpregu.90552.2008
Larnkjaer A, Nykjaer A, Olivecrona G, Thogersen H, Ostergaard PB (1995) Structure of heparin fragments with high affinity for lipoprotein lipase and inhibition of lipoprotein lipase binding to alpha 2-macroglobulin-receptor/low-density-lipoprotein-receptor-related protein by heparin fragments. Biochem J 307:205–214
LaRosa JC, Levy RI, Herbert P, Lux SE, Fredrickson DS (1970) A specific apoprotein activator for lipoprotein lipase. Biochem Biophys Res Commun 41:57–62
Lass A, Zimmermann R, Haemmerle G, Riederer M, Schoiswohl G, Schweiger M, Kienesberger P, Strauss JG, Gorkiewicz G, Zechner R (2006) Adipose triglyceride lipase-mediated lipolysis of cellular fat stores is activated by CGI-58 and defective in Chanarin–Dorfman syndrome. Cell Metab 3:309–319
Lee D, Walsh JD, Mikhailenko I, Yu P, Migliorini M, Wu Y, Krueger S, Curtis JE, Harris B, Lockett S, Blacklow SC, Strickland DK, Wang YX (2006) RAP uses a histidine switch to regulate its interaction with LRP in the ER and Golgi. Mol Cell 22:423–430
Lee J, Goldberg IJ (2007) Lipoprotein lipase-derived fatty acids: physiology and dysfunction. Curr Hypertens Rep 9:462–466
Lee JJ, Smith PJ, Fried SK (1998) Mechanisms of decreased lipoprotein lipase activity in adipocytes of starved rats depend on duration of starvation. J Nutr 128:940–946
Levak-Frank S, Weinstock PH, Hayek T, Verdery R, Hofmann W, Ramakrishnan R, Sattler W, Breslow JL, Zechner R (1997) Induced mutant mice expressing lipoprotein lipase exclusively in muscle have subnormal triglycerides yet reduced high density lipoprotein cholesterol levels in plasma. J Biol Chem 272:17182–17190

Levak-Frank S, Hofmann W, Weinstock PH, Radner H, Sattler W, Breslow JL, Zechner R (1999) Induced mutant mouse lines that express lipoprotein lipase in cardiac muscle, but not in skeletal muscle and adipose tissue, have normal plasma triglyceride and high-density lipoprotein-cholesterol levels. Proc Natl Acad Sci USA 96:3165–3170

Li C (2006) Genetics and regulation of angiopoietin-like proteins 3 and 4. Curr Opin Lipidol 17:152–156

Ling C, Svensson L, Oden B, Weijdegard B, Eden B, Eden S, Billig H (2003) Identification of functional prolactin (PRL) receptor gene expression: PRL inhibits lipoprotein lipase activity in human white adipose tissue. J Clin Endocrinol Metab 88:1804–1808

Liu GQ, Olivecrona T (1991) Pulse-chase study on lipoprotein lipase in perfused guinea pig heart. Am J Physiol 261:H2044–H2050

Lookene A, Chevreuil O, Ostergaard P, Olivecrona G (1996) Interaction of lipoprotein lipase with heparin fragments and with heparan sulfate: stoichiometry, stabilization, and kinetics. Biochemistry 35:12155–12163

Lookene A, Nielsen MS, Gliemann J, Olivecrona G (2000) Contribution of the carboxy-terminal domain of lipoprotein lipase to interaction with heparin and lipoproteins. Biochem Biophys Res Commun 271:15–21

Lookene A, Zhang L, Hultin M, Olivecrona G (2004) Rapid subunit exchange in dimeric lipoprotein lipase and properties of the inactive monomer. J Biol Chem 279:49964–49972

Lookene A, Beckstead JA, Nilsson S, Olivecrona G, Ryan RO (2005) Apolipoprotein A-V–heparin interactions: implications for plasma lipoprotein metabolism. J Biol Chem 280:25383–25387

Lowe ME (2002) The triglyceride lipases of the pancreas. J Lipid Res 43:2007–2016

MacArthur JM, Bishop JR, Stanford KI, Wang L, Bensadoun A, Witztum JL, Esko JD (2007) Liver heparan sulfate proteoglycans mediate clearance of triglyceride-rich lipoproteins independently of LDL receptor family members. J Clin Invest 117:153–164

Makoveichuk E, Castel S, Vilaro S, Olivecrona G (2004) Lipoprotein lipase-dependent binding and uptake of low density lipoproteins by THP-1 monocytes and macrophages: possible involvement of lipid rafts. Biochim Biophys Acta 1686:37–49

Martin-Hidalgo A, Holm C, Belfrage P, Schotz MC, Herrera E (1994) Lipoprotein lipase and hormone-sensitive lipase activity and mRNA in rat adipose tissue during pregnancy. Am J Physiol 266:E930–E935

Masuno H, Blanchette-Mackie EJ, Chernick SS, Scow RO (1990) Synthesis of inactive nonsecretable high mannose-type lipoprotein lipase by cultured brown adipocytes of combined lipase-deficient cld/cld mice. J Biol Chem 265:1628–1638

Masuno H, Schultz CJ, Park JW, Blanchette-Mackie EJ, Mateo C, Scow RO (1991) Glycosylation, activity and secretion of lipoprotein lipase in cultured brown adipocytes of newborn mice. Effect of tunicamycin, monensin, 1-deoxymannojirimycin and swainsonine. Biochem J 277:801–809

Mead JR, Cryer A, Ramji DP (1999) Lipoprotein lipase, a key role in atherosclerosis. FEBS Lett 462:1–6

Mead JR, Irvine SA, Ramji DP (2002) Lipoprotein lipase: structure, function, regulation, and role in disease. J Mol Med 80:753–769

Meegalla RL, Billheimer JT, Cheng D (2002) Concerted elevation of acyl-coenzyme A:diacylglycerol acyltransferase (DGAT) activity through independent stimulation of mRNA expression of DGAT1 and DGAT2 by carbohydrate and insulin. Biochem Biophys Res Commun 298:317–323

Merkel M, Weinstock PH, Chajek-Shaul T, Radner H, Yin B, Breslow JL, Goldberg IJ (1998) Lipoprotein lipase expression exclusively in liver. A mouse model for metabolism in the neonatal period and during cachexia. J Clin Invest 102:893–901

Merkel M, Eckel RH, Goldberg IJ (2002a) Lipoprotein lipase: genetics, lipid uptake, and regulation. J Lipid Res 43:1997–2006

Merkel M, Heeren J, Dudeck W, Rinninger F, Radner H, Breslow JL, Goldberg IJ, Zechner R, Greten H (2002b) Inactive lipoprotein lipase (LPL) alone increases selective cholesterol ester uptake in vivo, whereas in the presence of active LPL it also increases triglyceride hydrolysis and whole particle lipoprotein uptake. J Biol Chem 277:7405–7411

Merkel M, Loeffler B, Kluger M, Fabig N, Geppert G, Pennacchio LA, Laatsch A, Heeren J (2005) Apolipoprotein AV accelerates plasma hydrolysis of triglyceride-rich lipoproteins by interaction with proteoglycan-bound lipoprotein lipase. J Biol Chem 280:21553–21560

Miyoshi H, Perfield JW 2nd, Souza SC, Shen WJ, Zhang HH, Stancheva ZS, Kraemer FB, Obin MS, Greenberg AS (2007) Control of adipose triglyceride lipase action by serine 517 of perilipin A globally regulates protein kinase A-stimulated lipolysis in adipocytes. J Biol Chem 282:996–1002

Morley N, Kuksis A (1972) Positional specificity of lipoprotein lipase. J Biol Chem 247: 6389–6393

Morley N, Kuksis A, Hoffman AG, Kakis G (1977) Preferential in vivo accumulation of sn-2,3-diacylglycerols in postheparin plasma of rats. Can J Biochem 55:1075–1081

Narayanaswami V, Maiorano JN, Dhanasekaran P, Ryan RO, Phillips MC, Lund-Katz S, Davidson WS (2004) Helix orientation of the functional domains in apolipoprotein e in discoidal high density lipoprotein particles. J Biol Chem 279:14273–14279

Neuger L, Ruge T, Makoveichuk E, Vlodavsky I, Olivecrona G (2001) Effects of the heparin-mimicking compound RG-13577 on lipoprotein lipase and on lipase mediated binding of LDL to cells. Atherosclerosis 157:13–21

Neuger L, Vilaro S, Lopez-Iglesias C, Gupta J, Olivecrona T, Olivecrona G (2004) Effects of heparin on the uptake of lipoprotein lipase in rat liver. BMC Physiol 4:13

Nielsen MS, Jacobsen C, Olivecrona G, Gliemann J, Petersen CM (1999) Sortilin/neurotensin receptor-3 binds and mediates degradation of lipoprotein lipase. J Biol Chem 274:8832–8836

Nilsson SK, Lookene A, Beckstead JA, Gliemann J, Ryan RO, Olivecrona G (2007) Apolipoprotein A-V interaction with members of the low density lipoprotein receptor gene family. Biochemistry 46:3896–3904

Nilsson SK, Christensen S, Raarup MK, Ryan RO, Nielsen MS, Olivecrona G (2008) Endocytosis of apolipoprotein A-V by members of the low density lipoprotein receptor and the VPS10p domain receptor families. J Biol Chem 283:25920–25927

Nilsson-Ehle P, Egelrud T, Belfrage P, Olivecrona T, Borgstrom B (1973) Positional specificity of purified milk lipoprotein lipase. J Biol Chem 248:6734–6737

Nordenstrom J, Thorne A, Aberg W, Carneheim C, Olivecrona T (2006) The hypertriglyceridemic clamp technique. Studies using long-chain and structured triglyceride emulsions in healthy subjects. Metabolism 55:1443–1450

Nye C, Kim J, Kalhan SC, Hanson RW (2008a) Reassessing triglyceride synthesis in adipose tissue. Trends Endocrinol Metab 193:56–361

Nye CK, Hanson RW, Kalhan SC (2008b) Glyceroneogenesis is the dominant pathway for triglyceride glycerol synthesis in vivo in the rat. J Biol Chem 283:27565–27574

Nykjaer A, Bengtsson-Olivecrona G, Lookene A, Moestrup SK, Petersen CM, Weber W, Beisiegel U, Gliemann J (1993) The alpha 2-macroglobulin receptor/low density lipoprotein receptor-related protein binds lipoprotein lipase and beta-migrating very low density lipoprotein associated with the lipase. J Biol Chem 268:15048–15055

Nykjaer A, Nielsen M, Lookene A, Meyer N, Roigaard H, Etzerodt M, Beisiegel U, Olivecrona G, Gliemann J (1994) A carboxyl-terminal fragment of lipoprotein lipase binds to the low density lipoprotein receptor-related protein and inhibits lipase-mediated uptake of lipoprotein in cells. J Biol Chem 269:31747–31755

Obunike JC, Sivaram P, Paka L, Low MG, Goldberg IJ (1996) Lipoprotein lipase degradation by adipocytes: receptor-associated protein (RAP)-sensitive and proteoglycan-mediated pathways. J Lipid Res 37:2439–2449

Obunike JC, Lutz EP, Li Z, Paka L, Katopodis T, Strickland DK, Kozarsky KF, Pillarisetti S, Goldberg IJ (2001) Transcytosis of lipoprotein lipase across cultured endothelial cells requires both heparan sulfate proteoglycans and the very low density lipoprotein receptor. J Biol Chem 276:8934–8941

Olivecrona G, Beisiegel U (1997) Lipid binding of apolipoprotein CII is required for stimulation of lipoprotein lipase activity against apolipoprotein CII-deficient chylomicrons. Arterioscler Thromb Vasc Biol 17:1545–1549

Olivecrona G, Hultin M, Savonen R, Skottova N, Lookene A, Tugrul Y, Olivecrona T (1995) Transport of lipoprotein in plasma and lipoprotein metabolism. In: Woodford FP. (ed) Atherosclerosis X. Elsevier, Amsterdam, pp 250–263

Olivecrona T, Bengtsson G (1978) Heparin and lipoprotein lipase – how specific is the interaction. In: Carlson LA, (eds) International conference on atherosclerosis. Raven, New York, pp 153–157

Olivecrona T, Olivecrona G (1999) Lipoprotein lipase and hepatic lipase in lipoprotein metabolism. In: Betteridge DJ, Illingworth DR, Shepherd J. (eds) Lipoproteins in health and disease. Arnold, London, pp 223–246

Olivecrona T, Bengtsson G, Marklund SE, Lindahl U, Hook M (1977) Heparin–lipoprotein lipase interactions. Fed Proc 36:60–65

Olivecrona T, Chernick SS, Bengtsson-Olivecrona G, Paterniti JR Jr, Brown WV, Scow RO (1985) Combined lipase deficiency (cld/cld) in mice. Demonstration that an inactive form of lipoprotein lipase is synthesized. J Biol Chem 260:2552–2557

Olivecrona T, Chernick SS, Bengtsson-Olivecrona G, Garrison M, Scow RO (1987) Synthesis and secretion of lipoprotein lipase in 3T3-L1 adipocytes. Demonstration of inactive forms of lipase in cells. J Biol Chem 262:10748–10759

Olswang Y, Cohen H, Papo O, Cassuto H, Croniger CM, Hakimi P, Tilghman SM, Hanson RW, Reshef L (2002) A mutation in the peroxisome proliferator-activated receptor gamma-binding site in the gene for the cytosolic form of phosphoenolpyruvate carboxykinase reduces adipose tissue size and fat content in mice. Proc Natl Acad Sci USA 99:625–630

Ong JM, Kern PA (1989a) Effect of feeding and obesity on lipoprotein lipase activity, immunoreactive protein, and messenger RNA levels in human adipose tissue. J Clin Invest 84:305–311

Ong JM, Kern PA (1989b) The role of glucose and glycosylation in the regulation of lipoprotein lipase synthesis and secretion in rat adipocytes. J Biol Chem 264:3177–3182

Ong JM, Kirchgessner TG, Schotz MC, Kern PA (1988) Insulin increases the synthetic rate and messenger RNA level of lipoprotein lipase in isolated rat adipocytes. J Biol Chem 263:12933–12938

Ong JM, Saffari B, Simsolo RB, Kern PA (1992a) Epinephrine inhibits lipoprotein lipase gene expression in rat adipocytes through multiple steps in posttranscriptional processing. Mol Endocrinol 6:61–69

Ong JM, Simsolo RB, Saffari B, Kern PA (1992b) The regulation of lipoprotein lipase gene expression by dexamethasone in isolated rat adipocytes. Endocrinology 130:2310–2316

Osborne JC Jr, Bengtsson-Olivecrona G, Lee NS, Olivecrona T (1985) Studies on inactivation of lipoprotein lipase: role of the dimer to monomer dissociation. Biochemistry 24:5606–5611

Osibow K, Frank S, Malli R, Zechner R, Graier WF (2006) Mitochondria maintain maturation and secretion of lipoprotein lipase in the endoplasmic reticulum. Biochem J 396:173–182

Ost A, Ortegren U, Gustavsson J, Nystrom FH, Stralfors P (2005) Triacylglycerol is synthesized in a specific subclass of caveolae in primary adipocytes. J Biol Chem 280:5–8

Osterlund T, Beussman DJ, Julenius K, Poon PH, Linse S, Shabanowitz J, Hunt DF, Schotz MC, Derewenda ZS, Holm C (1999) Domain identification of hormone-sensitive lipase by circular dichroism and fluorescence spectroscopy, limited proteolysis, and mass spectrometry. J Biol Chem 274:15382–15388

Osuga J, Ishibashi S, Oka T, Yagyu H, Tozawa R, Fujimoto A, Shionoiri F, Yahagi N, Kraemer FB, Tsutsumi O, Yamada N (2000) Targeted disruption of hormone-sensitive lipase results in male sterility and adipocyte hypertrophy, but not in obesity. Proc Natl Acad Sci USA 97:787–792

Otarod JK, Goldberg IJ (2004) Lipoprotein lipase and its role in regulation of plasma lipoproteins and cardiac risk. Curr Atheroscler Rep 6:335–342

Page S, Judson A, Melford K, Bensadoun A (2006) Interaction of lipoprotein lipase and receptor-associated protein. J Biol Chem 281:13931–13938

Paglialunga S, Fisette A, Yan Y, Deshaies Y, Brouillette JF, Pekna M, Cianflone K (2008) Acylation-stimulating protein deficiency and altered adipose tissue in alternative complement pathway knockout mice. Am J Physiol Endocrinol Metab 294:E521–E529

Park TS, Yamashita H, Blaner WS, Goldberg IJ (2007) Lipids in the heart: a source of fuel and a source of toxins. Curr Opin Lipidol 18:277–282

Parthasarathy N, Goldberg IJ, Sivaram P, Mulloy B, Flory DM, Wagner WD (1994) Oligosaccharide sequences of endothelial cell surface heparan sulfate proteoglycan with affinity for lipoprotein lipase. J Biol Chem 269:22391–22396

Patsch JR, Gotto AM Jr, Olivercrona T, Eisenberg S (1978) Formation of high density lipoprotein2-like particles during lipolysis of very low density lipoproteins in vitro. Proc Natl Acad Sci USA 75:4519–4523

Payne VA, Au WS, Gray SL, Nora ED, Rahman SM, Sanders R, Hadaschik D, Friedman JE, O'Rahilly S, Rochford JJ (2007) Sequential regulation of diacylglycerol acyltransferase 2 expression by CAAT/enhancer-binding protein beta. (C/EBPbeta) and C/EBPalpha during adipogenesis. J Biol Chem 282:21005–21014

Pedersen ME, Schotz MC (1980) Rapid changes in rat heart lipoprotein lipase activity after feeding carbohydrate. J Nutr 110:481–487

Peinado-Onsurbe J, Staels B, Deeb S, Ramirez I, Llobera M, Auwerx J (1992) Neonatal extinction of liver lipoprotein lipase expression. Biochim Biophys Acta 1131:281–286

Pennacchio LA, Olivier M, Hubacek JA, Cohen JC, Cox DR, Fruchart JC, Krauss RM, Rubin EM (2001) An apolipoprotein influencing triglycerides in humans and mice revealed by comparative sequencing. Science 294:169–173

Pentikainen MO, Oksjoki R, Oorni K, Kovanen PT (2002) Lipoprotein lipase in the arterial wall: linking LDL to the arterial extracellular matrix and much more. Arterioscler Thromb Vasc Biol 22:211–217

Peterfy M, Ben-Zeev O, Mao HZ, Weissglas-Volkov D, Aouizerat BE, Pullinger CR, Frost PH, Kane JP, Malloy MJ, Reue K, Pajukanta P, Doolittle MH (2007) Mutations in LMF1 cause combined lipase deficiency and severe hypertriglyceridemia. Nat Genet 39:1483–1487

Peterson J, Bihain BE, Bengtsson-Olivecrona G, Deckelbaum RJ, Carpentier YA, Olivecrona T (1990) Fatty acid control of lipoprotein lipase: a link between energy metabolism and lipid transport. Proc Natl Acad Sci USA 87:909–913

Picard F, Naimi N, Richard D, Deshaies Y (1999) Response of adipose tissue lipoprotein lipase to the cephalic phase of insulin secretion. Diabetes 48:452–459

Pilch PF, Souto RP, Liu L, Jedrychowski MP, Berg EA, Costello CE, Gygi SP (2007) Cellular spelunking: exploring adipocyte caveolae. J Lipid Res 48:2103–2111

Ploug M, Kjalke M, Ronne E, Weidle U, Hoyer-Hansen G, Dano K (1993) Localization of the disulfide bonds in the NH2-terminal domain of the cellular receptor for human urokinase-type plasminogen activator. A domain structure belonging to a novel superfamily of glycolipid-anchored membrane proteins. J Biol Chem 268:17539–17546

Preiss-Landl K, Zimmermann R, Hammerle G, Zechner R (2002) Lipoprotein lipase: the regulation of tissue specific expression and its role in lipid and energy metabolism. Curr Opin Lipidol 13:471–481

Pulinilkunnil T, Rodrigues B (2006) Cardiac lipoprotein lipase: metabolic basis for diabetic heart disease. Cardiovasc Res 69:329–340

Querfeld U, Hoffmann MM, Klaus G, Eifinger F, Ackerschott M, Michalk D, Kern PA (1999) Antagonistic effects of vitamin D and parathyroid hormone on lipoprotein lipase in cultured adipocytes. J Am Soc Nephrol 10:2158–2164

Rader DJ, Jaye M (2000) Endothelial lipase: a new member of the triglyceride lipase gene family. Curr Opin Lipidol 11:141–147

Ranganathan G, Pokrovskaya I, Ranganathan S, Kern PA (2005) Role of A kinase anchor proteins in the tissue-specific regulation of lipoprotein lipase. Mol Endocrinol 19:2527–2534

Rapp D, Olivecrona T (1978) Kinetics of milk lipoprotein lipase. Studies with tributyrin. Eur J Biochem 91:379–385

Razani B, Combs TP, Wang XB, Frank PG, Park DS, Russell RG, Li M, Tang B, Jelicks LA, Scherer PE, Lisanti MP (2002) Caveolin-1-deficient mice are lean, resistant to diet-induced obesity, and show hypertriglyceridemia with adipocyte abnormalities. J Biol Chem 277:8635–8647

Reinbold M, Hufnagel B, Kewitz T, Klumpp S, Krieglstein J (2008) Unsaturated fatty acids liberated from VLDL cause apoptosis in endothelial cells. Mol Nutr Food Res 52:581–588

Roh C, Roduit R, Thorens B, Fried S, Kandror KV (2001) Lipoprotein lipase and leptin are accumulated in different secretory compartments in rat adipocytes. J Biol Chem 276:35990–35994

Rojas C, Olivecrona T, Bengtsson-Olivecrona G (1991) Comparison of the action of lipoprotein lipase on triacylglycerols and phospholipids when presented in mixed liposomes or in emulsion droplets. Eur J Biochem 197:315–321

Ruge T, Wu G, Olivecrona T, Olivecrona G (2004) Nutritional regulation of lipoprotein lipase in mice. Int J Biochem Cell Biol 36:320–329

Ruge T, Svensson M, Eriksson JW, Olivecrona G (2005) Tissue-specific regulation of lipoprotein lipase in humans: effects of fasting. Eur J Clin Invest 35:194–200

Rutledge JC, Woo MM, Rezai AA, Curtiss LK, Goldberg IJ (1997) Lipoprotein lipase increases lipoprotein binding to the artery wall and increases endothelial layer permeability by formation of lipolysis products. Circ Res 80:819–828

Saario SM, Laitinen JT (2007) Monoglyceride lipase as an enzyme hydrolyzing 2-arachidonoylglycerol. Chem Biodivers 4:1903–1913

Saiki A, Murano T, Watanabe F, Oyama T, Miyashita Y, Shirai K (2005) Pitavastatin enhanced lipoprotein lipase expression in 3T3-L1 preadipocytes. J Atheroscler Thromb 12:163–168

Saiki A, Oyama T, Endo K, Ebisuno M, Ohira M, Koide N, Murano T, Miyashita Y, Shirai K (2007) Preheparin serum lipoprotein lipase mass might be a biomarker of metabolic syndrome. Diabetes Res Clin Pract 76:93–101

Santamarina-Fojo S, Dugi KA (1994) Structure, function and role of lipoprotein lipase in lipoprotein metabolism. Curr Opin Lipidol 5:117–125

Santosa S, Jensen MD (2008) Why are we shaped differently, and why does it matter. Am J Physiol Endocrinol Metab 295:E531–E535

Sasaki A, Goldberg IJ (1992) Lipoprotein lipase release from BFC-1 beta adipocytes Effects of triglyceride-rich lipoproteins and lipolysis products. J Biol Chem 267:15198–15204

Saxena U, Goldberg IJ (1990) Interaction of lipoprotein lipase with glycosaminoglycans and apolipoprotein C-II: effects of free-fatty-acids. Biochim Biophys Acta 1043:161–168

Saxena U, Witte LD, Goldberg IJ (1989) Release of endothelial cell lipoprotein lipase by plasma lipoproteins and free fatty acids. J Biol Chem 264:4349–4355

Saxena U, Kulkarni NM, Ferguson E, Newton RS (1992) Lipoprotein lipase-mediated lipolysis of very low density lipoproteins increases monocyte adhesion to aortic endothelial cells. Biochem Biophys Res Commun 189:1653–1658

Schaffer JE (2002) Fatty acid transport: the roads taken. Am J Physiol Endocrinol Metab 282:E239–E246

Schoefl GI, French JE (1967) Morphologic aspects of vascular permeability to fat. Bibl Anat 9:495–500

Schotz MC, Garfinkel AS (1965) The effect of puromycin and actinomycin on carbohydrate-induced lipase activity in rat adipose tissue. Biochim Biophys Acta 106:202–205

Schultz FM, Wylie MB, Johnston JM (1971) The relationship between the monoglyceride and glycerol-3-phosphate pathways in adipose tissue. Biochem Biophys Res Commun 45:246–250

Schweiger M, Schreiber R, Haemmerle G, Lass A, Fledelius C, Jacobsen P, Tornqvist H, Zechner R, Zimmermann R (2006) Adipose triglyceride lipase and hormone-sensitive lipase are the major enzymes in adipose tissue triacylglycerol catabolism. J Biol Chem 281:40236–40241

Scow RO (1977) Metabolism of cyhlomicrons in perfused adipose and mammary tissue of the rat. Fed Proc 36:182–185

Scow RO, Blanchette-Mackie EJ (1985) Why fatty acids flow in cell membranes. Prog Lipid Res 24:197–241

Scow RO, Olivecrona T (1977) Effect of albumin on products formed from chylomicron triacylclycerol by lipoprotein lipase in vitro. Biochim Biophys Acta 487:472–486

Scow RO, Chernick SS, Fleck TR (1977) Lipoprotein lipase and uptake of triacylglycerol, cholesterol and phosphatidylcholine from chylomicrons by mammary and adipose tissue of lactating rats in vivo. Biochim Biophys Acta 487:297–306

Semb H, Olivecrona T (1986) Nutritional regulation of lipoprotein lipase in guinea pig tissues. Biochim Biophys Acta 876:249–255
Semb H, Olivecrona T (1987) Mechanisms for turnover of lipoprotein lipase in guinea pig adipocytes. Biochim Biophys Acta 921:104–115
Semb H, Olivecrona T (1989a) The relation between glycosylation and activity of guinea pig lipoprotein lipase. J Biol Chem 264:4195–4200
Semb H, Olivecrona T (1989b) Two different mechanisms are involved in nutritional regulation of lipoprotein lipase in guinea-pig adipose tissue. Biochem J 262:505–511
Sendak RA, Bensadoun A (1998) Identification of a heparin-binding domain in the distal carboxyl-terminal region of lipoprotein lipase by site-directed mutagenesis. J Lipid Res 39:1310–1315
Seo T, Al-Haideri M, Treskova E, Worgall TS, Kako Y, Goldberg IJ, Deckelbaum RJ (2000) Lipoprotein lipase-mediated selective uptake from low density lipoprotein requires cell surface proteoglycans and is independent of scavenger receptor class B type 1. J Biol Chem 275:30355–30362
Shen Y, Lindberg A, Olivecrona G (2000) Apolipoprotein CII from rainbow trout (*Oncorhynchus mykiss*) is functionally active but structurally very different from mammalian apolipoprotein CII. Gene 254:189–198
Shen Y, Lookene A, Nilsson S, Olivecrona G (2002) Functional analyses of human apolipoprotein CII by site-directed mutagenesis: identification of residues important for activation of lipoprotein lipase. J Biol Chem 277:4334–4342
Shirai K, Fitzharris TJ, Shinomiya M, Muntz HG, Harmony JA, Jackson RL, Quinn DM (1983) Lipoprotein lipase-catalyzed hydrolysis of phosphatidylcholine of guinea pig very low density lipoproteins and discoidal complexes of phospholipid and apolipoprotein: effect of apolipoprotein C-II on the catalytic mechanism. J Lipid Res 24:721–730
Shu X, Chan J, Ryan RO, Forte TM (2007) Apolipoprotein A-V association with intracellular lipid droplets. J Lipid Res 48:1445–1450
Sivaram P, Choi SY, Curtiss LK, Goldberg IJ (1994) An amino-terminal fragment of apolipoprotein B binds to lipoprotein lipase and may facilitate its binding to endothelial cells. J Biol Chem 269:9409–9412
Skowronski MT, Lebeck J, Rojek A, Praetorius J, Fuchtbauer EM, Frokiaer J, Nielsen S (2007) AQP7 is localized in capillaries of adipose tissue, cardiac and striated muscle: implications in glycerol metabolism. Am J Physiol Renal Physiol 292:F956–F965
Smith SJ, Cases S, Jensen DR, Chen HC, Sande E, Tow B, Sanan DA, Raber J, Eckel RH, Farese RV Jr (2000) Obesity resistance and multiple mechanisms of triglyceride synthesis in mice lacking Dgat. Nat Genet 25:87–90
Soma MR, Gotto AM Jr, Ghiselli G (1989) Rapid modulation of rat adipocyte lipoprotein lipase: effect of calcium, A23187 ionophore, and thrombin. Biochim Biophys Acta 1003:307–314
Soupene E, Kuypers FA (2008) Mammalian long-chain acyl-CoA synthetases. Exp Biol Med (Maywood) 233:507–521
Spillmann D, Lookene A, Olivecrona G (2006) Isolation and characterization of low sulfated heparan sulfate sequences with affinity for lipoprotein lipase. J Biol Chem 281:23405–23413
Sprecher DL, Harris BV, Stein EA, Bellet PS, Keilson LM, Simbartl LA (1996) Higher triglycerides, lower high-density lipoprotein cholesterol, and higher systolic blood pressure in lipoprotein lipase-deficient heterozygotes. A preliminary report. Circulation 94:3239–3245
Stein Y, Stein O (2003) Lipoprotein lipase and atherosclerosis. Atherosclerosis 170:1–9
Stremmel W (1988) Uptake of fatty acids by jejunal mucosal cells is mediated by a fatty acid binding membrane protein. J Clin Invest 82:2001–2010
Stump DD, Zhou SL, Berk PD (1993) Comparison of plasma membrane FABP and mitochondrial isoform of aspartate aminotransferase from rat liver. Am J Physiol 265:G894–G902
Sukonina V, Lookene A, Olivecrona T, Olivecrona G (2006) Angiopoietin-like protein 4 converts lipoprotein lipase to inactive monomers and modulates lipase activity in adipose tissue. Proc Natl Acad Sci USA 103:17450–17455

Talmud PJ (2007) Rare APOA5 mutations – clinical consequences, metabolic and functional effects: an ENID review. Atherosclerosis 194:287–292

Talmud PJ, Hawe E, Martin S, Olivier M, Miller GJ, Rubin EM, Pennacchio LA, Humphries SE (2002) Relative contribution of variation within the APOC3/A4/A5 gene cluster in determining plasma triglycerides. Hum Mol Genet 11:3039–3046

Tan GD, Olivecrona G, Vidal H, Frayn KN, Karpe F (2006) Insulin sensitisation affects lipoprotein lipase transport in type 2 diabetes: role of adipose tissue and skeletal muscle in response to rosiglitazone. Diabetologia 49:2412–2418

Teusink B, Voshol PJ, Dahlmans VE, Rensen PC, Pijl H, Romijn JA, Havekes LM (2003) Contribution of fatty acids released from lipolysis of plasma triglycerides to total plasma fatty acid flux and tissue-specific fatty acid uptake. Diabetes 52:614–620

Thorne A, Aberg W, Carneheim C, Olivecrona T, Nordenstrom J (2005) Influence of trauma on plasma elimination of exogenous fat and on lipoprotein lipase activity and mass. Clin Nutr 24:66–74

Tornvall P, Olivecrona G, Karpe F, Hamsten A, Olivecrona T (1995) Lipoprotein lipase mass and activity in plasma and their increase after heparin are separate parameters with different relations to plasma lipoproteins. Arterioscler Thromb Vasc Biol 15:1086–1093

Traber MG, Olivecrona T, Kayden HJ (1985) Bovine milk lipoprotein lipase transfers tocopherol to human fibroblasts during triglyceride hydrolysis in vitro. J Clin Invest 75:1729–1734

Ullrich NF, Purnell JQ, Brunzell JD (2001) Adipose tissue fatty acid composition in humans with lipoprotein lipase deficiency. J Invest Med 49:273–275

van Bennekum AM, Kako Y, Weinstock PH, Harrison EH, Deckelbaum RJ, Goldberg IJ, Blaner WS (1999) Lipoprotein lipase expression level influences tissue clearance of chylomicron retinyl ester. J Lipid Res 40:565–574

van der Hoogt CC, Berbee JF, Espirito Santo SM, Gerritsen G, Krom YD, van der Zee A, Havekes LM, van Dijk KW, Rensen PC (2006) Apolipoprotein CI causes hypertriglyceridemia independent of the very-low-density lipoprotein receptor and apolipoprotein CIII in mice. Biochim Biophys Acta 1761:213–220

van Kuiken BA, Behnke WD (1994) The activation of porcine pancreatic lipase by cis-unsaturated fatty acids. Biochim Biophys Acta 1214:148–160

van Tilbeurgh H, Roussel A, Lalouel JM, Cambillau C (1994) Lipoprotein lipase. Molecular model based on the pancreatic lipase X-ray structure: consequences for heparin binding and catalysis. J Biol Chem 269:4626–4633

van Vlijmen BJ, Rohlmann A, Page ST, Bensadoun A, Bos IS, van Berkel TJ, Havekes LM, Herz J (1999) An extrahepatic receptor-associated protein-sensitive mechanism is involved in the metabolism of triglyceride-rich lipoproteins. J Biol Chem 274:35219–35226

Vannier C, Ailhaud G (1989) Biosynthesis of lipoprotein lipase in cultured mouse adipocytes. II. Processing, subunit assembly, and intracellular transport. J Biol Chem 264:13206–13216

Veltri BC, Backus RC, Rogers QR, Depeters EJ (2006) Adipose fatty acid composition and rate of incorporation of alpha-linolenic acid differ between normal and lipoprotein lipase-deficient cats. J Nutr 136:2980–2986

Verger R (1976) Interfacial enzyme kinetics of lipolysis. Annu Rev Biophys Bioeng 5:77–117

Vilaro S, Llobera M, Bengtsson-Olivecrona G, Olivecrona T (1988) Lipoprotein lipase uptake by the liver: localization, turnover, and metabolic role. Am J Physiol 254:G711–G722

Vilella E, Joven J, Fernandez M, Vilaro S, Brunzell JD, Olivecrona T, Bengtsson-Olivecrona G (1993) Lipoprotein lipase in human plasma is mainly inactive and associated with cholesterol-rich lipoproteins. J Lipid Res 34:1555–1564

Wagner EM, Kratky D, Haemmerle G, Hrzenjak A, Kostner GM, Steyrer E, Zechner R (2004) Defective uptake of triglyceride-associated fatty acids in adipose tissue causes the SREBP-1c-mediated induction of lipogenesis. J Lipid Res 45:356–365

Wang L, Fuster M, Sriramarao P, Esko JD (2005) Endothelial heparan sulfate deficiency impairs L-selectin- and chemokine-mediated neutrophil trafficking during inflammatory responses. Nat Immunol 6:902–910

Wang L, Gill R, Pedersen TL, Higgins LJ, Newman JW, Rutledge JC. (2008) Triglyceride-rich lipoprotein lipolysis releases neutral and oxidized free fatty acids that induce endothelial cell inflammation. J Lipid Res (in press)

Wang SP, Laurin N, Himms-Hagen J, Rudnicki MA, Levy E, Robert MF, Pan L, Oligny L, Mitchell GA (2001) The adipose tissue phenotype of hormone-sensitive lipase deficiency in mice. Obes Res 9:119–128

Watt MJ, Steinberg GR (2008) Regulation and function of triacylglycerol lipases in cellular metabolism. Biochem J 414:313–325

Weinstein MM, Beigneux AP, Davies BS, Gin P, Yin L, Estrada K, Melford K, Bishop JR, Esko JD, Fong LG, Bensadoun A, Young SG. (2008) Abnormal patterns of lipoprotein lipase release into the plasma in GPIHBP1-deficient mice. J Biol Chem. DOI:10.1074/jbc.M806067200

Weinstock PH, Levak-Frank S, Hudgins LC, Radner H, Friedman JM, Zechner R, Breslow JL (1997) Lipoprotein lipase controls fatty acid entry into adipose tissue, but fat mass is preserved by endogenous synthesis in mice deficient in adipose tissue lipoprotein lipase. Proc Natl Acad Sci USA 94:10261–10266

Williams CM, Maitin V, Jackson KG (2004) Triacylglycerol-rich lipoprotein-gene interactions in endothelial cells. Biochem Soc Trans 32:994–998

Willnow TE (1998) Receptor-associated protein (RAP): a specialized chaperone for endocytic receptors. Biol Chem 379:1025–1031

Wing DR, Fielding CJ, Robinson DS (1967) The effect of cycloheximide on tissue clearing-factor lipase activity. Biochem J 104:45C–46C

Viso A, Cisneros JA, Ortega-Gutierrez S (2008) The medicinal chemistry of agents targeting monoacylglycerol lipase. Curr Top Med Chem 8:231–246

Wong H, Yang D, Hill JS, Davis RC, Nikazy J, Schotz MC (1997) A molecular biology-based approach to resolve the subunit orientation of lipoprotein lipase. Proc Natl Acad Sci USA 94:5594–5598

Wong K Ryan RO (2007) Characterization of apolipoprotein A-V structure and mode of plasma triacylglycerol regulation. Curr Opin Lipidol 18:319–324

Wu G, Olivecrona G, Olivecrona T (2003) The distribution of lipoprotein lipase in rat adipose tissue. Changes with nutritional state engage the extracellular enzyme. J Biol Chem 278:11925–11930

Wu G, Olivecrona G, Olivecrona T (2005) Extracellular degradation of lipoprotein lipase in rat adipose tissue. BMC Cell Biol 6:4

Yen CL, Stone SJ, Koliwad S, Harris C, Farese RV Jr (2008) Thematic review series: glycerolipids. DGAT enzymes and triacylglycerol biosynthesis. J Lipid Res 49:2283–2301

Yokoyama M, Yagyu H, Hu Y, Seo T, Hirata K, Homma S, Goldberg IJ (2004) Apolipoprotein B production reduces lipotoxic cardiomyopathy: studies in heart-specific lipoprotein lipase transgenic mouse. J Biol Chem 279:4204–4211

Yokoyama M, Seo T, Park T, Yagyu H, Hu Y, Son NH, Augustus AS, Vikramadithyan RK, Ramakrishnan R, Pulawa LK, Eckel RH, Goldberg IJ (2007) Effects of lipoprotein lipase and statins on cholesterol uptake into heart and skeletal muscle. J Lipid Res 48:646–655

Young SG, Davies BS, Fong LG, Gin P, Weinstein MM, Bensadoun A, Beigneux AP (2007) GPIHBP1: an endothelial cell molecule important for the lipolytic processing of chylomicrons. Curr Opin Lipidol 18:389–396

Zambon A, Schmidt I, Beisiegel U, Brunzell JD (1996) Dimeric lipoprotein lipase is bound to triglyceride-rich plasma lipoproteins. J Lipid Res 37:2394–2404

Zannis VI, Chroni A, Krieger M (2006) Role of apoA-I, ABCA1, LCAT, and SR-BI in the biogenesis of HDL. J Mol Med 84:276–294

Zdunek J, Martinez GV, Schleucher J, Lycksell PO, Yin Y, Nilsson S, Shen Y, Olivecrona G, Wijmenga S (2003) Global structure and dynamics of human apolipoprotein CII in complex with micelles: evidence for increased mobility of the helix involved in the activation of lipoprotein lipase. Biochemistry 42:1872–1889

Zechner R, Kienesberger PC, Haemmerle G, Zimmermann R, Lass A. (2008) Adipose triglyceride lipase – and the lipolytic catabolism of cellular fat stores. J Lipid Res (in press)

Zechner R, Strauss J, Frank S, Wagner E, Hofmann W, Kratky D, Hiden M, Levak-Frank S (2000) The role of lipoprotein lipase in adipose tissue development and metabolism. Int J Obes Relat Metab Disord 24[Suppl 4]:S53–S56

Zechner R, Strauss JG, Haemmerle G, Lass A, Zimmermann R (2005) Lipolysis: pathway under construction. Curr Opin Lipidol 16:333–340

Zhang J, Chu W, Crandall I (2008) Lipoprotein binding preference of CD36 is altered by filipin treatment. Lipids Health Dis 7:23

Zhang L, Wu G, Tate CG, Lookene A, Olivecrona G (2003) Calreticulin promotes folding/dimerization of human lipoprotein lipase expressed in insect cells (sf21). J Biol Chem 278:29344–29351

Zhang L, Lookene A, Wu G, Olivecrona G (2005) Calcium triggers folding of lipoprotein lipase into active dimers. J Biol Chem 280:42580–42591

Zhao G, Souers AJ, Voorbach M, Falls HD, Droz B, Brodjian S, Lau YY, Iyengar RR, Gao J, Judd AS, Wagaw SH, Ravn MM, Engstrom KM, Lynch JK, Mulhern MM, Freeman J, Dayton BD, Wang X, Grihalde N, Fry D, Beno DW, Marsh KC, Su Z, Diaz GJ, Collins CA, Sham H, Reilly RM, Brune ME, Kym PR (2008) Validation of diacyl glycerolacyltransferase I as a novel target for the treatment of obesity and dyslipidemia using a potent and selective small molecule inhibitor. J Med Chem 51:380–383

Zimmermann R, Strauss JG, Haemmerle G, Schoiswohl G, Birner-Gruenberger R, Riederer M, Lass A, Neuberger G, Eisenhaber F, Hermetter A, Zechner R (2004) Fat mobilization in adipose tissue is promoted by adipose triglyceride lipase. Science 306:1383–1386

Index